Abdul Lateef Olanrewaju
Abdul-Rashid Abdul-Aziz

Building Maintenance Processes and Practices

The Case of a Fast Developing Country

Springer

Abdul Lateef Olanrewaju
Department of Construction Management
Tunku Abdul Rahman University
Kampar
Malaysia

Abdul-Rashid Abdul-Aziz
School of Housing Building and Planning
Universiti Sains Malaysia
Minden
Malaysia

ISBN 978-981-10-1378-2 ISBN 978-981-287-263-0 (eBook)
DOI 10.1007/978-981-287-263-0

Springer Singapore Heidelberg New York Dordrecht London

Softcover reprint of the hardcover 1st edition 2015

Printed on acid-free paper

Springer is part of Springer Science+Business Media (www.springer.com)

Indeed I have been blessed. My late father, Alhaji Abdul-Gafar Olanrewaju, was my mentor and my friend. He threw me to the wolves as early as he could. I am who I am because of him. This book is dedicated to his memory with love, affection, humility and respect. My father's love for his family was outstanding.

My father died about a week before my doctoral viva voce in Nigeria while I was still in Malaysia. On 7 July 2011, we lost him after he was diagnosed as having diabetes and hypertension. I sincerely pray to ALLAH (SWT) to forgive and reward him with Aljanat Firdaus.

My father was an amazing person who cherished education even though he himself was not well educated. He was not able to complete his own primary education because of his family circumstances and it was because of this that he pledged that his own children would all be formally educated. He believed in education and the one thing that made him happy was for us all to share his passion for education; and for that reason I feel sad that he did not live to see me complete my Ph.D. Knowing that I had received my doctorate would have given him the happiest moment of his life, even though he was on his sickbed. I wish to acknowledge the support of all the members of my family, who stood solidly by him while his sickness lasted.

Even though I had the desire to pursue graduate degrees, starting them when and where I did was due to my father's support, conviction and persuasion. For you to understand how passionate he was about education, I would like to share the following with you. At around 50 years of age, my father became keen to employ a teacher to teach him how to read, write and speak English. For years the teacher would come to our house in the evening for about two hours from Monday through to Friday. I was impressed with my father's humility and understanding of issues. He was such an amazing person that he was able to take a rational systemic perspective towards all issues. Although naturally, as human beings, we sometimes had different points of view, we both shared enormous respect for one another. Even today I always consider my father's actions and reactions whenever I have to make any decisions. I will continue to remember him always.

I also dedicate this book to aspiring writers who have interests in building maintenance management and the built environment.

Foreword

Facilities management (FM) is still at the nascent stage in Malaysia. The oft-repeated remark among construction players is that Malaysia is good at building 'world class' edifices but poor at maintaining them. Generally, from government bodies, corporations right down to the general public, the maintenance mentality is missing. But that is not to say, advances are not being made. The Construction Industry Development Board has developed an Asset and Facility Management guide and competency standard for facility management contractors. The Malaysian Association of Facility Management provides the platform for facility managers to come together and get their voices heard. International conferences on FM continue to be held in Malaysia to elevate research and discourse on the subject. In time, decision-makers in public and private bodies will come to realize the strategic importance of FM in its value adding capacity rather than burdensome cost element. When that happens, new economic opportunities would expand for contractors and professionals alike, spurring additional economic growth to the nation. This book is timely in that it adds to the literature on FM in Malaysia, which at the moment is lacking.

Abdul Hakim Mohammed

Preface

Inadequacies in the current approaches to building maintenance are leading to poor service delivery, unnecessary increase in maintenance costs and poor users' satisfaction. The current approaches are retrogressive to the building fabrics, the engineering services, the maintenance organisations and the building users. In order to sustain the corporate objectives of organisation, an integrated and dynamic maintenance management system is required. The problems with the current maintenance management strategies could be solved by better management—that is, by having a concise planning, controlling, directing, implementing and organising methodology. The other problems still remain because of value mismatch and misallocation of resources. Such problems are dominant in the service sector compared to the product-oriented sector. The last problem is uniquely fundamental in the maintenance service delivery. Maintenance management cannot be stand-alone, however, prefixing it with terms like strategic, proactive and systematic without active considerations of the value systems of both providers and users cannot bring specific user functional requirements into the mainstream maintenance management.

Current trend indicates that, in general, the complexities of buildings are on the increase and these are posing enormous challenges to maintenance organisations. Requirements are also constantly changing with increase in user experience and expectations. Most organisations are now placing much emphasis on outsourcing the maintenance aspect of their functions, instead of empowering their in-house maintenance organisations. Given that universities have a large buildings base (around 30 % perhaps), there is an ever-increasing need to develop and empower their in-house maintenance organisation to execute most of their maintenance demands. The in-house maintenance organisation should be robust, reliable, supportive and efficient. The majority of the problems associated with in-house organisations have been due to direct neglect, incompetency and poor planning by the parent organisation. The top management in the parent organisation should be more concerned with the building performance, productivity and user satisfaction. In this highly competitive environment, and increase in user expectations, it is now more compelling than ever before to make certain that maintenance service delivery is improved. A term that can readily factor in user requirement is 'value'. This will reduce or avoid reworks and service mismatches. Maintenance objective must be considered from the user value system rather than focusing on the building itself. Buildings are resources. In other words, a building is a means not an end. This concept may not be new or novel *per se*. In fact, value management in its current context has been a subject of interest for the last 50 years and it has been well established in the procurement of new buildings. But why it has not been applied in the maintenance management process remains difficult to answer. Some well-established contractors that have maintenance subsidiaries that provide outsourcing services could believe that they are utilising the methodology. However, they are not really considering it actively in the maintenance management process if at all. What these organisations are doing, is to apply the concept to building maintainability, but not for maintenance. Suffice it to say, many confuse maintainability, which is a design issue, with maintenance process. To consider the value management concept well, the value system of the building users must be well established using well-defined yet sufficient and simple methods at all stages. A value culture must be created. The value management concepts were used as a platform to develop the value maintenance management framework that forms the basis of this book. Value management makes explicitly clear the requirement of stakeholders and goes ahead to display how it should be delivered and at the same time provide mechanism for monitoring whether the services are provided, and to what extent, and what has to be done to improve service delivery.

Therefore, organisations can improve their profit margin, productivities and satisfy their customers by having a systemic maintenance management system. With the continuing increase in the proliferation of building materials and components, more and more maintenance works are required. Therefore, a reduction in maintenance expenditure will enable the availability of funds to be directed to other activities. The information in this book will be useful to maintenance personnel, facility managers, building users and organisations guiding them to ways towards managing their building infrastructure. It will be useful to academics.

In today's competitive world, maintenance is becoming a strategic tool and process, with organisation adopting maintenance as a means of keeping their business in operation. In most countries, a huge sum of money is spent on maintenance annually. The 'maintenance management manual' seeks to facilitate gathering of information and experience and make applications of information useful to other building projects. The manual as a tool will be used to communicate to the clients as to what is important in order for the organisation to achieve its corporate objectives. It aims at identifying, coordinating, monitoring, auditing and diagnosing the quality and quantity of the maintenance works required in order to keep the building in optimum operation.

Acknowledgments by Abdul Lateef Olanrewaju

Alhamdullilah. I am very thankful to Rabul l' Allamin (ALLAH (SWT)) for giving me the opportunity and strength to start and complete this project without hindrance. My humble thanks go to the Universiti Teknologi Petronas for sponsoring my Ph.D. My depest gratittudes to my mother, Mrs Fatimah. The supports and understanding of my father in-law-Alhaji Shobowale AbdulWahab-is immeasurable. I would like to thank my wife Kafayah and my dear sons, Harun and Haneef, for their relentless patience and understanding during the many long hours that I sat seemingly fascinated in front of my computers to complete this book. When the thought of publishing this book came to mind, the names of quite a number of Publishers came to mind, though we have actually been invited by some publishers to send our book proposal. However, as the writing and compilation of the book advances the number of potential Publishers that we could approach gets smaller as we consider our readers' interests in light of the benefits of the book. Concern of being turned down by the reputable ones is also a dominant one in our minds. Springer is one of the Publishers we were afraid to approach considering its reputation. But then, we just have to approach one or some of the giant ones. But, we settled for Springer and immediately an email was sent to three personnel about our intention to 'publish' a book with Springer. We sent to three staff because we were not quite certain as to who was directly responsible or would receive the email. A day later, we received a reply from Ms. Petra Jantzen. She acknowledged receiving our email and expressed her 'curiosity' to hear about the book. Based on the email, we were required to complete the book proposal form, which we promptly completed as required. Based on the form we completed, within a week we were assigned to the Asia Publishing Editor, for Applied Sciences, Ramesh Nath Premnath, Ph.D. Dr. Ramesh was passionate and supportive. Probably, if not for his encouragement and supports, the journey of this book would have been longer at best. We are grateful to him. We received the official acceptance of book proposal through Mr. Ravi Vengadachalam, Project Coordinator. Mr. Ravi proved to be a good coordinate, and we are happy with the way we were attended to at the various stages in the production process.

I greatly appreciate the good gestures of all the respondents who provided their time and other resources in order to complete the survey. I also thank anonymous commentators who offered useful suggestions in order to improve this book. My sincere gratitude to my colleagues at the Department of Construction Management, FEGT, UTAR. They include: Ar. Tan Seong Yeow, Assistant Professor Lee Lim Tat, Prof. Dr. Naato Mine. Thanks also to the following at FEGT, UTAR: Dr. Zafarulla Nisamani, Dr. Yap Vooi Voon and Puan Hasriani Binti Hassan who offered moral supports. I also acknowledge the supports of Mr Paul Anavhe. As a lecturer and mentor, I am blessed to meet many students, who ask various questions. Their questions and ideas triggered many challenges in my thoughts and understandings, all of which have influenced my writing. I am grateful to all other people and organisations too numerous to mention who provided me valuable time and ideas. I hope the readers find the book useful and interesting and I sincerely apologise for any errors or omissions that you might find as you use the book and I will be grateful if you could let me know these. Have a great read!

Acknowledgments by Abdul-Rashid Abdul-Aziz

I am grateful to Dr. Olanrewaju for inviting me to participate in this book project. As lecturers, we often refer our students to foreign textbooks, citing the lack of local textbooks as the problem. Yet, we are the cause of that problem in that we lack the drive to produce the textbooks that our students so badly need. This joint effort in coming up with this book is a small step in the direction of producing reading material that is relevant in the Malaysian context. I would like to thank everyone who have helped shape my career, most notably my father, Abdul-Aziz bin Basha, who as a teacher, inspired me to treasure knowledge. Also to my mother, Aimee binti Merican, who had nurtured me during my formative years, and even now never stops worrying over me and my other siblings. Special mention should go to my maternal grandmother, Che Puteh binti Ismail, who was really instrumental in inculcating good values in all of us. I am truly indebted to my wife, Anita Noraini Campbell, who has provided the backbone to the family. Whilst I was overly focussed on my work, she raised our children so well. I would like to express my love to all my children—Abdul Shahid, Abdul Zahir, Aliyah Salihah, Abdul Muhaimin, Ambar Summayyah and Ambrin Samiyah—who have been patient with me all this while.

Contents

Abbreviations

A number of acronyms were used in this thesis, it is instructive to identify them in order to aid in reading the thesis. Following are the key ones:

3R	Reduce, Reuse and Recycle
8th MP	Eighth Malaysia Plan
9th MP	Ninth Malaysia Plan
10th MP	Tenth Malaysia Plan
AHP	Analytical Hierarchy Process
AM	Asset Management
BN	Bank Negera
BMI	Building Management Information
CEO	Chief Executive Officer
CIOB	Chartered Institute of Building
CO_2	Carbon dioxide
COBRA	The annual RICS construction, building and real estate research conference
CSR	Corporate Social Responsibility
DOD	Department of Defence
DOHE	Department of Higher Education
DOSM	Department of Statistics Malaysia
EIU	Economic Intelligence Unit
etc.	Others
FEGT	Faculty of Engineering and Green Technology
FM	Facilities Management
GBI	Green Building Index
GDP	Gross Domestic Product
GFCF	Gross Fixed Capital Formation
GHGs	Greenhouse Gases
GLCs	Government-linked Companies
GNI	Gross National Income

GOM	Government of Malaysia
IPTA	Public Higher Learning Institution (Institusi Pengajian Tinggi Awam)
IPTS	Private Higher Education Institution (Institusi Pengajian Tinggi Swasta)
JCT	Joint Contract Tribunal
KBE	Knowledge Based Economy
KBI	knowledge-Based Index
K-economy	Knowledge Economy
KL	Kuala Lumpur
KPI	Key Performance Indicator
KTM	Keretapi Tanah Melayu
LNG	Liquefied Natural Gas
LRT	Light Rail Transit
M&E	Machinery and Equipment
MCI	Malaysian Construction Industry
MOE	Ministry of Education, Malaysia
MOHE	Ministry of Higher Education, Malaysia
MP	Malaysian Plan
MPM	Maintenance Performance Management
n.a:	Not Available
NAPIC	National Property Information Centre
NGOs	Non-Governmental Organisations
NHD	National Housing Department
NKEAs	National Key Economic Areas
NKRAs	National Key Result Areas
PEMANDU	Performance Management and Delivery Unit
PETRONAS	Petroliam Nasional Berhad
PFI	Private Finance Initiative
PM	Performance Measurement
PMM	Planned Preventive Maintenance
RICS	Royal Institute of Chartered Surveyors
RM	Ringgit Malaysia
RTC	Response Time to Complaint
SCORE	Sarawak Corridor of Renewable Energy
SEA	South-East Asia
SETARA	System Penarafan Institusi Pengajian Tinggi Malaysia (Rating System for Malaysian Higher Education Institutions)
SMART	Specific, Measurable, Achievable, Realistic and Time
SMEs	Small and Medium Enterprises
SMS	Short Message Service
SPAD	Public Land Transport Commission (Suruhanjaya Pengangkutan Awam Darat)
SPM	Malaysian Certificate of Education

STPM	Malaysian Higher School Certificate (Sijil Tinggi Pelajaran Malaysia)
SWOT	Strength Weakness Opportunity Threat
UiTM	Universiti Teknologi Mara
USBU	University Strategic Business Unit
USBU	Organisation Strategic Business Unit
UTAR	Universiti Teknologi Tunku Abdul Rahman
UTP	Universiti Teknologi PETRONAS
V2020	The Vision of the Malaysian to become developed country by the year 2020
VM	Value Management
VMMM	Value Maintenance Management Model

About the Authors

Abdul-Rashid Abdul-Aziz is a professor at the School of Housing Building and Planning, Universiti Sains Malaysia. He started his career as a quantity surveyor, but very quickly took up the offer to further his studies under the sponsorship of Universiti Sains Malaysia and the Malaysian government. He completed his M.Sc. in Construction Management at Brunel University, West London, England and thereafter returned to his alma mater, University of Reading, Reading, England, to pursue his Ph.D. His initial research interest was international construction. But over the span of more than two decades as an academician, his research interests have expanded to cover construction labour issues, construction development, privatisation, construction procurement and housing development. He is a registered quantity surveyor with the Board of Surveyors Malaysia, a member of the Royal Institution of Surveyors Malaysia and a fellow of the Chartered Institution of Chartered Surveyors. He is the co-editor of the Journal of Construction of Developing Countries. In addition to that, he is an editorial board member of several other journals. He has published more than 70 articles in international refereed journals, international books and book chapters. He has done consultancy research work for the Malaysian Construction Industry Development Board, Singaporean Contractors Association Limited and the International Labour Organisation.

Abdul Lateef Olanrewaju had his first employment in the construction industry as a quantity surveyor, about 15 years ago. He was involved in various construction projects from conceptual stages to completion. He possesses the capability and interest to conduct full-time research work and contributes towards enhancing and strengthening research work in the built environment. He gained his master of science in Built Environment from the International Islamic University Malaysia (IIUM) for his research in value management and his doctorate from Universiti Teknologi PETRONAS (UTP) Malaysia for his research in building maintenance management. He received his quantity surveying education at Kaduna Polytechnic, Nigeria. His academic activities centre on lecturing and research associated with quantity surveying, value management, building maintenance, research methodology, procurement and contemporary issues in construction, project management.

He has authored and co-authored several scientific papers. He is a referee for leading journals. He has served as panellist and keynote speaker on different topics relating to the construction industry. He is an incorporate member of the CIOB, member of the Quantity Surveyor International, member of Malaysian Facility Management Association. His current research focuses on sustainable maintenance management, ethical issues in construction and roles of construction sector in economic development and the procurements of construction projects through the Shariah method.

What this Book is Designed to do for you

This book contains part of a thesis which was approved by Universiti Teknologi PETRONAS for a doctoral degree. But the interest of this material goes far beyond academic orientation, and we felt obliged to rearrange it, with some omissions and additions in order to present it into a self-sufficient book which could be read without any special knowledge on management of building maintenance. The doctoral thesis from which this book was developed was designed to be an inclusive reference towards the best practice approach (the process, procedures and system which meets a high standard of professional and academic competence) of maintenance management, it closed the loops between the haves and the have-nots with respect to practices, principles, operations, strategies, experience and resources. It is intended to help you strike a balance by guiding you, and to level the playing field; if you are currently behind schedule, it will push you ahead faster if you are currently at or close to the best practices. Although the primary data was collected from the academic sector, the discussions are prepared bearing in mind various maintenance stakeholders, so that the book can be useful to many readers.

The book applies the general management literature and practice to explore the conceptual foundation and nature of building maintenance. It advocates a systemic consideration of the maintenance practices and processes before going ahead to develops the new management theory for maintenance principles and practices. Finally, the nature, characteristic, processes and practices, of maintenance are articulated into a management framework. The question is what value does 'management' bring to maintenance that each of the functions if considered separately will not offer? The value that management has on maintenance includes: determine maintenance objectives, establish strategic objectives, set maintenance standards, determine maintenance performance indicators, identify user and their value systems, set priorities, reduce 're-maintenance', deliver better service and maximise productivity and profits. But to achieve the above objectives, there is a need to correctly articulate maintenance functions and activities into a management theory and practice. Because focusing on one aspect without systemic consideration of the others would not add 'value to money'. But such studies do not provide the systemic/holistic link the planning function has with the other management functions. There is need to consider all the functions consecutively. Each of the functional stages should serve as a functional 'gate' facilitating connection with the next stage. If all functions are not consistently considered, the process chain will be disjointed and feedback loop cannot be maintained, thus it is punitive and regressive to maintenance. It will be difficult to measure the 'actual' importance. It is imperative that maintenance organisation and developers realize the importance to improve their maintenance performance in order to grow their business.

Therefore, if your organisation is an independent maintenance contractor or an in-house organisation, this book is still useful to you. It will help you to improve on your building performance, increase your productivity and profits and at the same time increase maintenance service user satisfaction and productivity. But it would not stop there. It will help you to maintain sustainability. Every aspect on your maintenance management system is given consideration. Some readers may still be wondering why it is we have added to the existing literature on building maintenance management. When much has actually been written, is there a need for more books? If we might answer that question point blank, then our reply is yes, there is a need for more books.

First, while the existing literature focuses so much on the tactical aspect of maintenance management, this book deals mainly with operation issues in building maintenance management. Much of the existing literature gives attention to increasing organisation and profits, while little attention is given to user satisfaction. Managing building fabrics, structures and services requires much more rigour than in the previous years. The goal of the manager of maintenance organisation is to continuously operate an ageing building optimally, while also meeting increasing regulations, laws, requirements and users' demand and optimising integrity cost. Industry currently has the basic methodology to solve simple and technical issues. Over the past decade, emphasis has been richly placed on the importance of cost issue. This is due in part to understanding the concern with the money–*building* consideration. Also, we felt that much of the existing literature is too theoretical for the readers this book is designed for. The book is also unique because the contents are not only about the ideas and opinions of the author, but perhaps because it is developed from a thesis that includes the understanding of various stakeholders.

We felt there was a need for a book which will demonstrate to the maintenance organisations how and what they might consider relating to building maintenance issues. The approach adopted here is therefore systemic rather than fragmented and theoretical. The choice of what service level to provide and the process to apply should be defined by the user value system and the organisation's mission and vision. The building maintenance: processes and practices focus on defining the value system of users and providers in order to provide improved service and increased productivity. A maintenance service delivery system defines the processes a maintenance service is organised in order to take it from need (i.e. established through user complaints or inspection) to the physical reality. The delivery system assigned duties, tasks, responsibilities, authorities and established relationships to the member within the maintenance organisation as well as the maintenance service users. It may be necessary to differentiate between processes and services, as both are not the same from all perspectives, however they are related. While services are results, another type of product, processes are sets of activities, which could lead to the performance of a service at any given level. It is our conviction that emphasis is required more on the process and procedure as compared to on the end products. However, we should not expect all readers to align with the ideas, actions and reactions being put forward in this book. However, if you are one of those who do not agree with where we stand, may we please suggest you should consider the ideas and thoughts you think could lead to the best practices.

Chapter 1
Overview

1.1 An Overview of Value Maintenance Management

The need for building maintenance is very high, and so it will remain. The prosperity of a nation is dictated by the size of its gross fixed capital formation. Buildings regardless of their purpose are gradually exceeding more than 50 % of any country's gross fixed capital formation. Therefore, there is the need to make certain that this asset is in optimal performance to keep the national capital consumption low. Buildings that are in a poor state of performance will increase the rate of capital consumption. The prosperity and well-being of a country therefore depend on buildings, because national building stock depends on the amount spent to maintain the buildings. Many developed countries not only spend more than half of their total investment in the construction industry to maintain their building stock, but also invest about 10 % of their GDP for maintenance of constructed facilities. Building maintenance both in practice and theory is service oriented, not product focused but it is technical in nature. Maintenance management involves making use of resources for the sustenance of the building performance optimally. In other words, it makes certain that buildings are in high performance during their operation phase. However, maintenance management can no longer be a standalone both as a term or concept due the extent and scope of the inadequacies. Buildings' users are becoming more experienced and demanding. Their experiences are being shaped and sharpened from the services they are receiving from other sectors. The automobile and telecom sectors are providing fascinating experiences to the buildings' users and they question why the maintenance sector could not provide customer service to them. Maintenance management practice relies on an extensive body of knowledge and information (derived from psychology, social, finance, management theory, bossiness principle, science and engineering). Maintenance management has been prefixed with terms including strategic, proactive or best practice. Yet, this has not allowed maintenance management to deliver best service delivery.

A.L. Olanrewaju and A.-R. Abdul-Aziz, *Building Maintenance Processes and Practices*, DOI 10.1007/978-981-287-263-0_1

In this book, value is prefixed with maintenance management in an attempt to bolster the scope and performance of maintenance management. This is peculiar since the objective of maintenance is to increase building performance and ultimately building user satisfaction and the maintenance organisation's productivity. Maintenance is service oriented because the maintenance activities mainly involve interaction with the building users and generally the quality of services provided is determined by the building users. The term "value" means different things to different people but it is not a buzzword. Its meaning is clear, even though it comprises both objective and subjective constructs. In the well-established service sector, stakeholders are aware of the elements in the value concept. However, in the engineering management disciplines such as building maintenance, its application has been slow and maintenance service providers are reluctant to inculcate this into their main stream delivery methodology. On paper in the service level agreement (SLA), everybody talks about value, but in practice, its achievement has become unachievable because of the methodology and poor understanding.

However, in some cases, the measuring of end service is easy but it is not so easy in some cases. For instance, in relating maintenance cost with the value to be derived, if the objective of maintenance is only to avoid or reduce accidents, obviously in this case, the amount invested to avoid or reduce the accidents can be related to the number of accidents that are reduced. Meaning that if the measurement metrics an organisation developed were not made, what is the equivalent of the damage to property and injuries to operatives or even the loss of lives. The replacement amount can be determined based on insurance and/or government. However, it is impractical to determine the value of maintenance service to building users in a straight forward manner. For that purpose, its surrogates are developed to elicit how satisfied or contented are the maintenance service users with the services they are receiving. In the phrase "maintenance management model", the term "model" is open to various interpretations. In this book, particular emphasis is paid to the decision-making process and procedure. Through the management of maintenance works, maintenance organisations have to make decisions of all sorts towards attaining the parent organisation's corporate missions and visions. The decisions are both technical and managerial in nature. What often comes to mind whenever the term model is mentioned or written is that the model is either mathematical or statistical. While this book contains both the statistical and mathematical models, much emphasis is placed on the process model (graphical). This is because, "figures are more easily and quickly interpreted, understood and memorised than words". Building maintenance practices prefer process models compared to mathematical models. The building operatives, and in fact the managers, have limited knowledge in *modelling*. In general, "…attempts by mankind to overcome the complexity of reality have involved, at some stage, a form of a graphical representation".

The model presented in this book is based on the real maintenance management practices in universities as well as in academic literature. The literature on building and maintenance management is found in journals, proceedings and texts across fields, such as industrial, manufacturing, and processing engineering in assessing maintenance management, we bridge the literature across fields to provide a directive and pointer to us. In addition, the model is intended for "mature" maintenance organisations. This is because it requires the maintenance organisations to be part of the parent organisation rather than outsourcing larger parts of the services. In this book, the attentions of the maintenance organisations are focused on aspects of maintenance services that increase their productivity and profits at the same time as increasing the productivity and satisfactions of the building service users.

With this book, we hope to offer guidance to maintenance organisations in their efforts to provide better maintenance services and at the same time increase their productivity and profits. Based on primary information gathered, the "innovative maintenance organisations" will make use of the model very often while the "traditional maintenance organisations" will make reference to it casually, probably when things go awry'. It should, however, be noted that this book is not written to be taken off the shelf and applied straight away; the book is to be used as template only. Further, while every effort is made to make certain the accuracy and correctness of the information and knowledge, no warranty is expressed or implied as to the opinions and documentation contained in the book. The information in the book carries no legal or professional liability.

1.2 Objectives of the Book

This book explores the systemic consideration of requirements of both the maintenance service providers and maintenance service users. The aim of the book, as will be explained later, was to develop a value maintenance management model for maintenance organisations. What determines the content of this book is the increase in service user satisfaction, user productivity, maintenance organisation productivity and profits. These are then translated to building maintenance objectives. The specific maintenance objectives were further developed into key processes and procedures. Therefore, the ultimate aim of the book was to provide building maintenance organisations with informative operational and procedure of maintenance functions, an understanding which will enable them to seek problems and systemically develop solutions.

While allocation for maintenance is on the increase, the amount of complaints and criticisms from those that have stake in quality of education and those that have concern about the government's expenditures keep on increasing unabated. Even though many of the maintenance organisations claimed that productivity and profits are on the increase, there is much evidence to conclude that the increase in productivity and profits is at the expense of poor users' satisfaction and a decline in building performance. Although maintenance costs increase with building age, a year or two could not make a difference except where the building is excessively used. This is abnormal though.

While some authors have identified the problem facing maintenance organisations, the solutions offered have not been adequate. In the recent time, particularly in the developing countries such as Malaysia, there have been continuous interests in research in building maintenance; however, the positive impact of the research from the practitioner's viewpoints are quite subjective. Virtually, all the studies only look at maintenance from theory and "mechanical" perspectives hence give very little or no consideration to the managerial and service users' perspectives. However, the writers and authors that gave their resources to carry out the researches demand recommendations, as a result of their renew interest, to improve building performance, increase productivity and user satisfactions.

However, in this book, a comprehensive overhaul was made of the original thesis to fit into other organisations other than universities or educational buildings. Building maintenance management has been previously researched, and there are books on the topic, however, the books are few in number. Furthermore, there is a shortage of books on building maintenance that specifically detail maintenance management that focuses on increasing maintenance service users' satisfactions from a systemic point of view. For that purpose, it is understood that readers do not have sufficient knowledge on maintenance management from value perspectives. Therefore, all detailed information and knowledge will be provided in this book to enable readers to understand the fundamental issues involved in building maintenance management.

The development of a value maintenance management model, which is the focus of this book, has as its concern with the multiple focus of increasing maintenance organisations productivity, increasing the building performance and life span as well as aiming at the user performance and satisfactions rather than building itself. The value maintenance management model contains a number of sub-models. Traditionally, building itself and maintenance service providers are the focus of maintenance management initiatives. Therefore, the model presented here aims to close the loop and has at its core science, the requirements to make explicitly known the value system of the building maintenance service users and the understanding and perceptions of the maintenance service providers. The model also formulates a quantitative model for schedule maintenance for both systemic and technical maintenance management. It is differentiated from previous theories and practices of maintenance management because of the significance that is attached to the influence that service users have on building maintenance. This is necessary because buildings are capital goods and maintenance is not product oriented. The focus of maintenance is towards the users rather the buildings, because building is a means not the end product itself.

1.3 General Organisation of the Book

The book contains a total of eleven chapters. Each contains a specific aspect of the book. Every chapter has its own specific learning outcomes, keywords and each starts with its own introduction, and summary reference. The book took into account the diversities involved in building maintenance. Because the book was developed from a thesis, it contains information on methodological issues, data analysis and discussion of findings. We acknowledge that research methodology and findings and perhaps discussion are not usually presented in the way we did in this a book. Yet these two or three contain enormous bodies of information and knowledge. A great disservice is committed if we do not include them where necessary. Doing that will create artificial gaps and will cause all sorts of misunderstanding and questions such as "how does he comes to this conclusion?", "is this necessary?", "this book is fragmented" or "it is difficult to arrive here". With the inclusion of this information and knowledge, we want to be clear at the onset about what we are not saying. In fact in quite number of cases, we did query some authors on why they have not included these information in their books. We are aware that in some cases, these are not required but in many cases they are required.

The overall structure of the book has been designed to make the information as usable as possible to the greatest number of maintenance professionals at every cadre of the organisation and maintenance management profiles. Therefore, it is intended to be used by all managers, directors, engineers, executives and professionals throughout the maintenance organisation. Members of maintenance organisations should thoroughly familiarise themselves with the contents of the book. Oversize materials (e.g. maps, figures, tables and charts) are presented by sectioning the original, beginning at the upper side and continuing from the downside. However, in some cases, the original materials are reduced to fit in while the eligibility of the content is visible. The desire to write this book is majorly mine. In order to do away with inept phrases or statement including "he or her" or "him or her" we have used "he or him", "they" or "maintenance organisation".

Although there have been an increasing number of women employed as maintenance managers, executives or engineers, the positions are predominantly occupied by men. For instance, in Malaysia, as of the time the first author carried out his doctoral research, only two of the about 50 of those contacted were women. We have often been irritated by writers and authors who use phrases such as "as described somewhere in this book" but without saying specifically where. For that purpose, we have included cross reference whenever necessary in order to facilitate understanding. However, we were more influenced to develop the doctoral thesis into a book by the motivations gathered from quite a number of authors.

1.4 Overview of Chapters

In Chap. 1, an overview of value maintenance management was presented. The problems with the current maintenance approaches were presented. This chapter also reviews various issues related to educational buildings and goes on to present the roles of maintenance in sustaining and enhancing the performance of the academic buildings. Chapter 2 is a review of the construction industry in Malaysia. The information on the construction industry is provided first to permit the readers to appreciate the state of the industry in general and the maintenance sector in particular. We also deem this necessary as there is no argument on the relationship between construction industry and the education sector. But just to remind some readers, the correlation between the education sector and construction industry is a positive one. Their interdependence is not a complex one. For instance, on the one hand, the innovations and development propelling productivity and performance of the construction industry hinge on the qualities of research and development churned out from the universities, research centres and other institutions. On the other hand, the universities, research centre and other institutions require the outputs of the construction industry in terms of buildings and other infrastructure. The chapter entails a lot of representative statistics on the Malaysian construction market. In Chap. 3, a review of the Malaysian educational system is offered. In addition, the roles of the governments and private investors in meeting the mission and vision of education are reviewed. Furthermore, the interrelationship between the buildings and delivering quality education is presented. But, readers will soon find that much of the information in this chapter is, however, not peculiar to education buildings only.

Chapter 4 describes in details various issues in building maintenance management. In this chapter, a review of how maintenance is linked to building performance is presented. What are contained in this chapter are elaborate explanations on building maintenance. Issues with the criteria in the building user value system were further put forward. Although readers will observe that the technical aspect of maintenance is not new per se, yet readers that have not considered the essence of maintenance to be users will find that the only way positive strategy to create "value" or even maximise profits is for the maintenance organisation to pay critical attention to user experience. Chapter 5 deals with the research design and methods used to analyse the research data. The research methodology describes the methods, process and design used for doctoral research, however, only the very basic information on the methodological issues is presented in this book. Primary, this chapter aims to connect reader with the succeeding chapters.

Similarly, in Chap. 6, the data analysis obtained from the survey questionnaire and interview conducted for the doctoral degree is presented but with some significant modifications to fit in with this book. Chapter 7 contains the discussions of the research findings. Based on the information from Chaps. 5 and 6, major observations and inferences were made. From the observations and inferences, the "shape" and "form" of the "value maintenance management model" is delineated. It is a case that this information and knowledge from the observations and inferences were "grounded" to develop the value maintenance management model. Very brief information on grounded theory technique is presented. The grounded theory forms the basis of Chap. 7. Grounded theory is a form of qualitative research tool. It is an excellent method to generate themes and ideas from data.

Chapter 8 contains the value maintenance management model. The model consists of graphs, tables, figures, numerical value and mathematical equations. Major activities at various stages of the model are defined, illustrated, explained or described. Chapter 8 includes the actual experience of several maintenance managers.

The experiences are not only limited to those gathered from maintaining academic buildings but other types of buildings. The expert survey method was used to validate the model. The purpose of the validation is to see the usefulness of the model developed. Chapter 9 draws together the major themes in previous chapters to develop the model. However, because of logistic, we are unable to apply the model in “real” situation. But few of the “maintenance managers” connected lately let us know that they are applying some the lessons learn during validations in the organisations. Chapter 10 contains a modified versions of two papers authored on sustainable maintenance initiatives. Todays, it will be inadequate to discuss about maintenance without considering its impact on sustainability development. In order to move close to sustainable development, there is a major need to move towards the construction industry to seek for “help”. However, to move towards sustainable construction, there are the business requirements to shift in the direction of building design, construction, operation and maintenance. Mainly, the arguments we made with the two papers are that without sustainable maintenance platform for the built assets, there is no way the construction industry could contributes to sustainable construction or else the aspiration to sustainable development would be curtailed. The first paper is concerned with a review of sustainable housing maintenance management model. It provides an argument for and the significance of housing maintenance to the sustainable development. The second paper presents a study that investigates how university maintenance organisations are approaching sustainable development. A summary and conclusion of all the major themes in the book were discussed in Chap. 11. This chapter also contains some of the recommendations.

These experiences are not only limited to those [illegible]

Chapter 2
An Overview of the Construction Industry

Abstract In this chapter the Malaysian construction market is described. This chapter is included because of the interrelationships that existed between the construction industry and quality of education. Their correlation is a positive one. The education sector has strong spill over implications on other sectors of the economy and has contributed directly to productivity enhancement directly, indirectly or both. But as we have illustrated in chapter one, the education sector requires infrastructural facilities including roads, buildings, services, that are produced by the construction industry. Likewise the human capitals needed by the construction industry are supplied by the education sector. Therefore, the learning outcomes the ability to: (1) Be able to describe and discuss the Malaysian construction industry, (2) Be able to explain the extent of maintenance market, (3) Understanding the problems facing the construction industry.

Keywords Construction industry · Maintenance · Private sector · Economic development

The construction industry is an economic investment and its relationship with economic development is well posited. Many studies have highlighted the significant contribution of the construction industry to national economic development (Myers 2013). Although, some argue that the extent of its contribution is not always clear. Others have argued that the construction industry can be used to regulate the economy, while others tend to differ. Some have argued that the contribution is limited to the short term. There are also those who argue that it is economic growth that drives the construction industry rather than the industry advancing the economy, i.e. that the construction industry is not a driver of economic prosperity rather it only follows the "path" defined by the total economic growth rate.

In short, there are numerous arguments and counter arguments regarding the true role of the construction industry in advancing economic development. This may be a source of confusion to many, however, we emphasise that irrespective of the position one takes regarding the relationship of the construction industry to economic growth, it does not in any way invalidate the importance of the construction industry in regards to providing the necessary infrastructures that stimulate economic

A.L. Olanrewaju and A.-R. Abdul-Aziz, *Building Maintenance Processes and Practices*, DOI 10.1007/978-981-287-263-0_2

development. As such, the industry is required for national development. It has been argued that for a country to have meaningful and sustained development, it urgently requires that it indigenize its construction industry, because if the construction industry is inefficient, it will be difficult for such a country to attain meaningful development.

Among the major economic sectors, the importance of the construction industry is unique regardless of whether the country is underdeveloped, developing or developed. For instance, the construction industry is subjected to quarterly and annual statements of national accounts. The construction industry appears more than once in the national accounts: GDP, GNI and GFCF. The outputs are measured by gross output, capital formation and added value. More than half of GFCF consists of construction outputs. The homes, offices, roads, factories, and shopping malls are all part of the outputs of the construction industry, among other capital or investment goods.

The following are some major uses of the construction industry:

- Strategic tool to achieving sustainable development
- Construction output as growth-initiating and growth dependent
- Contributes significantly to the economy; GDP, GNI
- Contributes to Gross Fixed Capital Formation (GFCF)
- Provide outputs to most industries and utilises the outputs of many industries
- Contributes significantly to the informal sector
- Income generation and re-distribution
- Employment generation.

2.1 The Malaysian Construction Industry

Malaysia is actively working towards achieving a high-income status by 2020. This involves intensive transformation of the economic structure. The government has outlined an economic road map to transform the country in order to be recognised as a developed nation. Since independence, the Malaysian economy has observed plans with five-year strategic thrusts. The strategic trusts are in line with the goal to become a high-income nation by 2020. Looking towards the 2020 target, the challenge is to sustain the impetus of robust growth. Specifically, this requires average growth of 6.0 % in GDP per annum during the Tenth Plan Period. To achieve this target, the economic sectors are to play significant roles. The construction sector is active and features prominently in terms of policy formulation and implementations. A comparison of the size of the construction industry with other countries suggests that its contribution has been consistent and stable (Table 2.1). As may be seen, among the countries cited, the contribution of the Malaysian construction industry, although not the highest performer, its contribution nevertheless remains modest.

Table 2.1 Percentage share of GDP for selected countries (production approach) for construction sector

Country	2005	2006	2007	2008	2009	2010	2011	2012	2013
South Korea	5.7	5.5	5.3	5.0	5.1	4.6	4.2	4.1	4.1
Hong Kong	3.4	2.9	2.7	2.9	2.7	2.9	3.3	3.5	3.5
Taiwan	2.8	2.7	2.6	2.4	2.3	2.3	2.3	2.2	2.2
Singapore	n.a	n.a	3.0	3.6	4.2	3.8	3.8	4.0	4.1
Thailand	2.4	2.4	2.4	2.2	2.2	2.2	2.1	2.1	2.1
Philippines	4.4	4.6	5.0	5.1	5.4	5.7	5.0	5.4	5.6
Indonesia	5.9	6.1	6.2	6.3	6.4	6.5	6.5	6.5	6.6
Malaysia	3.0	2.9	2.9	2.8	3.1	3.2	3.2	3.5	3.8

Source Department of Statistics Malaysia (2014a)

Table 2.2 is the contribution of the different industries to the GDP. The services industry is the largest in terms of size while construction is the smallest. However, it may be seen that the contributions of the construction industry is consistent and impressive. As Table 2.3 suggests for the period indicated, the construction starts with a moderate contribution but lags until 2008 contributing only 2.7 %. However, from 2009 until 2013, the contributions of the construction industry are on the increase. In fact, while the contributions of other major sectors plummet, that of the construction industry is increasing.

In 2013, Malaysia's economy grew at 4.7 % with the all sectors registering positive growth (Table 2.4). The Services and Manufacturing sectors remained the key engine in terms of supply. Consecutively, the Construction sector continued a double-digit growth by registering 18.6 %. The growth is mainly accountable by the strong growth in the residential sector (Table 2.5) coupled with the underlying strength in infrastructure and civil engineering projects. The Malaysian construction industry is classified into four sectors namely, residential buildings, none-residential buildings, civil engineering and the special trade sectors. The residential sector involves the construction of houses and condominiums. The non-residential construction comprises of all building construction other than residential. These include the construction of commercial and industrial buildings. Civil engineering pertains to the construction of public infrastructure such as bridges and highways.

The contributions of the residential sector have remained consistent from 2008 to 2013 (Table 2.5). However, the share of the non-residential sectors fluctuates through the period and peaked in 2010, at 44.6 %. Similar scenarios play for the other sectors. Value for construction work done includes for new work, conversions and maintenance which were carried out during the reference period. The value of the work completed in the first quarter of 2014 by the types of clients is contained in Table 2.6. In terms of construction activity by project owner, the private sector continued to dominate. The private client dominates in all the four sectors, except in civil engineering in which the private sectors contributed a little less than 50 %.

The increase in the shares of the private sector in 2013 as compared to 2008 is remarkable (Table 2.6). As is the case in the first quarter of 2014, from 2008 to

Table 2.2 GDP by kind of economic activity at constant 2005 prices—RM million

Country	2005	2006	2007	2008	2009	2010	2011	2012	2013
Agriculture	44,912	47,533	48,188	50,036	50,063	51,263	54,250	54,963	56,095
Mining and quarrying	72,111	71,276	72,758	70,996	66,386	66,182	62,607	63,243	63,680
Manufacturing	149,754	160,880	165,879	167,148	152,150	170,261	178,237	186,748	193,237
Construction	16,107	16,022	17,391	18,151	19,270	21,459	22,464	26,640	29,554
Services	254,322	272,555	299,933	325,712	335,027	359,829	385,550	410,339	434,460
Plus import duties	6,372	5,671	5,937	7,523	6,898	7,660	8,653	10,001	10,586
GDP at purchasers' prices	543,578	573,936	610,087	639,565	629,885	676,653	711,760	751,934	787,611

Source Department of Statistics Malaysia (2014a)

Table 2.3 GDP by kind of economic activity at constant 2005 prices—percentage share to GDP

Country	2005	2006	2007	2008	2009	2010	2011	2012	2013
Agriculture	8.3	8.3	7.9	7.8	7.9	7.6	7.6	7.3	7.1
Mining and quarrying	13.3	12.4	11.9	11.1	10.5	9.8	8.8	8.4	8.1
Manufacturing	27.5	28.0	27.2	26.1	24.2	25.2	25.0	24.8	24.5
Construction	3.0	2.8	2.9	2.7	3.1	3.2	3.2	3.5	3.8
Services	46.8	47.5	49.2	50.9	53.2	53.2	54.2	54.6	55.2
Plus import duties	1.2	1.0	1.0	1.2	1.1	1.1	1.2	1.3	1.3
GDP at purchasers' prices	100.0	100.0	100.0	100.0	100.0	100.0	100.0	100.0	100.0

Source Department of Statistics Malaysia (2014a)

Table 2.4 GDP by kind of economic activity at constant 2005 prices—annual percentage change

Year	Agriculture	Manufacturing	Construction	Services	GDP
2006	5.8	−1.2	−0.5	7.2	5.6
2007	1.4	2.1	8.5	10	6.3
2008	3.8	−2.4	4.4	8.6	4.8
2009	0.1	−6.5	6.2	2.9	−1.5
2010	2.4	−0.3	11.4	7.4	7.4
2011	5.8	−5.4	4.7	7.1	5.2
2012	1.3	1	18.6	6.4	5.6
2013	2.1	0.7	10.9	5.9	4.7

Source Department of Statistics Malaysia (2014a)

Table 2.5 Value of construction work done by type of construction activity and project owner, Q1 2014 (RM '000)

Types of construction	Private	Government	Public corporation	Total
Residential buildings	6,709,282	252,330	240,213	7,201,826
Non-residential buildings	6,471,322	1,122,923	458,106	8,052,351
Civil engineering	3,880,379	2,024,227	2,494,848	8,399,455
Special trades	1,057,150	147,728	114,830	1,319,708
Total	18,118,133	3,547,209	3,307,997	24,973,339

Source Department of Statistics Malaysia (2014b)

2013, the industry is led by the private sector. The expansion of private investment from 2008 to 2013 was almost 100 % overall and in all the sectors except in the special trade where expansion was less than 50 %. Whether we consider the sectorial construction individually of the industry as a unit during the period, the private sector is the dominant investor in the industry. This finding is contrary to the normal practice in which the public or government is the major client. In most developing countries, at all levels the government is the major client. The investments of the private sector in the 'special trade' are the least, being only 5 % in 2013. However, private investors are making massive investments both in the residential and the non-residential sectors.

For the government and public organisations, their investments tend towards the civil engineering sector. In comparison, both public corporations and government are investing an approximate 5 % in the residential sector, but government investment in non-residential buildings is quite high. The number of projects completed in the year cited is contained in Table 2.7. Except in 2009, the number of projects increased. This scenario will likely remain high as the government has outlined a number of ambitious packages to spur the construction industry, as Malaysia advances towards its 2020 target.

Among the major economic sectors, the contribution of the construction sector as a source of employment is high. For instance, in 2012, the industry directly employed approximately 1,028,000 (Table 2.8) people. This represents an estimated 8 % of the

Table 2.6 Value of construction work done by type of construction activity and project owners, 2008–2014 (RM '000)

Year	Residential buildings	Non-residential buildings	Civil engineering	Special trades	Total
Private					
2013	23,581,037	22,618,949	13,827,222	3,172,017	63,199,226
2012	19,624,697	19,386,220	10,948,244	3,062,682	53,021,843
2011	14,747,483	14,544,468	7,768,191	3,309,319	40,369,461
2010	10,631,163	12,021,759	4,640,156	3,181,929	30,475,007
2009	11,016,286	11,099,674	4,850,657	3,034,121	30,000,738
2008	12,957,437	10,963,133	6,076,449	2,666,011	32,663,030
Public corporation					
2013	520,776	1,525,073	8,873,459	459,460	11,378,768
2012	119,187	757,191	6,416,175	254,724	7,541,277
2011	94,210	552,864	1,123,505	482,609	2,253,187
2010	130,325	333,182	1,182,226	803,099	2,448,833
2009	240,565	270,923	1,264,874	794,811	2,571,173
2008	274,021	245,545	1,027,422	289,651	1,836,639
Government					
2013	861,413	5,159,365	9,600,307	675,825	16,296,910
2012	1,123,984	7,344,173	10,839,606	795,947	20,103,710
2011	1,214,648	10,119,190	9,412,826	889,029	21,635,694
2010	1,546,864	14,736,495	9,921,018	1,664,025	27,868,401
2009	1,840,696	12,949,060	11,090,137	1,562,992	27,442,885
2008	2,228,587	9,091,469	11,068,873	1,431,549	23,820,478

Source Department of Statistics Malaysia (2014b)

Table 2.7 Number of project completed of various projects

Year	2013	2012	2011	2010	2009	2008
Number of project	37,516	35,589	32,454	31,018	28,814	31,719

Source Department of Statistics Malaysia (2014b)

total employed workforce. The construction industry is the fourth employer of labour after agriculture, manufacturing, and services. However, based on estimates by the Department of Statistics, the construction industry was postulated to employ approximately 765,000 people in 2010. In fact, the estimate assumed that until 2015, the construction industry would not employ up to 800,000 people in a year. This remarkable performance was mainly stimulated and sustained by the projects under the second fiscal stimulus package such as the upgrading of roads, maintenance of schools and government quarters and the allocation by the Ninth Malaysia Plan that needed to be accelerated before the end of 2010. The projects under the Ninth Malaysia Plan were double tracking rails, Penang's Second Bridge, KLIA2, Mass

Table 2.8 Total number employed, 1964–2012

Year	1963	1964	1965	1966	1967	1968	1969	1670	1971
Number	–	47,799	48,014	51,056	60,162	45,383	44,231	51,702	60,285
Year	1972	1973	1974	1975	1976	1977	1978	19790	1980
Number	71,614	80,135	101,679	106,812	109,190	140,622	150,004	142,012	–
Year	1981	1982	1983	1984	1985	1986	1987	1988	1989
Number	243,603	282,337	306,412	297,811	294,065	250,172	195,632	201,576	250,481
Year	1990	1991	1992	1993	1994	1995	1996	1997	1998
Number	305,547	362,748	425,285	447,177	507,516	533,500	627,369	–	551,866
Year	1999	2000	2001	2002	2003	2004	2005	2006	2007
Number	–	458,580	–	455,663	–	449,944	551,775	–	595,139
Year	2008	2009	2010	2011	2012	2013	2014	2015	2016
Number	–	602,694	974,488	–	1,027,900	–	–	–	–

Source Department of Statistics Malaysia Construction (2013a) (survey was not carried out for reference years 1980, 1997, 1999, 2001, 2003, 2006 and 2008.)

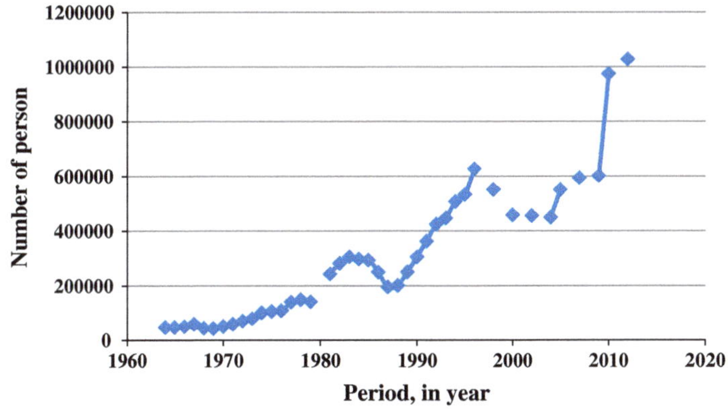

Fig. 2.1 Number of person engaged in the construction industry from 1964 to 2012. *Source* Department of Statistics Malaysia 2014b (constructed by the authors)

Railway Train, Iskander Malaysia and construction activities in the five Economic Corridors. The information contained in Table 2.8 is displayed in Fig. 2.1 for a pictorial representation.

The data from the Department of Statistics, indicates, that the distribution of the 1,027,900 people employed in the industry comprises of general workers (2 %), clerical (3.8 %), technical and supervisory (4.1 %), management and professional (3.7 %) and 86.3 % for operative staff. However, the 1,028,000 is not a static as it could increase or decrease due to the peculiar nature of the construction industry. However, it could be argued that the figure underestimates the performance of the industry. These are definition and methodological problems. The activities of the construction industry are considered solely by evaluating the value added to site activities. This would increase if the labour forces from the allied industries that produce, process and transport construction materials, components and services are added. The consultancies that are involved in the construction industry are also not considered. The statistics presented in this report refer to construction work done by registered contractors. Data for reference years 1963–1970 refer to the principal statistics of the construction industries in Peninsular Malaysia while data since reference year 1971 refer to Malaysia. Since reference year 1998, the survey covers establishments with the value of construction work done of RM500,000.00 and above. Therefore, if the workforces of those that are not directly working on sites are added, the contribution of the construction industry could be in the region of 10 % of the total workforce.

Table 2.9 reveals that, individually, the turnover of most (85.9 %) of the business units is less than five million. In other words, 86 % of the businesses delivered a combined output of RM17 billion or 16 % of the investments and employed approximately 240,000 persons. The large company with output size exceeding RM100,000,000 constitutes less than 1 % in terms of establishments, employs nearly 260,000 persons, and produced the combined output of RM32,000,000,000. The interpretation of these statistics is that as an industry, the construction industry is

Table 2.9 Number of establishments, gross outputs and employment size group 2012

Outputs size group	Establishment	Gross outputs		Employment	
RM million	% share	RM billion	% share	Number	% share
Total	100	110.1	100	1,027,900	100
<5	85.9	17.1	15.5	238,398	23.2
5 to <10	5.4	8.3	7.6	87.651	8.5
10 to <100	8.1	49.5	45.0	446.272	43.4
100 and above	0.6	35.1	31.9	255.609	24.9

Department of Statistics Malaysia (2012a)

fragmented and characterized by small units. Technically, the implication of this finding is that most units are subcontractors or trade contractors. While major contracting organizations employed approximately 200,000, their individual value added is RM235,000,000. This represents approximately 25 % of the total value added.

Table 2.10 represents the information of the gross outputs of the construction industry in 2012. The indicators designate that the value added in 2005 and 2012 are RM19 billion and RM327 billion respectively with an annual growth of around 15 %. The salaries and wages of the construction workers have expanded from RM11 billion to RM22 billion. This represents an approximate 100 % increment in five years. More than 85 % of full time employers in the industry are site operative. While less than 4 % are either professionals or managers, and 4 % are clerical staff. When the salary/wages of the professional/manager is considered, they are accountable for nearly 13 % while general operative costs account for 75 %. In 2013, construction industry was the main contributor of employment in the informal sector at 19.6 percent (Department of Statistics 2013b). The construction industry contributed about RM billion to the GFCF in 2012 which is more than 100 % compared to its contribution in 2005.

Figure 2.2 displays the contributions of major construction sectors in real terms. Most of the projects are new projects. However, while the sizes of new projects fluctuate, that of maintenance work (i.e. renovation, repair, and maintenance and upgrading) is uniform. A major implication of these results is that only a small amount of the investment in construction is channelled to maintenance, which could be interpreted to mean that most of the projects are still new. The situation is likely

Table 2.10 Key Indicators n Construction Sector, 2012

Indicator	2005	2010	2012
Value Of gross outputs (RM billion)	54	91.3	110.1
Value of intermediate inputs (RM billion)	35	59.4	73.1
Value added (RM billion)	19	31.9	37
Total person engaged (number)	551,755	974,488	1,027,900
Salaries and wages (RM billion)	10.9	19.8	22.4
Value of fixed assets (RM billion)	5.5	14.5	14.1

Department of Statistics Malaysia (2012a)

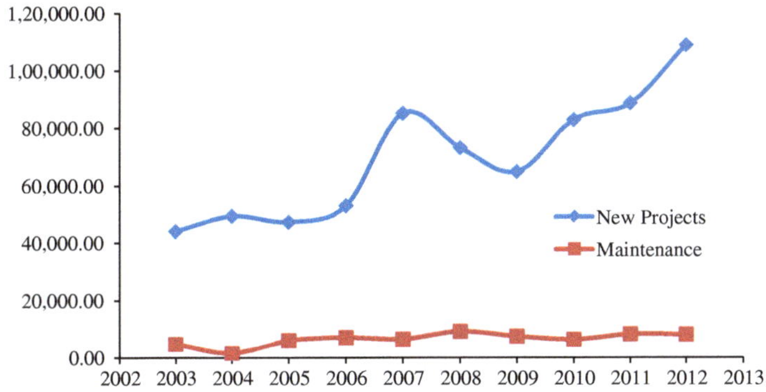

Fig. 2.2 Value added for construction works (RM million) Constructed based on data obtained from CIDB Malaysia

to remain like this for some time, as massive infrastructural projects have been unveiled to achieve the targeted developed nation status. However, the allocations to the maintenance sector are grossly inadequate, and even if the expenditure for maintenance is under reported, the government needs to invest in this sector as a large number of constructed facilities require maintenance.

The spill over impact from the projects of the Economic Transformation Programme (ETP), which started in 2010, was reflected in the performance of the construction sector for the year 2012. The year 2012 was remarkable for the industry, as the industry recorded highest growth and output of 18.5 % (Table 2.11) This high growth is the highest since 1995 (21.1 %). Correspondently, the GDP only grew at 5.6 % in 2012. The unprecedented expansion in the growth of the construction industry is explained by the civil engineering and residential sectors and reflects government efforts to expand and improve the roads and rail network, improve electricity generation and increase oil and gas outputs. In 2012, the construction sector recorded a value of gross output of RM110.1 billion with the

Table 2.11 GDP by kind of economic activity at constant 2005 prices—annual percentage change

Economic activity	2006	2007	2008	2009	2010	2011	2012	2013
Agriculture	5.8	1.4	3.8	0.1	2.4	5.8	1.3	2.1
Mining and quarrying	−1.2	2.1	−2.4	−6.5	−0.3	−5.4	1.0	0.7
Manufacturing	7.4	3.1	0.8	−9.0	11.9	4.7	4.8	3.5
Construction	−0.5	8.5	4.4	6.2	11.4	4.7	18.6	10.9
Services	7.2	10.0	8.6	2.9	7.4	7.1	6.4	5.9
Import duties	−11.0	4.7	26.7	−7.1	9.6	13.0	15.6	5.8
GDP at purchasers' prices	5.6	8.3	4.8	−1.5	7.4	5.2	5.6	4.7

Department of Statistics Malaysia (2014a)

intermediate input of RM73.1 billion. The value added was RM37.0 billion, while the total number of workers involved totalled 1,027,900 persons with the total salaries and wages paid at RM22.4 billion in 2012. Meanwhile, the value of fixed assets owned in 2012 were RM14.1 billion with the largest capital expenditure contributed by the machinery and equipment amounting to 50 % of the total capital expenditure during the reference year (Department of Statistics 2012a).

One of the prime movers of the SME is the construction industry. Construction recorded a strong growth of 15.3 %, while SME grew at 6 % in 2012 (Table 2.12). The construction industry contributed 2.8 % to the share of the SMEs. In 2012, the share of SMEs value added to the construction sector was 25.9 %. The construction industry maintains a consistent contribution to the SMEs (Table 2.13). This contribution is low, but with the nature of the construction industry is not surprising.

The contributions of women in the construction industry are small (Table 2.14). This refers to the establishment wherein at least 51 % of the equity is held by woman/women or the largest shareholder is a woman, or the establishment is run by a woman, or the Managing Director or Chief Executive Officer is a woman who owns at least 10 % of the equity. Women-owned establishments in the construction sector accounted for 6.2 % and generated output of RM4.4 billion (4.8 %) of the total sector in 2010. Approximately 54,000 persons were employed, with salaries

Table 2.12 Value added of smes by kind of economic activity at constant 2005 prices—annual percentage change

Economic activity	2006	2007	2008	2009	2010	2011	2012
Agriculture	8.3	3.3	2.3	1.2	5.9	6.2	1.9
Mining and Quarrying	4.9	12.7	1.6	1.8	3.5	5.5	14.6
Manufacturing	5.7	7.2	0.7	−7	11.2	7.6	6
Construction	0.3	12.4	2.5	6.9	16	4.6	15.3
Services	6.8	12.6	9.8	2.6	7	6.8	5.9
Import duties	−12.3	23	68.1	11.7	48	33.1	29.6
Total (SMEs expansion)	6.4	10	6.5	0.2	8.3	7	6

Department of Statistics Malaysia (2012b)

Table 2.13 Value Added of SMEs by Kind of Economic Activity at Constant 2005 Prices—Percentage Share to GDP

Economic activity	2005	2006	2007	2008	2009	2010	2011	2012
Agriculture	3.4	3.5	3.4	3.3	3.4	3.4	3.4	3.3
Mining and quarrying	0.1	0.1	0.1	0.1	0.1	0.1	0.1	0.1
Manufacturing	8.1	8.1	8.2	7.8	7.4	7.7	7.9	7.9
Construction	0.7	0.7	0.7	0.7	0.8	0.8	0.8	0.9
Services	17	17.2	18.2	19.1	19.9	19.8	20.1	20.2
Import duties	0.1	0.1	0.1	0.2	0.2	0.2	0.3	0.4
Total	29.4	29.7	30.7	31.2	31.8	32	32.6	32.8

Department of Statistics Malaysia (2012b)

Table 2.14 Key indicators of women owned establishment, in construction sector, 2010

Key Indicators	Total construction	Women-owned establishment	% share
Establishments (number)	22,140	1,371	6.2
Value of gross outputs (RM billion)	91,3	4.4	4.8
Value of intermediate inputs (RM billion)	59.3	2.8	4.7
Value added (RM billion)	31.9	1.5	4.7
Total person engaged (number)	974,488	53,816	5.5
Salaries and wages (RM billon)	19.8	0.9	4.5
Value of fixed assets (RM billion)	14.5	0.6	4.1

Department of Statistics Malaysia (2011)

and wages paid of RM0.9 billion (4.5 %). The value added to the fixed assets accumulated by this women-lead business was RM0.6 billion (4.1 %) in comparison to the total fixed assets (RM14.5 billion) recorded during the census period. The involvements of foreign contractors in the construction industry are minimal. In 2011 while only six government projects were undertaken by the foreign, 120 of the private projects has foreign involvement. In terms of value, 10 % of the projects awarded to foreign contractors are government while the remaining is from the private sectors (Table 2.15). However, in 2012, the involvement of foreign contractors in has dropped significantly. A similar situation is also experienced in 2013. In fact, the government does not award any contract to the foreign contractors in 2013. Regardless of whether the contracts were awarded to local contractors or foreign contractors, most of the works are awarded based on the traditional procurement method. For the last three years, more than 60 % of the contracts awarded to the foreign contractors were for civil engineering, or electrical engineering or mechanical engineering. The EPCC (engineering procurement construction and commissioning) contracts are normally refers to oil & gas projects only. Under the contract, the contractors designs the installation, procures the necessary materials and build the projects.

2.2 The Value of Maintenance Sub-sector

Table 2.16 list the number and value of projects awarded categorised in terms of specialisation. Apart from new projects, other activities are classified into upgrading, renovation, repair, expansion or maintenance. Again, due to methodological issues, data on the construction industry is underreported and maintenance works like repairs, renovation, and upgrading were considered exclusive items. However, both in practice and theory, this is not the case; the lines are not that clear. For this purpose, and is this text, maintenance includes repair, renovation and upgrading.

Table 2.15 Number and Value of Projects Awarded By Status of Contractors and Type of Contract as of December 2013

Type of Contract	Total Number of Projects	Total Project Value (RM M)	Local Contractors				Foreign Contractors			
			Government Projects		Private Project		Government Projects		Private Project	
			Number	Value (RM M)	Number	Value (RM M)	Number	Value (RM M)	Number	Value (RM M)
2011	7,605	99,461.61	1,905	21,408.19	5,574	59,451.14	6	1,610.74	120	16,991.54
Conventional	7,324	91,442.74	1,805	19,429.96	5,412	57,822.53	5	627.49	102	13,562.76
Design and build	159	4,251.60	64	1,670.79	86	954.18	1	983.25	8	643.38
Turnkey	99	3,502.31	34	300.84	57	419.65	–	–	8	2,781.82
Built, operate and transfer	5	41.11	–	–	5	41.11	–	–	–	–
Engineering, procurement, construction and commissioning	18	223.85	2	6.60	14	213.67	–	–	2	3.58
2012	7,650	123,600.42	1,940	18,116.04	5,567	91,216.52	1	144.86	142	14,123.00
Conventional	7,217	113,004.77	1,785	14,373.44	5,319	87,422.16	1	144.86	112	11,064.31
Design and build	247	6,778.39	103	2,767.27	124	1,731.33	–	–	20	2,279.79
Turnkey	114	3,052.90	38	867.67	67	1,429.36	–	–	9	755.87
Built, operate and transfer	42	505.71	8	32.80	34	472.91	–	–	–	–
Engineering, procurement, construction and commissioning	30	258.65	6	74.86	23	160.76	–	–	1	23.03
2013	5,985	92,304.06	1,481	19,915.09	4,423	62,451.30	0	0.00	81	9,937.67
Conventional	5,717	76,063.25	1,377	14,466.88	4,269	57,777.11	–	–	71	3,819.26

(continued)

Table 2.15 (continued)

Type of Contract	Total Number of Projects	Total Project Value (RM M)	Local Contractors				Foreign Contractors			
			Government Projects		Private Project		Government Projects		Private Project	
			Number	Value (RM M)	Number	Value (RM M)	Number	Value (RM M)	Number	Value (RM M)
Design and build	156	4,248.63	77	2,807.54	74	1,105.20	–	–	5	335.89
Turnkey	69	11,434.28	18	2,615.88	47	3,053.77	–	–	4	5,764.63
Built, operate and transfer	21	87.00	6	10.43	15	76.57	–	–	–	–
Engineering, procurement, construction and commissioning	22	470.90	3	14.36	18	438.65	–	–	1	17.89

Source CIDB Malaysia (2013a)

Table 2.16 Number and value of projects awarded by type of work and work specialization as of December 2006[a], 2010[b], 2013[c]

Work spsecialization	Total projects	Total value	New projects		Upgrading		Expansion		Maintenance		Repair		Renovation	
2003	**4,541**	**49,559.88**	**3,649**	**43,980.75**	**281**	**2,528.34**	**178**	**920.25**	**172**	**881.52**	**152**	**796.91**	**109**	**452.11**
Building	2,909	25,213.82	2,534	23,680.07	66	317.14	151	483.78	17	138.85	40	167.44	101	426.54
Civil engineering	1,404	16,899.50	960	13,416.21	193	2,158.71	23	53.47	121	653.68	10.	596.44	4	20.99
Electrical	116	7,054.58	78	6,608.39	11	23.96	2	369.95	21	48.18	1	0.59	3	3.51
Mechanical	112	391.98	77	276.08	11	28.53	2	13.05	13	40.81	8	32.44	1	1.047
2004	**4,881**	**52,694.36**	**4,117**	**49,257.87**	**214**	**1,280.05**	**147**	**565.86**	**125**	**708.85**	**155**	**666.07**	**123**	**215.66**
Building	3,325	29,221.00	2,964	28,112.21	66	162.58	118	434.81	20	224.80	45	82.98	112	204.17
Civil engineering	1,312	16,722.17	978	14,685.61	133	964.73	19	94.96	73	406.01	102	562.96	7	7.88
Electrical	132	6,303.91	95	6,066.09	9	139.41	6	24.99	19	36.02	5	3.04	1	1.36
Mechanical	112	447.28	80	360.96	6	13.33	4	11.10	13	42.05	6	17.62	3	2.25
2005	**5,565**	**54,277.08**	**4,686**	**47,183.64**	**289**	**3,066.34**	**107**	**1,162.68**	**89**	**865.90**	**216**	**1,416.06**	**178**	**582.46**
Building	3,691	35,896.84	3,279	33,643.65	80	372.07	86	989.11	20	147.11	66	218.79	160	525.26
Civil engineering	1,562	15,748.62	1,165	11,491.22	180	2,580.58	15	71.43	50	374.11	142	1,182.78	10	48.50
Electrical	179	1,311.48	144	1,128.80	18	86.72	2	80.90	9	8.74	3	3.80	3	2.52
Mechanical	133	1,320.14	98	919.97	11	26.97	4	21.24	10	335.09	5	10.69	5	6.18
2006	**5,854**	**60,926.99**	**4,888**	**52,896.23**	**3,160**	**4,145.83**	**137**	**1,055.48**	**125**	**972.90**	**188**	**1,219.67**	**200**	**636.88**
Building	3,736	36,374.37	3,284	34,005.92	77	470.72	108	490.26	20	576.30	61	203.57	186	583.48
Civil engineering	1,712	21,050.78	1,282	15,605.92	226	3,650.85	21	542.28	57	226.42	114	978.83	12	46.61
Electrical	236	1,893.66	188	1,720.05	6	10.82	4	7.36	33	114.54	4	9.33	1	1.57
Mechanical	170	1,608	134	1,520.22	7	13.77	4	15.59	15	25.64	9	27.94		5.02
2007	**7,358**	**93,294.20**	**5,845**	**85,081.63**	**448**	**3,124.03**	**351**	**1,848.03**	**183**	**1,138**	**268**	**1,303.99**	**263**	**798.23**
Building	4,719	51,658.31	4,009	47,813.18	99	605.24	305	1,693.38	27	272.87	60	566.77	219	706.54

(continued)

Table 2.16 (continued)

Work spsecialization	Total projects	Total value	New projects		Upgrading		Expansion		Maintenance		Repair		Renovation	
Civil engineering	2,150	37,057.53	1,463	33,108.88	321	2,429.77	33	143.08	108	606.03	192	694.24	38	75.54
Electrical	269	1,807.08	210	1,637.34	11	20.98	9	5.91	28	110.33	9	28.32	4	4.21
Mechanical	220	2,771.27	163	2,522.24	17	68.03	4	5.67	22	149.05	7	14.67	7	11.61
2008	**6,522**	**85,837.08**	**4,875**	**73,124.48**	**565**	**5,410.41**	**287**	**3,628.76**	**259**	**1,313.48**	**254**	**1,229.51**	**282**	**1,130.44**
Building	4,081	55,137.55	3,277	48,432.11	195	2,344.49	237	2,606.83	44	373.17	79	345.62	249	1,033.32
Civil engineering	1,947	24,770.75	1,284	19,448.75	311	2,720.80	35	960.73	134	726.68	163	847.23	20	1,033.32
Electrical	299	3,003.60	201	2,634.45	36	225.72	10	34.09	42	79.90	7	27.12	3	2.32
Mechanical	195	2,925.18	113	2,609.18	23	119.40	5	25.11	39	133.73	5	9.530	10	28.24
2009	**7,039**	**74,913.63**	**4,990**	**64,757.29**	**726**	**4,055.88**	**563**	**2,859.23**	**174**	**916.07**	**322**	**1,379.13**	**264**	**946.02**
Building	4,558	48,457.95	3,381	42,618.48	307	1,865.07	516	2,465.83	22	199.47	107	464.36	225	844.74
Civil engineering	1,926	18,358.93	1,213	14,604.18	379	1,995.32	35	347.34	82	500.23	193	843.30	24	68.56
Electrical	342	2,865.24	243	2,604.55	21	56.00	8	33.07	46	108.01	17	45.19	7	18.42
Mechanical	213	5,231.52	153	4,930.08	19	139.50	4	13.00	24	108.36	5	26.28	8	14.30
2010	**7,302**	**91,008.49**	**5,830**	**82,793.62**	**499**	**3,064.61**	**271**	**1,932.47**	**245**	**1,181.46**	**217**	**1,101.01**	**240**	**935.33**
Building	4,679	61,105.35	4,044	57,034.82	107	778.62	232	1,772.04	22	90.12	64	590.11	210	839.64
Civil engineering	1,926	16,652.15	1,294	13,402.49	336	2,070.61	27	99.01	110	542.07	136	467.42	23	70.55
Electrical	335	6,971.15	240	6,556.80	23	68.09	4	13.89	55	294.57	8	18.62	5	19.19
Mechanical	362	6,279.84	252	5,799.51	33	147.29	8	47.54	58	254.71	9	24.86	2	5.95
2011	**7,605**	**99,461.30**	**6,160**	**88,888.50**	**515**	**4,706.21**	**177**	**2,394.39**	**192**	**1,191.59**	**253**	**1,236.43**	**308**	**1,044.49**
Building	4,797	60,621.39	4,169	57,412.56	139	747.59	139	970.39	33	124.65	80	476.67	237	889.54
Civil Engineering	2,162	24,224.89	1,513	17,662.78	335	3,528.84	19	1,375.03	93	848.79	147	690.65	55	118.79

(continued)

Table 2.16 (continued)

Work spsecialization	Total projects	Total value	New projects		Upgrading		Expansion		Maintenance		Repair		Renovation	
Electrical	314	8,522.79	230	8,268.73	25	121.27	11	34.11	28	68.44	16	20.59	4	9.65
Mechanical	332	6,092.53	248	5,544.43	16	308.51	8	14.87	38	149.71	10	48.52	12	26.51
2012	**7,650**	**123,600.41**	**6,044**	**112,778.68**	**516**	**3,131.75**	**179**	**2,373.62**	**327**	**2,283.55**	**247**	**1,625.84**	**337**	**1,406.96**
Building	4,885	70,391.20	4,135	66,308.41	178	1,193.06	133	1,108.21	90	431.99	72	324.85	277	1,024.68
Civil Engineering	2,149	38,020.56	1,510	33,064.14	280	1,697.95	28	1,167.33	152	1,204.11	139	697.45	40	189.60
Electrical	327	8,950.20	215	8,301.67	35	137.65	15	94.17	44	372.91	11	23.15	7	20.65
Mechanical	289	6,238.45	184	5,104.47	23	103.09	3	3.91	41	274.54	25	580.39	13	172.04
2013	**5,985**	**92,304.04**	**4,648**	**84,086.64**	**486**	**3,306.12**	**114**	**1,021.99**	**189**	**1,070.62**	**317**	**1,349.21**	**231**	**1,469.45**
Building	3,548	60,916.69	3,069	57,976.48	113	602.43	81	604.01	12	56.79	86	466.92	187	1,210.06
Civil Engineering	2,003	19,108.00	1,287	14,334.01	328	2,572.93	22	312.90	118	848.08	213	805.75	35	234.35
Electrical	223	5,357.65	156	5,105.66	27	81.44	8	83.81	18	40.48	9	33.31	5	12.95
Mechanical	211	6,921.70	136	6,670.49	18	49.33	3	21.28	41	125.27	9	43.23	4	12.10

[a] *Source* CIDB Malaysia (2006)

[b] *Source* CIDB Malaysia (2009 and 2010)

[c] *Source* CIDB Malaysia (2013b)

You may wish to see the definition we provided for maintenance (see Sect. 3.2.4 page 43). But if for the purpose of argument and understanding, maintenance includes the activities we just considered in the last sentence; 'maintenance' is only accountable to 8.5 % of the total expenditure. But if we are conservative, the maintenance and repair are only considered, the amount is contribution is reduced to 2.96 % of the total investment in the construction industry. If we are more categorical, as the table indicates, the contribution of maintenance is reduced to a mere 1. 44 % of the total investment. In effect, the argument here is that there is insufficient allocation or investment for maintenance. Whether one considers maintenance as 1. 44 %, 2.96 % or even the 8.5 %, the amount is inadequate in many respects. The issue is that proper attention is not accorded to maintenance. In reality, the required amount for maintenance should be more than 10 % of the total expenditure considering the annual contribution of new buildings to the exiting stocks.

We considered that the government is not doing enough in terms of allocation to the maintenance sector. Most developed countries and some ambitious developing countries have observed the roles of maintenance and now allocate roughly 50 % of annual expenditure to the industry to the maintenance sector. It was also suggested that about 5 % of the replacement values of the built assets is also applicable. Table 2.17 presents the sectoral expenditure for maintenance. In terms of methods of procurement or types of contract, the traditional method is the dominant in the period selected for the different categories of works (Table 2.18).

2.3 Conclusion and Future Outlook of the Construction Industry

The outlook of the Malaysian construction industry in quite impressive though the global economic climate is volatile. The economy is postulated to remain on steady growth, with an expansion of around 6 % towards 2020 and beyond. Government spending is expected to reduce as the role of the private sector increases. Meanwhile, in line with a private sector led economy, government has outlined various measures to spur private participation to enable the private sector to be the engine of growth. Advancing towards the 2020 target, the government will adopt new privatizations policies. Furthermore, there will be equitable risk sharing and more collaboration particularly in high growth and strategies between the public and private sectors. Tendering processes will be more competitive particularly for engineering projects, and PPP system will dominate procurements and the construction industry will be dominated by the local contractors.

While the services and manufacturing sectors will dominate economic growth, the growth of the construction industry is postulated to increase and remain resilient mainly because of the implementation of major projects. Though many projects including the KLIA 2, Second Penang Bridge, The Iskandar Development Region (IDR), Northern Corridor Economic Region (NCER), East Coast Economic Region

Table 2.17 Number and value of projects awarded by type of work and project category, 2011–2013

Project category/ Year	Total number of projects	Total project value (RM M)	New project		Upgrading		Expansion		Maintenance		Repair		Renovation	
			Number	Value (RM M)	Number	Value (RM M)	Number	Value (RM M)	Number	Value (RM M)	Number	Value (RM M)	Number	Value (RM M)
2011	**7605**	**99,461.60**	**6,160**	**88,888.51**	**515**	**4,706.21**	**177**	**2,394.39**	**192**	**1,191.58**	**253**	**1,236.43**	**308**	**1,044.48**
Residential	2,257	24,642.92	2,141	24,229.97	21	30.19	12	53.09	22	60.61	30	106.74	31	162.32
Non residential	2,566	35,806.83	2,033	33,522.73	90	732.24	96	410.72	62	177.30	50	185.69	235	778.15
Social amenities	832	6,808.50	625	5,644.25	77	358.05	44	419.49	33	104.11	23	216.64	30	65.96
Infrastructure	1,950	32,203.35	1,361	25,491.56	327	3,585.73	25	1,511.09	75	849.56	150	727.36	12	38.05
2012	**7,650**	**123,600.40**	**6,044**	112,778.68	**516**	**3,131.75**	**179**	**2,373.61**	**327**	**2,283.55**	**247**	**1,625.85**	**337**	**1,406.96**
Residential	2,206	32,375.11	2,072	31,991.25	25	33.64	11	75.05	66	132.91	19	103.85	13	38.41
Non residential	2,636	39,263.42	2,055	35,075.87	115	842.10	98	834.13	78	675.69	43	700.43	247	1,135.20
Social amenities	894	7,492.19	608	6,409.13	102	407.52	39	186.53	42	160.66	46	178.69	57	149.66
Infrastructure	1,914	44,469.68	1,309	39,302.43	274	1,848.49	31	1,277.90	141	1,314.29	139	642.88	20	83.69
2013	**5,985**	**92,304.05**	**4,648**	**84,086.64**	**486**	**3,306.13**	**114**	**1,021.99**	**189**	**1,070.62**	**317**	**1,349.21**	**231**	**1,469.46**
Residential	1,636	27,693.99	1,566	27,322.58	11	19.45	13	39.37	10	108.79	22	159.32	14	44.48
Non residential	2,266	38,741.75	1,818	35,893.42	103	379.31	56	597.85	59	308.05	48	265.41	182	1,297.71
Social amenities	540	7,050.97	372	6,212.26	60	383.80	18	116.88	20	44.10	47	225.69	23	68.24
Infrastructure	1,543	18,817.34	892	14,658.38	312	2,523.57	27	267.89	100	609.68	200	698.79	12	59.03

Source CIDB Malaysia (2013c)

Table 2.18 Number and value of projects awarded by type of work and type of contract as of December 2013

Type of contract	Total number of projects	Total project value (RM M)	New project		Upgrading		Expansion		Maintenance		Repair		Renovation	
			Number	Value (RM M)	Number	Value (RM M)	Number	Value (RM M)	Number	Value (RM M)	Number	Value (RM M)	Number	Value (RM M)
2011	**7,605**	**99,461.61**	**6,160**	**88,888.50**	**515**	**4,706.21**	**177**	**2,394.38**	**192**	**1,191.60**	**253**	**1,236.43**	**308**	**1,044.49**
Conventional	7,324	91,442.74	5,953	82,531.20	491	4,265.97	165	1,424.81	187	1,170.39	240	1,057.03	288	993.34
Design and build	159	4,251.59	114	3,543.54	15	327.03	9	294.07	1	14.45	6	42.69	14	29.81
Turnkey	99	3,502.31	80	2,672.24	7	27.09	3	675.50	1	4.60	5	119.93	3	2.95
Built, operate and transfer	5	41.12	3	29.04	1	0.68	–	–	–	–	–	–	1	11.40
Engineering, procurement, construction and commissioning	18	223.85	10	112.48	1	85.44	–	–	3	2.16	2	16.78	2	6.99
2012	**7,650**	**123,600.41**	**6,044**	112,778.68	**516**	**3,132**	**179**	**2,373.62**	**327**	**2,283.55**	**247**	**1,625.83**	**337**	**1,406.97**
Conventional	7,217	113,004.76	5,718	102,957.99	476	2,839.73	158	2,224.90	320	2,200.97	234	1,490.55	311	1,290.62
Design and build	247	6,778.38	180	6,318.54	25	194.88	14	41.68	–	–	6	110.22	22	113.06
Turnkey	114	3,052.90	94	2,868.67	10	85.10	4	85.34	–	–	5	12.64	1	1.15
Built, operate and transfer	42	505.71	31	459.04	2	1.99	3	21.70	3	9.92	2	12.42	1	0.64
Engineering, procurement, construction and commissioning	30	258.66	21	174.44	3	10.06	–	–	4	72.66	–	–	2	1.50
2013	**5,985**	**92,304.04**	**4,648**	84,086.64	**486**	**3,306.12**	**114**	**1,021.99**	**189**	**1,070.62**	**317**	**1,349.21**	**231**	**1,469.46**
Conventional	5,717	76,063.24	4,452	68,415.73	463	3,099.55	103	907.82	183	1,044.45	296	1,165.04	220	1,430.65
Design and build	156	4,248.62	109	3,825.58	16	126.46	6	88.07	3	10.17	17	170.85	5	27.49
Turnkey	69	11,434.28	54	11,314.74	4	76.03	5	26.10	–	–	2	8.66	4	8.75
Built, operate and transfer	21	87.00	14	75.19	2	3.32	–	–	1	1.26	2	4.66	2	2.57
Engineering, procurement, construction and commissioning	22	470.90	19	455.40	1	0.76	–	–	2	14.74	–	–	–	–

Source CIDB Malaysia (2013d)

(ECER), Sabah Development Corridor (SDC) and Sarawak Corridor of Renewable Energy (SCORE), that spur economic previous expansion in the industry have been completed or already at advanced stages, there remains many mega projects that are expected to stimulate growth in the industry in the years to come. These projects include the MRT, the HSR (the High Speed Rail connecting Malaysia with Singapore), expansion of the KTM rails, LRT extension, Pahang-Selangor Raw Water Transfer, Melaka LNG Regasification, and the Pengerang projects. As at the end of 2009, a total of RM 245.49 billion was secured for Iskandar Malaysia, Northern Corridor Economic Region, Eastern Corridor Economic Region, Sabah Development Corridor and Sarawak Corridor of Renewal Energy (Government of Malaysia 2010)

The HSR, which is expected to commence next year is worth an approximate RM40 billion. The MRT and HSR apart of the cost of construction will spur development in the major cities in which the trains have a stopover. The Pengerang integrated projects is worth more several billion. Another development that sustains growth in the industry is the Greater KL project. The Greater KL programme will impinge highly on the demand for both residential and non-residential buildings.

The industry will continue to be dominated by the private sector. There will also be a reduction on the dependency on foreign labour in the industry. The applications of the Industrialised Building System (IBS) will increase as the government has granted accelerated capital allowance for up to 3 years for companies with purchase on mould for the production of IBS components. Similarly, government emphasis on the consideration of sustainable development will be intensified. To this end, government has introduced the Green Building Index (GBI) and tax exemption of close to 100 % for expenditure incurred in obtaining the first GBI certificate. Chapter 10 of this book contains some of the government's efforts towards sustainable development (see Table 10.1 on page 300). Based on our observation, while the demand for new projects will increase, we expect a surge in maintenance expenditure. Maintenance expenditure will more than double its current value. Initiatives like The National Asset and Facility Management (NAFAM) convention need to be revived. The previous convention in our opinion offered awareness and constituted a platform for the maintenance of built assets in Malaysia.

In summary, there are many opportunities for local contractors and professionals to export their services abroad. However, while the future of the Malaysian construction industry is optimistic, the industry is also facing a number of challenges. For instance, there are problems of instabilities of supply and prices of construction materials and components. Prices of construction materials and components are one of the highest in the region. These problems are leading to the poor performance of the construction deliveries. Another problem facing the players pertains to the Stamp Duty. According to the Master Builders Association Malaysia, a payment default is considered a consistent problem facing the industry. For instance, RM5 is to be paid for every RM1000 or part therefore effective 2009 for loan agreement and service agreement instruments except for education loans. This will affect the profit and productivity of the industry. The number of litigations due to claims and disputes are also on the rise. There are also problems relating to labour. There is a lack of competent labour to meet the demands of the industry. While we are of the understanding that for a country to have meaningful and sustained development, it must indigenize its construction industry, but to increase healthy competition in the construction, foreign involvement is encouraged. Local contractors should form alliance or joint venture with some foreign construction companies to deliver some the projects. There are some advantages with the involvement of foreign collaborations like project financing and knowledge transfer. In addition, more investment is required from both the private and public sector for research and development (R&D), and more funds and required from banks and other financial institutions. The contributions of the industry to the GDP and GNI at around 3 % are comparatively low. A significant improvement is required to increase the contributions of the industry. Proactive methods are required in disposal of waste. The Malaysian construction industry is synonymous to waste generator and the situation is further worsening and complex due to lack of suitable landfills. The waste that the industry produces create avenue to the some stakeholders (NGO, environment etc.) to see the industry as major cause of pollution. While the large construction business has good waste disposal programme the small and medium classes do not. However, dumping the waste into dump sites is not an adequate way to waste management. We suggest change in attitudes of the stakeholders. Emphasis should be given to waste minimisation and where possible waste avoidance over waste disposal no matter how effective this could be. The attitudes of the industry toward waste management are mainly to meet regulatory requirements. The industry should educate the clients on reuse and recycle of 'waste' materials and components. The government efforts towards climate change and green technology also pose greater challenges to the construction industry as the have significant impacts on how the constrction industry designs and constructs constructed facilities. To address the problems pose by the climate change /gboabl warming, Malaysia as adopted the Low Carbon Cities Concept (LCCC). Earlier on, a Low Carbon Cities Framework (LCCF) was formulated through governments collaborations to serve as guide for stakeholders including developers, contractors, designers local councils and town planners to reduce the levels of carbon emission in cities.

References

Construction Industry Development Board (CIDB) Malaysia. (2006). Construction statistics quarterly bulletin number and value of projects awarded by type of work and work specialisation, 2003–2005. https://www.cidb.gov.my/cidbweb/images/pdf/buletin/2006/B2_Q4_06.pdff.

Construction Industry Development Board (CIDB) Malaysia. (2009). Construction statistics quarterly bulletin. Number and value of projects awarded by type of work and work specialisation as of December 2009 (Table 2.17). Accessed on July 2013 from https://www.cidb.gov.my/cidbweb/images/pdf/buletin/2009/B2_Q4_09.pdf.

Construction Industry Development Board (CIDB) Malaysia. (2010). Construction statistics quarterly bulletin. Number and value of projects awarded by type of work and work specialisation as of December 2011 (Table 2.17). Accessed on July 2013 from https://www.cidb.gov.my/cidbweb/images/pdf/buletin/2011/BahagianKeduaQ42011.pdf.

Construction Industry Development Board (CIDB) Malaysia. (2013a). Construction statistics quarterly bulletin. Number and value of projects awarded by status of contractors and type of contract as of December 2013 (Table 2.18). Accessed on 23 February 2014 from http://www.cidb.gov.my/cidbv3/images/pdf/buletin/bhg%202.pdf.

Construction Industry Development Board (CIDB) Malaysia (2013b) Construction statistics quarterly bulletin. Number and value of projects awarded by status of contractors and work specialisation as of December 2013 (Table 2.16). Accessed on 23 February 2014 from http://www.cidb.gov.my/cidbv3/images/pdf/buletin/bhg%202.pdf.

Construction Industry Development Board (CIDB) Malaysia, (2013c) Construction statistics quarterly bulletin. Number and value of projects awarded by type of work and projects category as of December 2013 (Table 2.20). Accessed on 23 February 2014 from http://www.cidb.gov.my/cidbv3/images/pdf/buletin/bhg%202.pdf.

Construction Industry Development Board (CIDB) Malaysia. (2013d). Construction statistics quarterly bulletin. Number and value of projects awarded by type of work and type of contract as of December 2013 (Table 2.19). Accessed on 23 February 2014 from http://www.cidb.gov.my/cidbv3/images/pdf/buletin/bhg%202.pdf.

Department of Statistics Malaysia. (2011). *Economic census 2011, Construction.*

Department of Statistics Malaysia. (2012a). *Reports on the survey of construction industries*, ISSN 0128-3493.

Department of Statistics Malaysia. (2012b). Small and medium enterprises national accounts 2005–2012.

Department of Statistics Malaysia. (2013a). Construction. Accessed on 07 August 2014 from (http://www.statistics.gov.my/portal/download_Economics/files/DATA_SERIES/2013/pdf/17Pembinaan.pdf).

Department of Statistics Malaysia. (2013b). *Informal sector workforce survey report*, ISSN 2289-4691. Accessed on 23 September 2014 from http://www.statistics.gov.my/portal/index.php?option=com_content&view=article&id=2080&Itemid=169&lang=en.

Department of Statistics Malaysia. (2014a). *Annual Gross Domestic Product* 2005–2013, ISSN 1955-0751.

Department of Statistics Malaysia. (2014b). *Quarterly construction statistics, First Quarter 2014.*

Government of Malaysia. (2010). *Ten Malaysian Plan 2011–2015*. Putrajaya: Economic Planning Unit Prime Minister's Department.

Myers, D. (2013). *Construction Economics: A new approach* (3rd ed.). New York: Routledge.

Chapter 3
Maintenance of Academic Buildings

Abstract This chapter discusses maintenance issues in the university buildings in Malaysia. Building requires maintenance to perform optimally. While there commitments to raise the performance of the university building little emphasise is placed on maintenance. There is a need to tie maintenance of buildings the productivity of the academic sector. In this chapter, problems facing maintenance organisations are identified and further elaborated. The specific objectives and learning outcomes of this chapter include: (1) Be able to define building maintenance and its purposes. (2) Understand the problems facing maintenance organisations. (3) Have a feeling of the connection between building performance and quality of education. (4) Be able to observe the maintenance expenditure.

Keywords Value maintenance management model · Building maintenance · Malaysia

3.1 Introduction

This chapter starts with an outline of the Malaysian education system as background to the research and goes ahead to outline the significance of buildings to the universities. A correlation linkage is established between quality education, university buildings and maintenance. The role of maintenance in the building life cycle is concisely discussed. Buildings are procured to perform different roles. The roles that buildings are procured to perform depend primarily on functional requirements with technical solutions to the clients/user needs and wants. The needs and wants or value systems are of a form conveyed by the client's needs/wants when such needs/wants are shaped, configured and aligned by society, norm, perceptions, experiences and/or personality. University buildings have emblematic, cultural and moral implications on the users and the society at large. Building regardless of its purpose exists within communities and alongside other existing community value. University buildings must be places that reflect value placed on the students' learning, safety, aesthetic, comfort and convenience in order to fulfil

A.L. Olanrewaju and A.-R. Abdul-Aziz, *Building Maintenance Processes and Practices*, DOI 10.1007/978-981-287-263-0_3

their functional performance. The reason to maintain a building is that the building performance is significantly determined by the how, why and when the management of the building maintenance is conducted.

Buildings and their management need to be considered as investments rather than as liabilities. Buildings are factors of production to the universities, like human resources, capital, land, technology and information. In other words, buildings are not wanted for their own sake but for the service or product they help to provide or create. For instance, the purposes of the university building are to stimulate and sustain teaching, learning and research as well as to instil good values in the mind of the users. However, the value of a building decreases unless maintenance is administered on the building. In economic terms, when organisations invest more to maintain the building in relation to the rental income, the capital consumption is high. If expenditure is higher than the revenue, such organisations will run at a loss and will close shop for business. In order to sustain and improve building performance, maintenance is introduced. Maintenance interventions seek to delay defects in building and go ahead to provide optimum user satisfaction and better productivity. However, the management of the building and its maintenance requires a multi-disciplinary approach, from engineering, technological, commercial, cultural, economic and social perspectives.

The premise of this book is the move from the building and its management as the main focus for the maintenance management process, towards the value added initiative (to enhance productivity and user satisfaction as the main focus). The value drive is the focus on the systemic strategy of creating value to maintenance organisations and building users. This concept or idea is based on the value model approach. Fundamentally, it considers the value systems of both the service providers and service users in creating a maintenance service model. The value system is a complex set of criteria that explains the basis of the client/user *accepting* the project as the *solution* to their *problem.*

The model that is explained in this book is a philosophy that facilitates maintenance organisation to deliver satisfactory services and products to building users, reduce maintenance backlogs, reduce maintenance costs and increase maintenance organisation's productivity. Simply explained, value maintenance management practice entails making managerial decisions and organisational practices taking into account the complex algorithms that influence maintenance, behavioural issues and organisation objectives. Attentions are directed to meeting the users' maintenance needs effectively and efficiently by paying critical and careful attention to user value system, maintenance needs, criteria that influence maintenance decision making and resources.

3.2 An Overview of Malaysian Education System

Education is very significant and it is an integral part of the Malaysian Strategic Thrust. Education is one of the National Key Economic Areas (NKEAs). The NKEA is the government initiative unveiled in 2006 to propel Malaysia towards 2020. The NKEAs are the drivers of high economic activity that have the potential

Table 3.1 Development allocations for the educational and training sectors (RM Million)

Sectors	Six Malaysian plan	Seven Malaysian plan	Eight Malaysian plan	Nine Malaysian plan
Education	7,469.8	17,948.5	18,660	40,356.5
Training	615.4	2,237.3	4,000	4,792.6
Total	8,625.2	20,185.8	22,660	45,149.1

Source Government of Malaysia (1996, 2001, 2006)

to directly and materially contribute meaningfully to Malaysian economy. These are oil and gas; palm oil and related products; financial services; wholesale and retail; tourism; information and communications technology; electrical and electronics; business services; private healthcare; agriculture; and Greater Kuala Lumpur. The economy of Malaysia has, since the independence, been planned on 5 year strategic plans. For each of the plans, the education sector features prominently in terms of value and policy implementations. For instance, there is an increase of more than 420 % on allocation for the education sector over the last 20 years. Table 3.1 contains the federal development allocations to the education sector from 1990 to 2010. The increase in the allocations is a confirmation of the government's commitment to improve the performance of education system in order to produce human capital capable of competing locally and internationally.

In tandem with the expansion in the allocations for the education sector at all levels, allocations at the tertiary level are also increasing substantially. The illustration of this can be cited. Under the Eighth Malaysian Plan, the sum of RM8,900 million was allocated to institutions of higher learning out of the total RM18,660 million to the education sector (Government of Malaysia 2001). However, under the Ninth Malaysian Plan, allocation to institutions of higher learning increased to RM16,069 million (Government of Malaysia 2006). This represents an increment of more than 80 %. As a percentage of the total public expenditure, 4.4 % was attributable to higher education and education constitutes 2.1 % of the Malaysian GDP (MOHE 2006). This is massive compared to most countries as most developed and developing countries allocate only about 1 %. Table 3.2 contains the annual expenditure for education.

The percentages for education as federal allocations are 23.9, 21.9, 19.6, 19.8 and 17.1 in 2003, 2004, 2005, 2006 and 2007, respectively. Table 3.3 contains distributions of government expenditure to the public universities over a period of 5 years. The table indicates that a total of RM14,375,256,345.33 has been invested to the public universities for new construction and maintenance.

Higher education in Malaysia is a source of employment and foreign exchange and it is gradually becoming a major contributor to the economic sector as well as becoming a strong export source (MOHE 2006). Malaysia is fast emerging as an educational hub and a destination for nationals from different countries, particularly those from Africa, Asia Pacific, Europe, America and the Middle East. Malaysia's current global market share is about 3 % of total internationally mobile students and

Table 3.2 Principal statistics of education services 1971–2010, Malaysia (RM million)

Year	1971	1972	1973	1974	1975	1976	1977
Amount	15,985	23,385	27,751	32,614	38,312	45,805	53,282
Year	1978	1979	1980	1981	1982	1983	1984
Amount	58,537	67,275		95,095		123,317	155,339
Year	1985	1986	1987	1988	1989	1990	1990
Amount	169,539	187,690	201,203	213,863	239,956	275,476	329,633
Year	1991	1992	1993	1994	1995	1996	1997
Amount		426,126		626,383		974,491	
Year	1998	1999	2000	2001	2002	2003	2004
Amount		1,567,426			2,814,008	3,167,036	
Year	2005	2006	2007	2008	2009	2010	2011
Amount	4,727,323		5,424,138		5,705,293	8,508,074	

Source Service (2010)

Table 3.3 Public higher education institutions expenditures (2006–2010) (RM million)

Year	2006	2007	2008	2009	2010
Development and maintenance	1,711.8	2,804.0	1,718.4	4,310.9	3,830.9

Source MOHE (2010a)

aims to attract 150,000 international students by 2015 (Government of Malaysia 2010). To this ends, Ministry of Education is now aggressively promoting the *Edu-tourism* agenda to woo foreign students to study in Malaysia.

3.2.1 Roles of University in National Development

The Government of Malaysia has identified and recognised that universities have a major role to propel Malaysia into the innovation led high-income economy. Under the current Malaysian Plan, 10 MP (2011–2015) universities are expected to contribute significantly to the Malaysian high-income economy status. This entails the universities producing graduates that can and will compete nationally and internationally and at the same time for the universities to be able to commercialise their research discoveries and innovations.

The government is also set to grant autonomy to the public universities by reducing their direct control. Under this new arrangement, the government will only serve as a regulatory agent. The independencies of the universities in taking strategic decision are important to the latest government policy (the New Economic Model or NEM), of achieving educational excellence, innovation and internationalisation. At the moment, Five universities have been granted the Research Status. This is recognition of the concerted efforts that these universities have made in the areas of research and innovation. As requirements, these universities have special financial allocations for research and development from the government and corporate bodies. Malaysia's quest to transform into a knowledge-based economy (K-economy) where science, technology and engineering are integrated into the production process is high. That is, Malaysia desires to be a society, where the main prominent sources of national prosperity and wealth, i.e. creativity, imagination and design capability are embodied in well-educated skilled workers (Government of Malaysia 2006 and Malaysia and the Knowledge Economy 2007). This is to move the economy from production based to innovative based by making the GDP service led. The government is now critical of improving the service sector. Table 3.4 contains the sectoral distribution of allocation to key sectors of the GDP.

In order to achieve this objective, however, university assets must be adequate and appropriate in all respects. This involves a holistic overhaul of the performance of university systems, though. In fact, the government has stressed that Malaysian universities must excel in all aspects if really they must be an engine that will spur Malaysia towards being an innovation led high-income economy as well as being the major factor in defining, shaping and guiding the nation's workforces.

Table 3.4 Distribution of the GDP in terms of sectors

Amount	RM million (in constant 2,000 prices)						
	Actual				Est.	Target	
Sector	2006	2007	2008	2009	2010	2012	2015
Agric, forestry, livestock and fishing	37,701	38,177	39,828	39,992	41,497	4,42,121	48,812
Mining and Quarrying	42,030	42,881	41,831	40,246	41,522	42,093	43,875
Manufacturing	147,154	151,257	153,171	138,809	147,406	165,885	194,823
Construction	14,640	15,707	16,366	17,321	18,187	19,775	21,813
Services	247,099	272,406	292,555	300,154	320,089	366,704	452,221

Government of Malaysia (2010)

3.2.2 Academic Building as a Factor of Production

On the one hand, university assets are finance, technology, human capita, buildings, equipment and plant. On the other hand, university education is labour intensive, therefore human resource is considered its most significant resource. However, apart from the human resource, buildings are the most significant asset of a university institution. Therefore, while the methods of teaching, studying, learning and doing research require improvement, the building's fabric, structure and engineering services must be in optimum functional condition and high performance standard at all times.

Over the years, it has been recognised that there are positive correlations between performance of educational buildings and quality of education. The example of this can be referred. Based on a study conducted in the USA, it was found that children could do 5–11 % better on standardised tests if the physical condition of their schools improved (USA Weekend 1991). Even though this example is based on school children, the conditions and performance of university buildings affect students and faculty members' productivity and performance too. Similarly, report by California State University found that apart from competitive financial compensation, faculty members regard the quality of their buildings when taking up employment (California Postsecondary Education Commission 2007).

From the foregoing, it is not difficult to argue that university buildings are capital goods. This implies that the buildings are procured to create suitable, conducive, and adequate environments that support, stimulate and encourage learning, studying, teaching and scientific innovations. A university contains unique, complex, and sophisticated buildings used to perform various types of activities and functions. In fact, university buildings mirror the buildings in small cities. University buildings are also unique due to multiple occupancies: many students having varying and conflicting requirements and always in transition. Perhaps, more than in any other sector, the occupancies of the university institution are more complex and sophisticated-*educated and enlighten users*.

Buildings serve certain purposes, whether for accommodation, inspiration or otherwise. More than 90 % of our activities are conducted in or around the buildings. The example of this can be provided. Students and faculty members spend most if not all of their productive time in or around classrooms, laboratories, theatres or and tutorial rooms. It is through the university buildings that future leaders, captains of industry, entrepreneurs, professionals and scientists are produced (Mat et al. 2009). Universities need functional building to operate. Even the virtual universities require some minimum amount of buildings to carry out their business operations.

In order to explain the importance of buildings to quality education, we present a statement attributed to the technologies guru; Bill Gates in his bibliographies: "one cannot totally communicate through the technology; one needs to meet face to face to be able feel what is being uttered". The implication of this statement is that, the need for educational buildings will only intensify. More buildings are required to cater for soaring student enrolments. In monetary terms, the values of buildings accounts for some 30 % of university assets. Apart from the staff salaries, the costs of owning or leasing the buildings are the prominent issue on university balance sheets. These prominences are due to their significance. The cost of the property assets in comparison with the turnover of the university is high. This could be around 10 %. Therefore, any inadequacies in the buildings will serve as an obstacle towards achieving the university's corporate mission and vision.

Universities these days use the nature, design, condition and performance of their buildings to woo students. In Malaysia, universities that in the past relied on the applications sent to them through the Ministry of Higher Education now use their buildings as variable to attract students. This is also the case elsewhere, a visit to any university's website; the first image that appears is that of buildings.

Universities have different options to address their space requirements issues. Universities either rent or construct new buildings for use. But considering the universities are building based organisations, most universities construct theirs in the long run. Often private universities start by renting their take-off building before moving/relocating to a permanent campus. On the other hand, the governments have a tradition of allocating some of their buildings to the public universities to start from. However, constructing new buildings helps in upgrading educational facilities and providing better quality education. But building components have a predetermined life span. For this purpose, buildings cannot remain new throughout their design lives span. Before a building is completed, nature begins to destroy it systematically, and gravity, winds and seismic movements continuously test the stability of the buildings (Allen 1995/2005).

3.2.3 Roles of Maintenance in Building Performance

Birth, growth, maturity, decline, decay, death, and rebirth are fundamental stages in all natural cycles. This also applies to a building, although humans like to keep the cycle under control through maintenance until its *death* suit human purpose (Allen 1995/2005). The business objectives of universities have placed buildings in a strategic position. University buildings are facilitators and enablers to the university's mission and vision. The building provides value not only to the university organisation, but also to the students, faculty members, parents and other users and stakeholders. However, the buildings are not new factors of production per se, but the relative significance of the building performance in comparison with other resources of the university is significantly increased. A building's condition and performance constantly affects everyone's life because people's comfort and productivity is relative to the performance of the building they live, learn, conduct research and work in (e.g. home, offices, schools, university and markets), not to mention the impact it has on the social fabric and economic growth. Thus, any inadequacy in the building performance is a loss in value to all that have a stake in quality education.

There are usually several factors, or combinations of factors, that cause buildings to fall in value. A building may not perform satisfactorily as a result of inadequate design, poor workmanship, defective materials and components, wrong installations and applications and failure to provide the required maintenance. But, even if the building was well designed, quality of workmanship was high, materials and components selected were of high quality and properly installed hence any inconsistency from the predicted service life is attributed to maintenance.

But assumption apart, that is if the above variables were right, maintenance is still fundamental. Thus, without maintenance, the full service life of buildings cannot be attained. The need for maintenance will only intensify as the value of the building's structure and the engineering services must be preserved and sustained for them to be meaningful to the users and clients. Expenditure on building maintenance is high and will remains so or even escalates in the years to come. For instance, about 70 % of the building operating costs is attributed to maintenance (Rendeau et al. 2006; Booty 2006) and considering the fact that more than 90 % of the life time of a building project requires maintenance work. But in actual reality, the life cycle of a building starts from the initiation of its space need that is the moment a decision is made to build.

The increase in land costs is also leading to the unprecedented growth in maintenance works. But, it is practically impossible to replace, refurbish, convert or rebuild all organisation buildings at a time. This is an illustration; in order to replace 1960s buildings in English universities alone, it is estimated to cost £11 billion (Rawlinson and Brett 2009). Compared to other initiatives like refurbishment, alteration, conversion and reconstruction, maintenance is unique and prominent.

Based on the data obtained from the Ministry of Higher Education (MOHE) Malaysia, maintenance expenditure for university buildings had expanded by nearly 40 % from 2005 to 2009. For instance, expenditure on maintenance was nearly 470 million in 2005 but increased to more than 643 million in 2009. See Fig. 3.1 for the breakdown. However, comparing these amounts with the total expenditure on education implied that roughly only about 1 % was invested on university buildings maintenance (see Table 3.1). However, the exact expenditure for 2010 is not

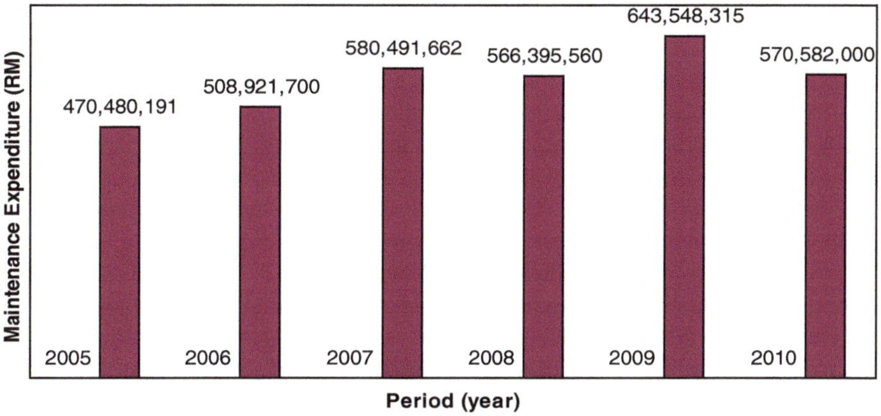

Fig. 3.1 Expenditure on maintenance of public university buildings (Million) (*Source* Adapted from data obtained from the MOHE 2010b)

available, because the date was collected at the middle of the year. However, the ministry will not provide additional funds in case universities require more than that. In the opinion of the MOHE, universities should be more effective and efficient with their allocations.

Despite the increase in the maintenance expenditures, empirical studies revealed that university buildings have not been well maintained because users were not satisfied with building conditions and performance. In fact, in recent times, there are many complaints and criticisms in the media on the conditions and performance of university buildings by the students, parents and other stakeholders (NST 2010).

This is happening when the governments are facing serious financial crises in the face of various pressures. Universities are now operating in highly competitive and turbulent environments. These also coincide with the times governments are cutting budgets to universities and the buildings are ageing. This is also happening at a time when students and their parents are becoming more demanding and concerned about building performance. Thus, the only way today's universities can survive and progress, is to be more effective and efficient in the face of the pressures facing them. These requirements impinge on all aspects of the university operations, and notably among the aspects is the management of the constructed facilities.

3.2.4 What Is the Scope of Building Maintenance?

Maintenance means different things to different people. The term has been defined and redefined by different authors. To illustrate, Pitt (1997) defined maintenance as "task of providing a repair service that protects the fabric and the use of the building to a standard that represent value for money" Best and de Valence (2002) labelled maintenance as "control engages to make sure that asset remains functional." According to BS3811 (1964), "maintenance is a combination of any actions carried out to retain an item in or restore it to an acceptable condition". In a review of maintenance management practices in local authorities in Malaysia, Aziz et al. (2010) considered the main purpose of maintenance management as that of combating the physical condition of the facilities as the mission.

Booty (2006) extends the definition a bit further as "the technical and administrative actions that are required in order to retain a building in a state in which it serves its required functions efficiently and effectively." From the definitions provided above, it is clear that maintenance revolves around the buildings. But of major concern is that these definitions believe that maintenance is initiated when failure is manifest. However, these definitions are not decisive (Jones and Sharp 2007). These classical definitions are inadequate and in fact unfortunate. These definitions put the physical conditions of the building at the centre of the maintenance services as against user satisfactions.

Buildings are not initiated for the sake of the building itself rather, for the value they contribute to the organisation mission and vision. Seemingly, this is where a complete and correct definition should stem from. The definitions also fail to link maintenance with the building performance. And if even if building is the object of maintenance, these definitions/understanding can only lead to correct maintenance. The physical condition of the building cannot in any way serve as a basis for initiating maintenance services. What is critical to the building users or client-occupiers of a building is the ability of the building to support the performance of the activities within and around the building and not necessarily the physical condition of the building per se (Chapman and Beck 1998; Jones and Sharp 2007). The problems of basing maintenance demand on physical condition are well documented, however.

Therefore, a definition is presented as "the required processes and services taken to preserve, sustain, protect and care for a building's fabric, structure and engineering services after completion, repair, refurbishment or replacement to current standards to enable it to serve its intended functions throughout its entire life span without drastically upsetting its basic features and use." From this definition, maintenance does not include activities like refurbishment, alteration, conversion, extension, and reconstruction, but it does include repairs, rehabilitation, renovation, renewal, restoration, reformation, redecoration and the general cleaning and services. The major concepts in this definition are explained in details in the Chap. 5 of this book(see Sect. 5.2 on page 80).

Maintenance is unique in the building life cycle as compared to initiatives like refurbishment, conversion and extension. There are illustrations. As soon as a building is refurbished, converted, altered, extended, completed or reconstructed, maintenance sets in, if not earlier. While maintenance is a must in a building's life cycle, other initiatives are usually one off if at all they are required throughout a building life cycle. Clients can decide not to refurbish or convert their buildings throughout the buildings' economic and or functional life span. But, this is not possible with maintenance as building components, material or designs are not maintenance free. A building that will not require maintenance will almost certainly compromise its life span and performance. Whereas only less than 5 % of a nation's building stock is refurbished, converted, replaced or altered, more than 95 % of any nation's building stocks must be maintained in a year. Often it is only when maintenance could not "rescue" the asset that, other initiatives are applied.

3.3 Issues in the Present Maintenance Practices

The failure in the operations of the buildings is a loss in value to the university, users and other stakeholders. If buildings fail, they lose their economic and functional values and hence could not represent value for money as that capital cannot be accumulated but consumed. The inabilities of the current maintenance systems to improve service delivery have placed a new demand on maintenance organisations (practitioners) and researchers. There is lack of logical, holistic and consistent reference point among the universities for maintenance management frameworks. University stakeholders may not attach much significance to their buildings as in the case of the hotel and recreation sector, yet the role of building is prominent in delivering quality education. Traditionally, parents, students and faculty members might pay much attention on the availability of latest and relevant reading, learning and research materials. However, times have moved on, there are greater concerns and pressures on the university to ensure optimum performance of the buildings in order to be competitive, attractive and innovative. Time is changing users' experience, perceptions and expectations very fast. A greater understanding of the building performance is shifting the whole lots of pressure to the maintenance departments. This wave is not only limited to universities but all organisations that own or rent their buildings. Universities are also realising that even with experienced workforce and high technological facilities, optimum value cannot be obtained if the academic environment is not enabling and supportive to leaning, teach, live and work. This requirement is more pronounced with labour intensive service like the university education.

While some of the university maintenance organisations have developed their methods of managing maintenance services, the weaknesses of these methods have been well documented. In those respects, the methods are objects of persistent criticisms. These criticisms include the missing connection between buildings with university corporate mission, focus on short term perspectives, disregard for user experience and focus on building conditions and placing much attention on financial benefits and the lack of capable workforces that understand and promote value cultures. In addition, the methods developed in-house are often based on the trial by error. The learning curves associated with the interactive processes are too steep in today's competitive environment. The achievement of best value can never be based on accident, rather it is the outcome of deliberate intention, committed efforts, and knowledge based and informed implementation.

In other regards, the growing significance of building maintenance has generated concerted efforts and increasing interests by authors and researchers to develop various maintenance management models to improve the performance of buildings through maintenance management. However, while such systems offer potential to improve the performance of maintenance management practices, the models were reactive, hypothetical, complex, and theoretical and conditional based. Many of the models suffer similar or same criticisms with those developed in-house. The models centre on the building as the subject and take a financial perspective, rather than focusing on users performance requirements.

Users are considered as feedback loops and are described in relation to the design. Maintenance is not related to the building performance. Users' participations in the maintenance management are grossly inadequate and opportunities for engaging and empowering them are mainly disregarded. While most of the models are conditional based, the focus of today's business and processes are performance or more appropriately value based. In part, most of the models are too academic, making it unusable to the practitioners. But, the major weakness and deficiencies with the models mentioned above are not viewed as barriers but progress to make good. Thus, the existing building maintenance management models are considered for the development of an all-inclusive model.

The need to integrate, simplify, amplify and extend the existing theoretical, hypothetical bias and tactical models in order to develop a prototype framework of that takes into accounts the current state of knowledge regarding maintenance management decision is compelling. The solutions to the shortcomings are not to avoid their applications but to design a framework which is user supportive and friendly to facilitate positive decision making. This is the case because many issues that reside in the building maintenance management involve investigations using theories and methodologies from varying disciplines including economics, law, engineering, science, culture, mathematical and statistical information and knowledge. These issues are critical to the systemic management of building maintenance that led towards the value creation initiatives. The continuous administration of the current maintenance strategies without a well defined supporting structure is grossly inadequate in today's market place, particularly for organisations that pride themselves on a source of knowledge.

3.4 Steps to Develop the Value Maintenance Management

In order to develop value maintenance practices for organisations, the following steps require critical examination:

1. Identify the purposes of university buildings
2. Identify and examine the purposes of maintenance
3. Identify and examine the value system of university building users
4. Review existing maintenance practices

5. Determine the required extra capacity to be achieved
6. Evaluate the feasibility of the development alternative model
7. Establishment of the nature and form of the proposed model
8. Engage and empower the users
9. Identify and examine the types of models
10. Eliminate complex mathematical models
11. Eliminate the use of complex specialised software.

However, it is most unlikely that a complete solution can be achieved without an increase in allocation to the maintenance sector. During the data collection for my doctoral thesis, many of the maintenance organisations were of the understanding that governing bodies of the universities should grant more funds for them to improve building performance and meet users' satisfactions. Nevertheless, it cannot be concluded that an increase in the maintenance allocations and expenditures will raise the building performance and user satisfactions and service delivery. This is because with increase in the allocations alone, maintenance could only be given tactical attention rather than value-based consideration. The dynamics of the complexity require a more rigorous analysis of the problems and to identify solutions, both short and long term. What is required is a step change in the way maintenance is managed.

From literature and a series of discussions with those responsible with building maintenance, it led to the premise that information and knowledge of the criteria that influence maintenance of university buildings, characteristics of defects in university buildings and criteria within the users' value system can be significantly explained or is positively correlated with how the maintenance is being managed. It is considered that equipped with such information and knowledge, maintenance organisations will make informed, sound, realistic and unbiased decisions. These are pertinent to increased productivity, attaining users' satisfaction and for enhanced services delivery. Simply put, the lack of this knowledge and information would imply that the scarce resources will not be strategically and holistically managed. As an illustration, poor maintenance management systems will only lead to unnecessary increase in maintenance costs and poor user satisfactions. Unnecessary costs are costs that neither provides use, nor life, nor quality, nor aesthetic, nor comforts or stakeholders features that meet the provider and consumer satisfactions (Kelly and Male 2001; Kelly et al. 2004). If maintenance is initiated, without forethought, the value of the building and investments will decrease.

It was on the basis of these backgrounds that a value-based maintenance management model or VMMM was proposed. The VMMM is required in order to align maintenance resource to users' satisfactions and to also align performance with university corporate strategy. This is a viable alternative that the current approaches can shift towards a value adding maintenance management initiative. The VMMM could drive out the unnecessary maintenance costs and drive in opportunity for higher productivity, user satisfactions and total service delivery. It is a supporting structure or processes required in facilitating decision making. Thus, the value-based maintenance management model is comprised of four interconnected components:

1. Criteria that influence building maintenance management
2. Criteria that influence users' satisfactions in buildings maintenance services
3. Defects in buildings
4. Maintenance performance metric for maintenance organisation.

Value-based management hinges on a collaborative approach towards creating value to stakeholders. Unlike the traditional maintenance management principle that is fixated on cost and on the provider's interests, only value-based management contains both financial and non-financial measures to assess strategic performance. The two perspectives of service providers and consumers must be combined for meaningful analysis and service delivery. Maintenance organisation (if they are in-house)/main organisations are not the only constituents with value expectations other stakeholders may also have them. The interests of all parties concerned with the organisation's success and failure must be aligned towards holistic value creation. Universities operate and survive because they provide meaningful services to their customers and for that service to be meaningful and useful, the perceptions and expectations of the users must be met (see Tucker and Smith 2008), after all, the students pay for the services they are receiving directly or indirectly.

A maintenance demand model (MDM) recognises that the role of users in the maintenance services delivery should be considered. The model requires that any maintenance demand must be commenced upon consulting with the users in order to factor in the user experience, perception and expectations. Although, and again, these ideas seem demanding to the maintenance organisations, but certainly the investment is worthwhile. The maintenance managers must be able to evaluate the quality of the maintenance service as well as establishing the minimum acceptable condition and performance of the buildings and of the maintenance services. Without the maintenance performance model or MPM, control functions cannot be optimised and it is difficult to make improvements since without improvements, users' satisfaction and productivity cannot be measured. The MPM enables continuous consideration of the user value systems through feedback and feed forward processes.

On our discussions above on the steps taking to develop the VMMM, we pointed out that we eliminate complex mathematical models and avoid the use com complex specialised software, will make some clarification in furtherance to that. We like to include here that readers will surely find out that this book involves mathematical and statistical models, but much emphasis is placed on graphical models/ process models. Because, figures are more easily and quickly interpreted, comprehend and memorised compared to words, and it is frequently understood that graphical models are easily and quickly applicable (Pourret et al. 2008). This consideration is very significant in building maintenance management. Because most of the maintenance organisations have little understanding behind complex mathematical models (Sherwin 2000). In real life, building fabric, structure and framework do not lend them to complex model application. Hence, it comes as no surprise that existing mathematical model for maintenance is yet to see practical applications in practice.

3.5 Significance of a Fresh Look

The needs for maintenance work are growing because, it is not cost effective to demolish, reconstruct or to rebuild new facilities to replace the existing ones. The increase in land costs has also lead to the unprecedented growth in maintenance works. Without doubt, there is no organisation, or even a nation no matter how wealthy that can afford to reconstruct all its building stocks at one time. This is not even practical aside from the economic consideration. However, it is extremely pleasing, but almost not practicable to produce buildings that are maintenance free. No matter how durably a building was designed and constructed, it must be maintained in order to continue to sustain its optimal performance. Therefore, it is becoming imperative that buildings are well maintained so that they serve their useful purpose within their live span. With the substantial amount of investments for the maintenance of the building's fabric, structure and services, any reduction in the amount invested will translate to a visible effect on the educational sector and to the nation's economy at large.

The well-being of the users will also be improved. Therefore this book is significant since the benefit of this research is to sustain the availability, convenience, conformability, reliability and safety of the buildings efficiently and effectively with minimum resources. It is envisioned that with the application of the manual, less resources will be required for maintenance, buildings will be maintained proactively, energy will be conserved, down time will be avoided or reduced to the barest minimum and recipients' satisfactions will be maximised.

The book seeks to guide decision makers towards formulating their maintenance policies. It will be of particular interest to maintainers, facilities managers, property managers, real estate investors and building users. It will also be useful to academia. To the academia it would serves as impetus for continuous research in order to improve and achieve best practices in building maintenance management, building performance and users' satisfactions.

3.6 Summary and Observations

Government has aimed to transform Malaysia into a high-income nation. Simplify put, the government wants Malaysians to have a better quality of life through better payment, but in order to achieve this aim there is the ultimate need to produce a quality human resource; that is producing well grounded graduates that can compete nationally and internationally. The business of a university is to transmit and disseminate education, knowledge and culture, and to teach and conduct scientific research. The current developments in all sectors of the economy are indeed outcomes of universities and research centres across the globe. Scientific knowledge and researches are vital in the development of new products and services. The scope and size of universities are growing. On the one hand, new universities were created and also colleges and polytechnics have been converted into universities. On the other hand, apart from the faculty members, buildings are the most significant assets of the universities. Universities need functional buildings to operate. However, the buildings have to be well maintained to ensure their optimum operable performance. There is no well defined guideline for the university building maintenance programme, however. This is detrimental to the building fabrics (and associated services), the university, the students and staff and other users. There is a need for a step change to ensure the consideration of user value systems to set the context for maintenance management practices, processes, procedures and implementation.

The value maintenance management model (VMMM) presents a promising guideline for the management of buildings. Under the VMMM, the maintenance management organisation considers complex set of criteria that influence maintenance management, the prevailing defect, user value system (i.e. functional requirement, user satisfactions) impact of maintenance on the university's core business objectives together with the corporate mission and vision to initiate maintenance. The prime objective of the VMMM is to help define and describe (the scope and quality of the maintenance services), plan, control, organise and monitor maintenance activities.

References

Allen, E. (1995/2005). *How buildings works: The nature order of architecture* (2nd edn.). Oxford: Oxford University Press.

Aziz, F., Ali, H. M., Khan, A. H. R., Jivasangeeta, S., & Jibril, P. (2010). Review on current maintenance practices within Iskandar Malaysia (case study: Iskandar Malaysia). In *Proceedings of 4th International Conference on Built Environment in Developing Countries (ICBEDC 2010): A Better Tomorrow with a Greener Today* (pp. 281–291). Universiti Science Malaysia, Pilau Pinang. December 1–2, 2010.

Best, R., & De Valence, G. (Eds.). (2002). *Design and construction: Building in value*. Woburn: Butterworth-Heinemann.

Booty, F. (Ed.). (2006). *Facilities management* (3rd ed.). Oxford: Butterworth-Heinemann.

BS3811 (1964) *British standard glossary of maintenance terms in terotechnology*. British Institute, London.

California Postsecondary Education Commission (CPEC). (2007). *California State University at the Beginning of the 21st Century: Meeting the Needs of the People of California* (p. 77). The academic senate of the California State University, February.

Chapman, K., & Beck, M. (1998). Recent experience of housing associations and others registered social landlords in commissioning stock condition surveys. In *COBRA 1998 Conference. RICS Research*. ISBN 1-873640-23-4.

Government of Malaysia. (1996). *Seventh Malaysian Plan 1996–2000*. Putrajaya: Economic Planning Unit Prime Minister's Department.

Government of Malaysia. (2001). *Eighth Malaysian Plan 2001–2005*. Putrajaya: Economic Planning Unit Prime Minister's Department.

Government of Malaysia. (2006). *Ninth Malaysian Plan 2006–2010*. Putrajaya: Economic Planning Unit Prime Minister's Department.

Government of Malaysia. (2010). *Ten Malaysian plan 2011–2015*. Putrajaya: Economic Planning Unit Prime Minister's Department.

Jones, K., & Sharp, M. (2007). A new performance based process model for built asset maintenance. *Facilities, 25*(13/14), 525–535.

Kelly, J., & Male, S. (2001). *Value management in design and construction*. New York: E & FN Spon.

Kelly, J., Male, S., & Graham, D. (2004). *Value management of construction projects*. Oxford: Blackwell.

Mat, S., Sopian, K., Moktar, M., Hashim, S. H., Abdul Rashid, A., Zain, M. F. M., et al. (2009). Managing sustainable campus in Malaysia—organizational approach and measures. *European Journal of Social Science, 8*(2), 201–214.

Ministry of Higher Education. (2006). *Towards Excellence Report: Report by the Committee to Study, Review and Make Recommendations Concerning the Development and Direction of Higher Education in Malaysia*. Shah Alam: University publication centre (UPENA), University Technology Malaysia (UiTM).

Ministry of Higher Education (MOHE). (2010a). Regular private communication.

Ministry of Higher Education. (MOHE) (2010b). Regular private communication.

New Strait Times. (2010, July 15). Surely UM can do better (p.19).

Pitt, T. J. (1997). Data requirements for the prioritization of the predictive building maintenance. *Facilities, 15*(3/4), 97–104.

Pourre, O., Naim, P., & Marcot, B. (2008). *Bayesian networks: A practical guide to application*. England: Wiley.

Rawlinson, S., & Brett, L. (2009). Cost model universities. *Building Magazine*. Accessed on June 15, 2010 from http://www.building.co.uk/data/cost-model-universities/3143057.article.

Rendeau, E. P., Brown, R. K., & Lapides, P. D. (2006). *Facility management* (2nd ed.). New Jersey: Wiley.

Sherwin, D. (2000). A review of overall models for maintenance management. *Journal of Quality Maintenance Engineering, 6*(3), 138–164.

Tucker, M., & Smith, A. (2008). User perceptions in workplace productivity and strategic FM delivery. *Facilities, 26*(5/6), 196–212.

USA Weekend. (1991, August 23–25). *If teens had money they'd buy decent buildings* (p. 8). USA Weekend. New York: Gannet Co., Inc.

Chapter 4
Malaysian Education System and University Buildings

Abstract The education system in Malaysia is presented in this chapter. How education is funded and managed is also discussed. Furthermore, the connection between the performance of buildings and quality education was outlined. This chapter and the next chapter underpin the previous chapter. In part, this chapter consolidates and expands parts of the issues in the previous chapters. It outlines the significances and roles of buildings. Initially, it provides the background of the Malaysian education systems. This is followed by detailed information on educational buildings before focusing on the university buildings. Following are the objectives and learning outcomes of the chapter: (1) understand how education is funded, (2) understand the levels of user satisfaction with building performance, (3) appreciate the purposes of university buildings, (4) understand the implications of failure to maintain buildings and (5) appreciate the need to move beyond building.

Keywords Education system · Malaysia · University buildings · University building users · Performance · Satisfactions

4.1 Introduction

Universities have developed and grown into significant national assets and while there have been significant periods of new provisions, many original buildings are still in use. The government of Malaysia is committed to transforming the universities into dynamic and responsive institutions which can hold their place internationally. However, because of shortage of revenue coupled with the national debts, government is cutting costs to meet other responsibilities. Restrictions on public spending have coincided with ageing and deteriorating building stocks in the education sector. These are creating new responsibilities for educational institutions to manage their buildings maintenance. The Malaysian government and private promoters are investing heavily in the university sector. However, when benchmarked against international standards, universities in Malaysia are experiencing

A.L. Olanrewaju and A.-R. Abdul-Aziz, *Building Maintenance Processes and Practices*, DOI 10.1007/978-981-287-263-0_4

inadequate financing. This is affecting allocations to the universities. The governing bodies of private universities are also cutting allocations to the privately funded universities in order to meet other business deals. Although the consequence of cutting allocations to the universities means that the allocations to building operation will reduce, maintenance management processes are poor.

4.2 Malaysia: An Overview

Malaysia is situated in South East Asia, consisting of 13 states lying between 2° and 7° north of the equator. The country is located at latitudes 3° and longitudes 101°. Eleven of these states are in Peninsular Malaysia, in the southern part of the Kra peninsula, having Thailand at the north and the island of Singapore to the south. The remaining two states of Sabah and Sarawak are located on the Island of Borneo sharing with Brunei. Malaysia shares borders with Thailand, Singapore, Borneo Island and Indonesia. The climate is hot and tropical located within the tropical belts (23.5° N and 23.5° S). There are rains in all the seasons and temperatures range between 21 and 32 °C (72 and 99 °F) with little rain variation throughout the year. The humidity is comparatively high. The annual rainfall varies from 2,000 to 2,500 mm. In general, Malaysia is characterised with high temperature, heavy rainfall, bright sunlight and high humidity. The humidity can be as high as 80 %.

The Malaysian capital, Kuala Lumpur, is a separate Federal Territory, as is the island of Labuan and the newly developed administrative capital of Putrajaya. The Supreme Head is a monarch (or Yang di-Pertuan Agong) elected for a 5-year term. The head must come from the hereditary rulers of nine of the thirteen states. Malaysia is a rapidly developing country, with its economy depending mainly on agriculture, forestry, petroleum and manufacturing. The overviews we provide here is specifically to enable the readers appreciate maintenance issues. For instance, weather has profound implications on decay and of on how it can be maintained.

4.3 Malaysian National Educational Systems

The education sector is very significant and integral part of the Malaysian Strategic Thrust. The economy of Malaysia has, since the independence, been planned on 5 years strategic plans. For each of the Malaysian Plan, education sector features prominently in terms of policy formulation and implementation. Education in Malaysia has been British since inception apparently as a result of the British incursion during colonial rule. However, over time, Malaysia has been able to develop its own brand of education systems though not without the British fragments. Over the years, access to education has increased significantly at all levels. For instance, enrolment to tertiary education in 1990 stood at 16.6 %, while by 2005, it grew to 29.9 % (Government of Malaysia 2006). Although there is no

Table 4.1 Student's enrolment in public education institutions, 2000–2010

Level of education	2000	2005	2010
Preschool	539,469	702,897	784,200
Primary	2,907,123	3,044,977	3,195,977
Secondary	1,964,607	2,093,847	2,306,478
Post-secondary	94,544	199,636	277,904
Teachers education	23,740	34,672	45,899
Total	5,529,483	6,076,029	6,610,458

Source Government of Malaysia (2006)

compulsory education, the government provides free education at the primary and secondary levels.

Primary and secondary educations are funded by the government though there are only a small number of privately founded primary and secondary schools. However, in the case of tertiary education, there are many privately owned colleges, polytechnics and universities offering various degrees to both overseas and local students. In fact, the size of privately owned university is gradually surpassing that of the government owned. The range of tertiary educations includes certificate courses, diploma courses, undergraduate and master degrees as well as doctoral degree courses. Government's commitment to education and the large-scale investments in the sector have paid off substantially. For instance, students' enrolment at public education institutions increased by 9.9 % during the 2000–2005 periods and it is expected to further expand by 8.8 % between 2005 and 2010 (Table 4.1). Table 4.1 contains the enrolment growth at the public education institution at the different levels from 2000 to 2010.

Educational system in Malaysia is classified into the following:

1. Preeducation (2–3 years)
2. Primary school (5–7 years)
3. Secondary education (5–6 years)
4. School extension (6 years), or preuniversity education classes (1–2 years)
5. Higher education—college students course (3 years)—postgraduate course (1–5 years).

4.4 National Education Philosophy

The ministry of education, later the ministry of higher education and now again the ministry of education, is saddled with the responsibility of supervising higher education policy. The National Accreditation Board (Lembaga Akreditasi Negera) evaluates the programmes offered by the private institutions. The national bodies that deal with the accreditation are Ministry of Education, Malaysian Association of Private Colleges and Universities (MAPCU), National Association of Private

Educational Institutions (NAPEI) and Persatuan Kebangsaan Institusi Pengajian Tinggi Swaata Bumiputera Malaysia (PKIBM). Education in Malaysia is ongoing effort towards further developing the potential of individuals in a holistic and integrated manner, so as to produce individuals who are intellectually, spiritually, emotionally and physically balanced and harmonic, based on a firm belief in and devotion to God.

Governments affirm commitments to produce Malaysian citizens who are knowledgeable and competent and also possess high moral standards. At the same time, responsible and capable of achieving high level of personal well-being as well as being able to contribute to the harmonious and betterment of the family, the society and the entire nation. Tertiary education is the education beyond the secondary education until higher education level. This is offered in community colleges, polytechnics, public and private training institutes, private higher education institutes and universities. University education emphasis is on professional training and brings about the awards of diploma and degrees or their equivalent. There should not be any argument, concerning the roles and contributions of university to economic prosperity of any nation. University is the viable means of facilitating economic development and progress.

4.5 Higher Educational Institutions in Malaysia

The mission of the Malaysian Government in respect to higher education is "to develop and put in place a higher education environment that encourages the growth of premier knowledge centres and individuals who are competent, innovative with high moral values in order to meet national and international needs" (MOHE 2008). Furthermore, the eighth objective of the Ministry of Higher Education is "to make available adequate and quality higher education infrastructure facilities comparable to international practices". To this, the Malaysian higher education system is playing a very significant role in producing a dynamic workforce that is highly qualified and excellent at fulfilling the country's needs towards achieving its vision. According to Barnett (1994), the primary purpose of higher education is to pass on the social, cultural, scientific, political and technology heritage to the next generation who test, reject and improve part of it (Milliken and Barnes 2002). With respect to Malaysia, the prime objective of the higher education is to produce skilled and trained professionals in order to meet the strategic thrusts of Vision 2020 and beyond. Malaysia is aware of developing its human capital to achieve the high-income nation status.

There are approximately 630 tertiary institutions in Malaysia educating more than 1 million students in an academic year. Currently, there are more than 50 established universities in Malaysia. The number of university academic is around 90,000 cutting across different categories. In 2012, there are 29,769 academics in the public universities and 2,151 of them are expatriates (MOE 2012). Enrolment for public and private institutions of higher learning is expected to reach 1,610,408

Table 4.2 Students enrolment in tertiary education institutions, 2000–2010

Level of study	Estimate		Target			
	2010		2012		2015	
	Public	Private	Public	Private	Public	Private
Certificate	40,204	85,885	33,733	103,166	25,927	135,822
Diploma	149,702	225,997	162,853	268,507	184,778	347,724
First degree	287,354	227,764	313,507	278,105	358,791	375,229
Master	51,133	15,690	66,335	21,588	98,021	34,842
PhD	17,425	2,810	24,592	4,279	41,233	8,041
Total	545,817	558,146	601,021	675,646	708,749	901,658

Source Government of Malaysia (2010)

by 2015. Table 4.2 contains the enrolment statistic for university. Financing public higher education in Malaysia has been the responsibility of the Government. However, supports were also provided by corporate organisations such as financial institutions, manufacturing companies and some multinational companies in the form of scholarships and grants. Notwithstanding, some states' governments have also set up foundations to operate higher institutions like the community colleges, polytechnics and universities.

The dedication to higher education and the significant investment in that sector have equally paid off to a large extent. Student enrolments in PhD, masters, first degree, diploma and certificate programmes at public higher education institutions have increased by 24.6 % during the 2000–2005 periods, and it is expected to expand by 118.65 % from 2005 to 2010. Similarly, in the private universities and university colleges, enrolments increased by 30.8 % during 2000–2005 and by 38.5 % during 2005–2010 (Government of Malaysia 2006).

4.6 University in Malaysia

Higher education is a major catalyst in generating and sustaining knowledge, skilled and competent human capital to meet the need and strategic vision of a nation. University education is created to contribute to the sustainable development of the society as a whole. It aims to produce highly qualified graduates who are able to meet the needs of all sectors of human activities. Advancing, creating and disseminating knowledge through research and innovations underpinned sustainable development. Universities are increasingly being recognised as a powerful force for national competition as well as a force for a world peace, global integration, mutual understanding, and geographical stability. It is therefore expedient that such organisations are well understood.

The origins and growth of the higher education system in Malaysia can be seen from four basic stages: first, the implementation and the subsequent development of a higher education system in Malaysia and Singapore before Malaysia's independence in 1957; second, the establishment of the University of Malaya (UM) in 1961; third, the establishment and growth of three national universities and the International Islamic University after 1969; and finally, the upgrading of the Agriculture and Technical Colleges in 1971 and 1972, respectively, to fully fledged university status (Selvaratnam 1985).

The University of Malaya, the Malaysian premier university, was established based on the recommendations of the commission that was set up by the Government of Malaya and Singapore to study the feasibility of its establishment after the Malayan independence. The objectives of establishing the UM were twofold to facilitate the replacement of expatriates with Malaysians and to cater for the ever increasing demand for high education for the qualified school leavers (Selvaratnam 1985; Bajunid and Jantan 2008).

However, for Malaysia to play its economic role competitively, it must transform into a knowledge-based economy (KBE). But this involves improving the university systems. Malaysia is committed towards transforming its universities into dynamic and responsive institutions which can hold their place internationally (Malaysia and the Knowledge Economy 2007). Concurrently with the view that universities are some of the best engines to advance a society towards a knowledge economy, the government of Malaysia carefully laid the foundation for the expansion of the system in its Ninth Malaysia Plan (2006–2010).

Under the current Malaysian Plan or 10 MP (2011–2015), universities are expected to contribute significantly to the Malaysian high-income economy status. This will entail the universities to produce graduates that can compete meaningfully nationally and internationally in economic and technological advancements and at the same time for the universities to be able to commercialise their discoveries. Although, the Malaysian government is investing heavily on the universities sector, however, when benchmarked against international standards, universities in Malaysia are experiencing inadequate financing (Malaysia and the Knowledge Economy 2007) and in the least, this is affecting the performance of the buildings.

4.6.1 Public Universities

Public education is provided by government for students. Public universities are government institutions. They are mainly financed by the government sector. Therefore, they are not for profit making rather they are parts of the government social responsibilities to the citizens. The public universities do not have stakeholders who are expecting a return for their investments. The management and operation of the public universities are subject to the various Education Acts and Regulations.

Public universities in Malaysia are either established under the Universities and University Colleges Act 1971 or the Acts of Parliament. However, only the International Islamic University Malaysia was set up under 1965 Companies Act. There are 21 public universities with a total enrolment of about 601,021 students in

2012 which is expected to rise to a million by the end of 2020. Table 4.3 contains the outputs of higher learning institution in Malaysia.

4.6.2 Private Universities

The liberalisation in the provision of education has encouraged private sector participation in education and training services. The purpose of private sector participation in higher education is to complement the government efforts in the provisions of higher education to the citizens. This further receives a boost with the enactment of the Private Higher Education Institutions Act 1996 (Government of Malaysia 2001). The Act also enables foreign universities to establish their branch campuses in Malaysia. The private institutions provide the same courses as their public counterparts, and they award degrees, diplomas and certificates. The Government has attempted to encourage more foreign universities to establish campuses in the country to improve standards and reduce the cost of sending Malaysian students abroad to study.

Privatisation of higher education in Malaysia, in a general sense, refers to the reduction of public funding in higher education provision, and parallel encouragement to the private sector to provide higher education (Mei 2002). The private higher education is also playing an active role in attracting foreign students to fulfil the government aspiration of the country as an education destination. The private education providers are given considerable recognition. This is explained by the fact that the president, vice president or CEO of many of the private universities sit on the National Higher Education council, NHEC (Bajunid and Jantan 2008). The NHEC is the highest decision-making body on all matters concerning higher education in Malaysia. The total output from public and private tertiary institutions at all levels increased from 130,161 in 2000 to 252,730 in 2005, out of which about 60 % (58.5 %) was produced by the private tertiary institutions (Government of Malaysia 2006).

The output of private higher institutions at all levels is expected to double by the year 2010. For instance, it was 33,111 in 2005 which is expected to reach 61,776 by 2010 (Government of Malaysia 2006) and the average annual growth is also expected to increase from 12.1 % between 2001 and 2005 and 13.3 % between 2006 to 2010. Most of the products from the private institutions are specialised in the areas of information and communication technology and electrical engineering. For instance, for ICT, public university, 903, 1,016 and 1,853 in 2000, 2005 and 2010, respectively, whereas in case of the private institutions it was 7,520, 11,844 and 12,886 in 2000, 2005 and 2010, respectively. For electrical engineering, while the public institutions are expected to reach 19,828 in 2010, private institutions are expected to expand to 33,498 (Government of Malaysia 2006). However, one of the problems with the private higher education in Malaysia is with regard to the recruitment of faculty members with the highest academic qualification (Morshidis 2006).

Table 4.3 Output of higher institutions in Malaysia, 2000–2010

Level of study	2000			2005			2010		
	Public	Private	Total	Public	Private	Total	Public	Private	Total
Certificate	23,816	81,754	105,570	37,931	94,949	132,880	141,290	143,480	284,770
Diploma	91,398	117,056	208,454	98,953	131,428	230,381	285,690	188,680	474,370
First degree	170,794	59,932	230,726	212,326	110,591	322,917	293,650	134,550	428,200
Master	24,007	2,174	26,181	34,436	4,202	38,638	111,550	5,770	117,320
PhD	3,359	131	34,490	6,742	140	6,882	21,410	270	21,680
Total	313,274	261,047	574,421	390,388	341,310	731,698	853,590	472,750	1,326,340

Source Government of Malaysia (2010)

4.7 University Administration

The administration of universities is contained in the Universities and Universities College (Amendment) Act, 1996, which states "The bill aims to promote the administration of public universities. Improving the administration of public universities and giving them greater financial autonomy is deemed conducive for their development and for them to accomplish more excellent academic achievements" (MOE 2012). The administrative structures of the universities are broadly divided into two parts, namely the academic and non-academic matters. However, universities are allowed to make internal decisions independently so far as it is not in conflict with the national aspiration. For instance, universities can make independent decisions about their mission, governance, hiring of their academic leaders, academic and non-academic staff, selecting students, and introducing new programs and courses. Though the universities must receive the approval from the MOE to engage faculty members, some universities will prefer to employ first before commencing the detail documentations for the MOE's approval but in most cases, particularly for the expatriates, the approval is sort before the member is allowed to arrive. Depending on the particular university's internal process and verification, the MOE often give their consent to employ or not to employ within a month upon submitting the correct and complete documents to the MOE.

The authorities of universities are the Council, the Senate, the Majlis, the Faculties (or Kulliyah as it is sometime referred), the Boards of studies, the Boards of selections, the Board of students affairs and other Bodies as may be prescribed by the statutes as authorities of the universities. The Ministry of Education is responsible for the general policy direction of the universities and administration. Therefore, the appointment of the Vice-Chancellor or Rector and Deputy Vice-Chancellor or Deputy Rector for the public universities is vested with the Ministry of Education. By this, the government has considerable stake in the administration of the public universities. Thus, the government does not only found the universities but also monitor and coordinate the activities of the public universities. The administration of the private university is not very much different from the publicly owned except that a governing council or board of governors appoints the management members.

The broad policies of universities are determined by the most senior administrative body, usually the council excepting academic matters, which are the responsibilities of the senate. The council appoints the Vice-Chancellor or Rector as the executive officer of the university. Although public universities are granted autonomy, the boundaries of responsibilities, when it comes to management of maintenance services, are little bit confusing and complex, however, since the government continues to have the major input and control of the funding of the maintenance services. Each university has to depend on their allocation on maintenance and failure or delay of which will affect the maintenance planning and invariably the performance of the buildings and ultimately the students' academic standard. To implement a sound maintenance management system, however, that is values based, requires more pragmatic attention from the governments and university administrators, and the allocation of funds needs to be made on a sound basis.

4.8 The Knowledge-Based Economy (KBE) Issues

KBE is an economy system that depends on knowledge. Knowledge engineering (and management) is used as a tool in the production processes. In an effort to progress actively towards a knowledge-based economy, knowledge-based index (KBI) was developed to assess Malaysia's readiness to become a KBE. The overall KBI expanded by 591 points from 2,413 in 2000 to 3,004 in 2005. This represents an increase of 25 %. In terms of KBI by countries, Malaysia remained at 17th position in 2005 (Government of Malaysia 2006).

The assessment of KBE of 1,819 firms was conducted through a survey comprising of 21 variables that reflect knowledge enablers and actions; it was found that all the industries that were involved in the survey had built a certain level of knowledge competency and capability as well as having embarked on some form of knowledge acquisition, generation and sharing activities (Government of Malaysia 2006). Major drivers of a knowledge-based economy are computer infrastructure, info-structure, education and training as well as research and development and technology. Table 4.4 revealed that information technology services, chemical, telecommunication, financial service and tertiary education exhibited better readiness.

4.9 Funding University Education

Although, public universities in Malaysia are founded by the government, each university is given considerable autonomy as each university is allowed to draw up its course contents and employ staff, among other things, so long as it does not contradict and contravene the national objectives. The Government finances each of the public universities to the tune of approximately 90 % of their total operational cost. Meanwhile the funding of the public universities is tied to their performance.

Table 4.4 Knowledge-based economy development index: Malaysia 2000 and 2005

Sectors		2000	2005
1	Education and training	1,250	1,536
2	Info-structure	756	814
3	R&D and Technology	324	408
4	Computer infrastructure	83	246
Total		2,413	3,004

Source AbdulKadir (2002)

The APEX (Accelerated Programme for Excellence) and SETERA (A Rating System for Malaysian Higher Education Institutions-*Sistem Penarafan Institusi Pengajian Tinggi Malaysia*) are used by the government to monitor motivate universities to improve their performance. However, in recent times, public universities are entrusted with the responsibility of utilising tax payer money more effectively than it has been the case. An interpretation of this is that the universities must improve their expenditure pattern and manage their resources prudently. The allocations to the public universities were made for various purposes including for new construction, renovation, maintenance, repair, extension orconversion.

However, public universities are already facing shortages of allocations from the government, while at the same time, they must meet the ever increasing users' expectations on the facilities, including buildings. The expansion of student numbers alongside the aim of widening accessibility to higher education has expectedly created a number of problems for universities particularly in terms offunding. Universities are compelled to develop a progressive commercial view of their facilities. Therefore, like most places, university systems in Malaysia are in transition as they continue to accommodate varying stakeholders' requirements and the needs of the buildings in terms of sustaining their efficient operation through maintenance.

Government is cutting allocations to the universities to meet other commitments. Fortunately or unfortunately, this is happening when there are surges in students enrolments; university buildings are ageing, decaying, deteriorating and failing, and users are more demanding as a result of their high expectations and perceptions. Therefore, universities have no choice but to be market-driven in order to generate sufficient revenue in addition to the government interventions if they must remain operational. In that regard, the competitive environment demands the university to apply sophisticated approaches. This is calling for a need for step changes.

Since the 1990, the government has realised that public universities were consuming a large share of the public revenue without a corresponding return for investment. As a result, government is pressuring the public universities to be self-funding. In order to meet the requirements for quality education, government is already partnering with the private sectors so that government can maintain the control of the public universities. A private finance initiative was signed for the construction of an additional six campuses for the UiTM. The initiative is expected to save government RM500 million.

Though, a complicated arrangement for funding the universities involving the various agencies including the Ministry of Finance, Ministry of Education and each of the universities has meant that the university generally can only have corrective or at best preventive approach to maintenance. Financial constraint has called for an effective and efficient decision-making system. Almost certainly, this could be achieved through a maintenance philosophy that is value-based, that is designed for functional performance of the buildings, which moves away from the corrective and preventive maintenance strategies. Since to provide or produce quality product or service, with a high degree of users' satisfaction is not sufficient, the cost of producing or providing the product of service must be efficient so that the long-term effect on the business or organisation is a desirable one (Oakland and Marosszeky 2006).

The provision, operation and maintenance of the university buildings are subject to government intervention. However, to achieve the objective of transforming Malaysia's economy to the KBE requires holistic overhaul. The pragmatic transformation requires improving the overall effectiveness of the university systems. This includes improving the functional performance of the university system, and not only in terms of lecturers, methods of teaching, studying, learning and doing research but also the physical infrastructures, notably the management of building systems must be improved. Therefore, building maintenance management is progressively becoming an instrument of measuring the performance of the university services. Building and maintenance management is one of the major indictors that contribute to improving the performance and for enhancement of the services delivery of the universities' delivery systems.

This is an example to illustrate the above understanding; 50 programme managers of five universities agreed that buildings and maintenance management is second in terms of reliability regarding whether it could be used to measure the performance of public universities (Mohd Anuar 2008). Students also considered the conditions of the university they intend to study in. In fact, universities in Malaysia these days use the nature, design and condition of their buildings to woo students. Those universities that in the past relied on the application sent to them through the Ministry of Higher Education now rely on the condition and performance of their buildings as a variable to attract students.

4.10 Universities and the Tenth Malaysian Plan (10MP)

Government has aimed to transform Malaysia into a high-income nation. Simply put, the government wants Malaysians to have a better quality of life through better payment. Under the current Malaysian Plan or 10MP (2011–2015), universities are expected to contribute significantly to the Malaysian high-income economy status. The government will also grant autonomy to the (public) universities by reducing their direct control. Under the new arrangement, the government will only serve as a regulatory agent.

The independence of the universities in making strategic decisions is important to the government policy (the New Economic Model or MEM) on achieving educational excellence, innovation and internationalisation. Since, the cardinal requirement of granting universities the autonomy status is for the university management to have prime responsibility in the governance of the university. It is very likely that the management of the buildings will be more efficient. Currently, the universities have to rely chiefly on government intervention for all major decisions. Among other thing, this practice is reducing maintenance to corrective and thus encouraging poor management.

The success of a university must not only be measured by the quality of its students and faculty members, but also on how they manage resources towards achieving their business objectives. Buildings are a factor of production. The building and its management needs to be considered as an investment rather than as a liability. Buildings are factors of production to the university institutions like human resources, capital, land, technology and information. For instance, an Audit Commission in the UK recommended that business cases for funding should encompass a more explicit link between buildings and educational performance (Audit Commission 2003).

University education is gradually becoming a source of income in Malaysia. Malaysia is gradually becoming an education hub and destination for all nationals. As a part of the MOHE (now MOE) and re-organisation and re-branding exercises, the Education Malaysia has been established to woo foreign students. 19 centres in different countries including in Thailand, Kazakhstan, Jakarta, Amman, London, Wellington, Perth, Sydney, Chicago, Los Angeles, Washington DC, China and Vietnam have been established for these purposes. The Education Malaysia would act as a one-stop centre for international students, institutions and stakeholders interested in Malaysian higher education.

4.11 Asset Management (AM)

Assets are the resources owned by an organisation or country. They are the total capital base of an organisation or a country. Assets can be tangible or intangible. Asset management is the management of the assets/resources and processes to optimise organisation's business objectives or charter. AM aligns organisational physical infrastructures, people and environment with product or service for the achievement of organisations corporate objectives. It is also concerned with the procurement, operation, maintenance and renewal of the resources. AM extends to all the critical resources and processes within the environment that contribute to the performance of the resources and ultimately to the performance of the business.

Assets include buildings, bridges, hospitals, roads, barracks, vehicles, work-forces, plants and equipment, information telecommunication network, software and data. The assets of a university organisation include students, faculty member, capital and information. Other things to be included are technology, buildings, plants, vehicles, infrastructures and other supporting facilities. Apart from human resource, which is the only repository of knowledge organisation, buildings are the second largest asset on a university's balance sheets.

4.12 Facilities Management (FM)

Facilities are properties where people work and are accommodated, or where an organisation conducts its businesses (Seeley 1996). However, this is not a conclusive definition, since facilities could be a place where people work, study, conduct business, rest, learn and teaches, etc. In fact, the term facilities is not limited to only buildings as it includes all fixed properties of an organisation such as plants and equipment. Facilities contribute significantly to the enhancement in productivity, profit abilities and the service quality of an organisation. In effect, facility management involves managing land, storage, or living space, learning and teaching spaces within premises. It also involves acquiring, constructing, fitting-out, managing, maintaining, protecting, transferring, and disposing of land and buildings.

Facilities management involves the management of all the services that support *core* business of an organisation. Before moving on, I will need to mention that *core* is subject to various interpretations. There is the need to redefine what is meant by *core* for building-based enterprises like a university; it is not correct to consider the management of its buildings as a non-core function. Similarly, it will be wrong for those in the hotel sub-sector to consider the performance of their building as *non-core* activities. FM focuses on meeting organisation's performance in terms of relationship between operational facilities and business outcome. Therefore, facility management involves the coordination of the physical workplace (i.e. commercial and institution buildings), study place (i.e. university buildings) and the users (i.e. students and lecturers). Organisations are progressively recognising that sustaining facilities efficiently is critical to the success of their businesses. Besides the employees' salaries, expenditure on facilities is dominant on most university balance sheets. From the foregoing, FM is an aspect of assets management. However, FM is often confused with maintenance management as it is often used to describe maintenance management. In South East Asia (SEA), FM rather than AM is the preferred concept.

Based on a study that compared the level of development of FM in the SEA and the UK, it shows that FM in SEA is at the infancy stage (Moore and Finch 2004). The research, however, outlined that compared to countries like Hong Kong and Singapore where progress in the development of FM is well established, its development is mixed in Malaysia. The nature of services provided by the FM organisations in the SEA include maintenance, property management, space management, real estate service and IT management. The dominant service offered by the FM organisations in the SEA is more towards the management of buildings and invariably on maintenance of the buildings.

4.13 Building Facilities

Shelter is a necessity. All buildings serve certain purposes, such as accommodation, inspiration or otherwise. Buildings are complex facilities with various systems intended to provide pre-established functionality and respond to legislated and environmental demands, potentially accommodating several users with varying demands upon the building systems, fit-up and space usage requirements (Kyle 2001). These are particularly true for university buildings because the building users are constantly in transition. Buildings are made of various kinds of components and systems. For many people, "building" could generally be defined as a container or shells that facilitate users' activities required to filter external condition from internal needs (Watt 1999). The containment could be simple or complex depending on the functional performance of the buildings and the users' value system.

However, not all buildings are expected to perform the "containment" or "envelope" function per se. Buildings are wanted for other purposes. The provision of a physically comfortable shelter is not the only or even the major reason for a building as a building could be required to give an established place or social and religious identity: expression of it must indicate culture, status and mood, while creating the humanised space in which to learn, experience and carry out normal daily function (Osburn and Greeno 2007).

Buildings have become comprehensive life support systems to users. In other words, buildings facilitate human activities. As an illustration, university buildings are study place that facilitate the core business of learning, teaching and innovation. For it to succeed and survive, in performing its required functions, it must be dynamic, strong and resilient. Buildings lives and deaths depend on how they are designed, used and managed during their life cycle, however. Much has been written on the design theory and practice of buildings, the purpose of this book is not on how buildings are procured, designed and built, rather it is on how buildings are used and managed—*theaftercare services*. It is well established that the value of a building will depreciate with time. However, its use and maintenance will affect the rate of depreciation and performance.

A building has the following phase—birth, life and death. The value of a building is determined in relation to its current ability to provide user functional requirements, the current market value and the building condition and performance rating in comparison to that of a new building. Buildings appear on both the private and public sectors balance sheets.

Confusions sometime arise, however, as to what is building condition or building performance or what is not of each of the concepts. While performance is a qualitative measure, condition is not. The condition of the building is quantitative. They can be seen and touched. Performance is related to behavioural issues and cannot be seen nor touched but can be felt (in fact with any of the sense organs). However, both are related and are in fact sometimes mistakenly used interchangeably. For instance, the defects (cracked walls, damaged windows, etc.) could affect users' psychological instincts and as well affect the user well-being. Aesthetics in a building could affect the building performance. While some aspects of the university buildings are expected to be attractive to learning, some part might not be. For instance, medical or hospital buildings affect the psychological well-being to patient.

As a further illustration, the peeling or fading away of paints on the building structure is not only unsightly but irritating. While unsightly is a condition issue, irritating is a performance issue because its extent cannot actually be seen not quantified but can be felt by the person concern. It can affect the user's psychological well-being, safety, comfort and productivity. In addition, lack of maintenance can cause growth of mould in the building due to faulty air conditioning units. This could lead to sick building syndrome. This is also the case with decay of wooden components and or materials in the building. The decay can affect the indoor air quality. Bad illumination can also occur due to faulty lighting unit.

4.13.1 Functions of Buildings

A building serves to enclose a space and modify climatic conditions. It also serves to protect and maintain privacy as well as serving as a source of income. It is a facilitator of human activities. Like the human skin, buildings are layers that provide shelter between our body and the environment. Buildings are becoming more and more complex. The functions that are expected of buildings are becoming progressively more intricate. For efficiency, human activities must be physically housed in a building that is well designed, constructed and maintained. However, building users and organisations are interrelated to the extent that a failing in one link of the chain will substantially affect the overall building performance (Amaratunga and Baldry 1998) and ultimately the activities taking place in and around the building. A building once completed, must be able to be operated, maintained and changed in a useful and economic manner.

A building is an expensive item. It is expensive to own and to use and it is usually static in nature. Generally, it cannot be moved from one place to the other. But the immobility is over space and not over time (Douglas 1996). They can be extended in shape and form through processes like refurbishment and conversion. Buildings are not static over time since they deteriorate, decay and lose value economically and functionally as they grow from birth to old age. They respond to climatic factors such as sunlight, rain and wind. On top of that, building components or systems do not decay or deteriorate in the same rate. The building's systems and components have services lives that rarely coincide with one another yet the systems and components are expected to perform together satisfactorily throughout the building life cycle.

The list that follows includes the functions of the modern buildings (Allen 1995, 2005):

1. Provide most of the immediate necessities for human metabolism
2. Create the required condition for human thermal comfort
3. Create the require condition for non-thermal sensory comfort, efficiency and privacy
4. Control the entry and exit of living creatures of all kinds
5. Distribute concentrated energy to convenient points for use in powering various light, tools and appliances
6. Provide up-to-date channels of connections and communication with their external environment

7. Facilitate bodily comfort, safety and productive activity by providing useful surfaces (i.e. floors, walls, stairs and shelves)
8. Provide stable support for the users
9. Protect its own systems and components
10. Respond to changes due to climatic elements
11. Protect it users
12. Capable of being operated, maintained and changed in a useful and economical manner.

4.13.2 Types of Buildings

Buildings can be categorised by function, purpose, types of heating, number of stories, etc. Buildings consist of several types of structures such as private buildings, health facilities, public buildings, public housing, town halls, prisons, educational buildings and offices. Buildings have been classified differently by different authors including, Douglas (1996), Seeley (1996) and De-Chiara and Crosibie (2001). However, some of the classifications were defined narrowly, while the others were too broad. In this research, buildings have been classified as De-Chiara and Chrosibie (2001):

1. Commercial buildings
2. Hospitality buildings
3. Educational buildings
4. Residential buildings
5. Healthcare buildings
6. Cultural and entertainment buildings
7. Government and public buildings
8. Religious buildings
9. Transportation buildings, and
10. Industrial buildings (Douglas 1996).

Educational buildings are further divided into earlier childhood education facilities, elementary and secondary schools, college buildings, polytechnics building institutional buildings and universities buildings. University buildings can be classified into academic and non-academic buildings. The academic buildings include the classrooms, library and laboratories, while the non-academic buildings include hostels, staff apartments/quarters on and off campus. While the academic buildings may not be readily available for maintenance during the day time, the non-academic buildings can be maintained during the day time. Although many staff might not be willing or pleased to leave the maintenance operative behind while they go to work but in many cases, someone is always left behind to look after the house, while the maintenance operatives execute their works.

In the case of the hostels, usually students are obliged by the university law to provide access to the maintenance operatives at all time notwithstanding whether the students are in the room or not. In fact, the maintenance department has the right by law to enter into any of the rooms in the hostels to execute a reported maintenance case even when the students are not in the rooms or place that require maintenance. The maintenance department has spare keys to all the rooms in the hostels. Except for emergency maintenance works, hostels ought to be maintained during the day when students have already left for classes at least for ease of maintenance services. The interpretation of this is that students have great influence on the schedules of maintenance services. For maintenance departments to have a smooth maintenance programme flow during the working hours/time, proper planning and organisation is critical.

The use to which a building is put, the suitability of its design for the purpose, and the intensity and the way in which it is used will directly and indirectly determine its maintenance management system. The maintenance procedure of a building should stem from the functional value that the users attached to the building. For instance, the value of a building varies according to the use of the building and different standards may be adopted for the different types of building (Mills 1994; Chanter and Swallow 2007). Therefore, building maintenance policy has to be considered not only in terms of building function, but in relation to the users' perception of the building's condition and its relevance to his primary needs (Lacasse et al. 1997).

Thus, the users' requirements are an essential prerequisite for evaluating the value of building. Thus, it is only when these prerequisites are made explicit that maintenance can be planned, controlled, organised and coordinated proactively and systematically. The users or recipients are persons or people that have or will have influence on the service provided to them and can take action if their demand is or are not meant. Unlike during the design and construction stage when the users were not known, in case of the maintenance services, the users are basically known. If the performance criteria of those users are considered, the value of the maintenance services delivered will very likely attain the expectations, perceptions and satisfactions of the users.

4.13.3 University Buildings and Engineering Services

A university is an institution of critical thinking, where knowledge is developed, refined, rejected and disseminated as a source of value creation for the national economic development (Millikin and Barnes 2002; Mohayidin et al. 2007). A university has a very strong interrelation with a nation's economic development and growth. Universities own buildings to carry out various types of activities and functions; office, lounges, reception areas, conference/seminar rooms, storage, treatment rooms, workrooms for equipment (photocopiers, fax machines, printers, mailbox) classrooms, teaching and research laboratories, libraries, residence halls, cafeteria, places of worship (i.e. Mosque), tea rooms and athletic facilities.

Engineering services or M&E complement and enhance the roles and services of the buildings. For instance, air conditioning units are required to provide and maintain conducive temperature. Water is required for drinking, washing, in the laboratories, toilets and kitchens it is an essential requirement in a building. Lighting is required for good vision. Electricity is required to provide the lighting, power the air conditioning and other mechanical facilities.

University buildings create conducive environments for learning, teaching, research, innovation and other academic activities. They are factors of production. Unlike office buildings that are generic resources, university buildings are a critical and integral part of the university business. University buildings represent a substantial percentage of university assets and operational cost they account for about 30 % of most university expenditure and of course are the most valuable assets apart from the human resources.

The expansion of student numbers alongside the aim of broadening accessibility to higher level education beyond its traditional primary and secondary levels has not unexpectedly created a number of problems for universities, not least financial for the building and operation of university facilities (Milliken and Barnes 2002). Consequently, universities are experiencing intense turbulence as they struggle with the combined fundamentals of growth, relevant measures of output, less generous resources, increased competition, greater accountability and transparency.

Thus, it is expected for universities to be more prudent in the utilisation of resources. Universities have a wider range of differing buildings than most other organisations and the buildings are complex both in terms of designs and functional use. Despite the proliferation of technology in the areas of teaching and learning (Shabha 2000; Gabriel 2003), the future of university buildings will continue to grow. Undoubtedly, the world of education is changing particularly in respect of space and time management. Today, it is possible for students to obtain degrees without attending a (physical) class (thanks to "online university", virtual university, etc.) yet buildings are required for other purposes (wait a minute they convert their homes or offices to classrooms!). It should be stressed that most if not all of the virtual universities offer only management related courses.

Admittedly, this will influence space requirements since it will suppress the demand for teaching space. On the other hand, the massive expansions in students' enrolment into the universities make it imperative to increase demand for traditional learning–teaching–classroom methods. Degrees involving science, technology and engineering seemingly require laboratory and workshops to conduct other activities. Lecturers and researchers also require conducive space to carry out their academic responsibilities which cannot be conducive in their respective homes or offices. The idea of shifting emphasis from classroom to home is not always appropriate to most students due to family demands. This is also true with the lecturers and other faculty members. Research and teaching equipment require buildings to accommodate them. All organisations require a minimum amount of *space* to operate. Buildings are products of organisational changes. Essentially, university buildings must provide supporting and conducive environments that assist and promote the delivery of university core business objectives.

For some unexplained reasons, universities have a unique problem in identifying maintenance as a value-added or profit-generating function, however. University management often considered maintenance as an inevitable cost or a devil that costs what it costs. Making it difficult to define meaningful objectives and hence difficult to make effective long-term plans for maintenance. A university has its buildings scattered across its campus.

Many universities, particularly publicly owned ones, also have residential buildings that house their staff outside their various campuses. University buildings include academic buildings, administrative buildings, student hostels, staff residential buildings, Mosques, hospitals, staff clubs and other supporting facilities. All these buildings are part of the assets of the university. A university that does not own their own may rent apartments for their staff. So whether the university is IT-based university or not it requires buildings to operate. Requirements for space in universities are not a problem peculiar to less developed or developing countries. I will sight an example: Stubberfield (1996) opined that just a handful of educational institutions in the developed countries like the UK and USA are dipping into the pool of the "Virtual Campus" concept.

The basic questions that need looking at if the concept of the virtual campus is to be embarked on are as follows:

1. But what are the likely consequences if the University campus has to be transformed into a virtual campus?
2. What type of activities and functions can be incorporated within it?
3. If the students of the future are surfing the Net from their bedrooms, receiving and transmitting information relevant to achieving a degree or other qualification?
4. What will happen to the university buildings themselves? (Shabha 2000).

The spatial organisation of the buildings can be changed to always suit any unique requirements of the building users. Though electronic environments may enhance methods of learning, they cannot replace the intellectual society of the university. A "physical" space is required for effective teaching and learning to take place. For instance, the laboratories and workshops of the virtual universities will not be in the space in anyway. To reiterate what has been mentioned before, according to Bill Gates, "one cannot totally communicate through the technology; one needs to meet face to face to be able to feel what is being uttered". Therefore, from the foregoing, it is concluded that though "Virtual Campus" will have impact on space requirements, it is unlikely to lead to any noticeable overall reduction in university buildings.

4.13.4 Conditions and Performance of Buildings

A university has a more diverse and wider range of buildings that require complex operational needs than probably any other organisations. Buildings contribute significantly to the quality of education. University buildings provide a stimulating and supportive environment for students and staff alike in addition to supporting teaching, learning and other academic related activities. The performance of a university can be improved by improving the performance of the buildings wherein all activities of learning, studying, researching and teaching are taking place. University's funds are mainly spent on staff salaries though the universities still need to attract more students, which are attracted not only by the staff academic standards but by the facilities that accommodate university activities. It is common knowledge that universities are often identified by their facilities such as buildings. Students often consider the conditions of the university buildings while making their choice of the university. It was in 2009, while attending a conference, in a university in Malaysia, The VC, after delivering his Keynote address/Welcome Speech, said, "Please go around and enjoy the beauty of our buildings".

However, the roles of buildings in facilitating this change cannot be over emphasised since buildings provide the users with the required temperature, humidity, lighting and ventilation for people to live, work and learn productively. As a result of understanding the complex role buildings play towards the delivery of the complex mission of higher education institution, the Higher Education Funding Council for England has outlined that educational institutions should integrate building into their plans rather than listing buildings in their wish list (Wood and Worthing 1996). Therefore, proactive attention must be given to the operation of the building.

In the executive summary to a well-attended seminar in Australia that was attended by experts from different countries including Malaysia, entitled "Places and Spaces for Learning Seminars", as parts of the Focus Issues (FI) after the presentations, participants were asked to reflect on eight key questions. The fourth question put forth to participants is "How can we use building space to change/enhance the education experience of students?" The summary of the thoughts, feedback and reflections on this question is "The importance of management and maintenance of spaces: there is a need for strategies for ensuring new spaces continue to be effective, as a 'degraded' space is a powerful disincentive to student engagement" (The Carrick Institute for Learning and Teaching in Higher Education 2007).

There is no conclusive published literature on the satisfaction of the university buildings in Malaysia. However, comments from users suggest that the conditions of the buildings are generally below expectations. Comments such as "please change the maintenance officers in our block" and "they do not take good care of the buildings" and "our toilets are always dirty" suggest the buildings conditions are poor. The most common identified items or areas that require maintenance is door locks, toilet and doors. In fact, a well-rated university in Malaysia was recently been described by students as "nice environment, nice buildings, lousy maintenance, and lousy service". More quotations by dissatisfied university buildings users can be found in Appendix.

The environmental condition where students learn/study is a reflection of the nation's well-being and university's buildings worth considerable amount, both in terms of monetary and social assets (Turrell 1997). In fact university buildings are in themselves significant national assets which have been built up over a number of years many of which are still in use. In Malaysia, many of the universities were established more than 30 years ago, with the first two public universities established in the 1950s. But if the condition of these buildings are surveyed together with many intense building periods following various government policies with the decline in quality of materials components and labour, it will perhaps come as no surprise that there is a significant maintenance need in most of those buildings.

Malaysia is confronted with university buildings that are **ageing**, **decaying**, **deteriorating** and therefore requiring maintenance in order to improve their performance. Unfortunately, this is happening at the time the government is reducing its allocation to the university sector. However, when the operation budgets of the university are reduced, maintenance allocations are vulnerable to slashing. This is probably because building performance issues are not discussed at the board room until when everything goes wrong.

On the other hand, since the 1960s several studies have indicated the need to balance capital costs against the subsequent maintenance costs of the buildings. A perceived saving today could lead to high maintenance costs in future. However, older buildings constructed of well-established materials and components and using well tried and tested techniques (and which in all probability have been updated and improved) may not be subject to heavy maintenance works or expenditure. Typically maintenance management seeks to increase the productivity of the maintenance organisation and their profit but at same time increase building users satisfactions and productivity. Specifically, it facilitates achieving the organisation's corporate mission and vision through maintenance.

University stakeholders may not attach the same importance to the building as in the case of the hotel sector, but the role of the building towards achieving quality education is prominent. For instance, the parents, students and faculty members might pay much attention to the availability of latest and relevant reading, learning and research materials. However, times have moved on, there are greater concerns and pressures on the university to ensure optimum performance of the buildings in order to be competitive, attractive and innovative. A greater understanding of the building performance is shifting much pressure to the maintenance department. There are continuing realisations that the maintenance department should be considered as a university strategic business unit or USBs. This is the only way the maintenance department can provide value-added services.

4.14 Changing Dimensions of Students

The essence of a university is to develop quality students that possess analytical and problem-solving skills and interpersonal understanding as part of their learning achievement to be able to contribute to the national goals and aspirations. The number of students acquiring university education is on the increase globally and nationally. There is the continuous recognition of the importance of a university education. University students come from wider ranges of backgrounds and age groups. In the past, university education was for the privileged members in the society. The greater prior experience individuals have will have great influence on how they observe and analyse their environment and including the university environment. Universities in the future must allow for a range of learning demands and allow students to learn and nurture in the way that is conducive for the students' particular requirements (Shabha 2000).

The major challenge facing a university today is that students and faculty members represent a new, empowered and enlightened breed of consumers that are familiar to outstanding services in a variety of venues. Therefore, new breed consumers required multidimensional services as compared to how they use to be. They demand greater experience with responsiveness, customisation, variety, quality and convenience. The supply side needs to response to be competitive. For effective utilisation of university resources, there is the need to better understand the interactions that exist between students, staff and the buildings.

In university, students and staff are the typical end-users, while the purpose of enhancing building performance is to improve students' learning (Leung and Fung 2005). It is to also enhance teaching, innovation and studying. Customer intimacy demands that the university must have sufficient and adequate knowledge of their customers' needs and wants. The service provider must look far beyond the immediate objectives of the products or services to the users. The providers must provide service that has a wider ends than the customers realised was possible (Bartholomew 2008).

With Malaysia becoming an educational hub and destination for students from the Middle East, Africa, Caribbean, Asia Pacific, America and even Europe, there will be a need for the supply and operation of university buildings in order to provide better service for money. There are about 1.5 million students in Malaysian universities. Indeed, this is enormous. Even though students are the main occupants of the university buildings, academic staff are also affected by the condition and performance of the university buildings. Often, more than 70 % of students live on campus. Modern students consider the condition and performance of university buildings for enrolment and retention purposes.

4.15 How Satisfy are the Students with Buildings?

Satisfaction is an elastic concept and means different things to different people. Here, it means optimum satisfaction or customer delight. It is normally measured as the difference between experience and perception on the one hand and achievement on the other side of the scale. The requirements of the value systems of current building users particularly those that use the university buildings are complex and multidimensional. In other words, the users are looking for services that go beyond the basic to meet their unique needs and in fact exceed their expectations (see also Smith and Wheeler 2002). Sarshar and Pitt (2009) noted that users do not really care about the details of how services are designed but are aware of the process quality that they can observe in the purchase of services. Based on available results regarding the satisfactions of the students on the university building in Malaysia, the users are not satisfied with buildings condition and performance of the buildings on their campuses (see for example Abu Bakar and Abu Bakar 2008). However, and again, previous studies tell us that users' satisfaction with the environment they are studying will improve with new or recently upgraded facilities (Schwede and Davies 2008).

4.16 Summary

University buildings are long-lived capital assets, and a duration of 100 years is typical. The building provides values not only to the university, but also to the students, faculty members, parents and other users and stakeholders. University buildings are facilitators and enablers as they are not wanted for their own sake, but for the services, they provide and supply to the building users. Thus, a failure in the supply of the required services is a loss in value to the universities, users and other stakeholders. A condition that might be acceptable for resident buildings could require immediate maintenance in the case of university buildings.

University buildings constitute the most difficult collection of large buildings to maintain because of their complex engineering services and their heterogeneous nature. Today, especially in the industrialised economy, universities (and indeed the publicly owned) are considered corporate bodies. A university now takes the characteristics of the private sector (not least in their management approaches). In other words, universities have grown to the stage, that they are managed as modern profit-based organisations. A university is comparable to the manufacturing sector, wherein the different organisations within the system are mutually dependent (Miller 2007). There is the need to consider a university in terms of a single enterprise; that is, the performance of one unit is considered to have direct impact on the outcome of the others in order to succeed and progress. The performance of the maintenance organisation has positive correlations with the productivity and profitability of the university as an entity.

In order to earn value for money invested for construction and operation of the expansive, multifaceted and diverse nature of buildings in university campus, there is the ultimate need for a capable maintenance organisation. Although, considerable investment is made to develop and operate university buildings, the building condition and performance have not been recognised and identified as part of the major issues and challenges of higher education.

References

AbdulKadir, S. M. S. (2002). Human resource development for k-based economy. In H. A. D Jamaluddin, H. A Zulkurnain, & S. Naidu (Eds.), *Proceedings: K-based economy: Forging ahead for national transformation. Organized by National Institute of Public Administration (INTAN) and Administrative and Diplomatic Service Association (PPTD).* pp. 121–135 KL: INTAN ISBN; 9839933-083-0.

Abu Bakar, N., & Abu Bakar, N. (2008). Students' Satisfaction towards the University Accommodation and Public Amenities. In *Proceedings of International Conference on Construction and Building Technology (ICCBT.)* 2008-B-(28), pp. 307–320 Kuala Lumpur 16–20 June 2008.

Amarantunga, D., & Baldry, D. (1998). Appraising the total performance of higher educational buildings: A participatory approach towards a knowledge system. *COBRA 98RICS* (Construction and Building Research Conference). ISBN 1-873 640-23-4.

Audit Commission. (2003) PFI in schools: the quality and cost of buildings and services provided by early Private Finance Initiative schemes. In K. Alexander (2010) (Ed.), *Proceedings: Usability of Learning Environments. CIB W111 (Research Report, 330): Usability of Workplaces-Phase 3 CIB. Netherlands: International Council for Research and Innovation in Building and Construction (CIB)*(pp. 5–16). ISBN: 978-90- 6363-061-4.

Allen, E. (1995). *How buildings works: the nature order of architecture*. Oxford: Oxford University Press.

Allen, E. (2005). *How buildings works: the nature order of architecture*. Oxford: Oxford University Press.

Bajunid, I. A., & Jantan, A. H. (2008). Higher education in Malaysia: challenges to the scared and sacrosanct ivory tower. In I. A. Bajunid (Ed.), *Malaysian from traditional to smart schools: The Malaysian education odyssey*. Selangor: Oxford Fajar Sdn Bhd.

Barnett, R. (1994). *The limits of competence, knowledge, higher, education and society*. Buckingham: Open University Press.

Bartholomew, D. (2008). *Building on knowledge: Developing expertise, creativity and intellectual capital in the construction professional*. Oxford: Blackwell Publishing Limited.

Chanter, B., & Swallow, P. (2007). *Building maintenance management*. Oxford: Blackwell Publishing.

De-Chiara, J., & Crosbie, J. M. (2001). *Time-saver standards for buildings types* (4th ed.). Singapore: McGraw-Hill International Edition.

Douglas, J. (1996). Building performance and its relevance to facilities management. *Facilities,14* (3/4), 23–32.

Gabriel, G. C. (2003). Decentralising asset management in a university environment using Web enabled technology. *Facilities,21*(10), 233–243.

Government of Malaysia. (2001). *Eighth Malaysian Plan 2001–2005*. Putrajaya: Economic Planning Unit Prime Minister's Department.

Government of Malaysia. (2006). *Ninth Malaysian plan 2006–2010*. Putrajaya: Economic Planning Unit Prime Minister's Department.

Government of Malaysia. (2010). *Ten Malaysian Plan 2011–2015*. Putrajaya: Economic Planning Unit Prime Minister's Department.

Kyle, B. R. (2001). Toward effective decision making for building management. In *Proceedings of APWA International Public Works Congress, Philadelphia, Sep. 2001 NRCC/CPWA/ IPWEA Seminar Series Innovation in Urban Infrastructure* (pp. 52–69). Available at www.nrc.ca/irc/uir/apwa.

Laccase, M. A., Vanier, D. J., & Kyle, B. R. (1997). Towards integration of service life and asset management tools for building envelope systems, In *Proceedings of 7th Conference on Building Science and Technology: Durability of Buildings Design, Maintenance, Codes and Practice* (pp. 153–167), Toronto, 20–21 March.

Leung, M. & Fung, I. (2005). Enhancement of classroom facilities of primary schools and its impact on learning behaviors of students. *Facilities,23*(13/14) pp. 585–594.
Malaysia and the Knowledge Economy: Building a World-Class Higher Education System. (2007). Report No. 40397–MY Human Development Sector East Asia and Pacific Region. The World Bank. Available at: http://english.cpiasia.net/index.php?option=com_content&view=article&id=88%3Amalaysia-and-the-knowledge-economy-building-a-world-class-higher-education-system&catid=85%3AEducational+Reform&Itemid=169. Accessed on 21 Sept 2008.
Mei, A. T. (2002). *Malaysian private higher education: Globalization, privatisation, transformation and market places*. London: Asean Academic Press.
Miller, B. A. (2007). *Assessing organizational performance in higher education*. New Jersey: Wiley.
Milliken, J., & Barnes, L. P. (2002). Teaching and technology in higher education: student perceptions and personal reflections. *Computer and Education,39*(3), 223–235.
Mills, E. D. (Ed.). (1994). *Building maintenance and preservation; a guide to design and management*. Oxford: Butterworth-Heinemann.
Ministry of Education. (MOE) (2012) National Education Statistic: Higher Education Sector. Available at http://www.mohe.gov.my/web_statistik/Perangkaan_SPT_2012.pdf. Accessed on 26 Aug 2014.
Mohayidin, M. G., Azirawani, N., Kamaruddin, M. N., & Margono, M. I. (2007). The application of knowledge management in enhancing the performance of Malaysian universities. *The Electronic Journal of Knowledge Management,5*(3), 301–312.
Mohd Anuar M. (2008). The Balanced Score and service delivery systems for public universities in Malaysia. MAM-AC-Lanhkawi-4-7-June-2008-Service Service Delivery System in Pus. Retrieved on December 3, 2008 from http://www.uitm.edu.my/news2008/paperworkconference/DrAnuar Marzuki.pdf
Moore, M., & Finch, E. (2004). Facilities management in South East Asia, *Facilities*, 22(9/10), 259–270 ISBN 0263-2772.
Morshidis, S. (Ed.). (2006). Enhancing quality of faculty in private higher education institutions in Malaysia. Higher Education Research Monograph. USM, Penang: National Higher Education Research Institute (IPPTN).
Oakland, J., & Marosszeky, M. (2006). *Total quality in the construction supply chain*. Oxford: Butterworth-Heinemann.
Osburn, D., & Greeno, R. (2007). *Introduction to building* (4th ed.). New Jersey: Pearson.
Sarshar, M., & Pitt, M. (2009). Adding value to clients: learning from four case-studies. *Facilities,27*(9/10), 399–412.
Schwede, D. A., & Davies, H. (2008). Occupant satisfaction with workplace design in the new and old environments. *Facilities,26*(7/8), 273–288.
Seeley, I. H. (1996). *Building Economics: Appraisal and control of building design and efficiency* (4th ed.). New York: Palgrave.
Selvaratnam, V. (1985). The higher education system in Malaysia: metropolitan, cross-national, peripheral or national? *Higher Education,14*, 477–496.
Shabha, G. (2000). Virtual universities in the third millennium: an assessment of the implications of teleworking on university buildings and space planning. *Facilities,18*(5/6), 235–244.
Smith, J. & Wheeler, J. (2002). *Managing the customer experience*, Prentice-Hall, Harlow. In Sarshar, M & Pitt, M (2009) Adding value to clients: learning from four case-studies. *Facilities*, *27*(9/10) pp. 399–412.
Stubberfield, M. (1996). Virtual Campus (The Times, 16 October).
The Carrick Institute for Learning and Teaching in Higher Education. (2007). Draft report PLACES & SPACES FOR LEARNING SEMINARS. http://www.google.com/url?sa=t&rct=j&q=&esrc=s&frm=1&source=web&cd=1&ved=0CDYQFjAA&url=http%3A%2F%2Fwww.olt.gov.au%2Fsystem%2Ffiles%2Fresources%2FConsolidated%2520Report%2520P%2520%2520S%2520Seminars%2520without%2520slides%2520ref%2520v%25200%25205%2520_5_.pdf&ei=0lWdUvaNM4SNrQejqoGYDg&usg=AFQjCNE5XUCuY1afTGp47sGT8mUhZjQtBg

Turrell, P. M. (1997). Small is different: a strategy for effective management maintenance in non-profit making organizations, *COBRA 97*, ISBN 0-85406-840-6.

Watt, D. S. (1999). *Building Pathology: principles and practice*. Oxford: Blackwell Science Limited.

Wood, S., & Worthing, D. (1996). Post-occupancy evaluation of higher education buildings; A case study. The Cutting Edge, 2006 RICS Research, ISBN 0-85406-930-5.

Chapter 5
Building Maintenance Processes, Principles, Procedures, Practices and Strategies

Abstract This chapter focuses on maintenance management in terms of its operation, procedure, systems, strategies, procurement and structures. Within this chapter, the main components of management activity are discussed: organising, controlling, directing and planning. However, these functions are not dealt with separately. An interactive and holistic system was used. That is, the functions are related to maintenance technology and engineering in theory and how it should be practiced. The literature on the principle of value-based management is also provided. In this way, a common thread is provided by interweaving all the themes and subjects together. Because of the unique nature of building users in relation to the functional requirements of the buildings, each building type has peculiar maintenance management that is applicable in order to improve the building performance as well as increasing productivity, user satisfactions and profits. After studying this chapter, you should do the following: (1) Appreciate the unknown factors, issues and elements that are peculiar to maintenance management. (2) Be aware of the often omitted or over sighted functions of maintenance management. (3) Recognise the different methods of maintenance. (4) Be able to identity what type(s) maintenance technique to apply, considering the functional requirement of the building or engineering services. (5) Understand the theoretical functions involved in building maintenance management.

Keywords Building maintenance · Defects · Maintenance strategy · Building performance · Building condition · Value management · Maintenance management functions

5.1 Introduction

In the previous chapter, the roles and functions of buildings towards contributing to quality educations was discussed. It was established that buildings require maintenance in order to be a source of income and profit generators. In that regard, the maintenance services require systemic management to earn best value for money invested. Thus, the goal of this chapter is to discuss building maintenance and

A.L. Olanrewaju and A.-R. Abdul-Aziz, *Building Maintenance Processes and Practices*, DOI 10.1007/978-981-287-263-0_5

management principles. The various activities and issues in maintenance services are discussed. It is thus established that maintenance should be positively planned, strategically organised, proactively lead, holistically controlled and dynamically implemented if best value is critical. Many issues which reside in the building maintenance management often involve investigations using theories and methodologies from varying disciplines including economic, laws, engineering, science, culture, mathematical and statistical information and knowledge. Consequently, the roles of capable maintenance organisation are expedient.

There is the need to develop maintenance departments with forethought. The lack of this will mean the department is not considered as a major strategic business unit in an organisation. Organisation theory has described the importance organisation structure has on productivity and customer satisfaction. From the organisation point of view, the ways an organisation is structured and operates its business have significant effects on its output which consequently influences the environments it exists in. Put in other words, the university maintenance department has greater consequence on the university business. The maintenance department must have sufficient in-built flexibility to cater with every work arising from the university's activities. Institutions including the universities are now realising that greater saving is possible with an organised maintenance department. Emphases are now placed on the way maintenance is procured, and the competencies and experience of the members. Due to the complexities and significance of maintenance works, there is the need for a proper maintenance department.

The chapter introduces the *value* concept into the maintenance management practice and theory. Value is a generic term that denotes effectiveness and efficiency together with the recipients' satisfactions. It is a blend of environmental, cultural, economic and behavioural issues. Within this context, this book centres on a maintenance management model that is based around the concept of value rather than on the traditional approaches of managing building maintenance that focus on cost and condition concepts. Value maintenance management is the application of value management philosophy and principles to plan, coordinate, control and organise maintenance resources. In value management, the user/client value system is crucial in order to provide best value services. In other words, there is the need to identify the criteria within the user/client value system. Therefore, this chapter outlines that the essence of initiating maintenance should be due to the value the users and providers attach to the building rather than due to the physical nature of the buildings and availability of funds, and in this way, the amount invested will be maximised and recipient's satisfaction will be enhanced.

5.2 Ontology and Epistemology of Maintenance

Maintenance has been defined in chapter one of this book as "all the required services and activities provided in order to preserve, protect and care for a building's structure and form after completion or after any repair or replacement to current standards to enable it to serve its intended functions throughout its entire lifespan without upsetting its original features and use". However, the definition contains some key terms and concepts that require looking into. For instance, "to preserve" denotes that maintenance is an active process that never comes to end provided the building remains. Further, when carrying out maintenance, the strategy must take into account current advancement in building regulations, standards and technology.

"Care" implies that the performance of the building requires systemic consideration. "To serves its intended functions" suggests that maintenance aimed at allowing the building to perform the useful roles for which it was built. "Without upsetting its original features and use" denotes maintenance has no effect outside the boundaries created by the initial designs of the building. In another words, it implies that maintenance is caused due to depreciation from wear and tear—use function. Works caused due to changing in users' value systems cannot be associated with maintenance services. "To current standards" suggests that maintenance involves some degree of improvement, even if it is for the sake of accommodating new technology or latest codes, standards and/or regulations.

This definition recognises that there are progressively more realisations that maintenance must be viewed from the engineering, scientific, technological,

Table 5.1 Various maintenance definitions

Author(s) and date	Definitions
Pitt (1997)	The task of providing a repair service that will protect the fabric and the use of the building to a standard that represents value for money
Christensen (2005)	To preserve in a condition or state of quality equivalent to that which was designed, structured or used in the past
Best and De Valence (2002)	All control employed to make sure asset remains functional
Lofsten (2006)	Actions undertaken in order to reduce the adverse effects of breakdown and maximise the facility at minimum cost
Booty (2006)	All the technical and administrative actions that are required in order to retain a building in a state in which it can serve its required functions efficiently and as expected
CICA (1989)	Any intervention to make sure a facility reaches its optimal service life
Emmitt and Yeomans (2001, 2008)	The necessary actions that are carried out to prevent building failure and to extend the service life of the buildings
Burden (2004)	To provide upkeep, repair and care for the building's integrity and appearance after acquisition or after restoration at an acceptable level to enable it to perform its useful functions
Marsh (2003)	Services required so that the building can operate efficiently and effectively while taking into account environmental condition, reliability and economic issues
Ashworth (2004)	Works required to preserve building and other items or equipment in a state of fitness for purpose

political as well as business perspectives. In short, it must be considered from the commercial perspectives. The definitions also suggest that maintenance can be corrective (repair) or preventive (preserve) and more importantly value based, even though maintenance is not usually regarded as value added services. Buildings are to be maintained in order to support an organisation's business objective. See Table 5.1 for overview of various definitions of maintenance. However, these definitions are not explicit and comprehensive and hence the need to revisit and extend the definitions. The classical definitions are inadequate to cover the scope, size and prime objective of maintenance in its current context.

5.3 Analysis of Factors Leading to Maintenance

The extent to which a building is maintained is driven by the necessity to protect the building performance, increase productivity and increase user's satisfaction. Therefore, organisations should generally seek to establish a strategic maintenance policy to maximise the value of the buildings. Primarily, the decision to maintain should be driven by maximising user performance. Maintenance could emanate due to different factors, e.g. problems arising from poor design, bad workmanship, substandard materials and components, as well as the unaccommodating behaviour and nature of the users and occupants. Maintenance costs increase with the age of the building. However, older buildings constructed of well-established materials and components and using well-tried and tested techniques (and which in all probability have been updated and improved) may not be subjected to heavy maintenance works or expenditure. The major sources of maintenance are explained next.

5.3.1 Poor Design

This is a maintainability issue. Building maintainability is a design characteristic. It is concerned with ease, accuracy, safety and cost of building maintenance. Some buildings are designed without considering the maintainability aspect, thereby making maintenance difficult, expensive, risky, unsafe and tedious. A building should be designed/constructed so that its maintenance can be performed with minimum cost and time. Effective maintainability is mainly design related. Inadequate consideration of maintainability results in a lot of maintenance problems while the building is in operation. Maintainability entails maintenance time, labour/frequency of maintenance, maintenance costs and other related issues that facilitate maintenance.

Many maintenance works actually generate from the design stage, as poor design will lead to serious maintenance problems. However, no amount of maintenance can be adequate for a building that is poorly designed. The system, sub-systems, components, materials and equipment specified at the design stage have profound effects on the maintenance work required once the building is in operation. It is essential that adequate attention is given to the chosen components and elements. Practicable, versatile, maintainable and robust materials, components and systems should be preferred. The selected components should also be user-friendly and standardised in order to ease the maintenance or repair works. Reliability and durability require systematic and holistic thought while selecting alternative items to avoid interminable and expensive maintenance works, besides the disruption to daily activities and impeding productivity.

In spite of that, the design team, while pondering over the drawing board, usually gives insufficient attention to the maintenance issue, more so if the client is a developer that is not the end-user. A blame game normally arises at a later stage if a problem arises and as soon as a problem is reported a typical scenario is members of the design team pointing accusing fingers at one another. What more, the designers will also lay blame on the client or his agent of misuse, while the client on the other hand accuses the designers of negligence or professional incompetence. Suffice to say, design is defective if wrongly employed, e.g. flat roofing that is suitable or requires little maintenance for buildings in Europe, may on the contrary require more serious attention in Malaysia due to its high humidity and heavy rainfall. The under-listed questions should be addressed while selecting alternative designs:

(a) Are the parts readily available?
(b) Are the parts standardised and simplified?
(c) How expensive would the parts be?
(d) Is there a need for specialised labour?
(e) What is the duration of the likely maintenance works?
(f) What is the nature of the likely failure?

While considering alternative designs, it should be brought to mind that a cheap or simplified design might turn out to be expensive and difficult to maintain. This way, maintainability should be related with the maintenance process and the building life cycle cost. Feedback from a maintenance manager or maintenance engineer can be very useful at the design stage, although this rarely happens in practice. Designers hardly visit the building they designed while in operation to assess the post-occupancy satisfaction of the users and neither do they have the opportunity to interact with the maintenance personnel. However, the fact that each building is unique makes the matter more complicated. It will greatly assist if the maintenance organisations are consulted before new buildings are commissioned.

5.3.2 Poor Workmanship

Labour plays a very important factor towards reducing maintenance works. The consequence of labour on the performance of construction projects are multifaceted as the workforce factor starts from the concept and design stage and continues throughout the construction and building operation stages. The design and construction team as well as the maintenance personnel must display a high level of competency and be able to work in sync. It is not difficult to postulate that even with high-quality materials and components, the building will eventually fail if the workforce that put the building together are not good at the job. Poor workmanship could be the result of lack of know-how, know why, know when, know what, training and skill, or just plain negligence.

The construction team need to address the issues of technical know how, technical know what, technical know when and technical know why while putting the building together. Poor design also includes situation where the designer is not aware of local requirements. Many buildings suffer due to know-how issue. Even if most of the designers are local, but they are foreign trained, they may not have adequate knowledge of the local requirements before being allowed to partake in the design and supervision of projects.

5.3.3 Defective Materials and Components

Substandard and defective materials and components will lead to maintenance problems, if not at the commissioning stage then surely later, when the building is operational. In order to reduce the impact of the problem, durable and industry-proven materials and components should be given priority (see Emmitt and Yeomans 2001). Materials of good quality can also fail to make the grade if used for the wrong purpose or in an unsuitable environment. Each material has its own peculiar characteristics and qualities that the designer must not ignore. As for the contractor, a maintenance bond should be signed in order for the contractor to rectify all defects owing to the usage of unapproved or second-rate materials. The designers and other consultants should make available to contractors and key maintenance indicators that the contractors must comply with even though this rarely happens in practice.

5.3.4 Usage and Age

All built facilities, irrespective of materials, components or workmanship used will with usage and time, encounter structural and systemic deterioration and decay. Maintenance in terms of usage is hinged on, and very much related to, the culture of the occupants. However, a question that must be addressed is who is the design/building meant for? A suitable answer to the question is very important towards reducing the impact of maintenance due to usage and age of the buildings. Ahuja and Walsh (1983) suggested that maintenance programmes should be introduced early, i.e. as soon as the project is completed.

The personnel or department responsible for maintenance works should be encouraged to develop a maintenance inventory and record of all the equipment, plants and tools. Building users should be provided with the proper information in the form of a manual or guidebook, not unlike that provided to someone who purchases a washing machine or toaster. Since this practice remains a hallmark and a prevalent aspect of the manufacturing sector in helping to increase consumer awareness, the construction industry must emulate these simple deeds to enhance their service quality and earn the confidence of their customers. It is only natural that users compare the services they get from the construction industry with what they get from different providers such as the manufacturing industry. Hence, the construction industry players must embrace the culture of value added best practices as being observed by other industries.

5.3.5 Climatic and Environmental Factor

One of the characteristics of the construction industry's outputs is that they are remarkably different from the outputs of other industries. By and large, the products are exposed to atmospheric weather condition. Buildings are constantly subjects of sunlight, rain, frost action, humidity, condensation, wind and pollutions of all sorts. Buildings are affected by climatic and environmental factors. This is even if the architects and/or engineers designed the building correctly and the contractors constructed or built correctly. Hence, a major cause of maintenance problems in buildings is exposure to climate and weather conditions. Different building materials and components react differently to weather. Therefore, to minimise the maintenance expenditure, there must be a proper selection of materials and components, as well as alternative designs for different atmospheric conditions and geographical locations.

Once more, a wrongly designed roof, like a flat-roofed feature in a building that might be appropriate in the West, might be more expensive to maintain in the tropical countries such as Malaysia due to high humidity and heavy rainfalls. The effect of the climatic changes is also affecting our buildings. The materials and components that were suitable in the past are becoming unsuitable due to the effect of climatic changes. While increase in temperature with high humidity is increasing the rate of decay and deterioration, flooding is also making the buildings unreliable. One of the likely measures to improve maintenance activities is through simplifications of designs. Complicated designs are expensive to maintain, most especially for hot–warm humid regions such as Malaysia. Complicated design could increase the growth of lichen, algae and fungi and create dampness. Clearly if this does happen, it will lead to the increase in maintenance costs.

Building materials and components are exposed to different weather conditions for long periods of time, the reason why it is very important to determine the effects of sunlight, heat, humidity and other effects of climate changes in the properties of the materials, colour and gloss. Therefore, there is the need to understand these properties in the evaluation of the building performance. For instance, determining the colour of materials is important not only for aesthetic reasons, but also because any change in hue may be a sign of predecay alterations (Costa et al. 2010).

5.3.6 Nature of Users

In certain types of building, especially university, hospital and market buildings, because of the complex nature of the users, the users are the human enemies of these buildings, through various activities, namely vandalism, violence and arsonists. These elements are prevalent in the buildings mainly due to poor maintenance or dilapidation (Allen 2005). In university buildings, due to the intriguing nature of the users coupled with constant changes of users that could have diverse backgrounds, there seems to be high demand for maintenance or improvement. However, buildings, even if in a good state of repair, could suffer attack from vandals, arsonists and violence if they seems psychologically threatening to the users. For example, vandalism is caused by the physical environment of the building; its design, detailing and fittings may attract vandals (Mills 1994).

Female students in the university request more maintenance as compared to the male counterparts, and because of the user's age, maintenance is more compelling in the university buildings. Therefore, combinations of good designs and installation together with efficient maintenance could serve to prevent, or reduce maintenance as a result of vandalism and violence in the university buildings. Furthermore, the increased user loads on the building and lack of awareness of the user are also greatly contributing to maintenance.

5.4 Types of Maintenance

Maintenance is generally classified as planned and unplanned maintenance. Planned maintenance is further grouped into preventive and corrective maintenance. As the name implies, planned maintenance is when the building and its contents are maintained on a scheduled basis. The scope of the maintenance schedule is determined by the results of past experience and performance of similar facilities. On the other hand, unplanned maintenance is often corrective in nature. This is whereby remedies are initiated to restore an already defective or failing part of a building. Corrective maintenance is also often termed as emergency or normal response to a failure or breakdown (Ashworth and Hogg 2002). Empirical studies have suggested that maintenance of the university buildings in Malaysia is corrective, routine, preventive and proactive (Fauziah 2006; Zakaria and Ali 2006a, b; Zakaria et al. 2008). But it is not only the university buildings that are correctively maintained but most other building types are also based on corrective maintenance policy. The descriptions and explanations of these approaches to managing maintenance are provided next.

5.4.1 Preventive Maintenance

This approach is routine in nature. It is concerned with intervening in the life cycle of a building immediately before it can be expected to cause problems (Newman 2001).

Simply put, preventive maintenance is the replacement of building parts on the bases of certain criteria, usually time and usage, for instance, to replace door locks or hinges after 5 years whether the lock has failed or not. These requirements are often based on the manufacturer's instructions or experience or both. Preventive maintenance consists mainly of inspecting and cleaning the building and its contents. Essentially, it is to reduce the probability of failures or breakdowns before they emerge and to detect any existing malfunction before it becomes serious.

Some examples of preventive maintenance include repainting, minor repairs such as replacement of tiles, cleaning of windows and replacement. It also includes removal of stains (e.g. effervescence), removal of algae and vegetation, cleaning of surfaces (washing, abrasive washing and chemical treatment), resin impregnation, and treating of timber framework and studs. These problems could probably be as a result of poor workmanship or poor design. Problems of structural or functional deterioration can be avoided by introducing preventive maintenance before the building fabric fails. Preventive maintenance can reduce maintenance costs by about 15 % if properly introduced.

Preventive maintenance procedures should be incorporated into a building's operational programme. Naturally, it makes more sense to prevent bad events from happening than to look for remedies after they strike. Maintenance programmes that are formulated based on the preventive maintenance principle must relate to the organisation's policies and form part of its overall development objectives. In practice, preventive maintenance is most likely to be an appropriate maintenance strategy for buildings whose condition can be monitored and for which the cost of applying time-based preventive maintenance is less than the cost of applying corrective maintenance. Nevertheless, the major setback of the preventive maintenance practice is that it incurs unnecessary cost on replacement of components and systems whose conditions may be perfectly satisfactory. Often, the replaced items are still operable when they are replaced.

5.4.2 Corrective Maintenance

Corrective is what is generally referred to as repair. Therefore, when people mention repair, what they mean is to repair. Corrective maintenance is the type of maintenance that is initiated when a building, and/or its systems and components, has failed. However, in most cases, corrective maintenance is initiated as a result of lack of maintenance culture, whereby a building or its parts are allowed to fail before maintenance is initiated. Corrective maintenance is done to restore the building into its normal condition. Corrective maintenance entails the repair, servicing and replacement of any worn out, defective or broken parts, including the cutting-off of any decaying or crumbling component of a building. Corrective maintenance causes degrading of the building structure and makes the progress of future maintenance activities more difficult and expensive.

Therefore, if an alternative strategy is not put in place, the building structure will continue to deteriorate since lack of proper maintenance leads to further deterioration, decay and failure. Maintenance is contagious! It is like cancer. Corrective maintenance leads to remaintenance or multiple maintenances. Corrective maintenance is spontaneous and involuntary in nature unlike the more proactive variant in preventive maintenance. However, in practice, corrective maintenance might not be avoided in the building life cycle, for instance, the damage caused to buildings by natural calamities. Often, building fabrics are maintained on a corrective basis. Nevertheless, the emphasis is that corrective maintenance should be reduced to the barest minimum and it should be implemented with little disruption and disturbance to operations taking place in and/or around the building.

5.4.3 Predictive Maintenance

In predictive maintenance, modern sophisticated signal processing tools are used to monitor the condition of the buildings. For instance, non-destructive testing, transducers, vibration measurement, thermography, ferrography and spectroscopy make it possible to perform non-intrusive inspection in order to monitor the conditions of buildings. Predictive maintenance is also known as condition-based maintenance. This is because a building's components and systems are continuously monitored for their condition and reliability. Monitoring is usually done by techniques such as non-destructive testing, infrared, vibration analysis, noise detection, oil analysis and others. The benefit of this type of maintenance is the failure time of the components is known so their replacement is planned during shutdowns to avoid stopping the systems and components and also breakdowns while in operations.

However, this practice is more suitable for specialised building elements. It is most suitable in the manufacturing industry as compared to the building and construction sector. For instance, due to advancement in technology and communication, it is possible to embed sensors that incorporate wireless technology into a range of building components. But, this has great limitation especially in building fabrics where the applications of mathematical models prove difficult and unable to cope with the complexity of a real-life situation (Umeadi and Jones 2003). It is very expensive to install most of the sophisticated tools into many building components and systems. In most non-manufacturing industries, such as the building industry, its use is often limited to the electrical fittings, appliances and mechanical components.

5.4.4 Proactive Maintenance

Proactive maintenance is akin to the identification and elimination of defects in a building prior to the manifestation of any symptoms occurring in the building. It is very similar to the predictive maintenance and generally cheaper in the long run when compared with other strategies of maintenance. It involves concentrating on the root cause instead of the symptoms of the damage. Root cause failure analysis, which is the determination of the mechanisms and causes of building faults (see April and Abran 2008), is crucial in proactive maintenance.

The basic consideration in proactive maintenance is the elimination of the causes of defects from all sources. The fundamental causes of building failures can thus be corrected, and the failure mechanisms can be gradually engineered out of each building system in order to improve the efficiency of the building. Proactive maintenance is the combination of predictive maintenance and preventive maintenance. Table 5.2 contains the relationship and basic elements of the four maintenance strategies (Spare 2001).

Types of maintenance service could also be adaptive or perfective in nature (Forrester et al. 2010). The adaptive maintenance service involves adjusting or adapting the service system to changing to a different service delivery. An example of this is changing the maintenance service of residential building to academic building or changing from lecturer's requirements to student's requirements. Perfective maintenance service involves developing or acquiring additional service system or improving the operation capability of the service system. This should not, however, be confused with refurbishment work, as it does not involve changing the physical outlook of the building but only the service provided. However, building maintenance strategies, whether corrective, preventive, predictive or proactive, are most likely to be budget driven and adhere to the culture of fragmentation and correction. This means that maintenance is not carried out according to actual functional performance but it is dictated by financial priorities decided at a specific time together with the identification of defects in the buildings. While proactive maintenance can be an effective approach to maintenance, it fails to take into account the performance of the buildings, users' experience and culture of the users. These issues are very critical to the users since the overall purpose of maintenance management system usually entails how the building can efficiently enable the occupiers to perform their functions with minimum disruptions and disturbances and at the same time increase the provider's productivity.

Building occupiers often judge maintenance service directly as to how it affects them rather than on the actual quality of the repair (Wordsworth 2001). The proactive approach to maintenance is based on the culture of prevention instead of addressing the symptoms of the defect as is the case under the corrective, preventive and predictive approaches to maintenance, and the root causes of the defects and failure are

Table 5.2 The relationship between maintenance strategies

	Approaches	Logic
1	Corrective maintenance	Maintain when it failed
2	Preventive maintenance	Maintain it regularly
3	Predictive maintenance	Maintain it (just) before it failed
4	Proactive maintenance	Maintain the root causes as against the symptoms

properly addressed where possible. However, it is still initiated only when a problem is in existence. For effectiveness and efficiency, buildings ought to be maintained when defects hamper the organisation from achieving its business objectives.

Furthermore, according to Oakland and Marroszeky (2006), it is basically not enough to provide just a quality product or service to the customers. Rather the product or service provided or produced must be weighed against the price the customers must pay in exchange for the services. Therefore, mere providing of maintenance service to the users within the budget, with the highest degree of quality and when the users want them, is not sufficient. The service must also be efficient and satisfactory to the users. The service provider must also spend less money to achieve more.

5.5 Condition Survey

Building condition survey is the process of evaluating the conditions of buildings, systems and installations. This is to establish whether they are capable of continually supporting business needs and legal requirements. The survey can also determine the ability and capability of buildings regarding whether they can satisfy internal expectations and external benchmarks. Condition surveys are also a useful means of recording the nature and condition of a property for a number of other reasons, for example, prior to building works or other activities by a third party which might affect a site. In these cases, the condition survey provides clear records of the condition which underpin a claim for loss or damage.

Knowledge of how buildings behave is an important prerequisite towards choosing correct maintenance policy. All buildings suffer wear and tear by the effect of the wind, sun and rain and by the activities of occupants. Building components and services do not last forever but require replacement. As a building and its components age, it requires maintenance to avoid ongoing problems and potential health and safety risks. Sometimes, a building is constructed with inherent weakness that causes dampness, movement or other problems within the structure. The condition survey is an important technique to the maintenance managers in decision-making for property asset management. However, initiation of condition surveys has been generally reactive and uncoordinated. A condition survey is initiated if there is element of failure in building systems and elements.

Condition-based maintenance is an appropriate maintenance strategy for items whose condition can be monitored and for which online condition monitoring techniques are available and cost effective and/or for which the cost of applying condition-based maintenance is less than the cost of applying corrective maintenance. In developed countries, such as the UK, it is a requirement of HEFCE that universities carry out building condition surveys and that periodic strategic maintenance reports are presented to the university management. However, the effectiveness and shortcomings of the condition survey have already been seriously questioned (O'Dell 1996; Chapman 1998; Umeadi and Jones 2003).

5.5.1 Condition-based Maintenance

Condition-based maintenance is the maintenance initiated on the basis of information collected, processed and transmitted from the condition survey. Condition-based maintenance is a process that determines what must be done to ensure that a building continues to function in the desired manner. Condition-based maintenance predicts impending failure based on actual operating data instead of relying on traditional preventive maintenance, thus generally eliminating unnecessary maintenance performed on simple manufacturer's schedule recommendations. The benefits of condition-based maintenance include the following:

(a) Avoids unnecessary maintenance by performing maintenance when it is actually required.
(b) Provides early failure detection to increase asset availability, thus reducing costs and avoiding unnecessary downtime.
(c) Supports continuous improvement through an integrated work flow management capability, ensuring accurate and consistent response to developing conditions.
(d) Provides operations, engineering and maintenance with easy access to current asset information for better decision-making.
(e) Provides integration with the control, safety and maintenance environments.
(f) The solution enables organisations to turn data into actionable information to help reduce maintenance expense while simultaneously improving throughput.
(g) Enables the use and standardisation of work flows for other monitoring solutions and data collection required with point solutions.
(h) Enables tracking of equipment activities and statistics not normally maintained by asset historians.
(i) Provides a repository to capture the conditions, rules and other tacit knowledge applied by experienced maintenance and reliability personnel, thus allowing organisations to retain the intellectual property as the workforce changes due to ageing workforce and increased mobility.

However, the basis of condition survey puts the physical condition of the building at the centre of maintenance demand. This is wrong as it can only reduce the essence of building and its management to the building itself, particularly as the purpose of the building is to provide the required performance. To the extent that the building can support the required activities, the physical condition of the building is secondary. Buildings are capital goods. In other words, they are enablers and facilitators of human activities. In addition, coupled with the reduction of the government interventions in university management, the defective condition of the building will become more obvious. For one, the available budget will not be adequate to meet the maintenance backlogs not to talk of meeting the users' satisfaction. What more, in reality, condition-based maintenance has been based on mere virtual inspection. However, such condition assessment is insufficient to the real actual level of maintenance backlogs. The shortcomings of the condition survey have been well documented.

5.6 Performance-based Maintenance

One of the terms demanding definition in the context of building or constructed assets is "performance". The definition of the term in plain English is not sufficient to fit in. Nearly 20 dictionaries of building and architecture were checked to see what the term "performance" means. However, there is no clear definition for this term. What is available is definitions associated with terms such as performance requirements, performance curve, performance test, performance standard, performance bond and performance specification. While the terms are related to the "performance", the relationships are not clear-cut ones. Therefore, meaningful understanding of "performance" was not obtained. However, brief insight into the term is provided as "the specified, expected or actual behaviour of a building material, component or building while in use" in the dictionary of architecture and building construction by Davies and Jokiniemi (2008). The most unique element in this definition is "while in use". This definition is useful because it clearly tells us that the required issue relates to behavioural issues during the operation stage. To paraphrase the definition, building performance is the performance which defines the requirements that a building must provide, stimulate and sustain the required conducive environment to enable the user to perform their duties and responsibilities while the building is in operation.

While the definition provides clues as to what to include in measuring building performance, the definition does not indicate specifically what to measure and whether the criteria are qualitative or quantitative or both. However, building behaviours while in use can only be determined in relation to the user functional requirements. Functional requirements of the users entail both the qualitative and quantitative criteria. Building performance is hard to define in precise terms. Building performance is "beyond engineering". It comprises both quantitative and qualitative criteria that have a large impact on building users. It is easier to think of the building performance as a complex set of criteria and to use a model or framework to focus on determination or evaluation. While the quantitative criteria can easily be measured, the qualitative criteria do not lend them to exact measurements.

5.7 Building Defects

Building defect means different things to different people (Watt 1999). Within the UK construction industry, defects are now referred to as snagging (Sommerville and McCosh 2006). Words including deficiency, damage, default, deterioration and decay are often used to describe defect. In this book, all the terms are synonymous with defect and are sometime used interchangeably. Some defects are caused by the designers (latent defects), while some are caused by the constructor (patent defects), yet some are caused due to normal wear and tear and lack of maintenance. The latent and patent defects are often in relation to new buildings though. Defects occur throughout the completed building life cycle. Defect can be defined as a failure in building and services arising from factors including inadequate brief, faulty design decision, method of construction, user activities, maintenance and fair wear and tear (Wordsworth 2001). To some, it means the shortcomings with the design and construction practices, while to others, it implies the inadequacies that arise from the "normal wear and tear".

The primary causes of building defects are inevitably interconnected, and the focus in this book is on the maintenance caused and/or used caused defects. The design and construction lead defects constitutes another arena of specialist study. There are sufficient studies on how building defects can be reduced through brief, design and programming process. However, this does mean that the factors listed in this book to lead to defects are not related to design and construction issues, it is only that detail explanations and descriptions will not be provided. Building defects can be identified by the users themselves or by inspection by those concerned with maintenance management. Practitioners have the capability to identify defects that are not yet obvious, at least to users. The appraisal of defects is usually undertaken through a sequence of site procedures that provide the necessary information on which to make an assessment as to the condition and fitness for purpose. This book will not specifically dwell on maintenance technology. Readers interested in a detailed approach to fault finding through diagnosis and prognosis can refer to Watt (1999) and Williams (1993). These are excellent reading materials.

Building defects account for some 20 % of maintenance costs. The proper consideration of defect in total maintenance management cannot be emphasised enough if meaningful efficiency is paramount. The severity of a particular defect is judged against the effect the defect has, as well as those that it will have on the building and the associated services. Consequently, the influence that defects have on a building if it is not maintained depends on factors including the nature of the defects, the building integrity, the user needs and wants as well as the successive maintenance programme.

The effect of the defect on the condition, appearance and performance of a building depends on the functional requirements of the building. For instance, a condition that might be acceptable for a resident building might require immediate maintenance in the case of university buildings (Chanter and Swallow 2007). The degree of defects in a building is a measure to evaluate the functional performance of the building. Defects hinder the functional performance of the building and its services. Although buildings' defects can be generally attributed to lack of good construction practice, they are often caused and in fact aggravated due to lack of good maintenance practices and usage (Garlot and Sherma 2006).

The unique consequence of defects in a building is that it affects the building's capability to function optimally, technically, aesthetically and even otherwise. Building users are becoming more vocal in their requirements; they develop a cognitive map of what they want and often this at variance with the maintenance organisation's perceptions. What further makes defect assessment critical is that it sets up inner psychological tension in the users, if the defect is not rectified (Sommerville and McCosh 2006).

5.8 Repair and Methods of Cost Forecasting

When maintenance demands arise, a decision has to be made whether to incur the costs of repairing them or whether it is better to replace them. There are several factors that are considered when making the decision, including the age of the item, the cost of the repair, height of building, location of building, methods of construction, quality of materials, the likelihood that it will need to be repaired again in the near future, and the cost of a new one (see Lee 1987; El-Haram and Horner 2002; Lai et al. 2006). Making the right decision about repairing or replacing a building can save money, but various factors rather than just the cost of the repair must be taken into account before deciding if something is worth due consideration. Maintenance costs incurred on buildings differ. The difference can be as a result of the types, functions, age, building status, size, complexity and methods of construction of the building. Building structures are also composed of distinct, but interrelated components, materials and elements. This is contributing to the difficulty in maintenance allocations for building maintenance. Generally, these complexities have discouraged the use of the quantitative maintenance model.

However, because of the standardised conditions and close plants control, a future expense of a unit of productivity cost or schedule can be produced with higher precision (Bennett 2003). Building maintenance budgeting, by comparison, is relatively difficult to determine. The absence of standardisation and the complexities required in maintenance makes the forecast of exact or reliable maintenance a matter more of accident/probability than design (Jones and Sharp 2007; Bennett 2003). However, such economic analysis provides support for a management decision on the need to repair or replace a building element or component other than relying so much on quantitative process, and range of other issues (mostly qualitative) should be attached with considerable insight.

Various variables could be used to develop quantitative models. The variables include age of the building, gross floor area, condition of the building, degree of building failure, location, budget, performance index and intensity of use. These variables have a relationship with maintenance. For instance, age of building and the size of the building in terms of gross floor are discovered to have significant impact on maintenance cost. The performance of a building also has meaningful influence on cost of maintenance because a building that has high performance status will require more maintenance. Also, the higher number of occupants in a given space will affect maintenance costs. The basic reason for a model or any maintenance decision support system is to assist in the prioritisation of maintenance needs and ultimately the cost of maintenance. This will enable the maintenance organisation to identify which element should be included in the subsequent maintenance rolling programme. This is necessary as with the ever-increasing building maintenance needs, many maintenance demands are often being deferred due to inadequate maintenance costs.

University financiers are sceptical about the mounting requests from the universities on maintenance expenditure. The major concerns of the financiers are with respect to the methods of estimating the maintenance costs. This is pushing the need to develop a quantitative model that predicts maintenance cost to the nearest accuracy. Financiers want to have a fair idea of how to spend/allocate to aid budgeting. The essence of the quantitative model is to provide support to the qualitative assessment, because the qualitative assessments are subjective which sometime leads to inadequate allocation of maintenance costs. However, the essence of qualitative requirements sometimes is more important to the quantitative criteria.

5.9 Description of a Mathematical Relationship for Maintenance

Cost modelling is used to forecast costs for clients and to estimate resources costs for contractors. The cost model for constructed facilities is probabilistic rather than deterministic. Building maintenance costs are the cost required to keep the building and engineering services beneficially functionally, and they include labour, materials, parts, statutory fees and profits. Mathematical models are used in order to determine variables that best describe cost. Mathematical models for maintenance works are used to provide better forecast for maintenance service.

Researchers have developed quantitative methods for planned maintenance approaches. Such approaches include Roué's formula system, priority category matrix, point accumulation system and the multi-attribute approach. However, these methods are inadequate to provide reliable estimate for building maintenance. For instance, in order to improve the inadequacies in the existing methods, it was proposed to use a computer program. The major problems with such models are that they put the physical condition of the building at the centre of the decision-making process. By so doing the very requirements of the building itself, which is to provide a conducive environment for users, is not adequately addressed. Therefore, the current approach suggests that maintenance practice is not realistic because it involves a feedback loop.

Maintenance management decisions can also be based on the relationship between maintenance cost and value obtained. In general, any increase in maintenance expenditure should improve value. At the optimum, the ratio of increase in expenditure to improved value is 1 (i.e. $\Delta MC/\Delta V = 1$). Therefore, a ratio of more than one is a measure of poor service delivery. Such an analytical model assumes an accurate relationship between input and output. The model would provide a reliable decision, where there is accurate information on maintenance expenditure and revenue. That is, where input and output can be translated to a common unit. The model is reliable to the technical aspect of maintenance, but applying it to the operational issue would pose difficulty. Also, it cannot be applied to qualitative criteria such as user satisfaction that are not measured in terms of cost. The fact that building maintenance does not produce new capital makes the calculation of a reliable estimate for value difficult. According to Sherwin (2000), particular features of building maintenance suggest that such a model is not realistic for the building maintenance decision-making process.

One of the applicable methods to prioritise maintenance needs is the AHP, but the factors they used are insufficient and the factors considered were obtained through condition survey only (Shen et al. 1998). Apart from the fact that condition survey through stock inspection has been criticised for various inadequacies, the data are only from the provider's perspective. Thus, building performance is not the focus of the study. The goal of visual inspection is to identify defect (i.e. crack and leakage) in building fabrics, structures and engineering services. Maintenance management science must consider performance to provide best value to the users and providers. Failure to consider maintenance from both the users and providers perspectives will manifest itself in developing models that are not adding value to money invested (Wood 2009).

There are growing concerns and interests among building clients (and users) on the need to effectively communicate with building users and consider their concerns (alongside physical condition) when developing maintenance policy. Maintenance demand should be driven by user requirements rather than condition of the building. Building users are more concerned about building performance not about condition (Jones and Sharp 2007). The usefulness of any building maintenance model is very much related to the extent it considers user satisfaction. Furthermore, a model that is primarily developed based on quantitative data cannot predict the real-life situation because they are stochastic or deterministic (Sherwin 2000; Jones and Sharps 2007). In fact, only through a comprehensive strategy can the modelling of maintenance be deemed successful from a scientific point of view. The involvement of users is associated with much higher appreciation and total satisfaction of services provided as compared with the situation where they are not involved in the decision of issue that concerns them.

5.10 Maintenance Policy

Policy is a guideline for decision-making that links the formulation of a strategy with the implementation. Thus, organisations can use a policy to make sure that their workers make decisions and to make sure buildings are well maintained. Organisations could use the policy to determine maintenance strategy. For instance, an organisation policy can also favour corrective maintenance against preventive maintenance or whether to use in-house labour or external contractors to carry out maintenance services. In any maintenance programme, there must be a maintenance policy that defines the environment and ground rules for the effective delivery of maintenance management services. Maintenance policy is mainly a rule that guides the distribution of resources required to maintain a building while prioritising maintenance service.

In an organisation, outsourcing all or parts of the maintenance services is an issue of maintenance policy. However, some organisations might favour executing maintenance service in-house. The types of maintenance strategies that organisations choose to procure maintenance services are as a result of their maintenance policy. For example, it may be a policy of the organisation to repaint external walls every 4 years or to inspect roofs annually to check the conditions of the roofs. Similarly, it may be a policy of the organisation to service or repair locks every 2 years. Parts of the maintenance policy of an organisation are normally detailed in the maintenance performance indicators. As an illustration, it could be set that at least 85 % of maintenance complaints must be attended to within a few hours of receiving the reports. But to be definite, the time frame should be adequate but well defined in terms of say 3, 5, 10 or 12 h but it could be days or weeks. The time period should determine based on scope, procurement method, workforce, nature of the maintenance works required or corporate mission and vision. Maintenance policy aims to do the following:

(a) Provide a clear statement of the organisation's objectives for the maintenance of its buildings
(b) Explain how the identified objectives support and facilitate the delivery of the organisation services
(c) Identify the unit, department or section responsible for the management of the building maintenance.

Maintenance budgets are usually under attack; as a result, it is not uncommon to have insufficient allocation to the maintenance activities. Quite often, when organisation budgets are cut, the maintenance of buildings is often one of the first areas to suffer with maintenance being deferred to cut costs. Therefore, available funding is usually prioritised in the face of the array of maintenance needs. This becomes very serious if the organisation in question favours a condition-based maintenance programme. In condition-based maintenance, the available allocations are not always sufficient for the backlogs revealed by the condition survey. Therefore, maintenance policy forms an important aspect of maintenance planning which ensures that resources are used efficiently and effectively and that the organisation's objectives are achieved. While formulating maintenance policy, the following factors must be considered:

(a) The aims of the organisation
(b) The required maintenance standard
(c) Legal and statutory liability
(d) Procurement strategies for maintenance services

(e) The consideration of cost/benefit of analysis, particularly in relation to the four above.
(f) The condition of the buildings.

Thus, a sound maintenance policy is a fundamental requirement for any organisation, if an organisation is to fully implement value-based maintenance management. Every organisation should develop and state its policy on maintenance together with an arrangement for its implementation. The content of the policy should be made known to the all employees. This is contrary to what is currently in operation because only some management staffs have information on maintenance policy. The preparation and implementation of a systemic maintenance policy together with continuous monitoring make for smoother production or services operation, minimisation of error and waste reduction. Top management should be dedicated to the continuous improvement of building maintenance. These ideas should be set out in a maintenance policy that requires top management to do the following:

(a) Identify the end-user needs and wants
(b) Assess the ability of the organisation to meet the needs and wants economically
(c) Ensure that materials, components and labour meet the required standards of performance and efficiency
(d) Ensure that contractors and suppliers share the organisation's value and process goals
(e) Focus on proactive and holistic rather than preventive and/or corrective philosophy
(f) Educate and train staff and operatives for quality improvement at all level even among the contractors, subcontractors and suppliers
(g) Monitor performance and measure the users' satisfaction
(h) Review the maintenance policy periodically when and where the need arises.

The maintenance policy should be the concern of all employees, and its principles and objectives communicated as widely as possible to all so that it is understood at all levels of the organisation and within the subcontracting supply chain on maintenance services. Practical assistance and training should be given where necessary to ensure the relevant knowledge and experiences are acquired for successful implementation of the policy.

5.11 Appraisal of Existing Maintenance Management Model

This section reviews existing building maintenance management models. Terms including model, approach, system, framework and strategy are synonymous and interchangeable. This is in recognition that different authors have used different names and terminology. The terms are used based on prevalence but are sometimes arbitrary. However, the principal approaches to develop the models are quite contrasting even though they have the same common purpose which is to improve the status of maintenance managements. The growing significance of building maintenance has generated an increasing interest in formulating maintenance management models to improve the performance of buildings through maintenance. The reviews presented here involve latest models developed in Malaysia and elsewhere.

In an attempt to address the problems inherent in the existing approaches, Jones and Sharp (2007) proposed and promote the use of "a performance-based process model for built assets". They review a plethora of building maintenance models, but conclude that all the existing systems are not efficient. For that purpose, they were able to convincingly establish that the current maintenance approaches depend mainly on the conditions of the building to arrive at a conclusion. Instead of basing maintenance on condition base, the model established the need for a performance-based maintenance strategy. The development of the model is based on the outcomes of an action research, though. Specifically, it is developed for commercial buildings in the UK.

He also argued for a restructuring of organisational structure of the maintenance organisation. Based on his model, maintenance is made if decay is noticed through visual inspection and/or upon receiving a complaint from customers. Budget is made annually for all probable maintenance in the year ahead. All maintenance problems observed through the stock condition survey must be addressed with the budget in the given year. Excess budget is not allowed to be carried forward to next year. No matter what the question may be. In a way, the model is a kind of input-based maintenance practice.

Zailan (2001) described the central management system (CMIS), under the Property Management Division in the Prime Minister's Department. The model is very common among public organisations in Malaysia. The model is centred on condition survey. The CMIS was introduced specifically for the maintenance management for public buildings in Malaysia. The system entails the application of a computerised maintenance management system to perform various tasks such as building registry, scheduling and budgeting. The system seeks to integrate day-to-day maintenance, repair and restoration with the planned preventive maintenance. She also proposed the introduction of an emergency maintenance regime to the CMIS system based on what might be described as personal investigation.

Syamilah (2006) proposed maintenance management system through strategic planning for public schools in Malaysia. Three key elements are identified in the strategic process, namely "broad participation and commitment to the process, including internal staff, the end-users and the consultants; development of performance criteria for change; and concurrent development of a new management plan reflecting recommended implementation strategies". Her system emphasised the broad participation and commitment of stakeholders and the development of performance criteria [e.g. key performance indicators (KPIs)] as well as paying urgent attention to maintenance complaints. The respondents at each of the schools are the principal, headmaster or teacher. The model was based on the outcomes of survey questionnaires administered to school managements.

After investigating the maintenance management procedures of more than 50 schools, Zainal Abidin and Roslan (2006) developed the school building maintenance management process model (SBMMPM). The model was based on the principle of integration definition for function modelling (*$IDEF_0$*). *$IDEF_0$* is a modelling language in the field of software engineering. The model was comprehensive and descriptive. A critical control element in the model is the determination of the status of building through inventory survey, though building users were not identified as stakeholders, but various criteria were used as control elements.

Zawawi et al. (2010) assess the building maintenance practice in Malaysia, from which they developed the building maintenance management framework. The framework is developed through survey questionnaire on a total of 15 buildings from hospitals, offices and hotels with five buildings from each group. This is probably the latest framework so far. The survey covers both the clients, building users and maintenance staff. Though the number of questionnaires sent out was not revealed, the reports revealed that 79, 80 and 100 responses were received from hospital, hotel and high-rise office buildings, respectively. The framework is divided into five steps. The major activity in the first step is the appointment of maintenance personnel. The second step is concerned with keeping records of buildings. The third step deals with maintenance schedules. In the fourth step, maintenance is planned, and finally, in the last step, the maintenance programme is reviewed.

Shohet (2010) also developed the performance-based maintenance principle. The model was based on the fact that performance-based contract can provide an effective procurement platform for contracting maintenance out due to optimisation of the technical solution and savings in inspection and control of the contract. Initially, the author summarised maintenance standards in Israel, the UK and the USA. The study area is Israeli, and the focus of the study was six public office facilities in Tel Aviv. The performance model included six KPIs as follows:

- Building performance indicator—monitors the performance and fitness of use of the building and its systems.
- Age coefficient—indicates the effect of age of the building and systems on the annual maintenance expenditure.
- Occupancy coefficient—indicates the effect of occupancy on the annual expenditure for maintenance.
- Annual maintenance expenditure—represents the actual expenditure for maintenance in a particular year, and this includes costs of labour, materials, spare parts and costs of rehabilitation works required to maintain the building at its designated use.
- Maintenance efficiency indicator—indicates the efficiency with which the maintenance is carried out in terms of ratio between the performance of the facility and the annual maintenance expenditure considering the effects of age and occupancy of the facility.
- Normalised annual maintenance expenditure—expresses the annual maintenance expenditure neutralised from the effects of age and occupancy of the

facility and thus allows comparison between facilities at different service conditions.

On the basis of the foregoing, there are continued efforts to improve building performance through maintenance management. However, while such models offer potential to improve the performance of maintenance management practices, the models were reactive, hypothetical, complex and conditional based. It is these substantial limitations to the models that are the fundamental problems with the existing building maintenance management models causing their inabilities to improve the existing systems.

More so, the determination of the maintenance budget through stocks inspection has been questioned. Stock condition at best provides only a snap shot of the state of buildings during the inspection. It is difficult to assess the amount of risk posed by an identified non-critical problem to future operations and productivity. A non-critical problem during inspection could deteriorate further or becomes more serious during the actual implementation due to time lapse and error of measurement.

Similarly, the maintenance management system through strategic planning for public schools (Syamilah 2006) is also a condition-based maintenance policy because a basic consideration with this model is that it is complaint driven. What this means is that a building is maintained upon receiving a complaint on defect. If complaint is not received, maintenance would not be initiated. Another basic issue with the model is with regard to defects that are not visible to the users. While the author identifies key elements in the strategic planning and makes some summary arguments concerning the need for the elements in the plan, she offers very little in terms of specificity for the interaction of the predictors.

In other words, their inclusion into the strategic model is not given considerable value. Unfortunately, the building users were not even included in the development of the model, as the respondents were only headmasters and teachers. Certainly, the respondents cannot clearly and explicitly explain the value system of the main users. Therefore, the users' interests are totally ignored.

The SMMMPM (Zainal Abidin and Roslan 2006) is quite simple and descriptive. While the system recognised the consideration of varying criteria as the control element, it is also conditional based and also failed to consider the inputs of building users in the formulation of maintenance policy. However, the input of the users and users' occupiers are very important. Their inputs cannot be compromised if value is to be achieved. Seeley (1987) has questioned the validity of any maintenance policy that does not consider the users' participation in the formulation of maintenance management system.

For instance, many parties contribute in the decision-making process which governs the design, construction and the maintenance of buildings. But at the end of the line, the users are the parties that will be affected by such decisions. The clients are, on the one hand, legally responsible for maintenance, while, on the other hand, they may not be aware of the technical problems involved. However, the $IDEF_0$ model is useful in detailing activities for function modelling. Because, it uses input, output, control and mechanism, it provides a brief description of systems, by using the Inputs, Controls, Output, and Mechanism. The hierarchical nature of $IDEF_0$ allows the system to be easily refined into greater detail until the model is as descriptive as necessary for the decision making task. However, the $IDEF_0$ models might be so concise that only the domain experts can understand them. $IDEF_0$ models are often misconstrued as representing a series of activities. The abstraction away from timing, sequencing and decision logic leads to comprehension difficulties for the people outside the domain (Munive-Hernandez et al. 2004).

Furthermore, Zailan (2001) recommended that if an emergency maintenance regime is factored into the maintenance management in the Prime Minister's Department, the maintenance policy will be improved. However, that cannot be proactive and, what more, be systemic. Emergency maintenance is corrective and when maintenance is not proactively managed, often it leads to buildings that will be very expensive to own or operate in the long run. The condition-based maintenance policy does not address the root cause of a defect, whereas many different defects have symptoms with superficial resemblance.

Condition-based maintenance depends on the existence of "sickness" in the building before "treatment" is recommended. However, what is critical to the users or client occupiers of a building is the ability of the building to efficiently and effectively support the activities within and around the building and not necessarily the physical condition of the building per se (Jones and Sharp 2007). Again, given that the model is based on a case study of an organisation, the recommendations to other organisation are constrained and limited.

The performance-based process model developed by Jones and Sharp (2007) is a promising model, but it failed to define value explicitly. Their concept of value is narrowly defined and reduces value to income generation issue. This is not to say maintenance should not be construed as income generating activities, but the purpose of maintenance actually exceeds that. The interpretation of the variables in their model is that value is constrained to commercial value to the prospective clients or investors. The value systems of the users are not factored into the model.

In addition, the model's conclusion is based on information gathered based on commercial buildings in the part of UK. While the purposes of maintenance for commercial buildings are often related to the aesthetic appearance, the same cannot be argued for other types of buildings. The example of this argument is now illustrated. The clients of commercial and recreation buildings can easily relate their profit to maintenance service but this might not be the case with other types of buildings such as educational or hospital building. In other words, it is not very easy for universities to relate investment with profit.

The framework proposed by Zawawi et al. (2010) is also good a model, but the model is not in any way comprehensive. Many of the questions addressed to users are inadequate in terms of enhancing users' satisfaction. The questions that were addressed to maintenance staff do not elicit details of their work practices. For instance, the staffs were only asked about their age, gender, academic qualification and working experience. In fact, the questionnaire is biased because the questions that were supposed to be directed to the maintenance organisation are directed to users and clients. Another problem with the model is that it over-simplifies maintenance management practices. Again, the development of the model is not based on thought process per se; rather, it is based on perceptions. Building performance is not considered; rather, the main emphasis was on the condition of the buildings.

A major inadequacy with the maintenance performance model developed by Shohet is that it is too mathematical and rigid for application. The maintenance performance model is centred on the building as the subject of maintenance service, while users' participation in the process is largely ignored. In addition, it is solely concerned with maintenance contracting methods. Specifically, it dwells on choosing between outsourcing and in-house procurement strategies. Although it spells out how the calculation should be carried out, it would be difficult for the maintenance management to apply in practice because of its complexities. This makes the frameworks complicated and theoretical.

In general, the applications of these models are time consuming and expensive to implement. Mostly, the current models are hypothetical and speculative. Some of the authors tend to use complex and difficult computer programs (mathematics) and require data that are difficult to collect. In the names of *novelty*, complex mathematical models are developed to improve building performance, but most maintainers do not possess the required skills to interpret most complex mathematical models (Sherwin 2000; Lewis et al. 2010). This is particularly true for maintenance organisations that are involved with the management of specialist buildings. Some of the models are too rigid for maintenance managers to adopt, and the coefficients are too subjective and artificial to be accepted by the practitioners.

While the general approach to managing building maintenance is common to most types of (maintenance) organisations, each building has its definite characteristics and functional performances that should be optimised. For instance, the roles of schools are distinct as compared to universities. It is not surprising that university building users would have more complex requirements, given that these institutions typically enrol more students from different racial, ethnic and cultural groups and social classes. In that regards, the maintenance objectives must be customised towards the peculiar requirements and value of the buildings to the organisation.

As an illustration, the functional requirement of a police cell is utterly different from say, a room at the hospital where a patient is treated (See also Pitt 2010). Although the buildings in the university vary, the common purpose is to stimulate, encourage and sustain learning, teaching, studying and research. The student's residences must be considered as part of the university serving and as the accommodation buildings, and must be designed, built and maintained with the above objectives in mind.

Apart from the existing literature that focuses on the development of a model for managing maintenance services, some works have been carried out to investigate or identify maintenance practices for higher education buildings, for offices and also for hospitals. The following includes the list of the literature (Zakaria and Ali 2006b; Aris 2006; Ishak 2007; Zawawi and Kamaruzzaman 2009; Isnin et al. 2010; Aziz et al. 2010). In fact, the work by Khan et al. (2010) was a case study on the practices of maintenance in a public university. They noted problems with the maintenance practices. From their study, they recommended among other things, the installation of CMMS, improvement in maintenance planning and benchmarking and involvement of users in the decisions-making. From an audit assessment of a public hospital, Ali and Wan Mohammad (2009) also found various inadequacies with the way maintenance service has been managed on the basis of which they recommend some improvement. But the major issues with these listed models/frameworks are that they are condition based and the building users experience and expectations were not considered. Too much emphasis is placed on service providers' interest only. But the use to which a building is put, the suitability of its design for the purpose, and the intensity and the way in which it is used by the users will directly and indirectly determine the maintenance policy and plan and ultimately the maintenance management system. For meaningful comparisons and analysis, building maintenance management has to be considered not only in terms of building function, but in relation to the users' perception of the building's condition and its relevance to their primary needs. What this mean is that the users' requirements are the essential prerequisite for the evaluation of the value of the building, and it is only then that maintenance can be planned, controlled and programmed proactively and systematically. User experience and expectations are crucial in formulating maintenance service delivery strategy.

5.12 Maintenance Management

Management is a generic term. However, management principles are applicable to specific organisation. An organisation could be an industry, agency, department, institution or university. It could also be either private or publicly owned or a private–public organisation. For instance, a university could be a publicly owned university or a privately funded university. Based on an extensive literature review, it is easy to conclude that management focuses on the entire organisation from both a short- and a long-term perspective and goes beyond the organisation's internal operations to include the industry and the general environment in which the organisation operates. It also involves the process of forming a strategic vision, setting objectives, formulating a strategy and then implementing and executing the strategy. It includes recording and storing facts and information for future use or for others within or outside the organisation. In simple terms, management is about choosing an alternative approach or means of delivery.

Essentially, it aims to deliver results efficiently and effectively in line with recipients' expectations. Efficiency is doing things in the right way. It involves eliminating waste and optimising processes of accomplishing activities. Wastes that management seeks to eliminate or improve include time, material and expenditure. Effectiveness is concerned with doing the right things. Managers are people who make an organisation that achieved its objectives. Summarily, management aims to do the following:

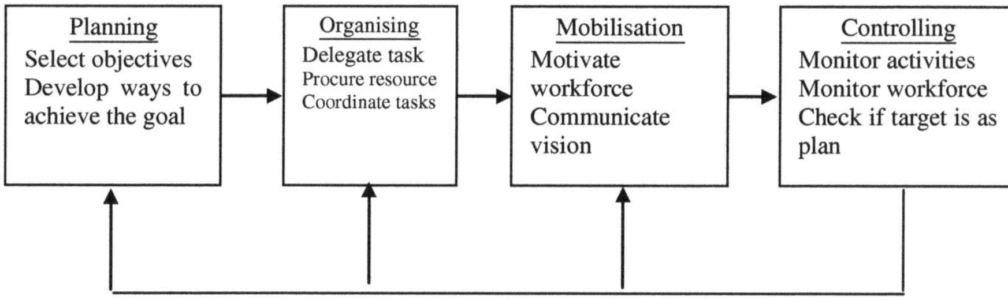

Fig. 5.1 Management process model (after Daft 2004)

(a) Enhance productivity,
(b) Improve users' satisfaction,
(c) Use resources efficiently and
(d) Use resources effectively.

The functions or duties of management vary and depend on the situation to which it is applied. The function that managers perform includes staffing, leading, coordinating, controlling, supervising, planning and motivating. However, they are generally confined to four main functions, namely planning, directing, controlling and organising. Managers are expected to have competences to perform these four broad functions irrespective of field of applications. It is not acceptable for a manager to have knowledge in three, two or one only.

However, the more specialised or complex the field of applications, the more the required skill is set away from pure management services. For instance, construction managers must be expected to be well grounded in construction service in addition to having knowledge on business and commercial perspectives. See Fig. 5.1, on the relationship between all the four functions of managers. However, each of the functions will be explained later in terms of maintenance management.

5.12.1 Value-based Management

In general management, there are two types of management, namely pure and applied management. Other types of management including strategic and value management are also present. This book is concerned with value-based management. The features and characteristics are explained next.

5.12.2 Theory and Concept of Value

Value means different things to different people. In general, value has many components including psychology and economical connotations. The user or consumer is the determinant of value in a product or service. Wilkinson and Kleas's book has provided sufficient information on the behavioural aspect of value (Wilkison and Kleas 2012). It is often associated with utility. In economics, both words are the same. In economics, utility means the benefits or satisfaction that consumer obtains from consuming or using good or services (Parkin 2012). This definition implies a one point. However, our concern is with servitisation. In the general service operation sense, this is thinking in terms of complete service delivering rather only at the point of consumptions (see also Johnston et al. 2012). In this book, and in our previous publications and discussions, it is the extension of this thinking that we refer to as *systemic*. This principle and practice we consider will add or create values to the maintenance service users. Furthermore, it is the extension of this principle and practice we argue to be considered in the maintenance management supply chain. By extension and as we have suggested, maintenance service should be contextualised into the various functions of management. This we consider will move the maintenance organisations closer to providing the services the users' wants and needs. From the maintenance service user point of view, value is created through the entire value chain. Because value is created through user's experience and outcomes, only strategic mean to add value to service delivery is through user involvements. Value is features or a characteristic that makes a product or service to sell more or that makes it more desirable to the users or clients (Stevens 1997). Value involves both the providers and users in order to set the scope of scheme, system or process, and it contains both hard and soft issues. While the hard issues are the technical requirements, the soft issues are subjective issues and involve aspects including cultural and behavioural issues. For the purpose of "value maintenance", in this book, we apply the reductionism to guide the scope of the "user value system". Some discussion on the reductionism is available in Wilkinson and Kleas's book. The user value system is concept or term that defines the aspects of maintenance services that the users required in maintenance service delivery. In other words, the factor the user wants and needs is maintenance service. User experience is an important issue in exploiting the value system. The meaning we attached to value in this book goes beyond the benefits the user of maintenance obtains from the services but includes all the "utilities" involved in the processing and deliveries of the services.

Maintenance management can no longer be a stand-alone both as a term/or concept due to the inadequacies with the extent and scope of maintenance. Thus, it comes as no surprise that maintenance management is prefixed with terms including strategic, proactive or best practice. Yet this has not allowed maintenance management to deliver best service. Therefore, "value" is prefixed with maintenance management in an attempt to bolster the scope and performance of maintenance management. This is peculiar since the objective of maintenance is to increase user satisfaction and organisation productivity through improving building performance. Maintenance management involves technical and operational issues, but the operational issues in the current definition are not sufficient to cover the context of maintenance (Sommerville and McCosh 2006).

The operational issues in the current maintenance management do not involve the building users in setting the determined by the users. As a result, the currente. The building users have little, if any, influence under the current maintenance practice. If the building users' value systems do not form the centre stage in setting the scope for maintenance, it will lead to complex problems such as misallocation of a resource to where it is not needed, or poor service delivery because the correct standard cannot be set (see also Parasuraman 2010). Also, there is the problem of lack of clear maintenance objectives with the current maintenance management system (Madu 2000). On the one hand, the inadequacies of the current schemes make the maintenance personnel focus too much on technical issues so as to have poor understanding on the organisational aspects. In the long run, all these greatly contribute to maintenance cost and yet deliver poor service. On the other hand, the basic factor that prompted the application of the value management principle to the construction industry is due to the inadequacies with the briefing process (or programming as it knows in the North America). Therefore, what is required is a method that supports building users' performance. A viable methodology is to consider the scope of maintenance in terms of the value systems of both the building users and maintenance organisations. The concept of value can mathematically be represented in Fig. 5.2.

The benefits should largely be defined based on the user values systems. Maintenance is service oriented because the maintenance activities mostly involve interaction with the current users and, generally, the quality of the services that are provided is determined by the users. As a result, the current building occupants have great influence on the maintenance management as this affects the way the maintenance services is planned, organised, controlled and directed. The users or recipients are a person or people who have or will have influence on the service provided to them and can take action if their demand is not met. Unlike during the design and construction stage, when the users were not often known, in the case of the maintenance services, the users are often known. In fact, the user determines when the service is to be executed and, most often, the services are provided when the users are still in the buildings. Therefore, the maintenance standards can be determined and inculcated into the maintenance programme. If the performance criteria of those users are considered, the value of the maintenance services that will be delivered will very likely attain the expectations, perceptions and satisfactions of the users.

$$\text{Value} \ \alpha \ \frac{\text{Benefits}}{\text{Investment}} \ \text{or} \ \frac{\text{Quality} + \text{Functions}}{\text{Total cost}}$$

Fig. 5.2 Mathematical representation of value

Based on conclusions drawn from analysis of over 40 companies, Treacy and Wiersema (1995) suggested that most successful companies achieve success by being leaders in one of three "value disciplines":

Operational excellence: this discipline is based on doing routine things in the most efficient manner. The discipline puts much emphasis on developing, recording, sharing and improving best practice approaches in order to minimise cost and to make the clients happy and trouble free.

Product leadership: this discipline places emphasis on innovation. This discipline capitalises on the principle that there is no need to develop standard process services (Bartholomew 2008). Therefore, creativity and lateral thinking is crucial for a successful value chain.

Customer intimacy demands: this discipline demands that companies must have sufficient and adequate knowledge of their customers' needs and wants. The service provider must look far beyond the immediate objectives of the products or services to the users. The providers must provide service that has a wider end than the customers realised was possible (Bartholomew 2008).

Value maintenance management can thus be defined as a comprehensive and structured approach to the long-term management of buildings as an enabler and facilitator for the efficient and effective delivery of benefits to the clients, users and the general public. A major emphasis of building maintenance management is to view a building as a means to an end, not an end in itself and to view the building as an asset or facility rather than as a burden and as an asset or facility that requires systemic consideration. Buildings require more care than they are currently accorded. Building users pay more attention to their cars and electronic gadgets than to buildings. However, our well-beings depend on the performance of buildings.

Theoretically, the major streams of value maintenance management are as follows:

(a) To identify the need for maintenance
(b) To establish the criteria of measuring efficiency and effectiveness
(c) To identify the policies to apply to maintenance as tasks
(d) To formulate maintenance procedures
(e) To make decisions regarding when to undertake maintenance.

Maintenance management can, however, be classified into maintenance technology, which is the tactical aspect of the process and services and the *main stream* management aspect of the process and services. The management aspect is strategic. It is in the management aspect that the expected quality, standard and quantity of the output for each of the maintenance activities are set as well as decisions regarding whether to maintain, and when to maintain, and for whom to maintain. The maintenance management, unlike the maintenance technology or engineering, cannot be readily outsourced and in fact should not be outsourced.

Maintenance management as shown in Fig. 5.3 transforms resources or input into output or value. Users' needs and wants are used continuously to modify or adjust the factors in the transformation process, and this usually affects the input.

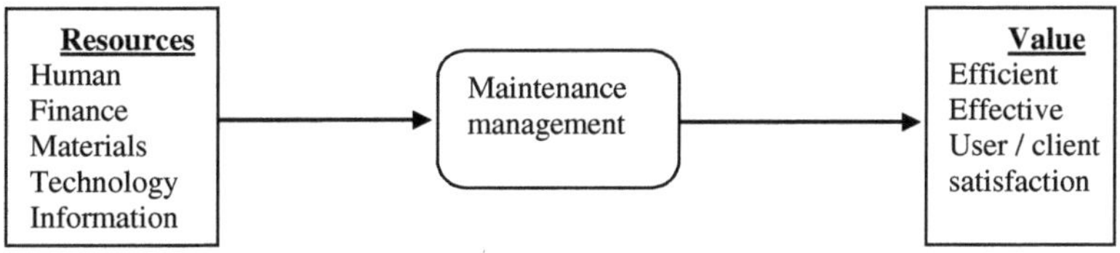

Fig. 5.3 Resources transformation (after Daft 2004)

However, the transformation process must be effective and efficient, i.e. the output must be greater than the input; thus, the essence of maintenance management is to create value and this is a value added process.

5.13 User Participations in Service Delivery

Service users are those individual or organisation that receives and benefits from the service or product that service providers deliver. Their primary objective is to obtain a reward or benefits that are defined by their *preferences*. We have argued in various pages of this book that user needs to be carried along in creating service for them. In fact, maintenance service users believe that they have greater role to play in service delivery. There is lots of literature on nature, functions and benefits that could be derived of customers in regard as "coproducer". The user or customers as they are often referred perform a number of roles in service delivery. Johnston et al. (2012) have categorised the benefits of customers' involvement in service delivery into three, namely inclusion, productivity and control. We will follow these meaningful categorisations to briefly explain the benefit of maintenance service users in service delivery.

Inclusion: If maintenance organisation engages the service in service delivery, a participatory environment is created and such users will be loyal and less complaint will be made. However, if users are not involved, often, the result is rework, dissatisfaction, anger, complaint and anger. In other words, there will be lack of "owner" from such users.

Productivity: If the user is involved, rework will be reduced to the barest minimal and important productivity will improve, as less resource will be used to achieve optimum satisfaction.

Responsibility and control: allowing user to be "coproducer" creates a sense of ownership and responsibility to the users. The users feel in control, and thus, this enhances their satisfactions and productivity.

5.13.1 Roles of Users

Here, we also adopt the classifications of Johnston et al. (2012) to explain the roles of maintenance service user in service delivery:

Coproducer—while this is not common in maintenance service delivery because of the technicalities involve, maintenance users could sometime perform some general work with them or perform maintenance service for other users. A simple example of this is the cleaning of the windows or doors or even screwing of nut and bolt of damage door without waiting for the maintenance department to carry out the repair. If there is cordial relationship and understanding, a number of maintenance users would do these themselves.

Specify service quality—one the functions that user could play is acting as service specifier. The quality of maintenance service is not straightforward as it seems. The users often have poor experience or information on the quality of service they required, and this therefore creates complexity for the maintenance department if the users' roles are not considered.

Quality inspection—this is peculiar to condition-based maintenance service delivery by user. The user makes complaints to the maintenance department regarding any defect in this building. Often, in more specific sense, quality inspection issue concerns the user to report the poor condition and performance of building to the maintenance department in order to initiate repairs.

Acting as the maintenance organisation—involves some users educating or advising their fellow users on the implications or the actions and reactions of them and other users' on the condition and performance of the buildings.

5.14 Functions of Maintenance Management

Whether management is general or applied, four fundamental functions are obvious. The sections that follow briefly examine each of the four broad maintenance management segments.

5.14.1 Maintenance Management of Planning

Management planning set the stage for the required actions and achievements (Bateman and Snell 2009). Maintenance planning is the first stage in maintenance management. During this stage, the organisation (i.e. university) sets specific maintenance targets it wishes to achieve and decides how it go about achieving the set objectives. It is concerned with determining what has to be done and how it has to be done prior to carrying out any physical maintenance. Planning should be an in-house function. Maintenance planning involves analysing the current condition of the building, anticipating future requirements and bringing the past into focus. Setting the value maintenance will add to the organisation functions and determines the standard and quality of required resources to achieve the planned target. It also defines the purpose of the maintenance activities, and it impacts on the organisation's corporate performance. In addition, it determines the quality maintenance activities required as well as answering questions on how and when to carry out maintenance activities.

Maintenance planning is a strategic function that should be performed by the higher level executive or management in an organisation. However, planning roles (i.e. tactical) can also be performed by lower- or middle-class managers. The prime aim of maintenance planning is proactive and systematic execution of maintenance process and activities in consonance with the organisation's corporate objectives. For instance, the organisation should ensure university's buildings should be maintained so that it provides flexible and cohesive support for learning, teaching and research.

5.14.2 Maintenance Management of Organisation

Organisation is the process of procuring and allocating resources to accomplish the planned target. Simply put, it is the maintenance management function of assembling and coordinating human, physical and financial information and other resources required to achieve the planned target. It also involves staffing the required workforce to carry out the maintenance activities as well as making decision on how resources can be maximised. This includes delegating duties and/or authorities to others (subordinate). This is usually displayed in the organisation chart. The competency of the person that heads and performs various duties is very important towards achieving the organisation's corporate objectives. The relationship of the maintenance employees with the higher level executive should be considered, and it should be ensured that maintenance employees are well acquainted with the organisation's mission and vision.

In order to achieve the objectives that were planned, resources are required and need to be assigned to respective functions in the organisation. For instance, efficient organisation of labour is crucial in maintenance management since maintenance is labour intensive as compared to undertaking a new build. Therefore, greatest potential of cost saving in executing maintenance service lies with efficient labour management. Poor planning and poor organisation will lead to poor maintenance management. Naturally, poor maintenance management leads to maintenance backlog.

5.14.3 Maintenance Management of Directing

This is sometimes called the management function of leading or mobilising. This is a very important function of managers because after making plans, creating a structure and employing the right personnel to do the various assigned tasks, someone must direct or lead the organisation in order to achieve the organisation goals. The main function that the manager is expected to perform here is to communicate the organisation's goal to the employee clearly. Also, the employees have to be motivated to perform their tasks highly as well as encouraging them to assume responsibilities in solving problems.

5.14.4 Maintenance Management of Controlling

This is another critical function of managers here. Organisations determine whether resources are used as planned and whether organisation is meeting its set target (Bateman and Snell 2009). For instance, the maintenance management function of controlling involves monitoring performance, making corrections and making continuous improvements. It also involves the determination of whether the organisation is achieving the level of quality set as well as whether maintenance activities are completing as schedule, with high reliability and safety and ultimately meeting the users' value system. To add value to the maintenance activities, target must be planned, resources must be organised, labour must be motivated and stimulated, and performance must be continuously controlled.

In the practice and principle, the maintenance programme can be initiated at all levels or at a time in the building life cycle with diminishing levels of return. The cost of implementation of the maintenance programme is very low if the programme is systemic, but the saving made possible by the productive intervention which is essentially high. As the maintenance programme proceeds through systemic to corrective approach, the possibility of proactive saving declines substantially and the cost to implement a maintenance programme increases. However, a proactive approach seeks the initiation of the maintenance programme before damage or failure occurs. This is made explicit through addressing the root causes of the problems as against addressing the systems as it usually the cases under the traditional approaches. It considers the physical performance of the building and its services as the only yardstick of measuring the intervention.

Under the proactive maintenance approach, the root causes of the problems are only initiated if problems arise. Maintenance is not initiated because it is linked with the organisation's business case or business objective rather than because of the superficial defect due to decay, deterioration and failure. At best, to be proactive in managing maintenance programme is to take the maintenance dimension of a buildings system and its services into account during the design and construction phase of alternative solutions.

However, in the value-based maintenance management, maintenance is initiated because the defect has a functional effect on buildings and because it will affect the users' satisfaction and productivity. In this way, the available resource will be efficiently and effectively optimised in the face of the wants of maintenance. However, this could only be ensured if maintenance is initiated on the basis of value to the users rather than on the basis of availability of funds and on its physical condition. In this context, to initiate maintenance is not a question of maintenance versus condition or maintenance versus business; rather, it is purely a business issue.

5.14.5 Maintenance Organisation

Irrespective of the nomenclature given to the roles of the maintenance organisation—maintenance department, property division, facilities department, administration department, logistics and building department or asset management department—the common aim is the same. Maintenance organisation is that department, section or unit within an organisation that is responsible for the management of buildings and engineering services. The size and scope of the department is factor of size, complexities, nature and geographical location of the buildings and size and nature of the main organisation of which they form part. The size of the maintenance department is also dependent on the types of procurement system that the parent organisation adopts. For instance, an organisation that prefers to outsource larger parts of its maintenance services will practically have a smaller maintenance department. The reverse is also true. In a small company, the functions may be performed by a staff member in addition to his/her other duties.

Organisation theory has described the importance an organisation's structure has on productivity and customer satisfaction. A number of studies have shown that the ways an organisation is arranged and operates its business have significant effects on its output which consequently influence the environments in which it exists (Mandal et al. 2008). The maintenance department has a great effect on the parent's business objectives. The maintenance department must concern itself with two basic functions, namely (1) it has the capability to provide quality services and (2) it is able to measure its performance against others (Chanter and Swallow 2007).

The department must be developed with forethought. A lack of foresight will mean the department is not considered as one of the (university strategic business units) USBUs. However, currently, maintenance departments evolve through informal interaction with building users, university top management, resources (materials, components and money) and other stakeholders. The structures of the department grow as the university grows in size, age and complexity. Certainly, at the initial stage of an organisation, it is quite natural (assuming the organisation starts operations with new buildings) that the structure of the maintenance department will be small. Organisations with old buildings will require a larger maintenance department.

The maintenance department must have sufficient in-built flexibility to cater for every piece of work arising from the demand. Strictly speaking, maintenance is a business and, in fact, a big business. Thus, the organisation should be strengthened by including it as one of the USBUs. Therefore, the department should be seen not as an engineering-led unit, but an entrepreneurship-led entity so that it can be creative, innovative and competitive. The Mission Statement (the underlying purposes and values) of the maintenance organisation should be to enhance users' satisfaction and to improve productivity.

The size and scope of the department is a factor of the size, complexity, nature and geographical location of the buildings. The size of the maintenance department is also dependant on the types of procurement system that the organisation adopts. Organisations that have sizeable buildings in their portfolio tend to favour using direct labour. Also, organisations that prefer to outsource larger parts of their maintenance services will, in practice, have a smaller maintenance department. The reverse is also true. In a small company, the maintenance functions are often performed by staff members in addition to their other duties (Wordsworth 2001). However, in a larger organisation, the staffs of the maintenance department perform dedicated functions. In other words, the staffs do not perform other duties aside from those they perform in the maintenance organisation.

More importantly, the structure of the maintenance department is indicative of the importance the parent organisation attaches to the performance of the buildings in the effort to achieve the organisation's business objectives. Organisations that place greater value on the buildings are more concerned with the nature and characteristics of the personnel and resources within the maintenance department. For instance, in the hotel sector, the performance of buildings is considered a direct generator of income and profits. In the same vein, almost certainly, a university cannot operate without functional buildings. Even the virtual university requires a minimum number of buildings to conduct its business. Thus, the need for high building performance is dictated by political, social, cultural and economic demands. A lack of a well-conceived maintenance department manifests itself in poor building performance as a result of poor service delivery.

Another critical issue that determines the structure of the maintenance department is with regard to budget allocations. In Malaysia, for instance, the maintenance budgets for public organisations depend greatly on the federal government's subventions. However, there are no provisions as to whether a university must use a particular procurement method or not. There is also no stipulation that a university can or cannot divert the maintenance budget it has been allocated for other purposes. The budget allocations from the Ministry of Higher Education are based on the request a university submits for maintenance and the availability of funds. However, more often than not, the subventions depend on government earnings. Many universities prefer to outsource what they often termed "non-core services". Non-core services are services that the organisation believes do not directly contribute to their corporate business objectives or a necessary evil, the costs of which must be borne come what may. In fact, the growth of outsourcing maintenance organisations could be directly responsible for companies wishing to contract their maintenance responsibility (Marsh 2003).

However, with large organisations such as universities that have expansive buildings in their portfolio, it is almost inevitable that there will be a need for a dedicated maintenance organisation. A university procures maintenance services from two diametrically opposite approaches, namely in-house and outsourcing. Though most organisations combine the two approaches to meet their needs, organisations that have sizeable buildings in their portfolio tend to favour using direct labour. Each of the approaches has it strengths and weaknesses. With the strategic functions of the buildings, outsourcing the larger part of the maintenance services will prove grossly inadequate for the university organisation. The buildings and their management need to be considered as investments rather than as liabilities. In other words, greater responsibilities are placed on the maintenance department. University buildings are not a new factor of production per se, but their relative significance is considerably increasing. The demand for a dedicated maintenance department for the university is critical if the provision of value added services is critical to the university.

Following on from the functions and size of the maintenance department as illustrated above, the structures of the maintenance department are typically presented in a chart. Figure 5.4 displays the organisation chart or organogram within a large private university in Malaysia. The organogram shows the span of control, responsibilities and the delegations of authority. The maintenance organisation is headed by a general manager, who is assisted by five executive officers, one each for the mechanical, electrical and civil property and landscaping units. The property executive is responsible for the general management of all the university's physical assets. The other four are responsible for the maintenance of the building structures and services and the landscaping. The general manager is the link between the

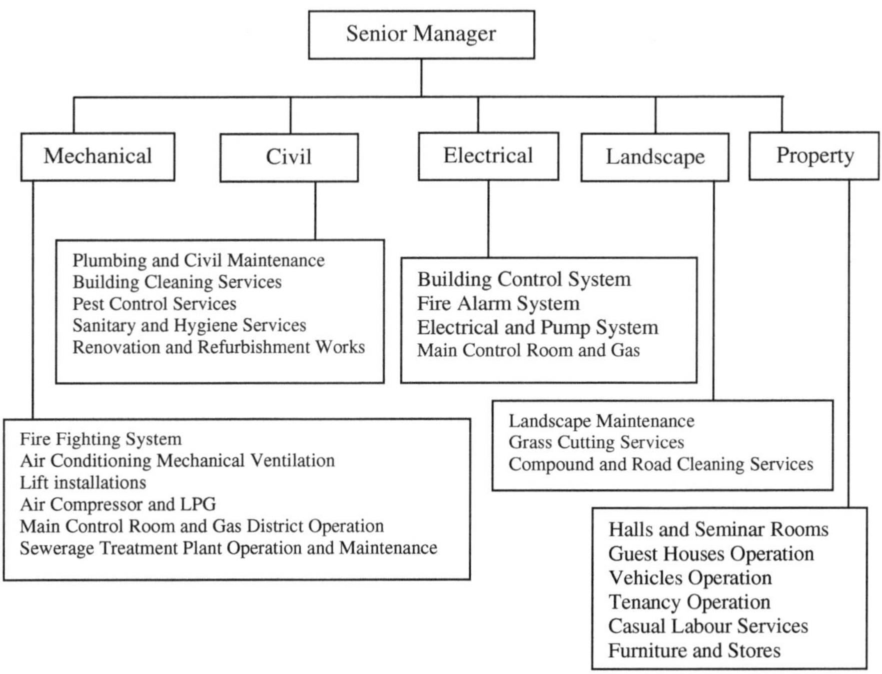

Fig. 5.4 Structure of the organisation chart

university top management and the maintenance organisation. Their roles are, therefore, very critical to service delivery.

In order to attain efficiency in service delivery, a university should specifically create building maintenance department. This will promote better results in terms of quality and independence (documentation, formalities, transition, specialisation of changes and over all employees' satisfaction). This is important, because, where building management is under the purview of either FM or AM, building maintenance is not accorded the required attention. It is not clear, however, at what organisation level it becomes preferable to switch to total outsourcing.

However, with large organisations such as a university that has expansive buildings in its portfolio, it is almost inevitable there will be a need for a dedicated maintenance organisation particularly, considering its significant contributions to their corporate business. No doubt a building only performs a supporting role to the university mission but it is almost inevitable that a university cannot operate without functional buildings. It deserves more attention than it is currently accorded by the university institutions. Its status needs to be elevated to the rank of human resource, since without functional buildings, a university cannot operate. Therefore, in order to earn optimum value for money invested, maintenance organisation must be intrinsically linked to the university corporate mission and vision. This is indicative of positive policy (Chanter and Swallow 2007) and thus recognises buildings as capital goods.

5.15 Maintenance Information

Information is generally defined as fact, detail or knowledge learned about a situation or an event. Information is becoming more essential in everyday activities than it ever was. Information is an intellectual asset (Flanagan 2007). Information is required in order to make any informed decision. Perhaps it is even safer to conclude that maintenance is about information because information management is crucial to maintenance management. Lack of sufficient and adequate information makes maintenance decisions difficult and or expensive.

Maintenance managers require adequate information about the organisation's core objectives, about the conditions/performance of the buildings, the age of the buildings, the users and their value systems, the functions of the buildings, the value of the buildings and the organisation's maintenance policy. Without information, the maintenance manager cannot make concrete decisions; for example, a building's component might be replaced earlier than its actual economic and functional life span. Information and communication technology are playing prominent roles in information management, albeit the level of data and information that is generated is sometimes, if not often, too large, thereby requiring careful planning and control for efficiency. It is a case of "garbage in garbage out" if the information is not well managed.

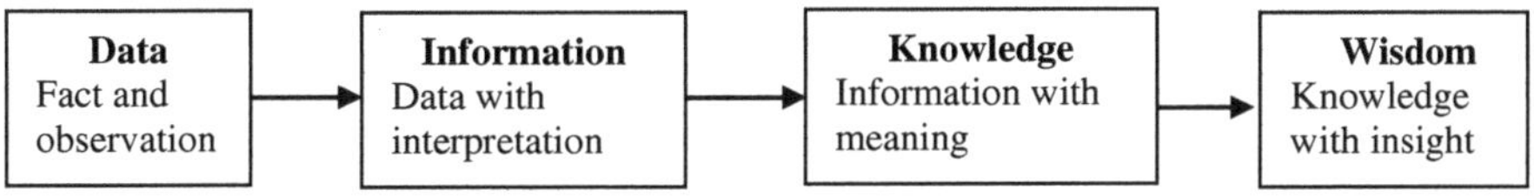

Fig. 5.5 Information model (after Flanagan 2007)

If a decision is based on wrong information, then such a decision can never be correct. However, comments, from notable facilities management practitioners in Malaysia, have shown that the nature of information on maintenance is diverse, unstructured, inaccurate and insufficient. Thus, this is making information sharing on maintenance very prohibitive. As a result, organisations are not utilising knowledge to the fullest to improve their performance. A fundamental reason adjudged for this deficiency is the poor management of the data and information they collect. However, it is important to distinguish between data, information and knowledge (Fig. 5.5).

5.16 Procurement of Maintenance Services

Maintenance procurement is concerned with the entire processes involved in the selection of consultants, contractors and other parties carrying out the maintenance works and includes all the actual execution of the maintenance works. However, in practice, procurement of maintenance service is often confusingly confined to issues relating to the implementation of maintenance activities, outsource or in-house. Although the market for maintenance work is increasing unfortunately, there are no standard procurement strategies developed for maintenance works. In fact, there is no any specific contractual arrangement for maintenance contracts (Marsh 2003). As a result, most contract arrangement for maintenance works is the modified version for capital projects.

The lack of specific procurement strategies for maintenance contracts is making the procurement of maintenance works more complicated and tedious compared to that for new works. Zavadskas and Vilutiene (2006) opined that due to the absence of specific procurement strategies for maintenance works, their procurement is usually based on lowest tenders' basis. To base selection of procurement strategy on the lowest tender price is, however, misleading particularly with the specialised requirements of maintenance contracts. Therefore, for a better outcome, the fundamental differences that exist must be distinguished while procuring maintenance works as compared to that for new procurement projects.

For one, maintenance work is complex and dynamic compared to new build work. These dynamics and complexities should be taken into account while selecting procurement strategies for maintenance works. Such consideration includes the complexity of the work involved, the working conditions, abnormal working hours, physical constraints, legal requirements, statutory obligations and unavailability of similar or same materials, equipment and components. Also to be considered is the availability of contract documents, and as unlike new projects that detailed working drawings, specifications, bill of quantities and other contract documents are often available, these are seldom available for maintenance works. In fact, the quantities and quality of the work cannot usually be determined prior to the actual execution of the work. For a traditional new project, works of such nature are executed under the day-work method or approximate quantities and the likes. For example, where the bill of dilapidation for maintenance is available, and it might not be entirely reliable due to the uncertainties and complexities associated with maintenance works (Ashworth and Hogg 2002).

However, to reduce the effect of these complications, experienced personnel should be assigned for such works. Adequate attention should be given to the nature and extent of the maintenance works involved. Perhaps a more suitable procurement approach that could be selected for emergency maintenance works could be the management contracting or construction management as the traditional approach will be very unlikely to be suitable for such works.

Industry-wide organisations currently favour two organisation structures. The first is the in-house maintenance structure, while the second structure prefers independent organisations to carry out the maintenance work. The merits and demerits of each of the structures have been previously explained elsewhere in this chapter; however, organisations usually mix the two approaches to ensure better outcomes. The extent of which the mix is made depends on the capacity and capability of the parent organisations and that of the maintenance organisation as well. The in-house organisation has been criticised for lack transparency, and their services are usually perceived as being of lower quality. The in-house method is also criticised for lack of proper documentation and leaves little room for flexibility for knowledge transfer. Based on analysis of the study conducted by building management information (BMI) in 1990, on occupancy expenditure on the main building of a university estate revealed that the university outsourced 70 % of the maintenance works (Seeley 1996). The remainder was maintained using in-house workforce.

On the other hand, typically, the outsourcing strategy does not ensure proper transfer of documentation from one FM company to another FM company. Usually, the contract terminates after a number of years which is subject to renewal as the client organisation deems fit. At the end of the contract period, if the outsourcer failed to win the bid, the problem that often arises is the problem of transferring the documents and experience from the FM organisation to another FM that won the current contract. Outsourcing also exposes the client organisation to external threats since the FM organisations will have access to some of the vital documents and processes that the client organisation uses. Therefore, it is suggested that it is in the best interest of large organisation to create an independent facility management. Outsourcing maintenance always reduces maintenance procurement to a lump sum; however, the risk and uncertainty involved with maintenance services dictate that the traditional lump sum is undoubtedly unsuitable for maintenance works. The lump sum contract demands that the work to execute can be measured (Kirkham 2007).

Another criterion that determines the procurement approach to be selected for maintenance work is the condition and performance of the existing facility (Marsh 2003). A detailed survey of the existing installations and structures must be carried out to ascertain the extent of and nature of work involved. The level of technological development and advancement should be taken into account, because some of the existing components and materials might have already become outdated. It is only natural that there will be improvement in technological and scientific advancement by the time the maintenance is to be executed compared to the time the building was initially built.

For new built, the contract is actually entered to cover legal matter in case of disputes among parties, in case of a maintenance contract, since no any new product is built or constructed, the maintenance contract is more that of instruction, as no new project is to be delivered, but only an improvement.

Procurement of maintenance contracts is loosely classified as fixed price, lump sum, price adjustment, cost plus percentage, cost plus fixed fee, cost plus fluctuating fee, target cost, shared savings or cost, bill of quantity, schedule of rates and package deal. It is also necessary to add to the list. The addition includes negotiated contract, day-work contract, specialist term contract, directly employed labour, fixed price maintenance and the JCT form for works of a jobbing nature. This is to accommodate the entire spectrum of maintenance, considering the complexities and intrinsic issue associated with maintenance works and the client value system in particular.

Different client requirements call for different approaches. As a starting point to identifying the best approach to use is the alignment of the user value system with the project's objective. For instance, contract arrangements such as those based on prime cost plus fee, schedule of rates and bills of approximate quantities are very suitable for emergency maintenance works. This is because for emergency maintenance works, time is of prime importance compared to other criteria.

Private finance initiative provides a good platform to consider maintenance of projects during design through to the operation stage. Expensive projects could also be let out under PFI, whereby a consortium of financiers, contractors and other services providers will team up to provide such services or projects that are usually the responsibility of the government to the public, at agreed benefits to the consortium. The consortium constructs the facilities and then operates and maintains the facilities throughout the concession period. The need to procure maintenance work under the PFI could also arise in some situations. For instance, where consortiums are required to carrying out maintenance and repair works on failed or abandoned facilities. They could be required to operate the facility on completion for an agreed period of time to recover the amount spent on the maintenance. The facility is then transferred to the client on the expiration of the concession period. This could sometimes be the case where the clients lack enough capital to finance the project perhaps due to liquidations or shortage of technical know-how. However, there must be a proper linkage between the procurement to be selected and the contractual arrangement that is intended to be used, because the contractual arrangement defines the performance standard as well as the benefits for the parties.

5.17 Maintenance Standards

Buildings are required to meet certain standards relating to sanitation, safety, quality, reliability, perception satisfaction and security. The degree of which depends on the functional requirements of the buildings. Similarly, maintenance service has the same standards that must be attained; for instance, while carrying out inspection, the rating which a component or system is required to meet before maintenance is initiated. This is usually measured on a preestablished scale. Seeley (1987) outlined that maintenance standard is concerned about determining the optimum level of maintenance work required on a building in order to preserve an acceptable environment in the building under the maintenance manager's care. Usually, a maintenance standard is determined for each of the various building components and systems.

Traditionally, maintenance is usually defined in terms of acceptable standard: "a combination of any actions carried out to retain an item in, restore it to an acceptable condition" acceptable condition implies acceptable standard. This underscores the importance of the standard of maintenance required in a building. In this context, however, standard means what occupiers perceive as acceptable which will vary from building to building. For university buildings, due to their diversity concurrent with the diverse backgrounds of the users, the maintenance standards are usually very high. The female hostels in most of the cases require maintenance of a higher standard compared to male hostels.

Maintenance standards can also be used in allocating a budget to the maintenance department. This is rooted in the ability of the maintenance managers to justify their case objectively and clearly by demonstrating the impact of the defect on the organisation's business objectives in the face of other activities that require more funding allocation. Maintenance standard defined the standard of care that the building and operation require in order to meet the business objectives. The value of the building is determined by the function of the services it renders to the users in combination with other factors of production. Therefore, if the function that the building offers is not required, there is no need to set maintenance standards for the building in the first place (Wordsworth 2001).

The minimum maintenance standards are usually in relation to legal requirements. Normally, this cannot be compromised as failure of which carries heavy penalties from the supervising agencies. However, other maintenance standards besides the legal requirements are normally dictated by the nature of the business that the building is supporting. However, for organisations such as public universities that are not for profit making, maintenance standards and performance cannot easily be measured since it cannot be related to profit or turnover. In fact, maintenance performance for public universities is also very slow and gradual because the level of fees the major recipients (students) pay is relatively small compared with university investment on maintenance. In a profit-making organisation, for example the private universities, the recipients can generally go elsewhere if they are not satisfied with the quality of services provided (Turrell 1997).

5.18 Labour Management

The construction of a building is labour intensive. Building maintenance is more labour intensive with the majority of the costs of maintenance arising from the operatives' wages. The operational costs of building are for salaries, utilities, contracts and maintenance. Labour management is concerned with the planning, controlling and reporting on the performance of workers to ensure that correct and competent workers or operatives are appropriately specified in a timely manner are obtained at a reasonable price and are available when necessary. For instance, a decision has to be made as to whether to use in-house labour or an outside contractor to provide the maintenance services or not.

About 40 % of the construction industry's operatives are employed for maintenance; almost 75 % of all painters and decorators are employed on maintenance (Chanter and Swallow 2007). A maintenance engineer in a public university once narrated his experience to us whereby four operatives spent more than half of their productive day to fix just a roofing slate (measuring about 0.4 m^2) on a damaged roof arising from termite attack that caused serious leakage in a university's library. Study shows that 65 % of the maintenance costs are contributed by the labour cost (Sliteen and Catarina 2010), 14 % for materials and remainder of 21 % for utilities costs.

From the foregoing, substantial cost saving can be derived from efficient management of the maintenance workforce. The overall essence of building maintenance management is concerned with striking a balance between maintenance needs and the available resource (i.e. time, labour and capital).

Labour management also involves the establishment of the organisation's labour force if the maintenance services are carried out in-house. However, if the maintenance service is procured through outsourcing, it is not the responsibility of the organisation to manage the labour. In the input-based maintenance services, the client organisation specifies the number of work office they want for maintenance on their buildings. This is a common practice in the management of the public offices in Malaysia. Maintenance labourers require detailed management, and in maintenance, a worker is required to possess knowledge in more than one trade. Labour must be effectively managed in order to control traffic and accidents on site. A decision has to be made whether or not to use in-house labour, or to outsource the labour, or their combination.

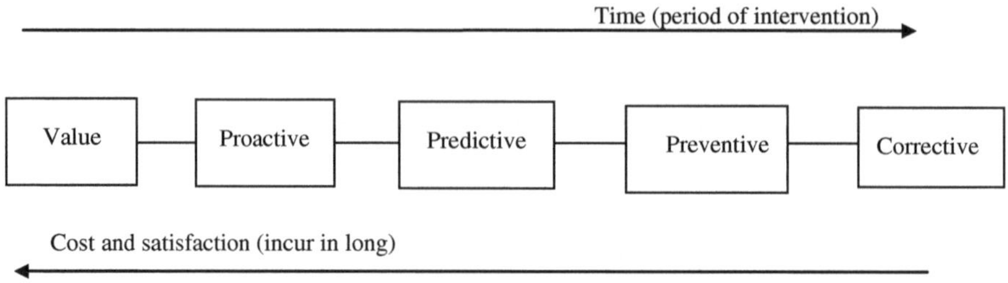

Fig. 5.6 Maintenance time–cost model

5.19 Time Management

Time is crucial in the building maintenance industry. Therefore, there is the need for an efficient way in which time is managed. Time management corresponds to the time that information is collected, processed, transmitted and retrieved. In other words, the time that a problem is reported by the recipients corresponds to the maintenance unit to the time the actual repair is made. Most projects are deemed desirable only if they are completed within a given period of time. In broad review of time management literature, Claessens et al. (2005) defined time management as behaviours that aim at achieving an effective use of time while performing specific goal-driven activities.

Strictly speaking, time cannot be readily managed; it is only the approach (i.e. processes) to time that can be managed. This is because time is not an accessible factor (Claessens et al. 2005). Time is very crucial in building maintenance services. Information must be collected, processed and released on time. The maintenance must be carried out on time to discourage worsening of the situation. This will reduce the extent of downtime and money that is required to carry out corrective maintenance (see Fig. 5.6).

From the maintenance time–cost model, it is shown that the earlier maintenance is introduced the lower amount that is required doing the maintenance work. Put in other words, maintenance costs increase as we move from the value to corrective maintenance paradigm. For instance, time usually reduces the quality of conditions survey to nothing, and time erodes fact. Maintenance leadership is under tremendous pressure to improve performance every year. This pressure puts heavy demands on the manager's, supervisor's or engineer's time.

5.20 Performance Measurement and Performance Management

Published literature revealed a wide range of opinion which tends to polarise either towards performance measurement or towards performance management. Authors distinguish between performance measurement and performance. The distinction is almost certainly required. Performance measurement is part of performance management, and ostensibly, measurement is required before it is managed in order to align with the organisation's corporate *mission* and vision.

Much has been written on performance measurement of business activities and the construction industry (for instance, Koplan and Norton 1993; Kulatunga et al. 2006). Indeed, the building maintenance sector is also a business after all. Managing buildings is a business. Performance measurement promotes the reduction or the elimination of unnecessary cost on the one hand and increased productivity and profitability on the other hand by exposing the size and extent of disruption and reworks in the supply chain process.

Performance measurement is one of the major tools used by the manufacturing sectors to improve market share. Traditionally, if cost of production is low, organisations seek high return. Prompting this is the shift from the traditional cost and time-based measurement systems. The application of performance measurement is beneficial to maintenance management organisations to understand its process and identifies opportunities for improvement or to review the performance of new processes (see Marosszeky and Karim 2002).

Currently, the criteria have not been explicitly linked to the services where value creating processes are taking place. The critical elements of a good performance measurement framework are as follows:

(a) Leadership and commitment
(b) Full employment involvement
(c) Good planning
(d) Sound implementation strategy
(e) Measurement and evaluation
(f) Control and improvement
(g) Achieving and maintaining standards of excellence (Oakland and Marosskeky 2006).

Almost certainly, if someone does not know how he or she is doing, there is no way he or she can know whether he or she is doing well or not (Cain 2004). Therefore, there is the need to measure how well one is doing. To know how well one is doing, one needs to measure one's performance which serves as the basis for benchmarking with others. Performance measurement and performance management are continual and dynamic rather than static. This is to provide opportunity for improving services to the users and to increase market strength and profit.

According to Oakland and Marosszeky (2006), in an effort to develop and use performance measurement instrument answers must be provided for

(a) Why measure?
(b) What to measure?
(c) Where to measure?
(d) How to measure?

The major challenge with performance measurement, however, is what to measure and how to measure performance. It is contended that a most practical way out is for the service industry, such as maintenance, to base any improvement on user satisfaction. No doubt the existence of any organisations depends on how satisfied are their customers (users) with the services. Accurate performance measurement is fundamental to any decision towards continual improvement (Cain 2004). Corroborating this view, Oakland and Marosszeky (2006) recommended, investigating the corporate goals of senior management, and problems that requires to be solved, opportunities that the organisation has, and what influence customers satisfaction. Even though a comprehensive list of what should be measured cannot be produced, they must reflect the true performance of the process in customer and/or supplier terms and continuous improvement.

Apart from, the problem of what to measure is also the problem of setting realistic and comprehensive measuring instruments. Achievable, measurable and realistic targets should be set. Martinez and Kennerley (2005) concluded that complicated and excess performance measurement often creates negative impacts on resources (i.e. time, money and labour) (Kulatunga et al. 2006). Complicated PM also limits managers' liberty and novelty. Kulatunga et al. (2006) argued that unrealistic and incorrect measuring instruments lead to misleading information.

A measurable and quantitative metrics that can be supported with facts are required. Therefore, in an effort to promote a basis for comparing maintenance organisations and maintenance services performance, it is necessary to produce the KPI for the maintenance organisation, here in after referred to as maintenance performance measurement (MPM). This is a set of criteria reflecting the performance of the maintenance service providers. The criteria entail aspects of the services that are related to the users of the maintenance organisation to be critical factors for the delivery of maintenance.

Such measurement indicators should be well balanced. What this means is that the criterion should not have unnecessary over bearing influence because it is the wholeness that represents value for money and ensures maximum user satisfaction. In order to manage performance accurately, organisations might find it necessary to measure effectiveness, efficiency, quality, impact and productivity (Oakland and Marosszeky 2006). The development of performance measurement should depend on the organisation's corporate objectives. For instance, the performance metric must be embedded into the client's organisation corporate mission and vision. Furthermore, useful indicators must be developed with the active participations of those who use and learn from the indicators (Alexander and brown 2006).

Performance measurement in maintenance involves both tangible and intangible metrics. Tangible metrics are those criteria that can be measured such as time to respond to complaint while an example on intangible ones is user satisfaction. Clearly, formulating performance measurement metrics for service providers such as the building maintenance sector is difficult, but critical examination with consultations with those concerned with building management and the end-users will reveal that it is practical to do so. Undoubtedly, one of the metrics of building maintenance services is the time to respond to a maintenance complaint.

5.21 Summary

The conventional approaches to managing maintenance services have been questioned. Currently, the object of maintenance is to extend the service life of a building without necessarily relating the building performance to users experience and activities. However, a building is not procured for its own sake. Rather, it is procured because of the services it provides to the users. Maintenance should be positively planned, strategically organised, proactively lead, holistically controlled and dynamically implemented if best value is critical. There is the need for a change in the way buildings and their maintenance are managed in order to provide better services, which is critical.

The building, and it management, needs to be considered as an investment rather than as a liability. Maintenance management requires a multidisciplinary approach: from the engineering, technological, commercial, economic and social perspectives. Strictly, maintenance is business. The maintenance department should be seen as a business unit. Its significant impact on the organisation can be illustrated by reviewing it in terms of the characteristics. It should be strengthened by including it in the Organisation strategic business units (OSBUs). This will also make it attractive and competitive. Therefore, the department should not be seen as an engineering-led unit, but entrepreneur led so that it can be creative, innovative and competitive.

References

Ahuja, H. N., & Walsh, M. A. (1983). *Successful methods in cost engineering; construction management and engineering*. New York: Wiley.

Alexander, K., & Brown, M. (2006). Community-based facilities management. *Facilities, 24*(7/8), 250–268.

Ali, M., & WanMohammad, W. M. N. (2009). Audit assessment of the facilities maintenance management in a public hospital in Malaysia. *Journal of Facilities Management, 7*(2), 142–158.

Allen, E. (1995, 2005). *How buildings works: The nature order of architecture* (2nd ed.). Oxford: Oxford University Press.

April, A., & Abran, A. (2008). *Software maintenance management: Evaluation and continuous improvement*. Hoboken, New Jersey: Wiley Interscience, Wiley.

Aris, R. (2006). Maintenance factors in building design. *M.Sc. Thesis, Universiti Teknologi Malaysia.*

Ashworth, A. (2004). *Cost studies of buildings* (4th ed.). Englewood Cliffs, New Jersey: Pearson Prentice Hall.

Ashworth, A., & Hogg, K. (2002). *Willis's practice and procedure for the quantity surveyors* (11th ed.). Oxford: Blackwell Science Limited.

Aziz, F., Ali, H. M., Khan, A. H. R., Jivasangeeta, S., Jibril, P. (2010). Review on current maintenance practices within Iskandar Malaysia (case study: Iskandar Malaysia). In *Proceedings of 4th International conference on built environment in developing countries (ICBEDC 2010): a better tomorrow with a greener today* (pp. 292–301). Pilau Pinang: Universiti Science Malaysia, December 1–2, 2010.

Bartholomew, D. (2008). *Building on knowledge: developing expertise, creativity and intellectual capital in the construction professional*. UK: Blackwell Publishing Limited.
Bateman, T. S., & Snell, S. A. (2009). *Management: Leading and collaborating in the competitive world* (8th ed.). New York: McGraw Hill/Irwin.
Bennette, L. F. (2003). *The management of construction: A project life cycle approach*. Oxford: Butterworth–Heinemann.
Best, R., & De Valence, G. (Eds.). (2002). *Design and construction: Building in value*. Woburn: Butterworth–Heinemann.
Booty, F. (Ed.). (2006). *Facilities management* (3rd ed.). Oxford: Butterworth–Heinemann.
Burden, E. (2004). *Illustrated dictionary of architectural preservation: Restoration, renovation, rehabilitation and re-use*. New York, USA: McGraw Hill Companies Inc.
Cain, C. T. (2004). *Performance measurement for construction profitability*. Oxford: Blackwell Publishing.
Chanter, B., & Swallow, P. (2007). *Building maintenance management*. Oxford: Blackwell Publishing.
Chapman, K. (1998). Dissatisfaction with stock condition surveys in social housing. *Structural Survey, 17*(14), 211–215.
Christensen, A. J. (2005). *Dictionary of landscape architecture and construction*. New York: McGraw Hill.
CICA. (1989). *Accounting and reporting for physical assets by governments*. Canadian Institute of Chartered Accountants, Toronto, Ontario, Canada. Cited in Vanier, D. J. (2001) Why industry needs asset management tools, NRCC-44702NRC, Publications Archive Archives des publications du CNRC.
Claessens, J. C., Van Eerde, W., & Rutte, C. G. (2005). A review of the time management literature. *International Journal of Personnel Review, 36*(2), 255–276.
Costa, R., Paiva, A., Rodriguez, O., De Rojas, M. I. S., & Frias, M. (2010). The influence of the colour behaviour of external envelope materials in the quality of the buildings. In P. Barrett, D. Amaratunga, R. Haigh, K. Keraminiyage & C. Pathirage (Eds.), *CIB 2010 World Congress Proceedings*. Salford Quays, United Kingdom: The Lowry, May 10–13, 2010.
Daft, R. L. (2004). *Management* (6th ed.). Ohio: Thomson (South–Western).
Davies, N., & Jokiniemi, E. (2008). *Dictionary of architecture and building construction*. Oxford: Architecture Press (Imprint of Elsevier).
El-Haram, M. A., & Horner, M. W. (2002). Factors affecting housing maintenance cost. *Journal of Quality in Maintenance Engineering, 8*(2), 115–123 (MCB UP Limited, 1355–2511).
Emmitt, S., & Yeomans, D. (2001, 2008). *Specifying buildings: A design management perspective* (1st and 2nd ed.). Oxford: Butterworth–Heinemann.
Flanagan, R. (2007). Vision 2020 for procurement in the construction sector. In *Proceedings of Quantity Surveying International Convention (QSIC)*. Jointly organized by Visit Malaysia 2–7, International Islamic University Malaysia, Board of Quantity Surveyors Malaysia, Institution of Surveyors Malaysia, JKR and CIDB. Held on September 4–5, 2007, Kuala Lumpur.
Forrester, E. C., Buteau, L. B., & Shrun, S. (2010). *CMMI for services: Guidelines for superior*. Boston: Pearson Education Inc.
Fuziah, C. C. A. (2006). *University asset management system (UniAMS)*. Unpublished Master's thesis, submitted to Universiti Teknologi Malaysia. Retrieved on January 02, 2009. Available at http://eprints.utm.my/5489/1/CheFuziahCheAliMFSKSM2006TTT.pdf.
Gahlot, P. S., & Sharma, S. (2006). *Building repair and maintenance management*. New Delhi: CBS Publishers and Distributors.
Ishak, N. H., Chohan, A. H., & Ramly, A. (2007). Implications of design deficiency on building maintenance at post-occupational stage. *Journal of Building Appraisal, 3*(2), 115–124.
Isnin, Z., Kamarudin, H., Ali, I. M., Baharuddin, M. N. (2010). User satisfaction on cleaning services in higher learning institutions. In M. Hisham Aariffin, E. M. Ahnuar, N. S. Syarifah, R. Takim & M. Bagdad (Eds.), *Proceedings of Management in Construction Researchers Associations (MICRA 2010)*. Paradigm Shift in Built Environment Research (pp. 19–30). Shah Alam: Universiti Teknologi Malaysia, December 1–2, 2010.

Johnston, R., Clark, G., & Shulver, M. (2012). *Service operations management: Improving service delivery* (4th ed.). England: Pearson Education Limited.

Jones, K., & Sharp, M. (2007). A new performance based process model for built asset maintenance. *Facilities, 25*(13/14), 525–535.

Kaplan, R. S., & Norton, D. P. (1993). Putting the balanced score card to work. *Harvard Business Review, 71*(5), 134–142.

Khan, A. H. R., Sharil, M. A. R., Ali, M. H., Sivanathan, S., Jivasangeeta, S., & Islamiah, S. (2010). Assessment of building practices in UTM for maintainability: Case study. In *Proceedings of 4th International Conference on Built Environment in Developing Countries (ICBEDC 2010): A Better Tomorrow with a Greener Today* (pp. 292–301). Pilau Pinang: Universiti Science Malaysia, December 1–2, 2010.

Kirkham, R. (2007). *Ferry and Brandon's cost planning of buildings* (8th ed.). UK: Blackwell Publishing Limited.

Kulatunga, U., Amaratunga, D., & Haigh, R. (2006). Performance measurement of construction research and development: Evaluation of performance measurement frameworks. In E. Sivyer (Ed.), *COBRA 2006 Proceedings of the Annual Research Conference of the Royal Institution of Chartered Surveyors*. London: University College London, September 7–8, ISBN: 978 -1-84219-307-4.

Lai, J. H. K., Yik, F. W. H., & Jones, P. (2006). Disputes arising from vaguely defined contractual responsibilities in buildings services maintenance contracts. *Facilities, 22*(1/2), 44–52.

Lee, R. (1987). *Building maintenance management* (3rd ed.). London: Blackwell Science Ltd.

Lewis, A., Elmualim, A., & Riley, D. (2010). Linking energy and maintenance management to improve knowledge transfer within building operation and management. In P. Barrett, D. Amaratunga, R. Haigh, K. Keraminiyage & C. Pathirage (Eds.), *CIB 2010 World Congress Proceedings*. Salford Quays, United Kingdom: The Lowry, May 10–13, 2010.

Lofsten, H. (2006). Measuring maintenance performance—In search for a maintenance productivity index. *International Journal of Production Economics, 63*, 47–58.

Madu, C. N. (2000). Competing through maintenance strategies. *International Journal of Quality and Reliability Management, 19*(9), 464–476.

Mandal, P., Venta, E., & El-Houbi, A. (2008). Business practices and performance in US manufacturing companies: an empirical investigation. *International Journal Business Excellence, 1*(1/2), 141–159.

Marosszeky, M., & Karim, K. (2002). Enterprise process monitoring using key performance indicators. In R. Best & G. De Valence (Eds.), *Design and construction: Building in value* (pp. 276–290). Oxford, UK: Butterworth–Heinemann.

Marsh, C. (2003). *Building services procurement* (1st ed.). UK: Spon Press.

Martinez, V., & Kennerley, M. (2005). *Performance management systems: Mix effects. Conference Proceedings EURAM*. Munich, Germany, May 4–7.

Mills, E. D. (Ed.). (1994). *Building maintenance and preservation: A guide to design and management* (2nd ed.). Oxford: Butterworth–Heinemann.

Munive-Hernandez, E. J., Dewhurst, F. W., Pritchard, M. C., & Barber, K. D. (2004). Modelling the strategy management process: An initial BPM approach. *Journal of Business Process Management., 10*(6), 691–711.

Newman, A. P. E. (2001). *Structural renovation of building: Methods, details, and design examples*. New York: MaCraw Hill.

O'Dell, A. (1996). Taking stock: A review of stock condition survey methods in the UK. In *User oriented and cost effective management, maintenance and modernization of building facilities*. ISBN: 951-758-358-3.

Oakland, J., & Marosszeky, M. (2006). *Total quality in the construction supply chain*. Oxford: Butterworth–Heinemann.

Parasuraman, A. (2010). Service productivity, quality and innovation: Implications for service-design practice and research. *International Journal of Quality and Service Sciences., 2*(3), 277–286.

Parkin, M. (2012). *Economics* (10th ed.). Harlow: Pearson Education Limited.

Pitt, T. J. (1997). Data requirements for the prioritization of the predictive building maintenance, *Facilities, 15*(3/4), 97–104.

Pitt, M. (Ed). (2010). Charging for non-core facilities services: The case of hospital car parking. *Journal of Retail & Leisure Property, 9*(4), 261–262.

Seeley, I. H. (1987). *Building maintenance* (2nd ed.). Houndmills: Macmillan Press Limited.

Seeley, I. H. (1996). *Building Economics: Appraisal and control of building design and efficiency* (4th ed.). New York: Palgrave.

Shen, Q., Lo, K., & Wang, Q. (1998). Priority setting in maintenance management: a modified multi-attribute approach using analytic hierarchy process. *Journal of Construction Management and Economics., 16*, 693–702.

Sherwin, D. (2000). A review of overall models for maintenance management. *Journal of Quality Maintenance Engineering, 6*(3), 138–164.

Shohet, I. M. (2010). Performance-based-maintenance of public facilities: Principles and implementation. In P. Barrett, D. Amaratunga, R. Haigh, K. Keraminiyage & C. Pathirage (Eds.), *CIB 2010 World Congress Proceedings*. Salford Quays, United Kingdom: The Lowry, May 10–13, 2010.

Sliteen, S., & Catarina, O. (2010). Benchmarking: Strategies to analyze operation and maintenance costs of healthcare facilities. In P. Barrett, D. Amaratunga, R. Haigh, K. Keraminiyage & C. Pathirage (Eds.), *CIB 2010 World Congress Proceedings*. Salford Quays, United Kingdom: The Lowry, May 10–13, 2010.

Sommerville, J., & MoCosh, J. (2006). Defects in homes: an analysis of data on 1,696 new UK houses. *Journal of Structural Survey, 24*(1), 6–21.

Spare, J. H. (2001). Building the business case for condition-based maintenance. In *IEEE*, 0-7803-7285-9/01.

Stevens, D. (1997). *Strategic thinking: Success secrets of big business projects*. Sydney: McGraw Hill.

Syamilah, B. Y. (2006). *Maintenance management through strategic planning for public school in Malaysia*. Unpublished Master's Thesis, University Technology Malaysia, Available at http://eprints.utm.my/4071/1/NoraziahAzminAbdLatifAzmiMAD2006TTT.pdf Reassessed on 05 September 2008.

Treacy, M., & Wiersema, F. (1995). The discipline of market leaders: Choose your customers, narrow your focus, dominate your market. In D. Bartholomew (Ed.), *Building on knowledge: Developing expertise, creativity and intellectual capital in the construction professional*. UK: Blackwell Publishing Limited.

Turrell, P. M. (1997). Small is different: a strategy for effective management Maintenance in non-profit making organizations, COBRA '97. ISBN: 0-85406-840-6.

Umeadi, B. N., & Jones, K. G. (2003). From condition to performance assessment of built components. In *COBRA*, 1-8419-148-9. University of Wolherhampton, September 1–2, 2003.

Watt, D. S. (1999). *Building pathology: Principles and practice*. Oxford: Blackwell Science Limited.

Wilkison, N., & Kleas, M. (2012). *An introduction to behavioural economics*. Hampshire: Palgrave.

Williams, A. R. (1993). *Domestic building survey*. UK: E & FN SPON.

Wood, B. R. (2009). *Building maintenance* (1st ed.). Oxford: Blackwell Publishing.

Wordsworth, P. (2001). *Lee's building maintenance management* (4th ed.). Oxford: Blackwell Science Limited.

Zailan, M. I. (2001). The management of public property in Malaysia. In *International Conference FIG Working Week 2001*. Seoul, South Korea, May 6–11, 2001.

Zainal Abidin, K., & Roslan, A. (2006). Maintenance management process model for school buildings: An application of IDEF0 modeling methodology. In *International Conference on Construction Industry 2006 (ICCI 2006)*. Padang, Indonesia: Universitas Bung Hatta. Available at www.fab.utm.my/download/conferenceseminar.icci200681pp12.pdf. Accessed on 23 May 2008.

Zakaria, Z., & Ali, N. M. (2006a). Assessment of implementation of maintenance works at Kolej Universiti Kejuruteraan dan Teknologi Malaysia (KUKTEM). In *Proceedings of the National Seminar on Civil Engineering Research 2006 (SEPKA 06), Universiti Teknologi Malaysia.*

Zakaria, Z., & Ali, N. M. (2006b). Maintenance management approaches in public higher learning institution in Malaysia: KUKTEM's perspective. In *Proceedings of the 1st Malaysia Technical Universities Conference on Engineering and Technology 2006 (MUCET 2006), Kolej Universiti Tun Hussein Onn.*

Zakaria, Z., Ali, N. M., Abd Hamid. M. R., Othman, N., & Haron, A. T. (2008). Exploring on issues of maintenance management at higher learning institution in Malaysia. In *Proceeding of the International Conference on Civil Engineering 2008 (ICCE '08): Challenges in Facing Natural Hazards for Future Engineering Practice, Universiti Malaysia Pahang.*

Zavadskas, E. K., & Vilutiene, T. (2006). A multiple evaluation of multi-family apartment block's maintenance contractors: I model for maintenance contractor evaluation and the determination of its selection criteria. *Journal of Building and Environment, 41*, 621–632.

Zawawi, E. M. A., Kamaruzzaman, S. N. (2009). Personnel characteristics of maintenance practice: A case of high-rise office buildings in Malaysia. *Journal of Sustainable Development, 2*(1), 111–116. www.ccsenet.org/journal.html

Zawawi, E. M. A., Kamaruzzaman, S. N., Ali, A. S., & Sulaiman, R. (2010). Assessment of building maintenance management in Malaysia: Resolving using a solution diagram. *Journal of Retail & Leisure Property, 9*(4), 349–356.

Chapter 6
Methodological Issues

Abstract The previous chapters have provided us with information on building maintenance management. This chapter sets out to provide descriptions and explanations on the methodological issues of the thesis from which this book was developed. We are aware that ordinarily this chapter could not be included in book; however, it is included for the readers to know the answers to how and also what primary data are collected, collated and processed. After studying this chapter, the reader should: (1) be able to describe the methods used to collect and analyse data used in developing this book. (2) understand why the method of data collection and analysis were adopted. (3) appreciate the science behind respondent selections. (4) appreciate the research design.

Keywords Data collection · Respondents · Survey questionnaire · Sample size · Qualitative data and quantitative data

6.1 Introduction

In Chap. 1 of this book, the outline of the methodology was presented. This chapter is therefore an elaboration of the outline. Thus, expanded versions of major issues presented in the outline are revisited here. It is therefore very much possible that reader will encounter some repetitions of facts presented earlier. For scientific research, it is uniquely necessary to explain *how* research aim, objectives, hypotheses and questions are conducted, achieved and answered. It is necessary to explain and describe how and why any decision is made and causes of certain actions that were taken. This chapter starts with discussion on the research problems, the research questions and as well as on the research aim and objectives. The methods of sampling as well as the scope and limitations of the research were also described.

A.L. Olanrewaju and A.-R. Abdul-Aziz, *Building Maintenance Processes and Practices*, DOI 10.1007/978-981-287-263-0_6

6.2 Recapitulation Problems Facing Maintenance Organisations

The value maintenance management model is required to align maintenance resource with users' satisfactions and to also align building performance with organisation corporate strategy and to ensure holistic consideration for maintenance management practices, processes, procedures and their implementation. The shortcomings and the failings of the current maintenance management processes have been identified and discussed. It was found that the building maintenance management system can only be improved with an alternative maintenance management framework that guides the decision making processes.

Maintenance per se cannot be blamed if things went wrong, rather it is the management that is requiring some step changes. On the one hand, there are many complaints and criticisms in media and research literature with how poorly university buildings are being managed in Malaysia. On the other hand, governments are already facing serious financial crises in the face of various pressures. As a result, universities are now operating in highly competitive and turbulent environments. This is also happening at a time when students and their parents are becoming more demanding and concerned about building performance. While there is a need to increase maintenance allocations, it is clear that lasting solutions cannot emerge without a step change in the maintenance management process. Organisations must improve their business practices to be competitive, market driven and to enhance their profitability shares.

Maintenance management requires a multi-disciplinary approach: from the engineering, technological, commercial, economic and social perspectives. This book is built around the argument that expectations and experiences of the building users and the structure of the maintenance organisation are crucial to deliver best service. But in order to do this, information and knowledge on the criteria that influence maintenance of university buildings, characteristics of defects in university buildings and criteria within the users' value system, are positively correlated with maintenance management policies, strategies and processes. The lack of this knowledge and information would imply that the scarce resources (material, time, etc.) are not deployed to best services and activities. A poor maintenance management system will lead to an unnecessary increase in maintenance costs and poor user satisfaction. Unnecessary costs are costs that provide neither use, nor life, nor quality, nor aesthetic, nor comforts or stakeholder features that meet the provider and consumer satisfaction (Kelly et al. 2004).

It was outlined that concerted efforts have been made towards improving the building maintenance management systems, such systems offer potential to improve the performance of maintenance management system; however, the inputs of building users were not factored in the formulation of maintenance policy. Furthermore, instead of considering performance issues, only condition issues were considered, and as can be seen, many of the models are for office and commercial buildings. The proposals were based on the culture of fragmentation and separation and the proposals still remain classical.

6.3 Design Strategy

Hitchcock and Hughes (1995) stressed that the choice of method to adopt for a particular research is determined by the chosen topic and the kind of data to collect. They also argued that the main trust in the method to use is "fitness for the purpose". As long as the research questions are answered and/or the objectives are achieved, the method is considered good. In fact there is no "best method", rather any approach that answers the research questions is appropriate. There are two broad types of research strategies, namely qualitative and quantitative strategies. The decision as to which to adopt depends on the purpose of the research and the type of available information required for the research.

There are two types of data; primary data and secondary data. Primary data are first hand in nature, they are collected from the field by the researchers and/or their agents. On the other hand, secondary data are collected from a secondary source. Secondary data include government processed information that is often in the form of a statistical publication. The IMF financial reports are an example of secondary data. Information obtained from literature reviews are also considered as secondary information. The Bank Negara Annual report also contains a lot of secondary information for this research. Seemingly, the compilation of these data depends greatly or mainly on primary data.

Primary data can be collected through:

1. Survey approach
2. Interview
3. Case study approach
4. Problem solving approach (action research)
5. Observation approach
6. Experimental approach.

6.3.1 Desk Work

Reviews of the related literature and theory are undertaken in order:

1. To provide background information on the maintenance practices in the Malaysian university; its past, its present and its future prospects.
2. To consider and comment on what others have already discovered and said about the problems faced by the university organisations in terms of maintenance management, the proposed solutions and existing situations.
3. To explore alternative solution(s) if any.

4. To develop the research aim and objectives.
5. To develop the research hypothesis for the research.
6. To confirm if the findings are consistent or not with the findings of other researchers in other places in relation to the solution to the maintenance management of university buildings.
7. To identify variables and concepts that are appropriate for this research.

6.3.2 Fieldwork

For the purpose of this book, data were collected through triangulation. The triangulation involved three data sources that cross examine the validity and reliability of one another. The three sources that were used for this research were questionnaires, interviews and literature reviews. Triangulation is used when there is a need to investigate same variable with two or more research methods. The essence of triangulation is to address or mitigate the weakness of one method with the others. However, before the primary data collection, pilot studies were conducted to test the wordings, ambiguities and ease of understanding of the questionnaire.

6.4 Survey Approach

Survey approach is useful when the required respondents' rate is relatively high and the research is expected to be completed within a limited time frame (Naoum 2007). The outcomes of the survey can be generalised to the main population. In the survey method, all the respondents are asked the same questions in the same circumstance (Bell 2005). The main emphasis of this method is on fact findings. Essentially, survey approach is adopted for this book because of the following reasons:

1. Flexibility: it is possible for the researcher to study wide ranges of research questions.
2. Situations can be described and relationships between variables can be studied.
3. The findings from surveys can be generalised to the entire population under consideration. Essentially, one of the main purposes of carrying out this research is to generalise the finding to universities in Malaysia as much as practicable.
4. Large amount of data can be gathered. The findings from this research were to be generalised, however. Therefore, it became imperative that a large amount of data had to be collected.
5. It was cheaper compared to other type of approaches like case study and experimental approach for this research.
6. Large areas can be covered, particularly with the postal questionnaire techniques.

6.4.1 Questionnaire Survey

The questionnaire approach was used because of the following reasons:

1. Questionnaires are a cost effective method of collecting data for survey research. Travelling to all the surveyed universities to conduct interview would involve a lot of expense. Conducting interviews through telephone or electronic media is also expensive compared to the postal questionnaire approach.
2. The electronic medium could not be used because many of the respondents do not have adequate access to the Internet and majority that have access are busy.
3. It is quicker to conduct the questionnaire approach.
4. Another reason for adopting the questionnaire approach was because of its familiarity to the respondents. This allowed respondents to complete the questionnaire at their own convenience, and provided them the opportunities to think about their opinions.
5. The questionnaire approach allowed large amount of data to be gathered within a short period. Hence make generalisations of results possible.
6. The findings could be expressed statistically and compared findings with previous findings.

6.4.2 Limitations of the Questionnaire Survey Approach

Irrespective of the process of administering the questionnaire, many of the limitations appeared to hold. However, the limitations associated with the questionnaire approach as they relate to this research are as described below:

1. Respondents cannot be probed. The outcomes from the questionnaire have to be taken on face value.
2. Questions about complex issues are difficult to examine. This is because respondents' opinion cannot be exhaustive.
3. With no control over the respondent, the researcher has little opportunity to ensure that the intended respondents complete the questionnaire. The target respondents might delegate completing the questionnaires to their subordinate.
4. Accuracy: the respondent could provide a general answer to a question whereas the intended response was specific.

6.4.3 Mitigating the Limitations of the Survey Approach

In order to ensure that targeted respondents completed the questionnaire, personal telephone calls were made to the organisations. This was for them to encourage the targeted respondents to complete the questionnaires. Furthermore, the questions were kept simple and short so that it was easy to be understood by the respondents. The prospective respondents were given approximately 2 months to complete the questionnaires this was to allow the willing respondents to complete the questionnaire at a time convenient to them. However, in relation to problems like this, Fallon (1971) stressed the importance of using available information against waiting for reliable and accurate information that would never come. This is because great opportunities are missed while the researcher is waiting too long for information that may never come after all.

Limitations of questionnaires can also be improved by avoiding leading questions (Cohen and Manion 1994). Leading questions are the questions that suggest to the respondents that there is only one acceptable way for him. However, in this research, there were "no one way", questions. The questions were also short. The total number of questions in the questionnaires was about a hundred. Although this number seems large, however, in many cases, a question was no more than five words. With this, we are of the opinion that the respondents would not be bored while completing the questionnaire. The questionnaires were largely closed ended questions, which do not require much thinking and consultation. On average, a set of the questionnaires takes about twenty minutes to complete.

6.5 Piloting

The questionnaire for this research went through a series of pilot studies. This was to correct ambiguities; checks of the wording ensured that respondents understood the questions in the contexts that it was meant it to be. A pilot survey is viewed as a prerequisite to collect meaningful data for research. According to Bell (1996), pilot study is "getting the bugs out of the instruments (questionnaire) so that the subjects in your main study will experience no difficulties in completing" (Green and Turrell 2005). Altogether, a total number of eight drafts were prepared before the final edition was achieved. The pilot survey took about two months before the final draft was finalised.

6.6 Reliability Tests

Reliability is defined as how reliable a research instrument or method is. Reliability of a measure is the degree to which respondents will consistently respond to the instrument or construct in the same way and under the same conditions. Therefore, the more reliable a method of data collection is, the more likely it will be that it gives the same or very similar results if the research is repeated. Reliability is an important consideration in survey research method. Reliability of an instrument could be checked in different ways, such as; test-retest (this is a situation where by the same instrument is re-administered on the same respondents again), the alternative form method (in this situation, an equivalent version of the same instrument is given and their results correlated) and the split half method (as the name implies, the instrument is split into two equal halves and the scores correlated) (Bell 2001).

However, these methods are not always feasible or necessary, and there are disadvantages and problems associated with all the three techniques. Testing for reliability should come at the stage of formulating questions, wording and the piloting of the instrument (Bell 2001) as the best way to test for reliability is through piloting a Crunch-Alpha reliability test conducted on the data where necessary. See the data analysis, in the Chap. 5 of this thesis for the details.

6.7 Validity Tests

In contrast to reliability that consistently measures the measure, validity measures the correctness and truthfulness of the measure. Therefore, to ensure this, the instrument must measure what it must measure accurately. An indicator is valid to the degree that it empirically represents the concept it purports to measure. Piloting is one of the ways of validating a research instrument in order for the instrument to measure what it supposes to measure correctly. Validity of instruments could be increased by recording data correctly and cross checking of data. The essence of triangulation is to counteract various possible threats to the validity of analysis. In order to improve the scientific rigours, a statistical validity test that was conducted on the data was appropriate.

6.8 Ethics of Research

Ethics of research reporting were preserved by observing the following rules and procedures:

- The names of the university and respondents were mentioned if they had been revealed in the public report, research literature or when they were allowed to be discussed by the participants in this research.
- Anonymity was employed otherwise, or when information shared may have been deemed disadvantageous to the participants.
- The possible consequence of the research was made clear to the participants prior to the survey and interviews.
- The confidentiality, anonymity, non-identifiably and non-traceability of the participants were guaranteed.

6.9 Main Database

A database comprising names and address of universities across Malaysia was compiled. The sample size was initially drawn from a published data base of the Ministry of Higher Education, but supplemented with information from the media (national newspapers) and catalogues. This was necessary as data from the ministry were not current. Herein referred to as the main database, it consisted of both private and public universities. The private universities were further divided into University College and University. There were several ways in which the respondents' organisations could have been identified. For instance, the respondents' university could be identified through the review of published data, reports of the ministry of educations and of the high education. It could also be identified through reading newspapers or visiting educational fairs. However, none of the information obtained from each of these sources was conclusive. Therefore, the entire sources were used to track the list of the universities in Malaysia. Due to time and cost constraints, it was not possible to interview or collect qualitative data from all the respondents across Malaysia.

6.9.1 Respondent Database

The survey questionnaire and interview were conducted on the maintenance organisations of the universities. Each university was represented by one person, "maintenance manager" to avoid duplications. Maintenance managers were appropriate as the respondents. The managers were involved in establishing the goals and objectives of the organisations. They were also responsible for setting goals for specific areas and departments (i.e. electrical unit, mechanical unit, building unit and landscape unit) in the organisation. They were responsible for making sure the day to day operation of the organisation ran smoothly.

The manager liaises with the chief executive officers/chief operating officer of the university. The manager set the direction that the organisation would follow towards meeting the university core function of teaching, researching and providing conductive environment for learning. The maintenance manager sat on various committees to deliberate on matters concerning the maintenance management of the buildings even though it was acknowledged that they were not directly involved in the technical and engineering aspect of the service.

Therefore, it is evidently believed that the "maintenance manager" possessed first-hand information, knowledge and experience to provide meaningful responses to the research. Consequently, the manager could offer independent, unbiased and meaningful responses. However, in some of the universities, the titles changed to names including maintenance executives, maintenance engineers or property manager as the case may be. However, whatever the name given to the roles of the maintenance manager-maintenance executive, asset manager, facility manager, property manager or general manager—the common intent was the same. The main issue was that one individual took the overall control and responsibility for managing the activities of the various personal.

6.9.2 Criteria that Influence Maintenance Management of Buildings

These criteria were necessary to develop an all inclusive maintenance management model and a pertinent variable requiring cogent consideration was the establishment of criteria that affected the maintenance decision making processes. The establishment of those criteria underlay the systemic consideration of maintenance management. Based on extensive literature and theory reviews, expert survey and series of case studies conducted on maintenance of buildings, criteria that influence maintenance management decisions were identified and addressed to the respondent to determine the degree of each of the criterion with maintenance management of university buildings in Malaysia. These criteria have significant impact on building maintenance management. Based on literature findings, a total of 56 criteria were identified but after rigorous analysis, they were reduced to 45. The 45 were later reduced to 31 by selected maintenance management experts during piloting. Empty spaces were provided for respondents to add additional sources they considered significant. The respondents were asked to grade each of the criteria on a five point Likert scale. The first of these categories, i.e. *strongly influential*, involved a situation where the respondent does not agree that the criterion could lead to influence maintenance decision in their previous or ongoing projects. *Strongly influence* indicated that the respondent believed the criterion had very high influence on maintenance decision.

6.9.3 Defects in Buildings

Both the maintenance organisations and the building users were asked to rate defects in the university buildings. Building defect means different things to different people. Some defects are caused by the designers (latent defects) while some are caused by the constructor (patent defects), yet some others are caused due to normal wear and tear and lack of maintenance. The latent and patent defects are often in relation to new buildings. Defects can be defined as failure in building and services arising from factors including inadequate brief, faulty design decision, method of construction, user activities, maintenance and fair wear and tear (Wordsworth 2001). To some, it means the shortcomings with the design and construction practices while to some, it implies the inadequacies that arise from the "normal wear and tear". Within the UK construction industry, defects are now referred to as snagging (Sommerville and McCosh 2006).

While it is acknowledged that the primary causes of building defects are inevitably interconnected, the focus here is on the maintenance caused and/or used that caused defects. The design and construction lead defects constitutes another arena of specialist study. There are sufficient studies on how building defects can be reduced through brief, design and programming process. Words including deficiency, damage, default, deterioration and decay are often used to describe defect. In this research, all the terms are synonymous with defect and are sometimes used interchangeably.

Building defects can be identified by the users themselves or by inspection by those concerned with maintenance management. If identified by the practitioners through the inspection or survey, it is usually presented in the bill of dilapidation. Practitioners have the capability to identify defects that are not yet obvious, at least to users. The appraisal of defects is usually undertaken through a sequence of site procedures that provide the necessary information on which to make an assessment as to the condition and fitness for purpose.

The severity of a particular defect will be judged against the effect that the defect has, as well as those that it will have on the building and the associated services. Consequently, the influence that the defect will have on a building if it is not maintained will be dependent on factors including the nature of the defect, the building integrity, the user needs and wants as well as the successive maintenance programme. The effect of defects on the condition, appearance and performance of the building depend on the functional requirements of the building. A condition that might be acceptable for a residential building might require immediate maintenance in the case of university buildings. The degree of defects in the building is a measure to evaluate the functional performance of the building. Defects hinder the functional performance of the building and its services. Although building defects can be generally attributed to lack of good construction practice, they are often caused and in fact aggravated due to lack of good maintenance practices and usage.

The unique consequence of defects in a building is that it affects the building's capability to function optimally, technically, aesthetically and even otherwise. Building users are becoming more vocal on the requirements; they develop a cognitive map of what they want and often this is at variance with the maintenance organisation's perceptions. Defects set up inner psychological tension in the users, if the defect is not rectified (see Sommervlle and McCosh 2006).

The identification of those defects contributes to the development of best maintenance management. As a result of extensive literature and theory reviews, expert survey and series of case studies conducted on maintenance of university buildings, 32 known building defects were identified and addressed to respondents to determine the degree of each of the defects that affect maintenance management of university buildings in Malaysia. The rationale for each of the defects relates to the rate of urgency required in maintaining each of the identified defects. Defects that influence/relate to the user value system are very critical to them and require urgent attention. Such defects influence their comfort (i.e. lighting quality,

Table 6.1 List of defects in buildings

No	Defect	No	Defect	No	Defect
1	Damaged window	12	Damaged roof structure	23	Damaged frames
2	Blocked water closet	13	Damaged walls	24	Faulty air conditioning system
3	Floor tile failure	14	Damaged door	25	Faulty electrical circuit
4	Wall tile failure	15	Faulty communications appliances	26	Lifts failures
5	Faulty shower	16	Failed furniture and fittings	27	Damaged staircases
6	Faulty sanitary appliances and fittings	17	Sink leakage	28	Faulty fire alarms
7	Faulty fans	18	Bad soap holders	28	Faulty floor drain
8	Damaged ceiling	19	Pipes Leakage	30	Faulty heat extractor
9	Damaged door locks	20	Faulty towel rail	31	Faulty fire extinguisher
10	Damaged floors	21	Faulty bulbs	32	Faulty smoke detector
11	Damaged taps	22	Damaged hot water dispenser		

temperature, humidity, odour, air quality and ventilation, security of the use of the buildings. The defects were contained in Table 6.1. Respondents were further asked to specify and rate any other defect they believed affected the building performance.

Severities of the defects were measured with a five continuum scale. The categories dictated when the work was to be undertaken, and allowed future works to be programmed and financed as part of the rolling programme. The severity of defects and the corresponding priority for maintenance are subject to consideration such as:

1. Statutory obligation
2. Health and safety requirements
3. Functional and operational requirements
4. Rate of deterioration
5. Lease or covenant obligation
6. Cost fluctuation
7. Value and utility of building and services
8. Desired and expected standards
9. Running or operational cost.

The defects addressed to the respondents were those that affect a building's fabric, services and structure and on the other hand, the defects addressed in this research did not include defects that relate to substructure works (foundations cracking, foundation bowing and foundation settlement). Substructure defects are beyond the scope of this research. Similarly, the defects addressed to the respondents were not comprehensive even to the building fabrics, structure or/and engineering service, but they were indicative. This is the case because it is not possible to address all the defects in buildings. In that respect, they are the object of many maintenance management problems.

6.9.4 Criteria of the User Value System (Factors that Enhance User's Satisfaction)

In general, the research that is specifically related to user satisfaction in building maintenance is remarkably scanty or non-existent. Studies that exist focus mainly on new buildings. To date there has been no any in-depth research into building maintenance user satisfaction though there is some research on post-performance evaluations. Thus, the fundamental issues of what the building user wants and expects from the buildings and the maintenance organisation must have, should have and nice to have (see also Sommerville and McCosh 2006) are ill researched. From the users' perspective, best buildings are often perceived as comfortable, safe, and healthy with built-in facility to respond rapidly when things go wrong.

Although, building users may not play any (active) role during the design and construction of the building they play active roles in managing them. In part, the users' value systems influence the decisions of those responsible for the management of the buildings. Maintenance managers and designers have different views on the needs and wants of the users. For instance, maintenance managers only tend to provide services of a "good enough basis" while the designers tends to oversimplify users' needs and wants (Leaman 2003). Often building users are satisfied with buildings when there is a rapid response capability when things go wrong (such as the speed and effectiveness of the help desk management), regardless of whether the buildings itself is scruffy or undistinguished architecturally (Leaman 2003).

6.10 Questionnaire for Building Users

Specifically, it is about the students and other building users including the staff, however, students were uniquely focused. This information is very relevant as students are the major occupants of buildings, the evaluation of their comments can provide valuable information and feedback to the maintenance organisations. Staff here means those other than those working in the respective maintenance organisation. Although staff and students are a different set of subjects and so considered as a corporate entity as such their needs and wants were considered as the same.

6.10.1 Criteria that Enhance User Satisfaction

Both the maintainers and building users were asked to determine and rank the criteria that influence users' satisfaction in maintenance delivery. These criteria intend to specifically determine and rank the criteria that building users know can enhance their satisfaction in the maintenance services. On why the user should identify the criteria even though the maintenance organisations also identified them is to determine the service gaps. Perhaps, it is possible that the maintenance organisations do not really understand what the users want. Often, what users need and want are not known, as providers tend to concentrate on the requirements that were never intended by the user and therefore lead to unnecessary maintenance cost and an increase in user dissatisfaction.

6.10.2 Defects in Buildings

Defect is a non-fulfilment of an intended functional requirement or an expectation, including that of safety and health. The same defects that were addressed to the maintenance organisation were also addressed to the students. The scale of measurement was also the same; the purpose of which was to compare whether there was a difference of opinion on the maintenance organisations and the users. Any difference discovered would guide the maintenance organisation towards a value based approach to maintenance management.

6.10.3 Method of Administering Questionnaires on the Users

Questionnaires are generally administered by hand delivery, postal questionnaire, and telephone and using electronic media. For the purposes of administering questionnaires to the building users, a non-random sampling technique was used. Specifically, the data collection combined stratified sampling and convenience sampling techniques. Stratified sampling is appropriate for population that occurs in groups or strata, for instance the private university and public university (Fellows and Liu 2008).

The population was initially stratified because it was felt that the population may have different parameters. Although convenient sampling is often used during the exploratory stage of a research project, it is probably the most appropriate alternative when accessible factors make it impossible to consider the required sample convenient, available, provides the needed information. Users should be involved and educated, so that they have realistic and achieved value system expectations and wants. This will address issues concerning unnecessary surprises on the aspect of the users and will also reduce unnecessary pressure on the part of the maintenance organisation.

6.11 The Analytical Hierarchy Process (AHP) Survey

The information provided in this chapter so far was on the general survey. The general survey was first undertaken to collect comprehensive data in building maintenance management. The purpose of the AHP survey was to collect refine and detail information in building maintenance. Essentially, it involved developing a mathematical model for building maintenance management. The research methodology and layout of the AHP are depicted in Fig. 6.1. The AHP survey commenced after refining and sharpening the results and findings of the general survey. AHP was used to prioritise factors and elements in the value maintenance management model. A pilot study was conducted prior to the AHP survey to test the suitability of the summarised criteria and factors from the general survey and to further scrutinise the comprehensibility of the questionnaire prior to sending it out.

AHP is a mathematical modelling technique for multi-criteria decision making. It was developed by Saaty, a mathematician in early 70s. The AHP method helps to specify numerical weights representing the relative importance of factors, elements, criteria or building maintenance management model. AHP allows both qualitative and quantitative approaches to solve complex decision problems. The technique structures/decomposes problems into a hierarchy of elements or factors influencing a system by incorporating levels. In quantitative aspects, the AHP can prioritise a set of attributes and distinguish in general the more important factors from the less important factors. The pair-wise comparison judgments were made with respect to the attributes of one level of hierarchy given the attribute of the next higher level of hierarchy (from the main criteria to the sub-criteria).

The AHP methodology has been adapted in various fields including engineering, military, management, manufacturing, politics, social, sports and even at the individual. It has been applied in areas including business administration, cost-benefit analysis, future planning, resolution of conflicts, determining requirements, allocating resources, measuring performance, designing systems and ensuring system stability. The wide applicability is due to its simplicity, ease of use, and great flexibility inherent in the technique.

AHP is particularly appropriate for complex decisions, which involve the comparison of decision elements which are difficult to measure. The technique is based on the theory that when faced with a multifaceted or complex decision, the natural human reaction is to group the decision elements according to their common characteristics. Particularly for maintenance management, the decision criteria may hold a different degree of preference or level of significance to the decision in the perspectives of the various stakeholders (users and maintenance organisations, top management and technical staff). However, AHP helps to bring consistency to choice made by the concerned stakeholders.

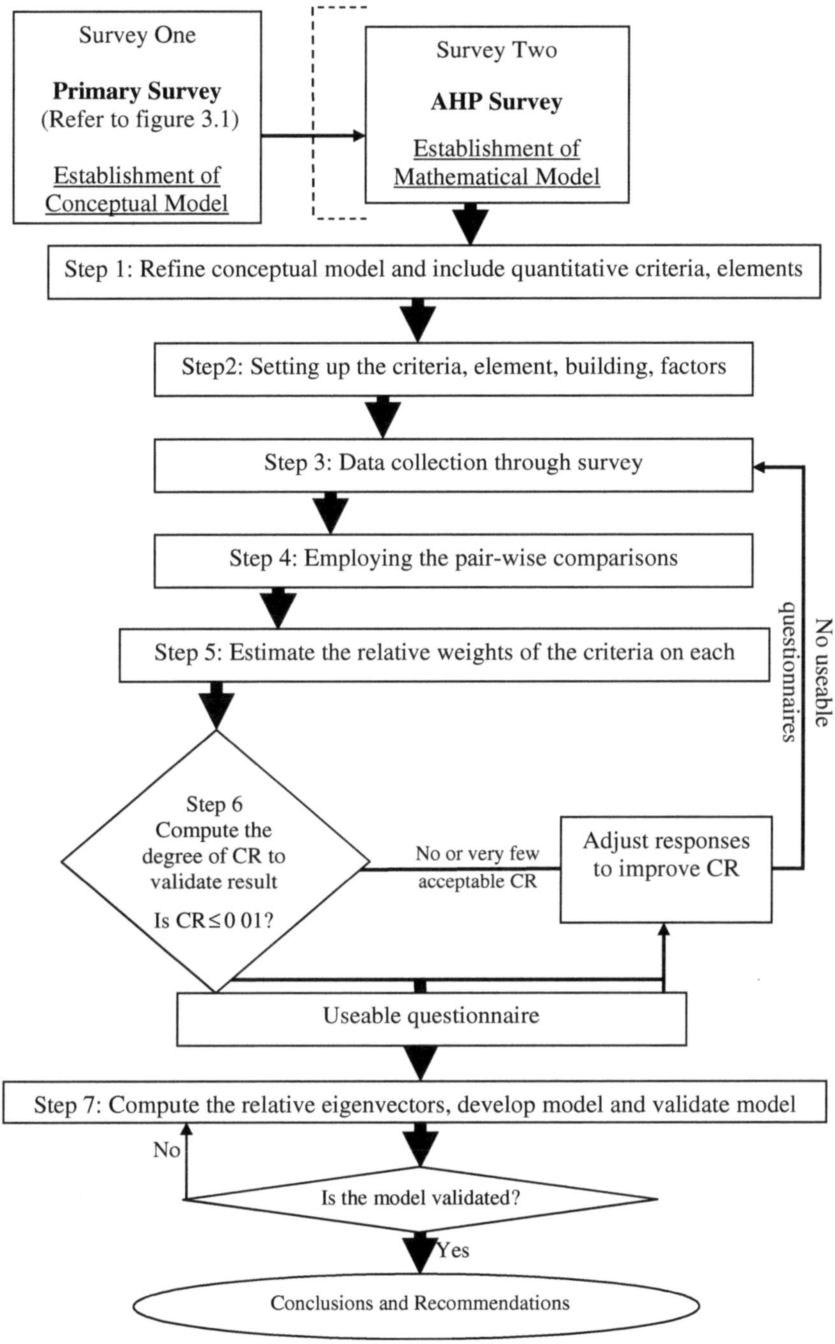

Fig. 6.1 Layout for the AHP Survey

The (manual) computational aspect of the AHP technique involves various steps as illustrated by Saaty (1980, 1990, 2003). However, the basic information of the technique is that the greater the weight or vector, the greater the importance of preference for the specific criteria, sub-criteria or alternatives.

The basic essence of the mathematics of the AHP technique is to construct a matrix expressing the relative value of a set of characteristics. For instance, what is the relative importance of a maintenance department of maintaining administrative buildings first as against maintaining the hostel under prescribed condition? Or what is the importance of maintaining roof structure of the building as against maintaining the walls? Or of spending much to maintain engineering services instead of roofs or walls? These are some of the multi-criteria decision making problems where AHP is required.

The mathematics of the technique demands that the factors or criteria should be relatively independent. A fundamental assumption of the technique is that if an element is much more important than another element, then it is therefore logical that the other element be of less importance to the first element. For instance, supposing both roof structure and walls required maintenance and are to be prioritised on a scale of 1–9; where 1 represents extremely less important and 9 denotes extremely important. Therefore, if roof structure is extremely more important than walls and ranked at 9, then walls must be extremely less important than roof and should be rated 1/9.

In this book, the AHP weighting is computed using the computer software application Expert Choice 2000 version. The software was developed by the inventor (Thomas Saaty) of the technique and has received widespread applications in different sectors. Expert Choice is a multi-attribute decision support software tool that is based on the AHP methodology. Expert Choice decision models follow the prescribed AHP module, a functional hierarchy with the broad overall goal at the highest level while the lower levels correspond to sub-criteria or alternatives.

The AHP model deals with prioritising of decision making by decomposing complex decisions to a series of pair-wise comparisons and then synthesising the results into degrees of importance. In addition, AHP gives room for sensitivity testing by computing a consistency ratio to do a check and balance on the consistency of the respondents to their subjective judgments. However, one of the problems of the AHP technique is that the technique assumes non-linearity, but the Expert Choice software has built-in mechanisms to adjust for this irregularity.

The questionnaire comprises of two main parts. The first part comprises the summary of the operation information or issue in the maintenance of buildings. With respect to the first part, seven main factors were identified and addressed to respondents. The factors were produced based on factor analysis performed on the criteria influencing building maintenance, criteria of the user value system and defects in buildings. These factors are responsiveness, safety of users, security of property, aesthetic, structure of maintenance department, financial and user comfort.

The second part focuses on quantitative issue in building maintenance. The part comprises cost-based criteria, element-based criteria and performance-based criteria.

The goal of the AHP questionnaire is to provide relative weights to the questionnaire items or categories of the items to generate mathematical relationship. Assigning relative weights to multiple items under multiple criteria in order to make a decision in AHP uses a multi-level hierarchical structure of objectives, criteria, sub-criteria and alternatives, and provides a quantitative computational method for generating priorities based on pair-wise judgment of the criteria.

With respect to the second part, building is divided into nine major elements. Building is broken down into: foundation, frames (staircases, columns, beams and floors), doors and windows, external walls, internal walls and partitions, roof structures, finishes and painting, lifts and escalator and engineering services.

Building element is a major component common to most buildings which usually perform the same function or functions irrespective of its design, specification or method of construction. Building components deteriorate at a different rate and it is very unlikely the entire building elements will be maintained at the same time. This will assist in prioritising maintenance needs.

Building is divided into nine basic elements, in accordance with the classification of the Building Management Information (BMI). Each element provides a more accurate estimate for maintenance purposes as compared to taking a building as a unit. However, the BMI divide building into 13 basic elements, however, from the maintenance perspective, it is very likely that the BMI classification requires modification. Furthermore, since the AHP is selected as the modelling technique, it is very necessary to reduce the element to a manageable size. This is because most humans cannot simultaneously compare more than seven objects (plus or minus two) (Miller 1956). Specifically, respondents were asked the contributions of each of the elements to their maintenance cost as well as how it affects developing value maintenance management.

Building maintenance is also prioritised in terms of gross floor area, age of building, building performance, number of users and number of buildings in an organisation (see El-Haram and Horner 2002; Shohet 2010; De-Marco et al. 2010). Usually, for maintenance organisations based their annual estimate on the above criteria with the addition of certain allowances to cover for increase in cost, inflation and errors. Most contractors or outsourcers also based their quotations on the above-mentioned criteria. In addition, building performance criteria including building condition, intensity of use, and effect of lack of maintenance on the elements/components significance of the element/components or building to users can also be used in prioritising maintenance repairs.

There are a number of steps to implement the AHP model. The first step is to describe the maintenance problem into an AHP decision hierarchy. It consists of an overall goal at the top, a group of options or alternatives for reaching the goal at the bottom, and a group of factors or criteria filling up at the middle which relate the alternatives to the goal. In most cases, the criteria are then decomposed into sub-criteria in some degree based on the needs of the problem.

Variation of the analytic hierarchical process to this study required identification of the objective, the criteria (the factors that affect the objective) and the alternatives. Those are as follows:

1. The objective to be achieved was to develop value-based maintenance management for maintenance organisations.
2. The criteria are:
   (a) Building elements
   (b) Operational criteria
   (c) Cost criteria
   (d) Performance criteria
3. The alternative strategies available were to maintain or defer the maintenance of the element or building. Specifically, it illustrated what element or building to maintain; first given the available resource before the others.

The hierarchy is illustrated in Fig. 6.2. The first level of the AHP hierarchy is the objective/goal to be achieved, which is developing value maintenance management model for maintenance organisation. The second level is concerned with criteria that affect the objective. The third level relates to the criteria considered as factors that the maintainers could consider in achieving the objective. Finally, the last level has to do with the main options or alternatives that the university had in order to achieve the objective. The priority vectors were derived by carrying out the comparison from the last level and ranking up to the first level.

The preparation of survey for the general survey and AHP survey are different. For the AHP survey, a common scale from (Saaty 1980) is adopted to prepare the questionnaire. Table 6.2 lists the classifications of the scale.

In AHP, before computing the weights based on pair-wise judgments, the AHP computes a weight for each decision element based on the pair-wise comparisons using mathematical techniques such as eigenvalue, mean transformation or row geometric mean. Inconsistency is measured by the inconsistency index (II). Perfect consistency implies a value of zero for II. In the pair-wise comparison, consistency ratio (CR) is performed to ascertain the consistency in data. The consistency ratio = CI/RI, where CI denotes consistency index and defined as $\mathrm{CI} = (\lambda_{\max} - n)/(n^{-1})$, RI = the random consistency index obtained by averaging the consistency index of a randomly generated reciprocal matrix from a scale of 1 to 9. It is not appropriate for the consistency ratio to be more than 0.10. Table 6.3 contains the consistency index against the size of the matrix (Saaty 1990).

6.12 Grounded Theory

The value maintenance management model was developed based on the principle of grounded theory. Theory is a set of interconnected statements, hypothesis or principles combined together to facilitate the explanation of a concept, another statement, principle or idea. Theories assist towards recognising, identifying, classifying, understanding explaining and/or relating variables, concepts or events. It provides a guide for research by putting previous and existing knowledge into

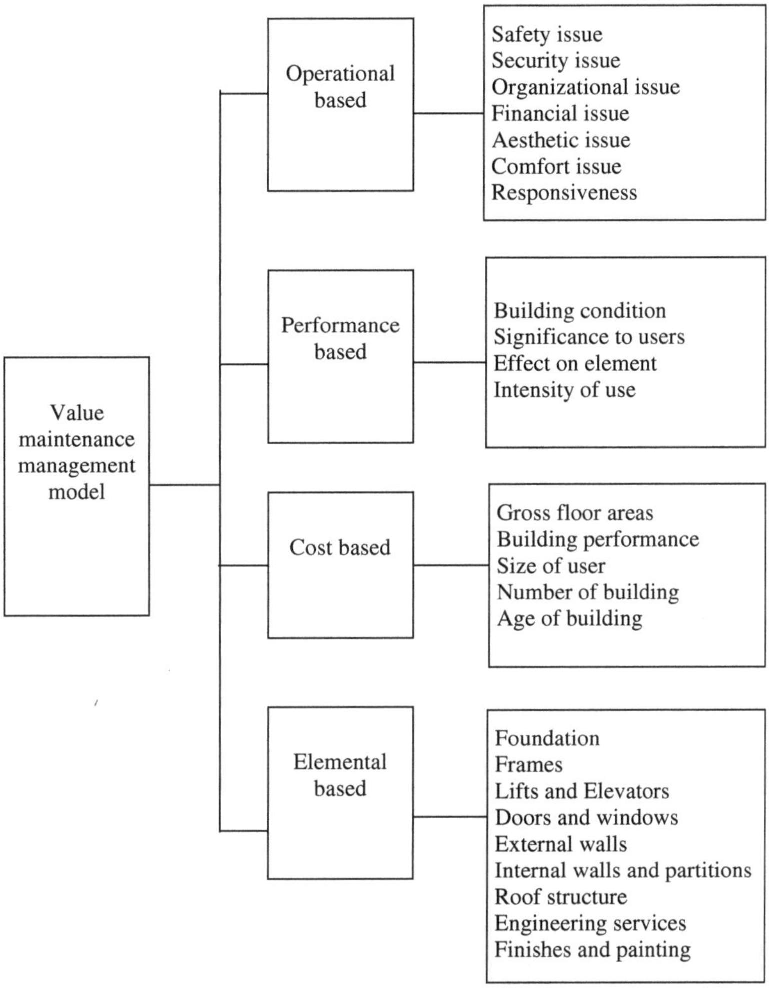

Fig. 6.2 Decision hierarchy of mathematical model for value maintenance management

context and characterised with factor, rational, concepts, causes, outcomes and assumptions.

Grounded theory is the method of organising or arranging data systematically to form concepts and themes. It actually gathers together all the isolated bits of empirical data into a coherent conceptual framework of wider applicability (Cohen et al. 2006). Though traditionally used in qualitative researches but its application in quantitative research has been recognised (Bell 2005). It is also commonly used in construction related researches. Reader interested in more details on the ground theory could refer to some of the books listed in end of this chapter.

Table 6.2 Interpretation of AHP measurement scale

Degree of importance	Definition	Interpretation
1	Equal importance	Two element making equal contribution to the goal
3	Somewhat more important	Moderate importance of element over the other element
5	Much more important	Essential or strong importance
7	Very much important	Very strong importance
9	Extremely important	Extreme importance
Scale, 2, 4, 6 and 8	Intermediate values	These are required when comprise between two adjacent judgment is needed
Reciprocals	If *v* is the judgment value when *i* is compared to *j*, then 1/*v* is the judgment value when *j* is compared to *i*.	

Table 6.3 Value of Random Consistency Index

Matrix Size	1	2	3	4	5	6	7	8	9	10
Average RI	0.00	0.00	0.58	0.90	1.12	1.24	1.32	1.41	1.45	1.49

6.13 Method of Validating Value Maintenance Management Model

Model comes in various forms. Model could be quantitative, graphical, mathematical, statistical, qualitative or tabular. However, regardless of the purpose or form of a model, it has one peculiar characteristic. It attempts to capture or represent a real life situation in a simplified form. In other words, a model endeavours to signify the complex of real-world situations in a more simplistic and potentially more easily understandable form. This is attained by developing the model so that it focuses on the key aspect of the situation and ignores the rest (Wisniewski 2006). There are no hard or fast rules in the developing of a model; inter alia, it should be developed based on the peculiar requirements of those that have a stake in the operation, application and effect of the model (Alexander and Brown 2006) and there are many ways to validate a model.

While some models lend themselves to be tested mathematically, some do not. Some can only be tested deductively, while others inductively. Whichever testing tool, is used, it must be meaningful to the research. In general, the approach to use could be an indicative of whether the research is exploratory, explanatory or hypothetical in nature. Research that depends solely on hypothesis testing would undoubtedly use a statistical tool in testing its hypothesis. While this current research combines explanatory, correlative and hypothesis, questions might be asked as to why not test the main research hypothesis statistically?

Ten maintenance managers were involved for their expert validations. Stakeholders from the MOHE were also involved with the validation. As the chief financier of the public universities and the main regulatory bodies for both the private and public universities, it is only natural that they were involved. And even if all the public universities in Malaysia were granted autonomy status (and which of course may be possible in say 20 years to come), the impact of the government will still be greatly felt, though less than how it used to be.

6.14 Summary

This Chapter has been able to describe and outlined the research design. Each of the methods of data collection was justified. The purpose of VMMM is to add value to buildings, increase user satisfaction and increase the maintenance organisation's productivity, quality and profits. It is one thing to be persuaded intellectually that better maintenance management could pay dividends, but it is another thing to believe that it is a strategic function. The top management must buy into it. The chief executive and other top management need to become convinced that value maintenance management is a top priority. Thus, a cost-benefit analysis calculation could be taken. However, many of the costs are hidden and many of the benefits are too intangible and uncertain. However, this does not mean that investment decisions must be based on faith and hope.

References

Alexander, K., & Brown, M. (2006). Community-based facilities management. *Facilities*, *24*(7/8), 250–268.

Bell, J. (2005). *Doing your research project: A guide for first-time researchers in education, health and social science* (4th ed.). England: Open University Press.

Bell, J. (1996). *Doing your research projects: a guide for the first time researchers in education and social science*. Buckingham: Open University Press.

Bell, J. (2001). *Doing your research projects: a guide for the first time researchers in education and social science* (3rd ed.). Buckingham: Open University Press.

Cohen, L., & Manion, L. (1994). *Research methods in education* (4th ed.). Oxon: RoutledgeFalmer.

Cohen, L., Manion, L., & Marrison, K. (2006). *Research methods in education* (5th ed.). Oxon: RoutledgeFalmer.

De-Marco, A., Ruffa, S., & Mangano, G. (2010). Strategic factors affecting warehouse maintenance costs. *Journal of Facilities Management*, *8*(2), 104–113.

El-Haram, M. A., & Horner, M. W. (2002). Factors affecting housing maintenance cost. *Journal of Quality in Maintenance Engineering*, *8*(2), 115–123.

Fallon, C. (1971). Value analysis to improve productivity cited in J. Kelly & S. Male.

Fellows, R. & Liu, A. (2003 and 2008). *Research methods for construction* (3rd ed.). Hoboken: Wiley-Blackwell.

Green, D., & Turrell, P. (2005). Schools buildings investment and impact on pupil performance. *Facilities, 23*(5/6), 253–261.

Hitchcock, G., & Hughes, D. (1995). *Research and the teacher: A qualitative introduction to school based research* (2nd ed.). London: Routledge.

Kelly, J., & Male, S. (2001). *Value management in design and construction*. New York: E & FN Spon.

Kelly, J., Male, S., & Graham, D. (2004). *Value management of construction project*. Oxford: Blackwell Publishing Company.

Leaman, A. (2003). User needs and expectations. In R. J. Cole & R. Lorch (Eds.), Building, culture and Environment: Informing localand global practices (pp. 157–176). Hoboken: Blackwell Science Limited.

Miller, G. A. (1956). The magical number seven plus or minus two: Some limits on our capacity for processing information. *Psychological Review, 63*, 81–97. In Ryu, S. Y (2005). Development of Usability Questionnaires for Electronic Mobile Products and Decision Making Methods, (Unpublished PhD Thesis, Virginia Polytechnic Institute and State University).

Naoum, S. G. (2007). *Dissertation research & writing for construction students*. Oxford: Butterworth-Heinemann.

Saaty, T. L. (1980). *The analytic hierarchy process*. New York: McGraw-Hill.

Saaty, T. L. (1990). *Multicriteria decision making: The analytic hierarchy process*. Pittsburgh, PA: RWS Publications.

Saaty, L. T. (2003). Decision-making with the AHP: Why is the principal eigenvector necessary. *European Journal of Operational Research, Issue, 145*, 85–91.

Shohet, I. M. (2010). Performance-Based-Maintenance of Public Facilities: Principles and Implementation. In Peter Barrett, Dilanthi Amaratunga, Richard Haigh, Kaushal Keraminiyage & Chaminda Pathirage (Eds.), *CIB 2010 World Congress Proceedings* (pp. 10–13). United Kingdom: The Lowry, Salford Quays, May 2010.

Sommerville, J., & MoCosh, J. (2006). Defects in homes: An analysis of data on 1,696 new UK houses. *Journal of Structural Survey, 24*(1), 6–21.

Wisniewski, M. (2006). *Quantitative methods for decision makers* (4th Ed.). England: Pearson Education Limited.

Wordsworth, P. (2001). *Lee's building maintenance management* (4th ed.). Oxford: Blackwell Science Limited.

Chapter 7
Analysis and Findings

Abstract This chapter is included in this book for you to have an understanding of some of the sources of information that go into the model. It is the information and knowledge gained in this chapter that is grounded into the value maintenance management model. Following are the objectives and learning outcomes of this chapter: (1) Understand the problems facing maintenance organisations. (2) Appreciate some detailed information on how university buildings are maintained. (3) Be able to explain know how and why maintenance are on the increase. (4) Be able to explain the process and strategies required to enhance methods and process of maintenance.

Keywords Data analysis · Results and findings · Criteria influencing building maintenance · Criteria of the user value system · Defects

7.1 Introduction

This chapter is concerned with the data interpretations and presentations of results. Data analysis was performed using three different computer packages: SPSS version 16.0, AHP and Microsoft Excel windows application package. These packages were combined to complement one another to facilitate analysis, interpretation and discussion. Whereas the SPSS was used to produce descriptive and inferential statistics, Microsoft Excel was used to produce illustrative statistics. The AHP was employed to demonstrate development of a mathematical model for maintenance cost distributions. Various questions were addressed to the respondents. Many of the questions or variables were control questions or variables. This chapter is presented in accordance with the questionnaire design and structured into two main segments. The first segment presents the findings of the maintenance organisations.

A.L. Olanrewaju and A.-R. Abdul-Aziz, *Building Maintenance Processes and Practices*, DOI 10.1007/978-981-287-263-0_7

In turn, the first part is organised into four parts, preceded by an introduction. The second segment dwells on the findings on the nature and characteristic of maintenance issues based on the users' perspectives.

Descriptive statistics were conducted on the data relating to the respondents profile and on other data that are concerned with achieving the aim of this research. Inferential statistics were performed on the data for further meaningful and insightful interpretations to provide robust significance for the research findings. Measures of goodness or fitness of the data were conducted through reliability and validity tests. A test score of more than 0.90 is statistically considered excellent; 0.70–0.90 is high; 0.50–07.0 is moderate, while below 0.50 depicts a low reliability or validity rate of the variables. Inferential statistics conducted includes cross-tabulation, chi-square, *t* test, ANOVA, MANOVA, regression analysis and factor analysis. These were to determine the significance of relationships between the variables or criteria or as the case may be.

The significance level is interpreted as an indicator of the strength of the relationships. The significance level for this research was set at 0.05 unless otherwise provided. For instance, if the calculated significance level is larger than 0.05, the null hypothesis is accepted. The lower the significance level, the stronger is the relationship between the variables or criteria. A test is considered extremely strong when the significance level of a test coefficient is $P < 0.0001$; it is very strong when significance level is $0.0001 < 1 < 0.001$, fairly strong when the significance level is $0.001 < p < 0.01$ and weak when $0.01 < p < 0.05$. When the significance level of the test coefficient is $p > 0.05$, the relationship is considered weak. The foregoing statistical parameters are consistent with the requirements of research in engineering and technology management.

Pareto principle was used to explain parts of the findings of this research. The principle is named after Wilfredo Pareto, a nineteenth-century Italian economist who found that wealth is not evenly distributed. He observed that few people have most of the money in a society. A corollary of this is that when several factors affect a situation, a few factors account for most of the impact. Quantitatively, the principle describes a phenomenon in which 80 % of variation observed in a process can be explained by about 20 % of the causes of that variation. A chart is used to illustrate Pareto principle. A Pareto chart is a compound of bars and linear graph. A Pareto chart can be used to show what the group agreed/disagreed, is important/not important or is influential/not influential practice or is behavioural in relation to a topic. Finding the 20 % of causes allows the most important cause of an issue to be addressed first. Example of the 80/20:

(a) Eighty percentage of maintenance expenditures can be explained by 20 % of defects.
(b) Eighty percentage of customer complaints arise from 20 % of the service delivery methods.

7.2 Summary of the Questionnaires from the Maintenance Organisations

A total of 50 questionnaires were sent out to the maintenance organisations of universities. Thirty-three useable questionnaires were received and analysed for this study. A plausible explanation as to why the entire completed questionnaires were useable was because the respondents were experienced. The fact that all the returned questionnaires were useable was also apparently because of the research design. Table 7.1 lists the distribution of the outcomes of the questionnaires sent and the ones returned.

The 33 received questionnaires represent a response rate of 66 %. This is considered very satisfactory for postal survey because 30 % is the common response rate for postal survey. For instance, Chan et al. (2009) drew valid conclusions from a response rate of about 2 % from a postal survey. Perhaps, this high response rate was possible because of the survey duration and the numerous reminders sent to the respondents.

7.2.1 *Characteristics of the Respondents*

Four questions were addressed to the respondents to elicit information on their profiles. The data analysis found that the majority (54.5 %) of the respondents that completed the questionnaires were engineers (Fig. 7.1). Many of them (30.3 %)

Table 7.1 Distribution of useable response rate of the questionnaire survey

Item	Total	Percentage
Sent	50	100
Total received	33	66
Total useable	33	66

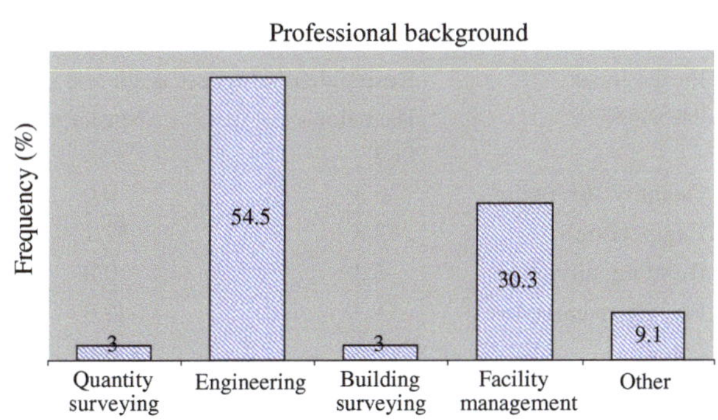

Fig. 7.1 Distribution of respondent professional background

were trained as facility managers. Only two of the respondents were quantity surveyors and building surveyors.

Seventy percentage of the respondents were degree holders (Fig. 7.2). In fact, close to half of the respondents held Bachelor degrees, while more than 21 % of the respondents held master's degrees. Theoretically, these results suggest that respondents have the capabilities to provide reliable feedback to the questionnaires. A considerable number of the respondents held other qualifications. The outcome of the survey revealed that 48.5 % of the universities were public, while the remaining 51.5 % were privately owned.

Table 7.2 displays the cross-tabulation between professional background and academic qualification. From the analysis, it was found that most of those possessing master's degrees were engineers. Similarly, most of those that had obtained bachelor's degree were engineers. It was interesting to find that many (42.9 %) of those that obtained master's degrees were facility managers.

A chi-square test was conducted to test the significance of the relationship between the respondents' professional background with the academic qualification. Chi-square test is appropriate for testing the significance of nominal data. However,

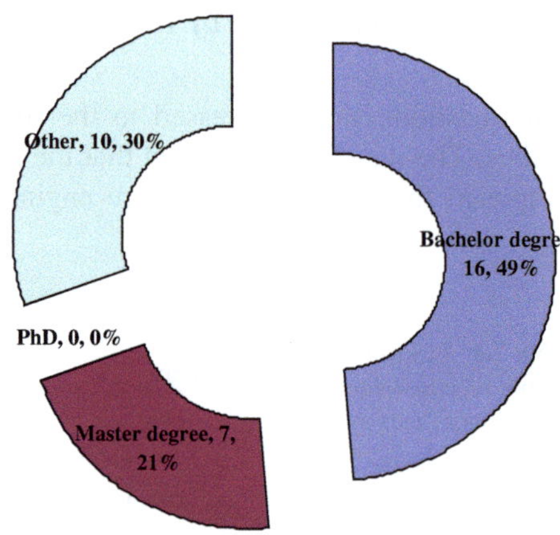

Fig. 7.2 Distribution of respondent highest academic qualification

Table 7.2 Cross-tabulation between professional background and highest academic qualification

Professional background	Respondent's highest academic qualification			Total (%)
	Bachelor's degree (%)	Master's degree (%)	Other (%)	
Quantity surveying	6.3	0.0	0.0	3.0
Engineering	43.8	57.1	70.0	54.5
Building surveying	6.3	0.0	0.0	3.0
Facility management	31.3	42.9	20.0	30.3
Other	12.5	0.0	10.0	9.1
Percentage of total	100.0	100.0	100.0	100.05

there is no evidence to conclude that the chi-square value of 4.483, with two degrees of freedom, is significant. In order words, professional background and academic qualification are not statistically related. To interpret, irrespective of the respondent's qualifications, he or she could belong to any of the professionals groups; however, these expected outcomes were found. It would naturally have been surprising if the research found any significant relationship.

On the basis of the respondent's position, 42.4 % of them were in other positions (Fig. 7.3). Among those of the other positions, who indicated their positions, were director of development and maintenance executives. Maintenance executive is synonymous to maintenance manager. Many of the respondents were maintenance managers (n = 10). Considerable sizes of the respondents were facility managers (30.3 %). Two of the respondents were general managers, and only one of the respondents was occupying the position of administrative manager. More than half of the respondents had being working for more than 5 years (54.4 %). About 46 % of the respondents had less than 5 years of working experience. Specifically, nearly 20 % of the respondents had more than 15 years working experience. Therefore, the respondents had satisfactory working experience to provide the required information.

Table 7.3 shows the cross-tabulation between the industrial experiences of the respondents with the respondents' positions in the organisation. The table indicates that half of the 10 maintenance managers had less than 5 years of working experience. It further shows that three of the facility managers had between 10 and 15 years of working experience. None of the general managers and the administrative manager had more than 10 years of working experience. Those occupying other positions also had their working experience spread to all the various levels.

Cross-tabulation analysis was conducted on the professional background and position of the respondents. Table 7.4 revealed that 10 of the 14 respondents that held a position other than those identified in the questionnaire actually had

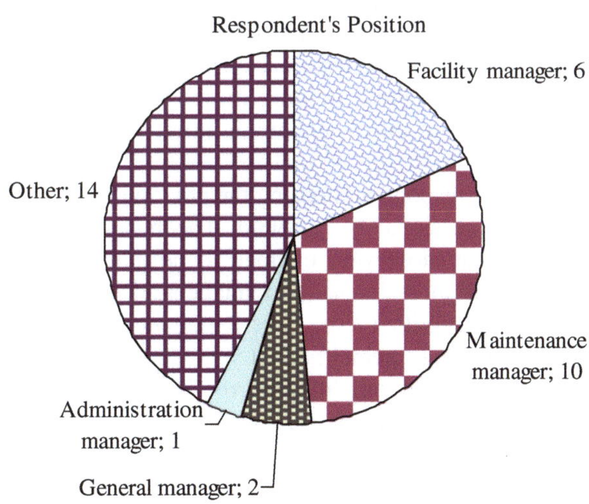

Fig. 7.3 Distribution of respondent position in the organisation

Table 7.3 Cross-tabulation between position and working experience

Working experience	Current position					Total
	Facility manager	Maintenance manager	General manager	Administration manager	Other	
Less than 5 years	2	5	1	0	7	15
5 years to 10 years	0	1	1	1	4	7
10 years to 15 years	3	1	0	0	1	5
15 years and above	1	3	0	0	2	6
Total	6	10	2	1	14	33

Table 7.4 Distribution of cross-tabulation between professional background and position

Respondent's latest professional background	Respondent's current position					Total
	Facility manager	Maintenance manager	General manager	Administration manager	Other	
Quantity surveying	0	0	0	0	1	1
Engineering	2	4	2	0	10	18
Building surveying	0	0	0	0	1	1
Facility management	4	4	0	0	2	10
Other	0	2	0	1	0	3
Total	6	10	2	1	14	33

engineering backgrounds. Interestingly, many (n = 4) of the facility managers had their professional background in facility management. Most (n = 4) of the maintenance managers had their background in engineering and facilities management. Also all (n = 2) those holding the position of general managers were engineers. Many (n = 2) of those occupying other positions had their background in facility management.

The only respondent holding the position of administration manager belonged to another professional background. The results were, however, interesting. It was expected that those who claimed to be maintenance managers would have their backgrounds in maintenance management or at best in either facilities or assets management. With these, their experiences in maintenance would be in doubt. Maintenance management is not yet a core course for the engineering programmes, at least at the undergraduate degree level. However, it is possible for the respondents to have attended some short courses or vocational training to augment their backgrounds.

ANOVA test was performed to determine whether the respondent's position depended on their academic qualification and professional background. The results indicated there were not any relations between the variables. In other words, the positions of the respondents were not determined by their academic qualifications and professional background. Theoretically, this was expected, however. It is possible to hypothesise that the position of a maintenance manager depends on his or her academic background and profession background. In this way, a round peg will be in a round hole. However, there is a moderate relationship between the three variables. This is indicated by the partial eta value of 0.311. This indicates a slightly more than 30 % relationship.

The value for the within-group sum of squares (intercept) is 0.672, which means that there is considerable variance in the respondents' positions. Variability of 67 % is high. Individually, academic qualification has a very weak effect, while the effect of professional background is modest. However, few universities in Malaysia now respond to the need of producing building maintenance graduates by incorporating into their curricula courses devoted to maintenance, and consequently, graduate engineers now actively involved in maintenance management problem solving have little training in maintenance.

7.2.2 Background Information

Many of the questions in this section were control questions. They were used to explain other questions or variables in the research. Specifically, this section collected information on the size of the built-up area, annual maintenance budget, method of estimating maintenance budget and ways of dealing with maintenance budget deficit. The variables also included maintenance procurement systems, current maintenance practices and information on user satisfaction survey. The other variables were the age of the buildings, size of the number of full-time employees, types of maintenance practices and method of receiving maintenance complaints from the building users.

Figure 7.4 revealed that about 50 % (n = 16) of the universities spent less than RM10,000,000 annually each on maintenance. Many (40.6 %, n = 13) of them

Fig. 7.4 Distribution of annual maintenance budget

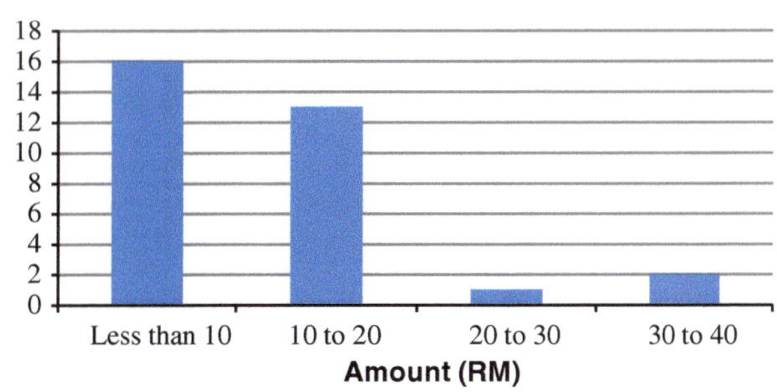

spent about RM15,000,000 to maintain their buildings. While one of the universities spent between RM20,000,000 and RM30,000,000, only two of them indicated spending between RM30,000,000 and 40,000,000 on maintenance. The mean score for the budget was found to be 1.66, while the standard deviation was 0.83 having variance of 0.68. A valid conclusion from the mean score value was on the average a university spent close to RM20 million in a year for maintenance for the previous 5 years.

ANOVA was performed to determine whether a relationship existed between the annual maintenance and size of the built-up area. It was expected that universities with large built-up areas would spend more on maintenance. While the post test was not performed for the size of the built-up area because one of the criteria had fewer than two cases, the table indicated there was no relation between the size of the built-up area and budget. However, the mean value is 95 % confidence, meaning the model value can significantly explain 95 % of the population value. However, a likely reason for this outcome is the sample size.

The 33 % response rate might be inadequate to provide the required effect for the test. Yet, theoretically, it would be expected that universities with large floor areas will absorb more funds compared to those that occupy smaller floor areas for maintenance. This is natural. However, this might not be so in the case of the buildings in large university that was newer and competently built and using quality materials and components. Nonetheless, considering the previous information, the last arguments might not be credible, however. There was no evidence to conclude that the materials and components were of the best quality and that the labourers engaged were the most skilled.

A cross-tabulation analysis was performed between types of university and annual maintenance budget, the outcomes of which were shown in Fig. 7.5. While we expect that the majority ($n = 9$) that spent less than RM10 million on maintenance annually are private, it was unexpected that most that invested between RM10 million and RM30 million would be private. The results also indicated that one private and one public university spent between RM30 million and RM40 million each on maintenance. The only university that spent RM20 million to RM30 million was a public university. However, a possible reason for these results could be that many or most of the private universities were operating in old

Fig. 7.5 Cross-tabulation university category and annual maintenance budget

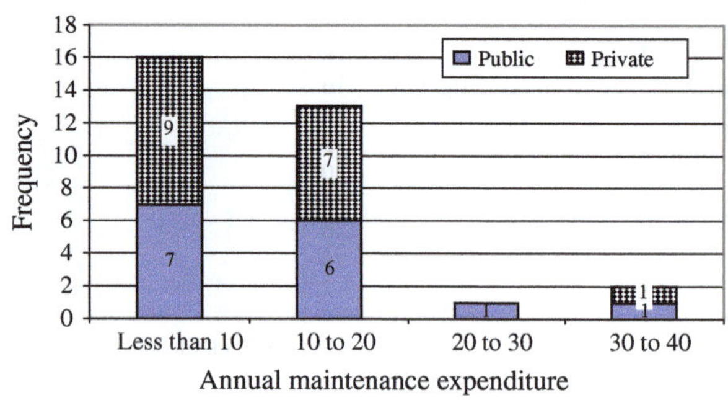

buildings or buildings that were not properly designed and/or built. Another reason could be that the maintenance systems of private universities were not effective and efficient. This is because previous analysis has shown that in terms of size of floor area the public universities were dominant.

Respondents were questioned regarding the methods they adopt in estimating their maintenance budgets. The results are listed in Table 7.5. The results indicate that estimates are based on the previous year's expenditure. Accordingly, for universities to determine their maintenance cost for the next year, they had to determine how much was spent in the current year and added a certain percentage, often in the region of 10–20 % to the preceding budget. Many of the maintenance organisations carried out an inspection to identify defects on the basis of which the budget estimate was established. A considerable proportion also based their estimate on how much the university actually gets.

But to base maintenance estimates on the previous year's expenditure is grossly inadequate to come up with realistic budgets. While it is also inappropriate to base maintenance estimates on the budget, it is also equally not very adequate to base the estimate on an inspection or condition survey. Unfortunately, only very few (5.45 %) of the respondents claimed to use other methods of estimate, although this group of respondents did not specify the method they claimed to use. The traditional approaches have been criticised for various shortcomings.

The only way modern organisations can survive and progress is to be more effective and efficient in the face of the pressures they have to confront. The competitive environments demand that the organisations must apply sophisticated approaches. Thus, the trend should be towards value-driven maintenance. Value is becoming a boardroom focus for organisations to be competitive and innovative. An organisation that puts *value drives* concept at the centre of its MS means that the organisation is viewing their processes from concepts through application and disposal from both the consumers' and providers' perspectives. Currently, maintenance and its management are treated as engineering issues. However, it should be entrepreneur led. This also underscores why the setting up of a maintenance department should be based on planning and foresight.

In Malaysia, maintenance expenditure for public universities was actually catered for by the Ministry of Higher Education. Maintenance expenditure does not depend on allocations in the real sense. This fact is corroborated by the results of how the maintenance organisations make up deficits (Table 7.6). More than 60 % of the respondents applied for funds from the management. Therefore, the result of this variable should be interpreted with caution.

Table 7.5 Distribution of method of estimating annual maintenance budget

Method	Frequency	Percentage
Based on previous expenditure	25	45.45
Carry out inspection	16	29.10
Based on budget	11	20
Other methods	3	5.45

Table 7.6 Distribution of method of making up maintenance deficit

Method	Frequency	Percentage
Apply for money from management	20	62.5
Cut money allocated for other purposes	10	31.3
Wait until next year's budget is released	2	6.3
Total	32	100.0

Most (66.7 %) of the maintenance organisations combined both in-sourced with outsourced work. However, many (21.2 %) of them actually outsource their maintenance services to external contractors. The percentage of those that used in-house services was 9.1 %. The number of universities that used only in-house labour for their maintenance services was 3. Based on the results of the cross-tabulation between procurement methods and category of university (Table 7.7), two of the three universities that used in-sourced services were private universities.

Interestingly, the majority of the universities that outsourced their maintenance services were private universities (Table 7.8). An equal number (n = 11) combined in-sourcing with outsourcing to procure maintenance services. One of the public universities claimed to use another method without mentioning the name of the other method. It is unfortunate, however, that most of the respondents procure maintenance by outsourcing. The shortcomings of outsourcing maintenance services have been well documented. Outsourcing maintenance always reduces maintenance procurement to a lump sum. However, the risk and uncertainty involved with maintenance services dictate that the traditional lump sum is undoubtedly unsuitable for maintenance works. This is the reality because the lump sum contracts demand that the work to be executed can be measured.

Table 7.7 Distribution of maintenance procurement systems

Method	Frequency	Percentage
In-house	3	9.1
Outsource	7	21.2
In-house and outsource	22	66.7
Other	1	3.0
Total	33	100.0

Table 7.8 Cross-tabulation on university category and maintenance procurement system

Respondent type of university	Maintenance procurement system preferred by respondent				Total
	In-house	Outsource	In-house and outsource	Other	
Public	1	3	11	1	16
Private	2	4	11	0	17
Total	3	7	22	1	33

Table 7.9 Distribution of average age of buildings

Age	Frequency	Percentage
Less than 10	10	31.3
10–20	13	40.6
20–30	6	18.8
30–40	1	3.1
40–50	2	6.3
Total	32	100.0

Based on the results on the age of the buildings, most (40.6 %) of their buildings are about 15 years old (Table 7.9). This is followed by those that are less than 10 years old (31.3 %). Fewer than 7 % of the buildings are between 40 and 50 years. While the standard deviation was 1.11, the mean score value for buildings' age was 2.1. These results were interpreted to imply that in reality some buildings were older, while some were relatively newer. The mean score value shows that on the average the age of all the buildings ranges between 20 and 30 years (i.e. 25 years).

The outcomes of this research indicated that virtually all the universities procured their maintenance services by combining in-sourcing with outsourcing (Table 7.10). Seven of the 10 universities with buildings less than 10 years old combined their procurement systems. Similarly, eight of the thirteen universities with buildings of about fifteen years old (10–20) also combined their procurement system. The same also goes for universities with buildings between 20 and 30 years old. However, most ($n = 4$) of the universities that outsource all their maintenance services were the universities with buildings between 10 and 20 years old. Only one each of the universities with buildings less than 10 years old, 10–30 and 30–40 years old carried out maintenance services using their in-house workforces.

It is interesting to find that universities with buildings older than 30 years did not outsource all their maintenance services. However, from the ANOVA test results, there was not statistical evidence to conclude that age of buildings depends on procurement system (F (4, 1) = 0.728, $p = 0.581$) or procurement systems depend on age of buildings (F (2, 1) = 0.373, $p = 0.773$).

Table 7.10 Distribution of cross-tabulation between ages of buildings and the procurement system

Age of building	Maintenance procurement system				Total
	In-house	Outsource	In-house and outsource	Other	
Less than 10	1	1	7	1	10
10–20	1	4	8	0	13
20–30	0	2	4	0	6
30–40	1	0	1	0	2
40–50	0	0	2	0	2
Total	3	7	22	1	33

A cross-tabulation analysis indicated that 70 % of the universities with buildings less than 10 years old were private universities. More than 60 % of the universities operating in buildings older than 10 years but less than 20 years were private universities. The results also showed that five of the six universities operating in buildings between 20 years and 30 years were public universities. One university each for private and public university operated in buildings older than 30 years but less than 40 years old. Interestingly, all the universities operating in buildings more than 40 years old were publicly owned. What this analysis pointed to was that private university's buildings were not that old. These are interpreted to mean that the private universities that claimed to be operating in buildings older than 20 years were not actually operating in their buildings. For instance, UTP purchased substantial part of its building facilities from USM. Those buildings were previously used as a satellite campus by the USM. They were more than 25 years old and apparently older than UTP itself.

A question was designed to capture size of the full-time employees in the maintenance organisation, the results of which are displayed in Table 7.11. The table shows that about 60 % of the universities employed fewer than 30 staff in the maintenance department. Nineteen percentage of the universities employed between 140 and 170 staff in the maintenance departments. Nearly, 6 % of the universities employed between 60 and 90 and 110–140 employees each, respectively. Fewer than 10 % of the universities employed roughly 45 staff. From the mean score values (2.41) and the standard deviation (2.05), it was determined that on the average, a university employed 75 staff and operatives in their maintenance department.

Taking into account the size of the built-up areas, these outcomes can be interpreted to mean that most of the universities actually outsourced their maintenance service rather than in-sourced it. Alternatively, this argument can be supported considering the annual maintenance expenditure. As an illustration, more than 70 % of the universities spent more than 5 million each for maintenance annually. In actual fact, half of the universities spent more than 10 million each for maintenance. If maintenance services were to be procured through in-sourcing, there is no way that the 30 or even the 60 full-time employees will be able to prudently use this amount. The amount is too much for them to be able to make use of in-house servicing. So therefore, it can be concluded that it is very likely that outsourcing is the most favoured procurement system.

Table 7.11 Distribution of full-time employees in the maintenance organisation

Employee	Frequency	Percentage
Less than 30	19	59.4
30–60	3	9.4
60–90	2	6.3
110–140	2	6.3
140–170	6	18.8
Total	32	100.0

The ANOVA values (F (4, 1) = 2.286, p = 0.86) with mean square of 1.342, 4 degree of freedom and F value of 2.286 indicates there is no statistical evidence to conclude that maintenance depends on age of buildings (0.086), meaning that regardless of the age of the buildings, the maintenance budget will not be significantly varied.

Table 7.12 also revealed the number of employees is not related to the maintenance budget. However, based on information contained in Table 7.13, the partial eta squared value for size of employees and age of building are 0.113 and 0.205, respectively. These show that the effects of sizes are modest. The value for the intercept is 0.78, which means that there is a lot of variance in the universities' budgets for maintenance. The table further demonstrates that the three variables cannot collectively explain the size of the maintenance budget that actually has considerable effect. The significant value of statistical analysis is actually a factor of

Table 7.12 Distribution of ANOVA test between maintenance budget and size of employees in organisation

Source	Type III sum of squares	df	Mean square	*F*	Sig.
Corrected model	2.709(a)	4	0.677	0.958	0.447
Intercept	37.044	1	37.044	52.380	0.000
Employee	2.709	4	0.677	0.958	0.447
Error	18.388	26	0.707		
Total	105.000	31			
Corrected total	21.097	30			

Table 7.13 Distribution of ANOVA test whether maintenance budget can be predicted by age of building, size of employees in the organisation and age of buildings on campus

Source	Type III sum of squares	df	Mean square	*F*	Sig.	Partial Eta squared
Corrected model	10.063(a)	17	0.592	0.697	0.760	0.477
Intercept	38.072	1	38.072	44.859	0.000	0.775
Size	0.308	4	0.077	0.091	0.984	0.027
Employee	1.399	4	0.350	0.412	0.797	0.113
Age of building	2.844	4	0.711	0.838	0.525	0.205
Size * employee	0.000	0	0.0	–	–	0.000
Size * age of building	0.020	1	0.020	0.023	0.882	0.002
Employee * age of building	0.000	0	–	–	–	0.000
Size * employee * age of building	0.000	0	–	–	–	0.000
Error	11.033	13	0.849			
Total	105.000	31				
Corrected total	21.097	30				

Table 7.14 Table cross-tabulation of full-time employees in the organisation and age of buildings

Size of employee	Age of buildings					Total
	Less than 10	10–20	20–30	30–40	40–50	
Less than 30	8	10	1	0	0	19
30–60	1	1	0	0	1	3
60–90	0	1	1	0	0	2
110–140	0	0	1	0	1	2
140–170	1	1	3	1	0	6
Total	10	13	6	1	2	32

the effect of size and sample size. Technically, the effect size measures the strength of a relationship between variables.

The ages of the buildings were also analysed with the size of the full-time employees in the maintenance organisations. The results of the cross-tabulation results are depicted in Table 7.14.

Nineteen of the universities that have less than 30 employees in their maintenance organisations were actually managing buildings between 10 and 20 years old. University employing between 30 and 60 employees is managing buildings less than 10 years old, 10–30 and 40–50 years old each, respectively. Most of the universities with employees between 140 and 170 managed buildings between 20 and 30 years old. What this means is that less numbers of employees managed buildings less than 20 years old. In other words, most of the universities actually outsourced their maintenance services. For instance, it is obvious from Table 7.14 that universities with buildings more than 20 years old actually have more than 110 employees each in the maintenance organisation. However, these results are not unexpected. Universities and probably many other places actually outsourced their maintenance to contractors. Universities often view maintenance as non-core services to their primary objectives, a perception which is unfortunate, however.

Furthermore, to determine whether there is statistical evidence to conclude that age of buildings and size of the number of full-time employees are related, the two variables were tested through ANOVA, the results of which are displayed in Table 7.15. The results revealed that a relationship exists between the ages of the buildings and the number of full-time employees in the maintenance organisations (0.003). The table further shows a moderate effect (0.434). This means that the age of the buildings can significantly explain about 45 % of the size of the numbers of employees in the maintenance organisations surveyed.

Figure 7.6 summarises the cross-tabulation between types of university with the size of maintenance employees. Expectedly, the majority ($n = 12$) of the universities that employed less than 30 employees were private university. Similarly, two of the three universities that employed fewer than 60 employees but more than 30 employees were private universities. All of the universities that employed 60–140 universities were publicly owned universities. Interestingly, while the majority ($n = 4$) of public universities employed more than 140 employees but fewer than

Table 7.15 Distribution of ANOVA test between size of employees and age of buildings

Source	Type III sum of squares	df	Mean square	*F*	Sig.	Partial eta squared
Corrected model	56.242(a)	4	14.060	5.167	0.003	0.434
Intercept	160.757	1	160.757	59.072	0.000	0.686
Age of building	56.242	4	14.060	5.167	0.003	0.434
Error	73.477	27	2.721			
Total	315.000	32				
Corrected total	129.719	31				

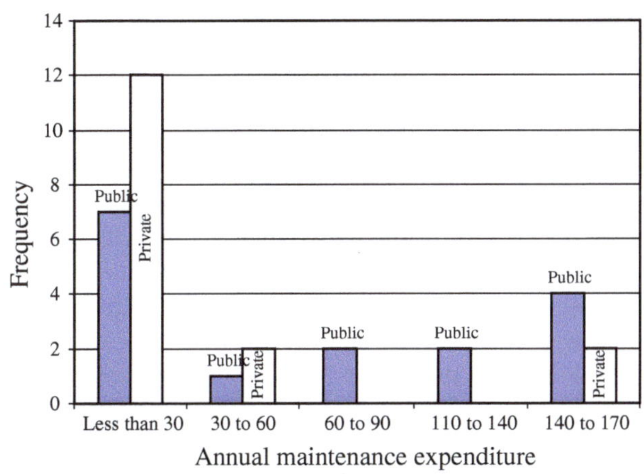

Fig. 7.6 Cross-tabulation on category and size of full-time employee

170, two private universities also employed the same number of employees. These findings were, however, not unexpected. It was expected that public universities considering the size of their buildings will employ more employees.

Another meaning to these findings was that, in reality, private universities actually outsourced their maintenance services to external contractors. This argument is valid as private universities, do not consider building maintenance as contributing to their corporate objectives. In other words, to them, maintenance is a non-core function. Private universities often had it at the back of their minds that maintenance is a 'necessary devil' and as such prefer to outsource it. However, this is a narrow perception.

Table 7.16 shows the maintenance practices. Most of the maintenance practices are corrective and cyclical. Most of them do not carry out condition and inspection-based maintenance. Predictably, an equal proportion of the respondents claimed their maintenance practices are inspection based. By the same token, the majority of

Table 7.16 Distribution of maintenance practices

Maintenance practice	Frequency	Percentage
Corrective/reactive based	8	24.24
Cyclical/planned based	7	21.21
Condition based	6	18.18
User based	5	15.15
Inspection based	7	21.21
Other	0	0.00
Total	33	100

the universities do not practice the user maintenance system. In most of the universities, corrective-based maintenance was found to be practiced. This outcome actually confirmed our undisclosed hypothesis that the maintenance practices employed by the universities are corrective, cyclical and inspection based. However, these maintenance practices are inadequate to lead to optimum service delivery. These practices reduce the objective of maintenance to that of the condition of the buildings rather than the buildings' performance. On the contrary, it is the performance of the building that the users are concerned about, not the condition of the building.

However, this is one of the fundamental problems facing the maintenance organisations. All be it, it is noted that organisations will carry out corrective maintenance on some selected building elements, but in a situation where most of the elements are maintained correctively is grossly inadequate. Under such arrangement, the ever inadequate funding would not stack up to the maintenance backlogs. Arguably, this will lead to poor user satisfaction and unnecessary increase in maintenance costs. In addition, this is one of the failings of outsourcing maintenance, particularly where it is not adequately managed. Where the outsourcers failed to manage the procurement process, the contractors in an attempt to maximise their profits suspend or carry out the maintenance till when it suits them. This is often the case, where the labour is not available to manage all the complaints requests.

The outcomes of the survey portrays that two of the three universities that prefer in-sourcing all their maintenance services rated their current practice as very good and good (Table 7.17). None of those that outsourced all their maintenance services rated their procurement systems as very good. However, the majority (n = 21) of those that combined in-source with outsource rated their system good (n = 11) and fairly good (n = 10). One of the universities claimed they preferred other procurement systems for maintenance. The conclusion that can be drawn from these results is that in-sourcing is the best option to procurement maintenance services, given that only one of the 22 universities that combined outsourcing with in-sourcing considered their practice as very good.

This research also seeks to determine how maintenance complaints are channelled to the maintenance organisations. The three most frequently used means of receiving maintenance complaints are by telephone, email and personal visits. Only a few of the complaints were mostly sent by fax and postal mail. Few users also claimed they sent their complaints by other methods, though these groups of the

Table 7.17 Distribution of cross-tabulation of between procurement and rating of maintenance practice as rated by the maintenance organisation

Maintenance procurement strategy	How organisation rate their current practice				Total
	Very good	Good	Fairly good	Do not know	
In-house	1	1	0	1	3
Outsource	0	2	5	0	7
In-house and outsource	1	11	10	0	22
Other	0	1	0	0	1
Total	2	15	15	1	33

respondents did not indicate the other methods used. Most (87.5 %) of the respondents did not send their complaints by fax. This outcome was, however, expected. The building users voiced a preference for methods that allow them to report their complaints simply. From theory and practice, users will prefer simple and user-friendly methods as compared to more complex methods. However, it was surprising that the respondents rank lodging complaints by email highly. We are sceptical about interpreting these findings. Further research is required to question the users themselves as to whether they actually believe making complaints through email is effective. Sending an email requires an Internet connection which is not available to all students, at least in their hostels.

It is natural for building users to send their complaints through phone calls, and personal visits. Those methods are perhaps easier and cheaper compared to sending messages through fax and the post. Users also report their complaints through the warden and designated person in charge of maintenance. A Web-based method of reporting is undoubtedly not widely practiced yet. For this method to succeed, there is a need for the users to be connected to the Internet. With the problem associated with Internet, it could be frustrating for users or even the maintenance organisation to access their mail. Many of the users do not even have access to Internet in their rooms and many that do not even own computers. In all likelihood, building users could not be expected to use cybercafes to lodge their maintenance complaints. This will be cumbersome and crude. This can only happen if that is the only alternative; otherwise, it would not be appropriate for frustrated building users.

A Pareto analysis performed on the methods of receiving maintenance complaints indicated that actually, five of the seven methods constitute 80 % of all methods (Fig. 7.7). Postal and other methods were least used to send maintenance complaints. Therefore, maintenance organisations could combine the five methods to optimise the complaints requisitions strategy.

The maintenance organisations were asked to rate their current maintenance practices. This is a self-assessment approach. It is intended to garner how the providers rate the quality of their services to the consumers. This is a control variable. Of the total respondents, only two rated their maintenance management as very good. Forty-six percentage of the universities rated their maintenance practices

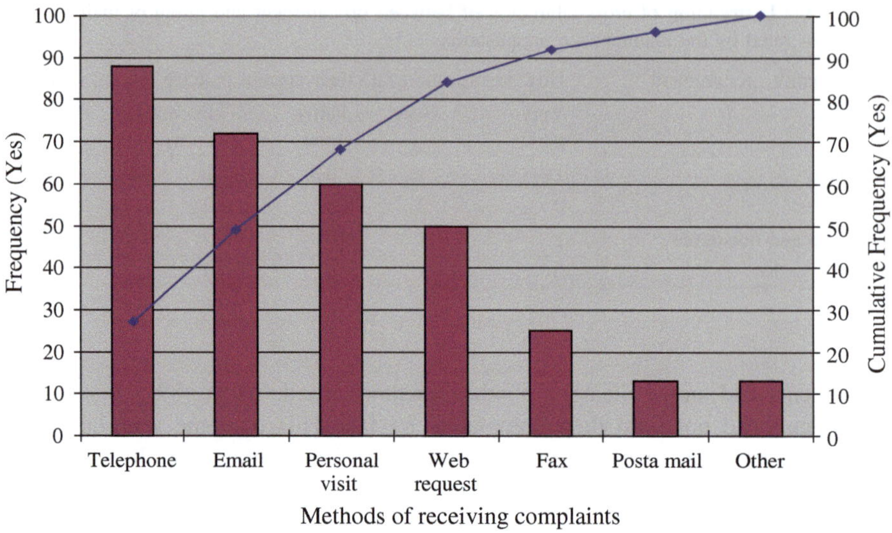

Fig. 7.7 Pareto analyses of methods of receiving maintenance complaints

good or fairly good. Only one of the universities did not know how to rate its maintenance practice (Table 7.18).

Figure 7.8 displays the results of universities that conducted their own users' satisfaction survey with respect to their maintenance practices. Consumers and service providers always have different perceptions with regard to value of a service. Put in other words, a perceived gap exists between the services that organisations provide compared with the services that end-users want. This is a control question; much of the discussions are dealt with in the second section of this thesis.

Consumers combine their experience with their expectations to define their wants. Need and want constitute a value. Through surveying users' satisfaction, improvement can be made to the maintenance practices. Apparently, it is only the users who know what they want. A successful maintenance management system depends on achieving value (for money invested) where value is characterised by cost (of production) and functional performance and satisfaction. Figure 7.8 revealed that the majority (54.8 %) of the universities ticked yes, meaning they had conducted the survey. Nearly 40 % of the universities had not conducted user satisfactions before. Two of the universities did not know whether they had conducted user satisfaction surveys before or not. Although, most of the universities had conducted users' satisfaction surveys, a number nearly equal size had not. However, this outcome is not surprising. Universities did not conduct satisfaction surveys often and those that did often asked questions that did not actually capture the real situation of their service. The questions asked were designed to suit the management.

Universities were asked to provide their opinions on how satisfied they considered their users to be with the maintenance services they provide to them. This was also self-assessment question and thus a control question. The results of the

Table 7.18 Distribution of rating of maintenance practices

Rate	Frequency	Percentage
Very good	2.0	6.1
Good	15.0	45.5
Fairly good	15.0	45.5
Not good	0.0	0.0
Do not know	1.0	3.0
Total	33	100.0

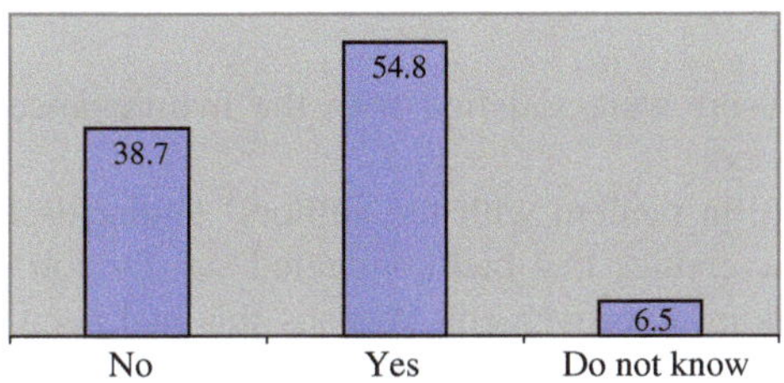

Fig. 7.8 Distribution on whether university conducts user satisfaction survey

survey are shown in Table 7.19. Sizeable (30.3 %) numbers of the respondents failed to respond to this question. However, the majority (78.3 %) of the universities that responded believed that the users were "satisfied". Only one of the universities that responded believed that the users were extremely satisfied with their maintenance practices. These outcomes are not unexpected. With the streams of complaints that the maintenance organisations received, this clearly suggests that the users were not very satisfied. This outcome could be regarded as honest opinions coming from the service providers. It is, however, believed that the users will probably rate the services they received less satisfactory.

This research found out that most (n = 15) of the universities that claimed their users were satisfied with the maintenance services they received had conducted user satisfaction surveys (Table 7.20). Only two universities each claimed to have conducted the survey and that their users were extremely satisfied and very satisfied with the maintenance service. Considerable size (n = 3) of the universities that

Table 7.19 Distribution of level of satisfaction of the users with service as rated by the maintenance organisation

Satisfaction	Frequency	Percentage
Extremely satisfied	1	4.3
Very satisfied	3	13.0
Satisfied	18	78.3
Do not know	1	4.3
Total	23	100.0

Table 7.20 Distribution of cross-tabulation between conducting users' satisfaction survey and degrees of user satisfaction

Whether organisation conduct users' satisfaction survey	Level of users satisfactions with service delivery			Total
	Extremely satisfied	Very satisfied	Satisfied	
No	0	0	3	3
Yes	1	1	15	17
Do not know	0	2	0	2
Total	1	3	18	22

indicated that their users were satisfied with the maintenance had not conducted user satisfaction surveys.

These results were in tandem with the authors' undisclosed hypothesis. It was expected that the universities that had conducted satisfaction surveys would have opportunities for making improvement. Perhaps this had been the case and thus is reflected in these outcomes. The outcomes of the users' satisfaction surveys were used to benchmark services that organisation provided. However, it is pertinent that the variables that facilitated user satisfaction and improved productivity were actually included in the survey. Two universities that claimed their users were very satisfied with the service did not know whether they conducted the survey or not. Often, where experts are not engaged to design questionnaires, the variables are often tilted to suit the organisation perception rather than reflect the reality. In that case, management is only told what they want to hear rather than what they need to hear.

In theory and in practice, building age, condition of building and size of floor area can predict maintenance costs. In order to test the hypothesis, a multiple regression analysis was done. The results show that the R^2 value is 0.022. The F value of 0.328 is not significant, however. Although, based on the results obtained, the variables are not good predictors to model maintenance costs. However, the outcomes are illustrative. From the results, it was clear that to reduce maintenance, the size of the floor areas must be reduced. However, the results indicated that to reduce maintenance, age of building is not factor. This aspect of the results is difficult to explain, however. Nevertheless, a likely reason could be attributed to the sample size.

7.2.3 Analysis of Criteria Influencing Maintenance Management of Buildings

This section was designed in order to achieve the first objective. That is to analyse the criteria that influenced the maintenance management of the buildings. The criteria in this section were designed to identify, evaluate and assess the criteria that maintenance organisation considered influential to building maintenance. The maintenance personnel were asked to grade 31 criteria in the order of their influence on maintenance. The criteria were major sources of maintenance-related issues or problems in the organisation. It was taken into account that the criteria were considerably capable of contributing to increasing service delivery, user satisfaction and reducing maintenance backlogs.

The requirements of each of the criteria have been described in Chap. 5 of this thesis. A total of 31 criteria were identified and addressed to respondents to either agree or disagree on the rate at which each of the criteria influenced maintenance management of the buildings. The list of the criteria was drawn from extensive literature and discussions with those concerns with the management of the buildings. The list of the 31 criteria may not have been exhaustive due to the complexities associated with maintenance management. The list covered the significant criteria that influenced maintenance management. Provisions were made for the respondents to include and rate any other criteria that influenced building maintenance. However, it was hypothesised that the criteria addressed to the respondents could significantly explain or influence maintenance management of the buildings.

Prior to analysing the criteria and the strength of the data, the "goodness of fit" was determined. In order to determine the strength of the data, reliability and validity tests were performed. There are many types of reliability tests that can be performed on data. For instance, Cronbach's alpha, split-half reliability coefficient, parallel form reliability and test–retest reliability can be performed. In case of validity of data, the following validity tests can be determined for data. For instance, criterion-related validity, convergent and discriminant validity are the types of available validity tests. For this purpose, Cronbach's alpha test was performed.

The reliability test results indicated that the Cronbach's alpha for all the criteria is very satisfactory (Table 7.21). However, the first three criteria were the weakest criteria because their correlation with the overall reliability score were in actual reality negatives. Their removal from the analysis would see the 0.923 figure rise to 0.934. Thus, the internal consistency reliability of the measure used could be considered excellent. The closer the reliability coefficient gets to 1.0, the better. The reliability results also revealed that the criteria were reliable measures. The values varied from 0.909 to 0.924 for the criteria. Average factor extracted of 0.8080 was found, which indicated a very good overall validity.

In order to assess, evaluate and rank the criteria influencing the maintenance management of the buildings, analyses were performed on the data. The degree of influence of each of the criteria was determined by the weight–age mean and mode score tests of each of the criteria. From the individual mean score tests, the average weight–age mean score was computed. Mean is the average or mid-score of a

Table 7.21 Distribution of reliability statistics for criteria influencing maintenance management

Cronbach's alpha	N of items
0.923	31

distribution. It is determined to explain the varying values or opinions of respondents. It is used to calculate the average of observations. The mean is the most stable of the three measures of average scores (Steinberg 2008). The standard deviation is also calculated to determine the level of spread of each of the individual values from the mean score. Therefore, if the mean score falls between 1.0 and 1.5, the criterion is considered not influential at all. See Table 7.22 for other distributions. This cut-off point is used, because the lowest possible mean score is 1. This also applies to the "extremely urgent" scale. The maximum possible score is 5.

Table 7.23 displays part of the descriptive statistics test scores for each of the criteria. It shows the frequency, mode scores, and the minimum and maximum scores. It further depicts the standard deviation and the variance scores and the skewness. Based on the information contained in the table, it was obvious that with the exception of few of the criteria that two respondents did not answer, all the other criteria were all answered. This can be verified from the frequency score. The maximum frequency score for all the criteria was 33, and all the respondents provided their viewpoints.

Apart from the problem associated with in-house procurement having a maximum score of 4, the maximum score for the other criteria was 5 each. The minimum score values for all the criteria were either 1 or 2. The maximum score value was either 4 or 5 even though, largely, 5 is actually dominant. These results were interpreted to mean that at best all the criteria could influence the maintenance of the buildings in Malaysia to a certain degree. The standard deviation score indicated that most of the criteria means clustered around the mean score value.

Experiencing problems with in-house staff, information on existing building performance, deterioration due to age of buildings and problems associated with outsourcing were, however, more close to the mean score application of building maintenance management software; involvement of maintenance personnel during design and lack of maintenance programmes were far from the average mean score. The standard deviation indicated that distribution was near to normal. With the exception of some few criteria, the standard deviations of most are less than one each. From Table 7.24, it was clear that most of the criteria were negatively skewed; the left tail was heavier than the right tail. This meant that the bulk of the criteria were very influential or extremely influential. Therefore, the mean and median mode scores in the data set were very similar.

Table 7.24 displays the weight–age mean tests for the criteria. It also displayed the overall mean score on the tests across the criteria. It further depicted the ranks of

Table 7.22 Distribution of index evaluation metric for criteria influencing maintenance management

Scale	Evaluation (interpretation)
1.00–1.50	Not at all influential
1.51–2.50	Not very influential
2.51–3.50	Influential
3.51–4.50	Very influential
4.51–5.00	Extremely influential

Table 7.23 Distribution of the descriptive statistics for the criteria influencing maintenance management

Criteria	Frequency	Minimum	Maximum	Sum	Mode	Standard deviation	Variance	Skewness
Use of the buildings	31	1.00	5.00	105.00	3.00	0.99139	0.983	−0.403
Budget constraints	31	2.00	5.00	126.00	4.00	0.98281	0.966	−0.454
Information on user performance satisfactions	31	2.00	5.00	97.00	3.00	0.78224	0.612	0.375
Inadequate training and development of staff	31	2.00	5.00	115.00	3.00	0.79535	0.633	0.449
Use of poor quality components and materials	31	2.00	5.00	123.00	3.00	0.88388	0.781	0.024
Lack of successful maintenance programmes	31	1.00	5.00	101.00	4.00	1.14399	1.309	−0.125
Bad budgeting control	31	1.00	5.00	103.00	4.00	1.11124	1.235	−0.398
Lack of maintenance performance indicators	31	1.00	5.00	102.00	3.00	0.94748	0.898	−0.425
Application of building automation system	31	2.00	5.00	97.00		0.93272	0.870	0.444
Lack of skilled personnel in maintenance department	31	1.00	5.00	115.00	4.00	0.97215	0.945	−0.498
Natural deterioration due to age and environment	31	2.00	5.00	119.00	4.00	0.86384	0.746	−0.349
Using in-house workforce	31	1.00	4.00	99.00	3.00	0.75000	0.563	−0.473
Frequent shortage of materials and spare parts due to absence of efficient inventory system	31	1.00	5.00	94.00	3.00	1.06423	1.133	0.320
Using outsource workforce	31	2.00	5.00	101.00	3.00	0.86384	0.746	0.498
Involvement of maintenance experts during design stage	31	1.00	5.00	118.00		1.19975	1.439	−0.479
Workers' motivation	31	2.00	5.00	104.00	3.00	0.87744	0.770	0.160
Lack of trust and confidence among staff	31	2.00	5.00	96.00	3.00	0.87005	0.757	0.129
Application of building maintenance management software (i.e. CMMS)	31	1.00	5.00	94.00	3.00	1.13970	1.299	−0.067
Lack of effective communications	31	2.00	5.00	104.00	3.00	0.91464	0.837	0.321

(continued)

Table 7.23 (continued)

Criteria	Frequency	Minimum	Maximum	Sum	Mode	Standard deviation	Variance	Skewness
Lack of maintenance manual	31	2.00	5.00	104.00	3.00	0.76200	0.581	0.000
Effective organisation structure	31	2.00	5.00	112.00	3.00	0.87988	0.774	0.000
Lack of maintenance standard	31	2.00	5.00	104.00	3.00	0.91581	0.839	0.269
Lack of top management support	31	1.00	5.00	113.00	3.00	1.01550	1.031	−0.682
Competent maintenance manager	31	1.00	5.00	111.00	3.00	1.10671	1.225	−0.754
Complexity of design	31	1.00	5.00	95.00	3.0	0.92864	0.862	−0.401
Availability of building and services register	31	1.00	5.00	101.00	3.00	1.01947	1.039	0.253
Lack of clear maintenance objectives	31	1.00	5.00	97.00	3.00	1.06208	1.128	−0.065
Information on existing performance of buildings	31	2.00	5.00	106.00	3.00	0.82060	0.673	0.090
High users' expectations	31	2.00	5.00	105.00	4.00	0.84370	0.712	−0.154
Improper response time	31	2.00	5.00	108.00	3.00	0.83280	0.694	0.246
Poor complaint reporting systems	31	1.00	5.00	108.00	3.00	1.03954	1.081	−0.287

Table 7.24 A summary of distribution of frequency, mean and ranking of the criteria influencing maintenance management

Criteria	Level of influence					Weight–age mean	Rank
	1 (%)	2 (%)	3 (%)	4 (%)	5 (%)		
Use of the building	6.3	9.4	43.8	31.3	9.4	3.2813	16
Budget constraints	0.0	12.1	21.2	39.4	27.3	3.8182	2
Information on user performance satisfactions	0.0	25	50	21.9	3.1	3.0313	29
Staff training/ development	0.0	6.1	51.5	30.3	12.1	3.4848	8
Quality of components and materials	0.0	3.1	37.5	31.3	28.1	3.8438	1
Maintenance programme	9.1	24.2	27.3	30.3	9.1	3.0606	24
Budgeting control	9.1	21.2	24.2	39.4	6.1	3.1212	20
Maintenance performance indicator	6.1	18.2	39.4	33.3	3.0	3.0909	22
Building automation system	0.0	34.4	34.4	25.0	6.3	3.0313	27
Lack of skilled personnel	3.0	12.1	30.3	42.4	12.1	3.4848	7
Deterioration due to age and environment	0.0	12.1	27.3	48.5	12.1	3.6061	3
Problem associated with in-house workforce	3.0	18.2	54.5	24.2	0.0	3.0000	30
Shortage of materials and components	9.1	27.3	42.4	12.1	9.1	2.8485	31
Problem associated with outsourcing	0.0	27.3	45.5	21.2	6.1	3.0606	25
Involving maintenance expert during design stage	6.1	12.1	27.3	27.3	27.3	3.5758	4
Worker motivation	0.0	16.1	41.9	32.3	9.7	3.3548	14
Trust and confidence among staff	0.0	29.0	35.5	32.3	3.2	3.0968	21
Application of maintenance management software	9.7	22.6	32.3	25.8	9.7	3.0323	26
Effective communication	0.0	16.1	45.2	25.8	12.9	3.3548	13
Maintenance manual	0.0	15.6	46.9	34.4	3.1	3.2500	18

(continued)

Table 7.24 (continued)

Criteria	Level of influence					Weight–age mean	Rank
	1 (%)	2 (%)	3 (%)	4 (%)	5 (%)		
Organisation structure	0.0	12.5	37.5	37.5	12.5	3.5000	6
Maintenance standard	0.0	21.9	40.6	28.1	9.4	3.2500	17
Top management support	6.3	3.1	37.5	37.5	15.6	3.5313	5
Competency of maintenance manager	9.4	3.1	34.4	37.5	15.6	3.4688	9
Complexity of design	6.5	16.1	45.2	29.0	3.2	3.0645	23
Building service register	3.1	21.9	43.8	18.8	12.5	3.1563	19
Clear maintenance objectives	9.4	15.6	46.9	18.8	9.4	3.0313	28
Information on existing building performance	0.0	15.6	43.8	34.4	6.3	3.3125	15
User expectation	0.0	16.1	35.5	41.9	6.5	3.3871	10
Maintenance response time	0.0	12.5	46.9	31.3	9.4	3.3750	12
Complaint reporting system	6.3	6.3	46.9	25.0	15.65	3.3750	11
Total average score	0.68	17.64	25.55	33.60	22.54	3.1834	

the criteria. For clear illustration, the outcomes were graphically presented in Fig. 7.9. From the analysis, the average mean score for all the criteria was 3.183. A Pareto analysis showed that the criteria were influential, very influential or extremely influential. Fewer than 20 % of the respondents believed that the criteria were not at all influential or not very influential. Eighteen criteria had their individual mean score test more than the average mean test.

This indicates that eighteen criteria were either very influential or extremely influential to the total building maintenance management. Only negligible (less than 1 %) part of the respondents considered the criteria as not influential at all, while nearly 18 % indicated that the criteria were not very influential. A majority of the universities (33.6 %) considered the criteria very influential, and many of them also measured the criteria as influential. A sizeable number (22.54 %) of respondents considered the criteria extremely influential. This research found out that the quality of materials and components was the most highly influential criteria that influenced maintenance management. Conversely, having materials and components in store was found to be the least influential criteria affecting maintenance consideration.

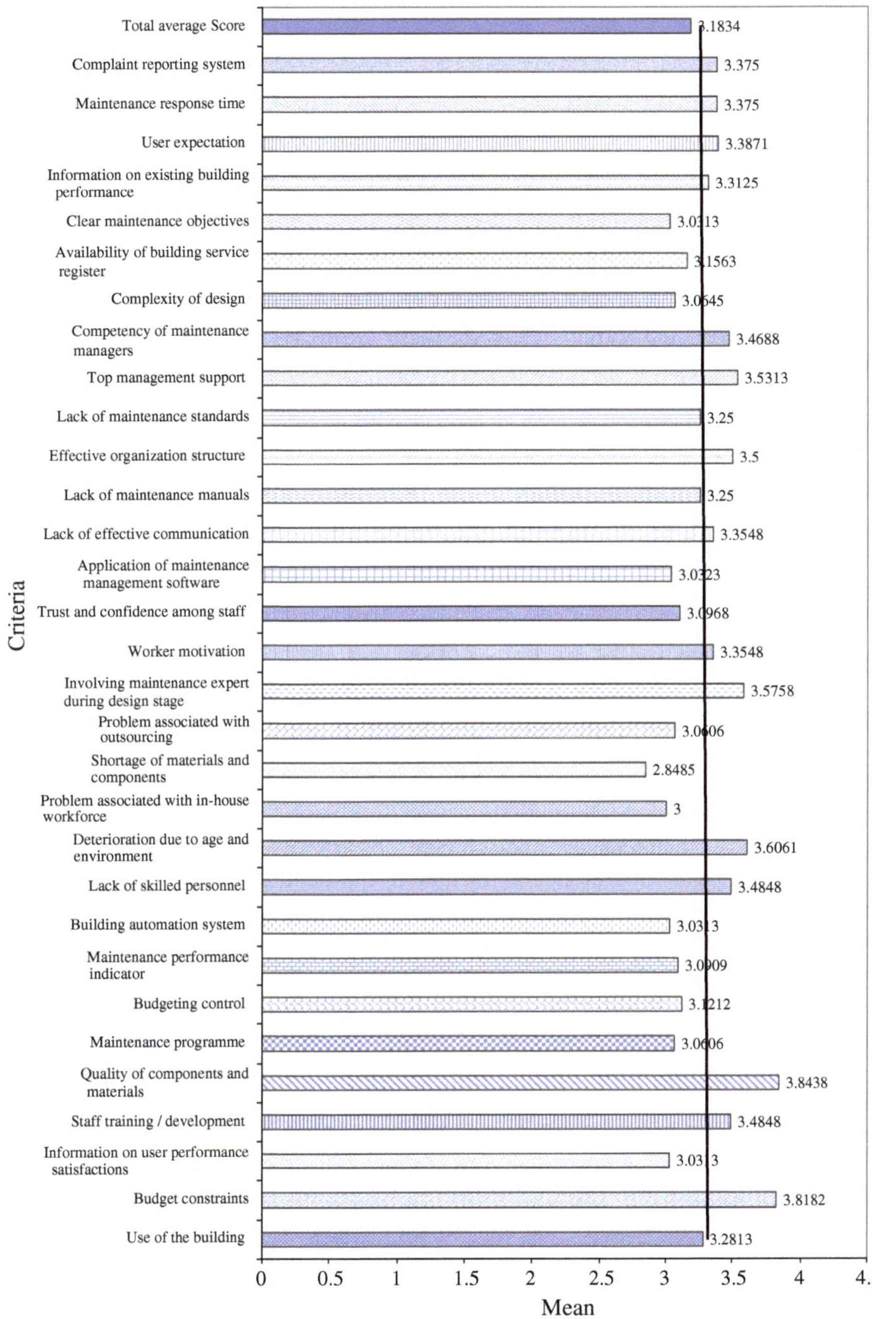

Fig. 7.9 Graphical illustration of the mean/ranking of the criteria influencing maintenance management

Virtually, 50 % of the maintenance organisations considered quality of materials and components used in maintenance as either very important or extremely important. Specifically, while 37.5 % of the maintainers viewed it as important, 31.3 % viewed it as very important, and 28.1 % of the respondents considered it extremely important. Only fewer than 4 % considered it as not very influential. These results were not unexpected. It is expected that quality of materials and components used during design and for maintenance will significantly influence the maintenance of buildings.

As would be hypothesised, the respondents ranked budget constraints overtly as very influential. While the majority (39.4 %) of the respondents considered budget constraint as very influential, many of them (27.3 %) considered it extremely influential. Many also believed it is influential. While none of the respondents did not considered it influential at all, about 12 % of them believed it was not very important, however. Budget constraint is a fundamental issue in maintenance management; arguably one of the requirements in maintenance management is the allocation for maintenance. Often if not in all cases, the available budget was inadequate to cater for the backlogs. Government has challenging universalities to be prudent on how they manage their allocations. This is placing high demand on universities for efficient utilisation of budgets. There is no doubt that government is also cutting allocations to their universities. However, with good management philosophy, budgets can be prudently utilised and justified.

Interestingly, deterioration due to age and environmental factors were highly ranked. Essentially, wear and tear defects are caused by normal use. Other factors in the wear and tear equation are the deterioration due to age and environmental factors. So this outcome is expected. Most (48.5 %) of the maintenance organisations concluded that deterioration due to age and natural environment can significantly influence maintenance management. Nearly 27 % of the respondents considered age and environmental factors as influential. Some 12.1 % of the respondents believed the combined criteria to be extremely influential. The same size of the respondents considered them not very influential.

The involvement of maintenance experts during the design phase of a new building is a great opportunity in maintenance cost reduction. In actual reality, maintenance consideration must commence from the design and construction stage. Maintenance experts' advice on maintainability issues is critical. About 82 % of the respondents found involving maintenance experts during design was influential, very influential or extremely influential. Nearly 12 % of respondents considered it as not very influential, and yet about 6 % of them believed that the criterion was not influential at all.

Seventy-five percentage of the maintenance organisations considered top management support in the maintenance management influential or very influential. Specifically, 37.5 % each measured it as influential or very influential. 15.6 % believed top management support was extremely influential for successful maintenance systems. Maintenance is not usually given top management support. Often, it is given operational consideration and is not given board room attention. It is only accorded with the attention it deserves when everything goes wrong. However, issues receive proactive and holistic consideration when brought to the attention of the "big boss".

Organisational structure was found to be highly favoured by the maintenance organisations. It was discovered that about 40 % of the respondents considered organisational structure as influential. An equal sized number of the respondents, however, measured it as very influential. Not even one of the respondents did not considered organisational structure influential. Interestingly, 12.5 % believed it is not very influential, even though 12.5 % considered it as extremely influential in maintenance management.

Unexpectedly, the mean score of storage of materials and components was below the mean score. However, budget constraint, quality of materials and components were found to be very influential. The maintainers did not see problems associated with in-house employees as so influential as to affect the maintenance management. They did not also see defining maintenance objectives clearly as influential. However, they considered time to respond to user complaints very influential. The method of reporting complaints was also found to be very influential. Surprisingly, they did not view trust and confidence among staff as influential.

The majority of the maintenance organisations (51.5 %) considered staff training and development as an influencing criterion. Many (30.3 %) of them measured staff training very influential. Nearly 12 % of the respondents believed that staff training was extremely influential. Fewer than 7 % measured it as not very influential, however. Continuous development is necessary for staff improvement. This is a very important consideration in maintenance management. In Malaysia, those managing building maintenance are not specialists in maintenance-related fields like facilities or property management. Even if they are, both operative and management staff need training or refreshment courses to update their knowledge on latest developments. In the past, the application of maintenance management software was not well recognised, but today, it is highly required in any maintenance organisation. Undeniably, the older generation did not think that an application such as software will become a fundamental issue now. Time is moving forward, and we also need to move with time; otherwise, we will become obsolete with the latest fashion.

The concept of employee empowerment, which is centred on involving employees in every stage of service delivery, stems from a well-trained operative. Competent employees will facilitate down–up initiative. Employee empowerment can also be built through building efficient communication network that includes the employee, delegated with responsibilities to operatives, and creates supportive team work as well as supportive organisation structure. Maintenance staff should be trained to have adequate technical and practical know-how, what and when, on issues related to maintenance management. Seminars, short courses and workshops provide platforms for maintenance staff to improve their skills and knowledge.

With regard to the competency of the maintenance manager, it was found that the majority (37.5 %) measured it as very influential, 34.4 % considered this as influential and 3.1 % of the maintenance organisations considered the competency of the maintenance managers not very influential. Contrarily, close to 16 % of them considered competency of maintenance managers as extremely important. The position of maintenance manager in the maintenance organisation is very sophisticated. He is responsible for ensuring the smooth operation of day-to-day activities of the organisation as well as liaising with top management. He or she needs to be competent in human resource management, labour management and versatile with methods of presentation. He or she must be able to convince the top management on the need to review and earmark adequate funding for maintenance.

The bulk (41.9 %) of the maintenance organisations measured user expectation as very influential in building maintenance management. Only fewer than 7 % viewed it extremely influential though. Yet a substantial (35.5 %) proportion of them saw it as influential. None of the respondents considered it not influential at all, while about 16 % of them, however, believed it to be not very influential though. These findings were in tandem with the author's expectations. User value systems, the needs and wants of service users, shape their expectations and perceptions. This in turn shapes how they assess the quality of service they receive. If their value systems are not met, albeit how costly it is in production cost, users attributed little or no value to the service. Therefore, for service providers to delight users, they must provide services that meet users' expectations. Arguably, some users' expectations are beyond normal. Thus, the involvement of users in service definition is critical.

Forty-seven percentage of the maintenance organisations believed the systems of reporting complaints are influential. However, 6 % of the respondents viewed as it not influential at all. However, some 40 % of the respondents measured it as very influential or extremely influential. These outcomes are not different from the author's expectations. The complaint reporting system is very crucial in maintenance management. For effectiveness, there is a need for a single point of reporting complaints. When maintenance needs arise, users are always anxious about whom to talk to. In this situation, they require a direct point of contact. Complex reporting systems complicate users' headaches and frustration. Perhaps users should be provided with a toll-free call to those responsible for maintenance directly.

Certainly, users will not prefer to have to make a personal visit to the maintenance organisation before their needs are attended to. In addition, users will not just want to be answered by machines. It is a common practice these days for maintenance organisations to install machines that attend to users. Some of the answering procedures are also frustrating. Users want to talk with people not machines.

All the respondents believed maintenance response time are influential. In fact, not even one of them did not agree that response time is influential. These outcomes are in order and for it to be rated next to complaint reporting system speaks volumes of these outcomes. Nearly, half of the respondents measured response time influential (46.9 %). Thirty-one percentage accepted it as very influential, although 12.5 % believed it was not very influential, and yet about 10 % measured it extremely influential. No doubt, the time taken for maintenance organisations to respond to complaints is critical to users. Users want their complaints to be taken care of as quickly as possible. Inasmuch as users prefer single reporting that is not complicated, another basic issue that makes users happy with maintenance service is when their complaint is responded to. To provide quality maintenance work is one thing, another thing is when such complaints are responded to.

It was interesting to find that effective communication comes next to maintenance response time. No doubt effective communication is crucial in maintenance management. Communication is pertinent even within maintenance organisation itself. An organisational goal in the form of a maintenance objective is critical. Maintenance objectives must be effectively communicated to all staff in the maintenance organisations. Operatives must have clear understanding of the objectives of the maintenance organisation. For instance, a maintenance objective of an organisation could be to satisfy their users. A performance indicator must be established to ease hassle. Thus, it is interesting to reveal that not even one respondent disagreed that communication is influential in maintenance management. Nearly 45 % of them though considered it as influential. Many (25.8 %) argued that it is very influential, while about 13 % of them indicated that it is an extremely influential criterion in maintenance management. However, participatory communication must be favoured to authoritative communication. Under authoritative communication, only senior management have a say. It discourages down–up flow of information.

From the outcome of the study, it is obvious that all maintenance organisations rated worker motivation as critical. While 16.1 % rates it not very influential, close to 42 % rated it as influential. Many, (32.3 %) of the respondents measured it as very influential. Ten percentage of the respondents considered it extremely influential. Effective communication among all staff is related to worker motivation. If workers are motivated, it will affect the quality of work they provide to users. Operatives must be equipped. Organisations must invest in their maintenance operatives, as they do for academic staff. Maintenance staff should be well motivated to achieve maximum output.

With regard to the availability or lack of available information on existing building performance, 44 % of the respondents measured it as influential; many (34.4 %), however, measured it very influential although none of the respondents believed it was not influential at all. About 6 % considered it extremely influential. All be it, though these outcomes were impressive, it was expected that the majority would consider it more influential, absolutely more than the observed outcome. Information on the existing building is crucial to maintenance management. It is when the existing building condition and performance are known that adequate improvement can be made on the building. This information is the basis for condition or inspection-based maintenance.

As it is expected, use of building comes near or next to availability of information on building performance. Forty-four percentage of the organisations actually considered use of the building as influential. But 6.3 % believed it is not influential at all to maintenance management. Ten percentage believed the contrary. Many (31.3 %) of the respondents submitted that use of the building is very influential. Buildings affect user attitude, and they need adequate care from the users as well. If a building is excessively used, carelessly used or neglected, it will undoubtedly require more than maintenance. It could require refurbishment, alteration or even conversion if not total reconstruction.

A very influential factor in maintenance management is maintenance standard. Forty-one percentage of the respondents opted that maintenance standard is influential. Many (28.1 %) believed it is very influential, and although 22 % believed it is not very influential, none of them believed it is not influential at all. About 10 % of them measured it as influential. Maintenance standards are critical in maintenance consideration. Maintenance standards should be tied to an organisation's mission and vision. This will also affect materials selection and quality of workmanship.

A maintenance manual was considered to be influential by 41 % of the respondents. About 34 % of the respondents measured it as very influential. Although none of the respondents did not believe that a maintenance manual is not influential, some 3.1 % of the respondents argued that it is extremely influential. About 16 % of the respondents accepted that a maintenance manual is not very influential in maintenance management. Specifications of system performance and functions, specifications of materials and components, mode of operation and usage constraint, the range of operating capacities, supply requirements, schedule of maintenance, permitted modifications, description of spares and alternatives, lists of test equipment and its checking procedure, method and instructions on disposal for hazardous materials, detailed drawings, photographs of the newly (or acquired) completed buildings, fire fighting methods, licenses and approval obtained floor areas and maximum bearing load capacity, the contents of maintenance manual are crucial to the building and the organisation's corporate objectives and their core business.

Closely related to a building maintenance manual is the maintenance register. The general information about organisation is located in maintenance register. The location, condition and performance of buildings are found in the building register. It also contains information on previous maintenance records of the buildings as well as possible time when maintenance is required next. Organisations that seek to succeed in maintenance management need a record of their buildings. Therefore, it was found that 44 % of the universities measured building registers as influential. A substantial number (22 %) of them did not believe it is very influential. While 18.8 % of them argued that it is very influential, 12.5 % further contented that it is actually extremely influential.

Budgeting control was found to be very influential by 40 % of the maintenance organisations. None of the responding organisations rated budgeting control as not influential at all. While 21 % believed it was influential, 12 % believed it was actually not very influential. Twenty-seven percentage of the universities, however, believed it was influential in maintenance management. Budgeting control can make a significant impact on maintenance management. Poor control of budget can result in unnecessary maintenance backlogs.

Trust and confidence was revealed to be very influential by 32.3 % of the respondents. While none of the respondents did not rate trust and confidence as not influential at all, substantial (29 %) part of the maintainers, however, consented that it was not influential. However, 36 % believe it was influential and 3.1 % also argued that it was extremely influential. Trust and confidence building among staff is a significant part of any successful organisation. Part of the strategies of motivating workers is through building trust and confidence. Maintenance operatives need to be trusted and confidence built in their capabilities. Staff should be confided with a reasonable amount of trust. This will go a long way towards boosting their morale and ultimately increase organisational productivity.

Around 40 % of the universities agreed that a maintenance performance indicator is influential. About 6 % of the respondents believed that it was not influential at all. Even though 18.2 % revealed that a performance indicator was not influential, 33.3 % believed that it was very influential in maintenance management. A small number (3 %), of the respondents saw it as extremely influential. It was also found that 6.5 % of the respondents believed that complexity in design was not at all influential. Close to 16 % further deviated noting that it was very influential. The majority (45.2 %) of the respondents, however, believed it was influential. About 30 % of them agreed that it was very influential and considerable, and 3.2 % of them believed it was actually influential.

About 10 % of the respondents considered that a maintenance programme was extremely influential. Conversely, 10 % of them consented to the contrary. About 24 % of the surveyed respondents agreed that a maintenance programme is not very influential. However, most (30.3 %) of the respondents claimed that a maintenance programme was very influential.

Almost 30 % of them also measured it as influential. A successive maintenance programme on buildings is critical to subsequent maintenance of the building. A successful maintenance management system depends on achieving value for users and the maintenance organisation. Much of the building equipment and plant requires scheduled or cyclical maintenance programmes. This is in order to avoid downtime if the plant or equipment is not maintained when due.

Maintenance procurement methods are very crucial in maintenance management. The types of maintenance procurement methods organisations choose have impact on the quality of maintenance service. Each of the procurements, however, has its own shortfalls. It is interesting to reveal that many (27.3 %) of the respondents disagreed that problems associated with outsourcing is not influential in maintenance management. While only 6 % of them considered it as extremely influential, 22 % believed it very influential. The majority (46 %) of the respondents measured it as influential.

About 32.3 % of the responding organisations measured the application of maintenance management software as influential. Twenty percentage of the respondents further considered it very influential. While 10 % of them considered it was not influential at all, 10 % believed it was extremely influential. However, 23 % of them believed it was not very influential. Coincidentally and next to the issue of building maintenance management software, using building automation systems was found to be overtly very influential.

It was found that 34.4 % of the respondents argued that it was influential. However, 34.4 % of them believed contrarily. About 25 % observed it as very influential and 6.3 % believed it actually extremely influential. These outcomes suggest that both criteria are closely related all be it serving different purposes in the building maintenance management. A building automation system is a form of performance-based maintenance management. Yet both are systems or ways to facilitate decision-making. In a way, these results were not expected, even though it was expected that both would be highly rated among the respondents.

To have a clear maintenance objective was discovered to be influential (46.9 %) among the respondents. Although a reasonably sized group (9.4 %) believed it not influential at all, about 10 % argued that it was extremely influential. Furthermore, while 16 % believed it was not very influential, 19 % measured it as very influential. This outcome was unexpected. It was expected that defining maintenance objective would be significantly rated. Achieving higher productivity and satisfaction hinges on having clear maintenance objectives. It is only when organisations have clear objectives that maintenance standards are established. It was revealed through the research that 50 % of the respondents opted that information on user performance satisfaction was influential. About 25 % of the respondents opted that it was not influential, although 22 % quantified it as very influential and 3.1 % believed it is extremely influential.

While 3 % of the respondents disagreed that problems associated with insourcing can influence maintenance management, 18 % also believed it was not actually very influential. Yet, close to 60 % of the respondents measured it influential and 24 % considered it very influential even though none of the respondents agreed it was extremely influential. Around 10 % of the respondents observed that shortage of material was not influential at all, but 10 % had contrary observations. A majority (42.4 %) of the respondents measured shortage of materials and components only influential, whereas 27 % believed it was not influential, and 12 % believed it was not very influential.

Summarily, on the basis of the data analysis, the criteria were categorised into the following functions or aspects:

- Planning function: this includes budget constraint, budget control, top management support, maintenance procurement strategies, maintenance objectives and standards
- Control function: this includes information on user performance indicators, shortage of materials and components, and shortage of labour,
- Organisation function: this includes organisation structure,
- Directing function: this includes the competency of the maintenance manager,
- Design function: this includes involving maintenance experts during design and design complexity
- Behavioural function: this includes user expectations, user performance satisfaction and use of buildings, worker motivation and trust and confidence in the organisation.

These categorisations will assist in guiding towards modelling the value maintenance management system. In part, they explained the various stages in maintenance management. However, it was hypothesised that the above criteria can significantly explain improving maintenance service delivery.

In order to confirm the hypothesis, a multiple regression analysis was performed, the result of which is listed in Tables 7.25 and 7.26. From Table 7.25, it was gathered that there is no statistical evidence to conclude that the criteria can explain service delivery.

These outcomes were not, however, unexpected. But in actual reality, this might not be the case. The influence of the above criteria is a matter of empirical questions, and several issues emerge over and beyond statistical problems. A very plausible reason for these results was because of the sample size. Sample size affects the significance level of scores. If the sample size is small as compared to the size of the variables, it will negatively affect the result, which will in turn provide unrealistic results. For this particular case, this could be an acceptable reason. For instance, wherein there were 33 respondents, the number of criteria was 31, which means it was on the ratio of 1–1. However, the confidence level for the data was determined at 95 %. Furthermore, the R^2 is 1. So in actual reality, these criteria are excellent predictors.

7.2.4 *Criteria of the User Value System: Providers' Perspectives*

This section seeks to analyse the criteria that enhance user satisfaction in maintenance service delivery. These criteria were addressed to both the maintenance organisations and the building users. However, this section is concerned with analysis of the criteria of the user value in terms of the provider perspectives (the maintenance organisation). Thus, the rating of criteria will be compared and contrasted with the same criteria as rated by the users themselves. There are usually "service mismatch" between what the user wants and needs and what the services providers provide. Consumers (i.e. building users) and service providers always have different perceptions with regard to value of service delivery.

Put in another word, a "perceived" gap exists between the services that organisations provide compared with the services that end-users want. Service outputs are

Table 7.25 Model summary of criteria influencing maintenance management

Model	R	R^2	Adjusted R^2	Std. error of the estimate
1	1.000	1.000	0.0	0.0

Table 7.26 ANOVA whether the criteria can explain service delivery

Model		Sum of squares	df	Mean square	F	Sig.
1	Regression	28.000	26	1.077	–	
	Residual	0.000	0	0.0		
	Total	28.000	26			

labour intensive and customised and are provided at the convenience of the users. Thus, human contact and its consequences are a crucial part of the process of producing services (Russell and Taylor 2006). They also found that service outputs are not easy to measure because they cannot be readily stored, felt, held and used again. The basic questions that the information in this section will provide answers to are the following:

(a) How do building users evaluate maintenance service delivery?
(b) What are maintenance organisations' perceptions about service quality?
(c) Do discrepancies exist between the views of building users and those of the maintenance organisation?
(d) Can users' and maintenance organisations' perceptions be combined into a common maintenance management model for service delivery?
(e) How can maintenance organisations improve users' satisfaction?

The determinants of service criteria include reliability, responsiveness, competence, access, courtesy, communication, credibility, security and understanding the user. While these qualities are established for classical and general consumers, they can be adapted for building users. However, the original Parasuraman's SERVUAL model is modified. The modifications are necessary because the classical dimensions of the service quality were inadequate to cater for the needs of varying consumers. The needs and wants of the building users are arguably different from those of other sectors. At the same time, the criteria are not static. They are dynamic and in transition. In fact, Russell and Taylor (2006) also identified dimensions of service quality to include timeliness, courtesy, consistency, accuracy, convenience, responsiveness and completeness.

Therefore, there is the need to modify the dimension to suit the particulars of the specific consumers. However, there is inadequate literature on the determinant services criteria for the building users. Arguably, buildings users are critical of reliable or dependable maintenance organisation and services and want easy access to maintenance service providers. Furthermore, building users want and need easy means of communication with the maintenance organisations. They also want to be treated with courtesy. It is also imperative that building users want to be attended to by competent operatives and that the places of living, studying and doing other activities within the university are safe and secure.

Users often attach different value to services they receive. The consumers combine their experience with their expectations to define their needs and wants. Need and want constitute a value. Need is tangible (objective), while want is intangible (subjective). But an extension gap also exists between what a member of a team believes he or she provides and what other members believe they provide. Naturally, questions hereby emerge such as "are the rating of the criteria within the user value systems the same as those rated by the users themselves?" If they are providing or not providing the said services, "what aspects of the services are they not providing?"

Evidence exists both in the literature and empirical findings that users are not satisfied with the maintenance services they are receiving in Malaysian universities.

Quotes from dissatisfied building users are indicative: "please change the maintenance department, the maintenance is too lousy, our buildings are not well maintained, the toilets in the building are always dirty, repair the defects without delay, improve the condition of toilets, staircases, classrooms and lecture halls, be alert of complaints, late to repair buildings, have incompetent and insufficient staff, be more efficient, staff are ill mannered, be effective and quick in maintenance service is important". These are just to mention a few.

This analysis is pertinent because maintenance organisations must take cognisance of the building user concern. The performance of a building is very much likely to be enhanced if the maintenance organisations are aware of these "gaps" and take them into consideration in the maintenance process in response to the user satisfactions. These criteria were addressed to the respondents to tick which criteria they considered important or not important in building maintenance management.

A total of 12 criteria were identified and addressed to respondents to rate the importance of each of the criteria with respect to meeting users' expectations and perceptions of service delivery. The criteria may not be exhaustive due to the complexities associated with maintenance management and customers and user satisfaction. However, the list covered the significant criteria that improve building users' satisfaction with maintenance service delivery.

Thus, it is hypothesised that the criteria can significantly explain meeting the users' satisfactions in maintenance service delivery. In other words, if the criteria are taken into account when making decisions on maintenance management, users will be satisfied with the outcomes of the maintenance service. In order to confirm the hypothesis, a multiple regression analysis was performed, the results of which are shown in Tables 7.27 and 7.28. From Table, 7.27, it is obvious that the criteria used can significantly explain meeting the user satisfactions. The model is moderately fit, however. Therefore, to meet users' satisfactions, the criteria need to be improved. The single most fitted criterion in the model is the quality of the maintenance services. Therefore, it can be concluded that these criteria are satisfactory measures.

Table 7.27 Distribution of model summary of the criteria of the value system

Model	R	R^2	Adjusted R^2	Std. error of the estimate
1	0.826(a)	0.683	0.471	0.72408

Table 7.28 Distribution of ANOVA test whether the criteria of the user value system can explain user satisfaction (based on maintenance organisation)

Model		Sum of squares	df	Mean square	F	Sig.
1	Regression	20.305	12	1.692	3.227	0.012
	Residual	9.437	18	0.524		
	Total	29.742	30			

Furthermore, reliability and validity tests were performed to test the strength of the data. The result of the Cronbach's alpha test for the reliability of the twelve (12) criteria is shown in Table 7.29. The average validity score was determined at 0.754. This therefore implies that the constructs were valid for this purpose.

Table 7.30 displays a descriptive statistic of the survey. With the exception of one criterion that was answered by 31 respondents, all other criteria were answered by 32 respondents. With the exception of 3 criteria, which have their mode scores as 3 each, other criteria each has 4 as their mode score.

The standard deviation suggests that the results were clustered to the mean scores. The skewness scores indicated that responding organisations were very critical of 7 of the criteria (i.e. those with negative scores). The minimum and the maximum scores suggest the respondents' rating spread across the five measurement scales. The sum scores vary from 115 to 136. This is pointing to the fact that the respondents spread apart in their opinions. Respondents are more critical of safety and health of users, users comfort and well-being, as well as convenience in making appointments. This is because the minimum scores for each of the criterion were 3 (3 is measured as urgent).

Table 7.31 displays the mode and the mean score tests for the criteria. The table further shows the ranking of each of the criterion calculated based on the mean score test. The mode score is for each of the degrees of measurement.

The average mean score test was 3.8196. This is a high score for a scale having 5 as its maximum possible score. More than 95 % of the respondents measured the 13 criteria as important, very important or extremely important. A negligible percentage (0.2449 %) of the respondents did not consider the criteria important at all. Fewer than 5 % of the universities claimed that the criteria were not very important. Many (32.44 %) of the universities, however, believed that to meet user satisfaction, these criteria are important. Nearly 40 % of the universities believed that the criteria were very important.

Quite a number (23 %) of the universities considered the criteria as extremely important. Six of the criteria have their individual mean score tests higher than the average mean score. Quality of maintenance service is the most highly ranked criterion followed by time to respond to maintenance complaints and reliable maintenance organisations. Aesthetic appeal is the least scoring criterion, with a mean score of 3.2183. All other criteria have their individual mean score close to the average mean score. See Fig. 7.10 for the illustrative statistics.

About 47 % of the respondents measured quality of the maintenance service as very important, while many (41 %) also considered it as extremely important. Only 3.1 % of the respondents considered it as not very important, but none of the respondents agreed that it is not important at all. But nearly 10 % believed it is

Table 7.29 Distribution reliability statistics of criteria of the user value system

Cronbach's alpha	N of items
0.919	12

Table 7.30 Distribution of descriptive statistics of the criteria of the user value system

Criteria	Minimum	Maximum	Sum	Mode	Std. dev.	Variance	Skewness
Optimum functional performance of buildings	2.00	5.00	115.00	3	0.79755	0.636	0.080
Safety and health of users	3.00	5.00	128.00	4.00	0.80322	0.645	0.000
Time to response to complaint	2.00	5.00	130.00	4.00	0.75935	0.577	−0.578
Aesthetic appeal	1.00	5.00	105.00	3.00	0.72887	0.531	−0.498
Security of property of user	2.00	5.00	121.00	4.00	0.79248	0.628	0.008
User comfort and well-being	3.00	5.00	119.00	4.00	0.73470	0.540	0.266
Time to complete the maintenance works	2.00	5.00	128.00	4.00	0.80322	0.645	−0.398
Efficient reporting systems	2.00	5.00	121.00	4.00	0.83219	0.693	−0.273
Reliable maintenance organisation	2.00	5.00	130.00	4.00	0.84003	0.706	−0.471
Convenience of appointment with user to make repair	3.00	5.00	119.00	3.00	0.72887	0.531	0.498
Friendly maintenance staff	2.00	5.00	119.00	3.00	0.81258	0.660	0.191
Quality of maintenance service	2.00	5.00	136.00	4.00	0.76200	0.581	−0.933

important. Time to respond to maintenance complaints was found to be important in the understanding or more than half (53.1 %) of the respondents. None of members of the respondent's group measured it as not important; 3.1 % did not see it as very important. On the other hand, many (28 %) of the respondents measured it as extremely important and another 16 % believed it is important.

Dependable maintenance organisations were measured to be extremely important by many (34.4 %) of the surveyed respondents. However, about 41 % actually believed it is very important. Quite a sizeable number (22 %) of the respondents also saw it as important. A small group (3 %) disagreed that it is very important, however. It is interesting to also reveal that the majority (47 %) of the universities measure time to respond to maintenance work very important. Moreover,

Table 7.31 Distribution of level of importance, mean and ranking of criteria of the user value system

Criteria	Level of importance					Weight–age mean	Rank
	Not at all important (%)	Not very important (%)	Important	Very important (%)	Extremely important (%)		
Optimum functional performance	0.0	6.3	40.6	40.6	12.5	3.5938	11
Users' health and safety	0.0	0.0	31.3	37.5	31.3	4.0000	5
Time to respond to complaint	0.0	3.1	15.6	53.1	28.1	4.0625	2
Aesthetic appeal	3.1	3.1	59.4	31.3	3.1	3.2813	12
Security user property	0.0	3.1	34.4	43.8	18.8	3.7813	8
User comfort and well-being	0.0	0.0	35.5	45.2	19.4	3.8387	6
Time to complete maintenance work	0.0	3.1	21.9	46.9	28.1	4.0000	4
Efficient reporting system	0.0	6.3	28.1	46.9	18.8	3.7813	7
Reliable maintenance organisation	0.0	3.1	21.9	40.6	34.4	4.0625	3
Convenience of appointment to make repair	0.0	0.0	43.8	40.6	15.6	3.7188	9
Friendly maintenance staff	0.0	3.1	40.6	37.5	18.8	3.7187	10
Quality of maintenance service	0.0	3.1	9.4	46.9	40.6	4.2500	1
Total average score	0.2449	4.1943	32.4329	40.3555	22.7725	3.8169	VI

VI very important

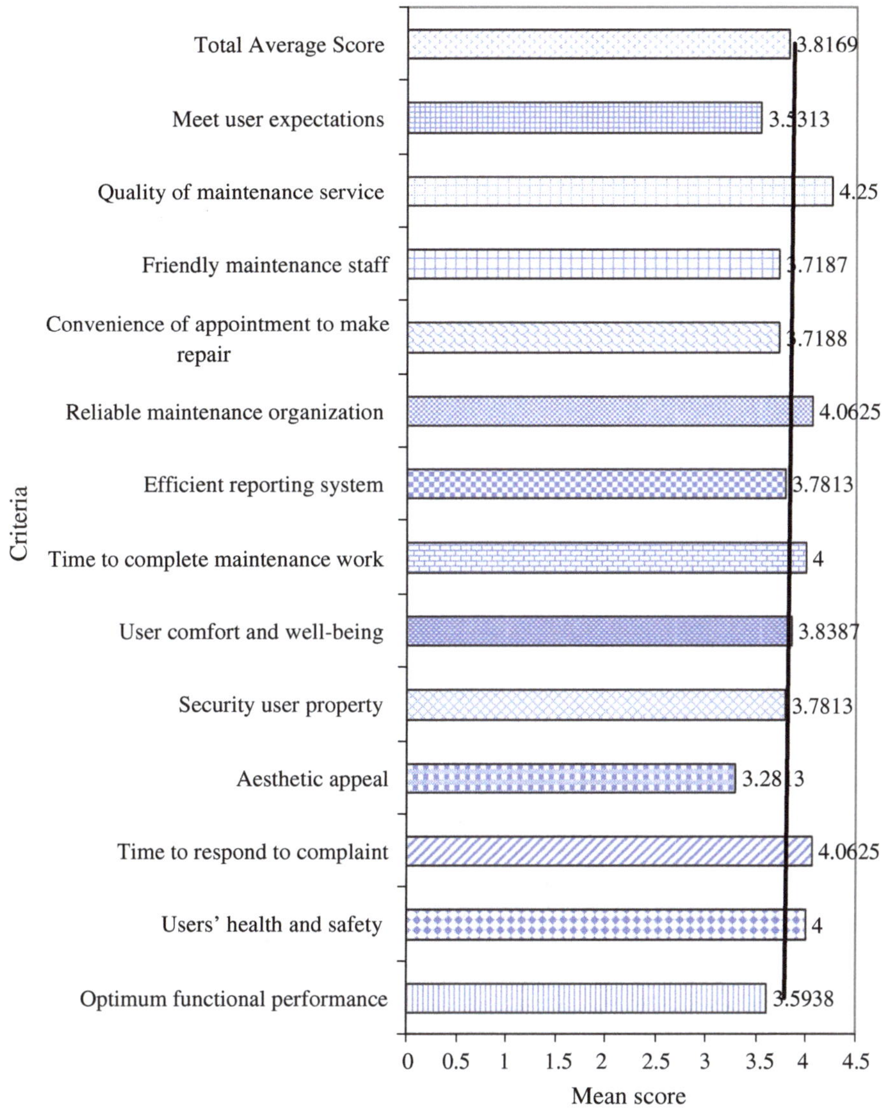

Fig. 7.10 Illustrative statistics for the criteria of the user value system

recognisable (28 %) respondents argued that it is extremely important. Twenty-two percent of them further believed it important, but 3.1 % think it is not important at all. Interestingly, none of the respondents agreed that user health and safety was neither important at all nor not very important. While 31 % of the respondents measured it as important, 38 % measured as very important and the remaining 31 % considered it as extremely important.

Similarly, not even one of the universities agreed that user comfort and well-being was neither not important at all nor not very important. In fact, 36 % of the respondents believed it is important, and nearly half (45 %) of the respondents measured it as very important. All be it, the remaining 19 % rated it extremely important. With respect to the maintenance reporting system, 6 % of the respondents considered it as not very important, 28 % believed it was important, 47 % rated it very important and 19 % of the felt it was extremely important. Security of the users' property was rated important by 34.4 % of the respondents. While only few respondents (3.1 %) believed it is not important at all, 44 % confirmed it is very important and 19 % measured it to be extremely important.

The responding organisations rated the convenience to make an appointment to undertake the repair was important, very important or extremely important. About 44 % said it is important, 41 % confirmed it as very important, and the remaining respondents (16 %) measured it as extremely important. The need for a friendly maintenance staff was measured by the majority (41 %) of the universities as very influential, however. Yet 38 % of the respondents believed it is very important, while 19 % judged it extremely important.

Optimum functional performance of buildings was rated low and below the author's expectation. This is unexpected though. As a matter of fact, 6.3 % of the respondents verified that it is not very important. Forty-one percent each measured it important and very important, respectively. Only 13 % of the respondents said it is extremely important. Another interesting finding was that the maintenance organisation rated meeting users' expectations low and even lower than building performance. In fact, a recognisably sized group of 19 % rated it as not very important even though another 19 % of the respondents evaluated it as extremely important. And while 34 % gauged it as very important, 28 % held it that it is important.

However, it is not surprising to note that aesthetic appeal or building appearance was rated the lowest criterion. In fact, only 3 % of the respondents believed it is extremely important. While 60 % of the respondents considered it as important, 31 % indicated that it is very important. Three percentage each of the respondents quantified it not important at all and not very important. Summarily, the finding of criteria within the user value systems based on the providers' perspective shows that the criteria can be divided into groups as follows:

- User comfort: this includes aesthetic appeal, comfort and well-being
- User health and safety: this includes security of user property
- Friendliness: this entails time to respond to complaint, to complete maintenance work and efficient reporting system
- Quality of service: this entails optimum functional performance and to meet user expectations and perception.

7.2.5 Analysis of Defects in the Buildings: The Providers' Perspectives

The defects in this section were designed to identify, evaluate and rank the defects that require maintenance urgently. However, the same building defects were addressed to building users. The maintenance organisations were asked to rate each of the defects in terms of how urgent they were in total maintenance management. An important segment of the total maintenance management is the identification, categorisation and evaluation of building defects. Building defects occur through normal wear and tear. Although building defects are inevitable, they can, however, be managed. Its analysis is critical if meaningful maintenance management is significant.

The following questions were created to guide this analysis:

(a) What are maintenance organisations' perceptions about defects?
(b) Do discrepancies exist between the views of building users and those of maintenance organisation on the building defects?
(c) Can users' and maintenance organisations' perceptions be combined into a common maintenance management model for service delivery?
(d) How can maintenance organisations improve users' satisfaction?

The degrees of a defect in building are measured to evaluate the functional performance of the buildings. The reliability tests for the defects were conducted, the results of which are listed in Table 7.32. The extracted validity also ranges from 0.692 to 0.916. Therefore, the reliability and validity defects are excellent.

It was hypothesised that the defects addressed to the respondents can significantly explain improving service delivery. In order to confirm the hypothesis, a multiple regression analysis was performed, the results of which are shown in Tables 7.33 and 7.34. From the Table 7.33, it was gathered that there is no statistical evidence to conclude that the criteria can explain service delivery. However, these outcomes were not unexpected, because of the sample size as well as the power effect.

Both power effect and sample size affect the result of inferential statistics. When the number of variables and the sample size are equal or almost equal, it affects the outcomes by producing results that might not be statistical reliable. The same is also the case if the sample size is small compared to the size of the variable. For better outcomes, the sample size should be as twice the variable size. However, confidence levels of the variables are reported.

Table 7.32 Reliability statistics for the defects in buildings as rated by the maintenance organisation

Cronbach's alpha	N of items
0.939	32

Table 7.33 Model summary of defects in buildings as rated by the maintenance organisation

Model	R	R^2	Adjusted R^2	Std. error of the estimate
1	1.000	1.000	–	–

Table 7.34 ANOVA whether concentrating on the defect can explain user satisfaction as rated by maintenance organisation

Model		Sum of squares	df	Mean square	F	Sig.
1	Regression	28.964	27	1.073	–	–
	Residual	0.000	0	–		
	Total	28.964	27			

Thirty-two wear and tear defects in buildings were identified and addressed to the respondents. The degree of urgency of each of the defects is determined by the frequency at which responding maintenance organisations agreed with each of the defects. For instance, where the mean score falls between 1.0 and 1.5, the defect is considered as not urgent at all. See Table 7.35 for other distributions. This cut-off point is used, because the lowest possible mean score is 1. This also applies to the "extremely urgent" scale. The maximum possible score is 5.

The defects were important sources of complaint that maintenance organisations received from building users in their respective campuses. Though the list of the 32 defects may not be exhaustive, due to the vast nature of the building maintenance concurrently with the problem of nomenclature, the list covers regular defects pertaining to buildings that users are regularly concerned about.

Defect considerations in buildings are important criteria in evaluating a building's performance. Twenty-one defects were ranked very urgent or extremely urgent. This in effect facilitates a comprehensive and clear understanding of the degrees of the various defects. Analysis of the degrees of the defects contributes to directing maintenance organisations on maintenance management. Defect classification is a strategic function of maintenance managers. For prudence, active efforts should be directed to the extremely urgent defects, and then the very urgent and finally to the ones that are not very urgent. Probably this is the only way the ever inadequate fund will be spent judiciously and users' satisfactions can be achieved and maximised accordingly. However, defect degrees in terms of urgency ought to be self-evaluation, supposedly by the users themselves.

Table 7.35 Average index evaluation metric for defects as rated by maintenance organisation

Scale	Evaluation (interpretation)
1.00–1.50	Not urgent at all
1.51–2.50	Not very urgent
2.51–3.50	Urgent
3.51–4.50	Very urgent
4.51–5.00	Extremely urgent

Table 7.36 Scale statistics of defects in buildings as rated by maintenance organisation

Mean	Variance	Std. deviation	N of items
112.83	265.11	16.28	32

Table 7.36 contains the total mean score and the average variance. The table also contains the average standard deviation of the 32 defects. A major conclusion that is drawn from the table is that 95 % of the respondents' mean scores will lie between 2 and 5.

The minimum and maximum scores were 1 and 5 (Table 7.37). The sum values range from 74 to 140. The model scores also range from 3 to 5. The standard deviation ranges from 0.066144 to 1.14647.

The frequencies of the degree of urgency of the different defects are depicted in Table 7.38. The graphical representation is displayed in Fig. 7.11. The table shows an overview of data obtained, the level of urgency and the mean scores. The mean scores indicate the degree of urgency for each of the defects. The standard deviation shows the variability of score found in the score tests. The average mean score is 3.42. Fifty percentage of the defects have their individual mean score more than the average mean score. Most (34.79 %) of the survey respondents considered all the defects as urgent, while a considerable group (20.78 %) considered the defects as extremely urgent. Just a few of the respondents viewed the defects as not at all urgent (2.08 %). Lift failure was the most (1) extremely rated defect followed by faulty electrical equipment, while the least most considered defect was faulty towel rail (32) after bad soap holder (31). Individual discussion on each of the defects is not provided.

7.3 Users' Perspectives

This section presents findings on the maintenance issues based on the building users' perspectives. It has been stated earlier that there are two main stakeholders in the maintenance management value chain. In order to develop a systemic maintenance management system, analysis of both parties' services are compared and contrasted. The two perspectives of service providers and consumers must be combined for meaningful analysis. When the two perspectives are separated, improvement in one will lead to deterioration of the other. Therefore, identifying what the users want will contribute to better service delivery. This is achieved through identifying and understanding the "performance gap". Specifically, the indices of the building defects and criteria of the user value system for both the maintenance providers and building users are compared and contrasted.

The sample size for building users was 500, though there are more than one million building users in the universities. This sample is chosen because of the

Table 7.37 Descriptive statistics of defects in buildings as rated by the maintenance organisation

Criteria	Minimum	Maximum	Sum	Mode	Std. dev	Skewness
Damaged window	2.00	5.00	112.00	3	0.966	0.430
Blocked water closet	2.00	5.00	129.00	3	0.861	−0.386
Floor tile failure	2.00	5.00	88.00	3	0.777	1.102
Wall tile failure	2.00	4.00	85.00	4	0.614	0.556
Faulty shower	2.00	5.00	100.00	3	0.952	0.632
Faulty sanitary appliances and fittings	2.00	5.00	121.00	3	0.890	−0.115
Faulty fan	2.00	5.00	102.00	5	0.765	0.287
Damaged ceiling	2.00	5.00	102.00	4	0.948	0.514
Damaged door locks	2.00	5.00	125.00	5	0.893	−0.114
Cracked floor	2.00	4.00	98.00	3	0.684	0.0380
Damaged taps	2.00	5.00	109.00	4	0.911	0.160
Damaged roof structure	1.00	5.00	140.00	5	1.032	−1.431
Collapse drains	2.00	5.00	116.00	4	1.034	0.047
Damaged/cracked walls	2.00	4.00	99.00	5	0.707	0.000
Faulty communications appliances	1.00	5.00	128.00	5	1.055	−0.598
Failed furniture and fittings	2.00	5.00	101.00	4	1.088	0.648
Broken doors	2.00	5.00	116.00	4	0.972	0.281
Sink leakage/ blocked	2.00	5.00	123.00	5	0.911	0.061
Bad soap holders	1.00	4.00	75.00	3	0.876	0.309
Pipes leakage	1.00	5.00	126.00	4	1.103	−0.511
Faulty towel rail	1.00	5.00	74.00	3	1.119	0.912
Damaged hot-water dispenser	1.00	5.00	90.00	3	0.839	0.569
Faulty bulbs	2.00	5.00	116.00	5	0.870	0.556
Faulty air-conditioning system	2.00	5.00	135.00	5	0.914	−0.449
Faulty electrical circuit	2.00	5.00	144.00	5	0.822	−1.148
Lifts failures	2.00	5.00	146.00	5	0.902	−1.525
Frames	2.00	4.00	98.00	3	0.684	0.038
Faulty fire alarm	1.00	5.00	135.00	5	1.071	−1.164

(continued)

Table 7.37 (continued)

Criteria	Minimum	Maximum	Sum	Mode	Std. dev	Skewness
Faulty heat extractor	1.00	5.00	106.00	4	1.203	0.058
Faulty fire extinguisher	1.00	5.00	124.00	5	1.146	−0.683
Faulty smoke detector	2.00	5.00	125.00	5	0.912	−0.630
Cracked staircases	2.00	4.00	99.00	3	0.661	0.000

guideline on the determination of sample. For population of several millions, 500 samples are adequately sufficient for all types of analyses. More so, it was also established that ten times the number of variables is statistically significant for the sample size. Inferential statistics also begin to lose its value if the sample sizes exceed 500. In fact, many writers have argued for 400 sample sizes for any population if inferential statistics is required.

7.3.1 Characteristics of the Buildings' Users

A total of 550 questionnaires were administered on the users. Fifty-eight of the questionnaires were not returned. Fifteen were not useable because they were not properly completed by the respondents. That left a total of 477 useable questionnaires. This was a response rate of 81 %, which is considered satisfactory. Fifty-four percentage of the respondents were female, while the remaining (46 %) were male (see Fig. 7.12). The analysis in Table 7.39 further suggests about 74 % of the respondents were from publicly owned universities. About 26 % of the students were from private universities.

The majority (40 %) of the responding users were in their third year or year three which follows closely with those in their second year (Fig. 7.13). Many (19.1 %) of the responding students were in their first year in the universities. Only 6.3 % were in year four, while less than 1 % were in the fifth year. The results further revealed that 15 % (n = 7) of the students were doing either master's or PhD degrees.

The above results indicated that on average the respondents had spent more than a year on the campus. This background is good considering that students have to spend a year or three semesters for the foundation programmes before moving to the first year. This is the case in all the public universities and most of the private universities. Therefore, the respondents were capable of providing valid responses to the condition/performance of their university buildings and the maintenance organisations.

A majority (73 %) of the students lived on the campus, while only 27 % of them lived outside the campus. See the distribution in Table 7.40. As such, the majority of the students had adequate information on both the academic and administrative

Table 7.38 Distribution of level of urgent, means score and ranking of defects as rated by maintenance organisation 00000

Defect	Frequency of respondents on degree of agreements (%)					Descriptive statistics	
	Not at all urgent	Not very urgent	Urgent	Very urgent	Extremely urgent	Mean	Rank
Damaged window	0	15.2	48.5	18.2	18.2	3.3939	18
Clogged water closet	0	3.1	25	37.5	34.4	4.0313	7
Damaged floor finishes	0	48.5	39.4	9.1	3	2.6667	29
Damaged wall finishes	0	48.5	45.5	6.1	0	2.5758	30
Faulty shower	0	33.3	39.4	18.2	9.1	3.0303	23
Faulty fan	0	21.1	51.5	24.2	3	3.0909	21
Damaged ceiling	0	30.3	39.4	21.1	9.1	3.0909	20
Faulty door locks	0	6.1	33.3	36.4	24.2	3.7879	10
Faulty towel rail	27.3	39.4	21.2	6.1	6.1	2.2424	32
Cracked floor	0	24.2	50.0	24.2	0	2.9697	27
Faulty hot-water system	3	39.4	42.4	12.1	3	2.7273	28
Faulty lighting	0	6.1	54.5	21.2	18.2	3.5152	14
Cracked walls	0	24.2	51.5	24.2	0	3.0000	25
Faulty air-conditioning system	0	3	27.3	27.7	42.4	4.0909	5
Faulty electrical circuit	0	3	12.1	30.3	54.50	4.3636	2
Lift failures	0	6.1	9.1	21.2	63.6	4.4242	1
Faulty taps/faucet	0	15.6	40.6	31.2	12.5	3.4062	17
Damaged roofing	3	3	15.2	24.4	54.5	4.2424	3
Collapse drains	0	18.2	33.3	27.3	21.2	3.5152	16
Faulty IT package/communications appliances	3	3	33.3	24.2	36.4	3.8788	8
Failed furniture and fittings	0	9.1	33.3	39.4	18.2	3.0606	22
Broken door	0	12.1	45.5	21.2	21.2	3.5152	15
Sink leakage/ blocked	0	6.1	39.4	30.3	24.2	3.7273	12
Bad soap holders	18.2	45.5	27.3	9.1	0	2.2727	31
Pipe leakage	3	6.1	33.3	21.2	36.4	3.8182	9
Faulty sanitary appliances and fittings	0.0	9.1	33.3	39.4	18.2	3.6667	13
Cracked frames	0	24.2	55.9	20.6	0	2.9697	26
Faulty fire alarm	3	6.1	15.2	30.3	45.5	4.0909	4
Faulty heat extractor	3.1	28.1	25	21.8	21.9	3.3125	19
Faulty portable fire extinguisher	3	15.2	15.2	36.4	30.3	3.7576	11
Faulty smoke detector	0	6.5	19.4	38.7	35.5	4.0323	6
Cracked staircases	0	21.2	57.6	21.2	0	3.0000	25
Total average index (%)	2.08	18.15	34.79	23.27	20.78	3.42	Urgent

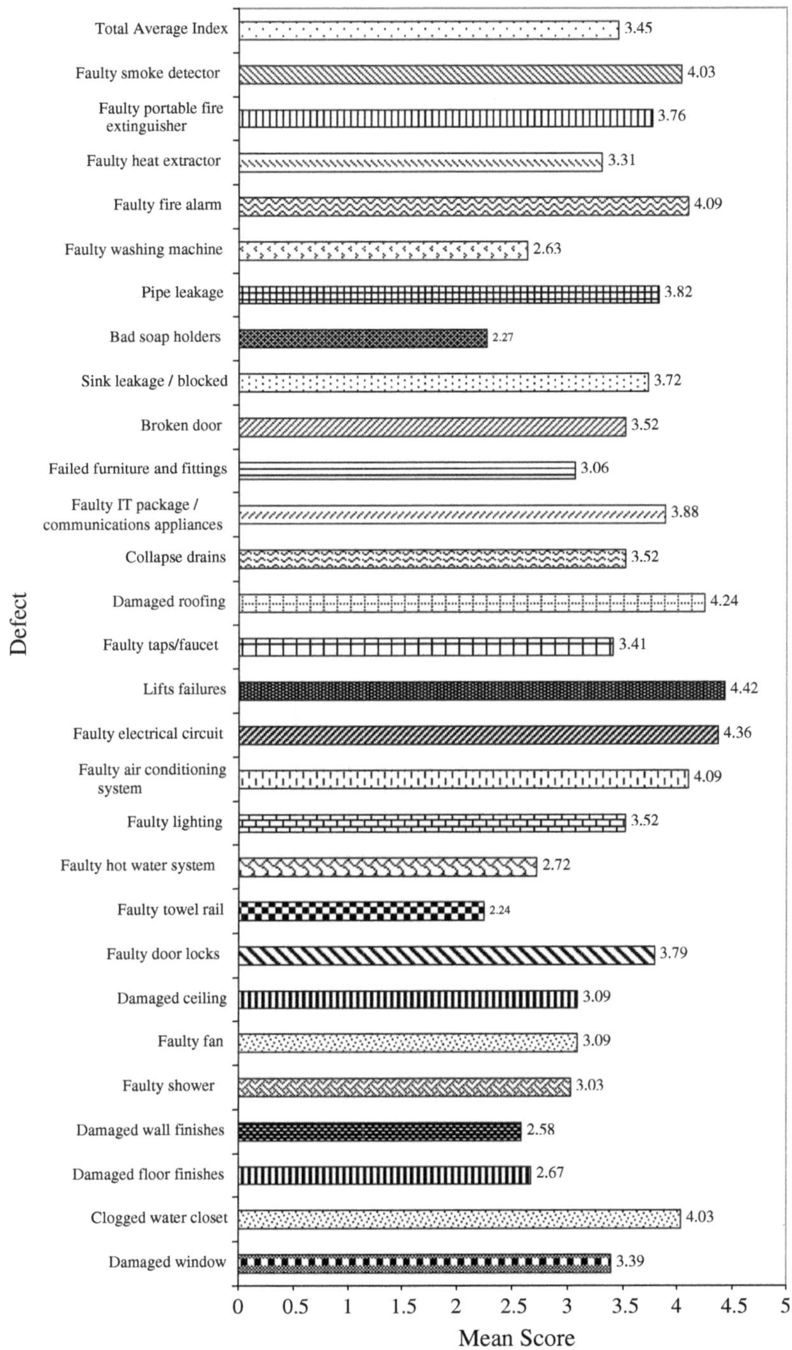

Fig. 7.11 Illustrative statistics for the criteria defects as rated by maintenance organisation

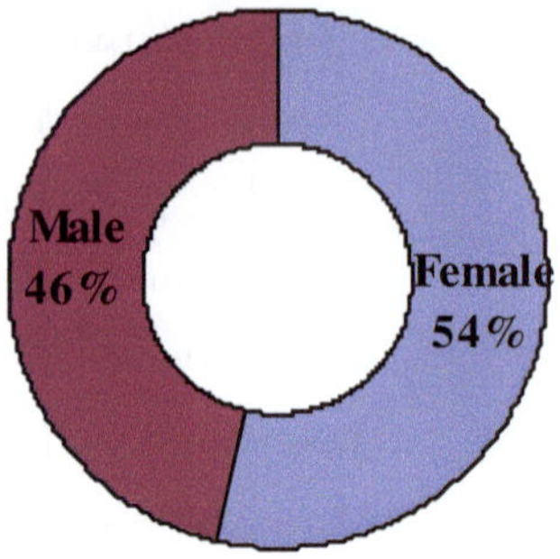

Fig. 7.12 Distribution of respondent's gender

Table 7.39 Distribution of respondent's type of university

Category	Frequency	Percent
Public university	348	73.70
Private university	124	26.30
Total	472	100.0

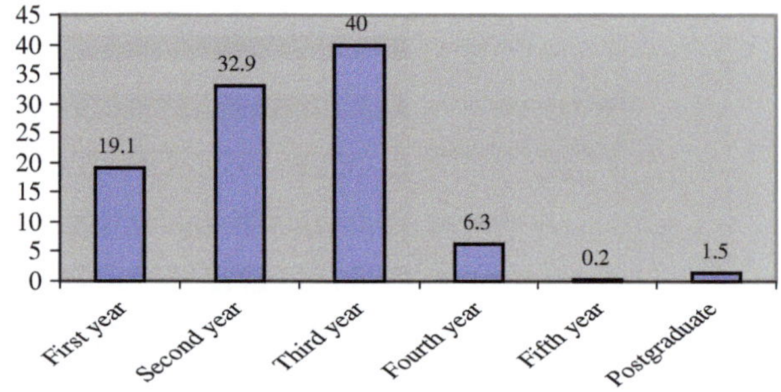

Fig. 7.13 Distribution of respondent's year of study

Table 7.40 Distribution of respondent residence

Residence	Frequency	Percentage
Outside campus	121	26.7
In campus	332	73.3
Total	453	100.0

sectors as well as the residential buildings. In fact, it is even possible that many of those living outside the campus once lived on the campus. Most students in their foundation/first year generally reside on the campus but opt to move outside the hostel as they "mature". Quite a number also move outside the campus as they get married while still studying. Although some universities provide couple

Table 7.41 Distribution of cross-tabulation between respondent's gender and respondent's residence

Count		Respondent residence		Total
		Outside campus	In campus	
Female	Count	70	174	244
	Percentage within respondent's gender (%)	28.7	71.3	100.0
	Percentage within respondent residence (%)	57.9	52.4	53.9
Make	Count	51	158	209
	Percentage within respondent's gender (%)	24.4	75.6	100.0
	Percentage within respondent residence (%)	42.1	47.6	46.1
Total	Count	121	332	453
	Percentage within respondent's gender (%)	26.7	73.3	100.0
	Percentage within respondent residence (%)	100.0	100.0	100.0

accommodations, most do not, but even the universities that do, the rent is bit high and the accommodation is not generally conducive to the couples. For these purposes, married couples mainly reside outside the campus, though close to their respective universities.

Gender and residence were crossed, the results of which are displayed in Table 7.41. The results indicated that 29 % of the female resided in their homes, while the remaining 71 % lived on campus. Out of the 158 male students that completed the questionnaire, 76 % of them took residence in the hostels. Only 24 % of the male students lived outside the university's hostels.

The study also determined the performance of the buildings. The results (Fig. 7.14) showed the majority (47 %) of the respondents measured the building good. About 42 % of the students measured the buildings fairly good. Only 5 % of

Fig. 7.14 Distribution of condition and performance of buildings on campus as rated by building users

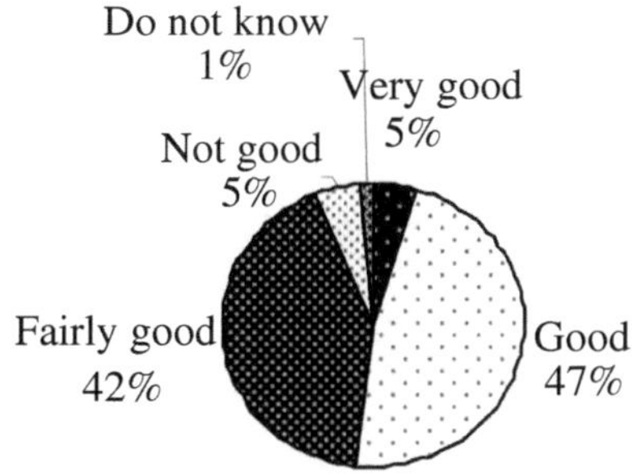

the responding students considered the performance of the buildings very good or not good each, respectively. Just 2 % of them considered the buildings not good at all.

A cross-tabulation analysis revealed that the majority of respondents that claimed that the performance of their buildings were very good (52.2 %) and good (61.9 %) were female (Table 7.42). Conversely, the majority of those that considered the performance of the buildings fairly good (51.2 %) and not good (73.9 %) were male building users. However, these outcomes were unexpected. It was expected that female students were more demanding and critical of building performance compared to male. Females are more concerned with appearances than males.

A chi-square test was performed to test whether there is a relationship with gender and building performance. The results of the tests are displayed in Table 7.43. The chi-square value of 16.833 with 4 degrees of freedom indicates there is a relationship between gender and building performance. The relationship was significant at a 0.002 confidence level.

Table 7.42 Cross-tabulation respondent's gender and the overall condition of buildings on campus

Count		The overall condition/performance of buildings on campus					Total (%)
		Very good (%)	Good (%)	Fairly good (%)	Not good (%)	Do not know (%)	
Female	Percentage within performance of buildings on campus	52.2	61.9	48.8	26.1	28.6	53.7
Male	Percentage within performance of buildings on campus	47.8	38.1	51.2	73.9	71.4	46.3
Total	Percentage within performance of buildings on campus	100.0	100.0	100.0	100.0	100.0	100.0

Table 7.43 Chi-square distribution of gender and building performance

Source	Value	df	Asymp. sig. (2-sided)
Pearson chi-square	16.833	4	0.002
Likelihood ratio	17.126	4	0.002
Linear by linear association	11.978	1	0.001
N of valid cases	477		

Table 7.44 Distribution of cross-tabulation between respondent's type of university and performance of buildings on campus

Count		The overall condition/performance of buildings on campus					Total (%)
		Very good (%)	Good (%)	Fairly good (%)	Not good (%)	Do not know (%)	
Public	Percentage within performance of buildings on campus	65.2	78.0	69.7	78.3	71.4	73.7
Private	Percentage within performance of buildings on campus	34.8	22.0	30.3	21.7	28.6	26.3
Total	Percentage within performance of buildings on campus	100.0	100.0	100.0	100.0	100.0	100.0

From the survey analysis, the majority of the users who considered the building performance very good, good and fairly good and not good were actually from the publicly owned universities. See Table 7.44. These can be interpreted to mean that public university buildings are in better condition than the private universities. However, in theory, this should not be so as private universities have sufficient funds for their maintenance compared to private universities. Public universities depend on governments' allocations to operate, and since the publicly owned universities are not for profits, the money accrued to them is limited in comparison with the private universities. Maintenance allocation and expenditure are closely related with earnings. It is common knowledge that the students in the private universities pay higher fees compared to those in public universities.

The chi-square tests displayed in Table 7.45 indicate there is no conclusive statistical evidence to conclude that a relationship exists between university category and building performance (chi-square = 4.884, df = 4, $p > 0.05$).

The majority of those that measured the performance of the buildings very good, good, fairly and not good were those that took residence in the hostels. See Table 7.46. While these results are difficult to interpret, it is those that live on campus that could provide valid feedback on the performance of the buildings. These groups of users do not only spend virtually all their time on campus, but they

Table 7.45 Chi-square distribution between university category and building performance

	Value	df	Asymp. sig. (2 sided)
Pearson chi-square	4.884	4	0.299
Likelihood ratio	4.869	4	0.301
Linear by linear association	0.440	1	0.507
N of valid cases	472		

Table 7.46 Cross-tabulation of respondent residence and the overall condition/performance of buildings on campus

		The overall condition/performance of buildings on campus					Total (%)
		Very good (%)	Good (%)	Fairly good (%)	Not good (%)	Do not know (%)	
Outside campus	Percentage within the overall condition/performance of buildings on campus	20.0	23.1	31.8	31.8	0.0	26.7
In campus	Percentage within the overall condition/performance of buildings on campus	80.0	76.9	68.2	68.2	100.0	73.3
Total	Percentage within the overall condition/performance of buildings on campus	100.0	100.0	100.0	100.0	100.0	100.0

also possess the experience in both hostel rooms and classrooms. However, chi-square tests of 7.217 with 4 degrees of freedom revealed that these results might be the case in reality (Table 7.47). In other words, there is a relationship between users' residence and a buildings performance.

The results of how students rate the current maintenance practice are displayed in Fig. 7.15. The survey results indicate that most (44 %) of the building users considered the current maintenance practice fairly good. Many considered the current practices good, however. A substantial (n = 66) group believed it was not good, but on the other hand, only 22 of the 473 responding users measured the current maintenance practices very good. Very few did not know how to rate the current maintenance management practice. That group of users opted for the "do not know" option.

Expectedly, most (58 %) of building users who considered the current practices very good were male (Table 7.48). But most (57 %) of users who believed the

Table 7.47 Chi-square tests on respondent's residence and building performance

	Value	df	Asymp. sig. (2-sided)
Pearson chi-square	7.217	4	0.125
Likelihood ratio	8.988	4	0.061
Linear by linear association	1.263	1	0.261
N of valid cases	453		

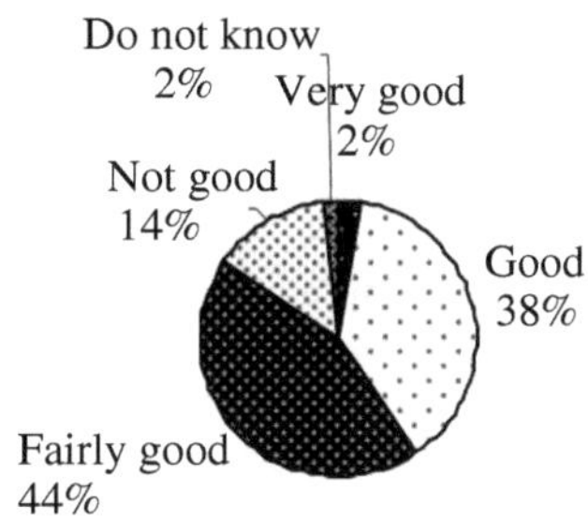

Fig. 7.15 Distribution on how students rate the current maintenance practice

Table 7.48 Cross-tabulation of respondent's gender and how students rate the current maintenance practice

		How students rate the current maintenance practice					Total (%)
		Very good (%)	Good (%)	Fairly good (%)	Not good (%)	Do not know (%)	
Female	Percentage within how students rate the current maintenance practice	41.7	57.2	56.5	39.4	37.5	53.7
Male	Percentage within how students rate the current maintenance practice	58.3	42.8	43.5	60.6	62.5	46.3
Total	Percentage within how students rate the current maintenance practice	100.0	100.0	100.0	100.0	100.0	100.0

practices were good were female. Contrary to expectation, most (61 %) of the users who considered current practices as not good were male. In theory and in practice, one will expect females to be the dominant group that would consider the practice not good enough, even so most of those that considered current practice very good, were male. The chi-square test was performed to establish whether a relationship existed between gender and maintenance practice. The results suggested there was statistical evidence to conclude that a relationship exits at 90 % significance level (Table 7.49).

As was mentioned earlier, buildings' users who could provide better opinions on the performance of the buildings were those residing on the campus. The results of the cross-tabulation are displayed in Table 7.50. However, based on the findings in

Table 7.49 Chi-square tests on genders and current practice as rated by building users

	Value	df	Asymp. Sig. (2-sided)
Pearson chi-square	8.537	4	0.074
Likelihood ratio	8.541	4	0.074
Linear by linear association	3.268	1	0.071
N of valid cases	473		

Table 7.50 Cross-tabulation of respondent residence and how students rate the current maintenance practice

		Very good (%)	Good (%)	Fairly good (%)	Not good (%)	Do not know (%)	Very good (%)
Off campus	Percentage within how students rate the current maintenance practice	20.0	20.7	30.7	29.2 %	28.6	26.4
On campus	Percentage within how students rate the current maintenance practice	80.0	79.3	69.3	70.8	71.4	73.6
Total	Percentage within how students rate the current maintenance practice	100.0	100.0	100.0	100.0	100.0	100.0

Table 7.51, there is not statistical evidence to conclude that such a relationship exists.

Most of those who considered the current maintenance practices as very good, good, fairly good and not good were building users from the public universities (Table 7.52). Two possibilities could trigger these outcomes. First, it is either that the public universities had a better maintenance management system or that the public universities have access to more funds or else the outcomes only happened by chance. However, the first reason is probably plausible since public universities depend on subvention from the government in addition to other avenues for generating funds. The public universities also have dedicated maintenance organisation though this often merges with development division. This is not the case with the private universities. Most do not even have a maintenance department as they rather choose to outsource their maintenance services. The chi-square test revealed there is no evidence to conclude the existence of any relationship statistically (Table 7.53). These indicate that there is no significant difference between the observed and expected frequencies.

Table 7.51 Chi-square tests residence and current practice as rated by building users

	Value	df	Asymp. sig. (2-sided)
Pearson chi-square	5.159	4	0.271
Likelihood ratio	5.262	4	0.261
Linear by linear association	3.194	1	0.074
N of valid cases	450		

Table 7.52 Cross-tabulation respondent's type of university and how students rate the current maintenance practice

		How students rate the current maintenance practice					Total (%)
		Very good (%)	Good (%)	Fairly good (%)	Not good (%)	Do not know (%)	
Public	Percentage within how students rate the current maintenance practice	66.7	72.7	75.2	72.7	75.0	73.7
Private	Percentage within how students rate the current maintenance practice	33.3	27.3	24.8	27.3	25.0	26.3
Total	Percentage within how students rate the current maintenance practice	100.0	100.0	100.0	100.0	100.0	100.0

Table 7.53 Distribution of chi-square tests of university category and maintenance practices as rated by building users

	Value	df	Asymp. sig. (2-sided)
Pearson chi-square	0.685	4	0.953
Likelihood ratio	0.672	4	0.955
Linear by linear association	0.157	1	0.692
N of valid cases	468		

The outcomes in Table 7.54 clearly indicate that most (22.33 %) of the building users prefer to channel complaints through telephone. This was followed closely by those that prefer to make complaints through Web-based request (21.8 %). Many of them also prefer to lodge complaints through email. Making complaints through personal visit comes next. Eighty-one percentage of the responding building users do not like to complaint through postal mail, and 87 % also do not appreciate sending complaints through fax. These findings are expected; building users want

Table 7.54 Distribution of preferable methods of sending maintenance complaints to maintenance organisation

Method	Yes	Percentage
Telephone	45.2	22.33
Email	41.7	20.56
Fax	12.7	6.30
Personal visit	38.9	19.13
Postal mail	19.5	9.60
Web request	44.3	21.80
Other	0.6	0.30
Total	203	100

easier methods of sending complaints. However, sending complaints through fax and postal mail are undoubtedly difficult. It will require much more of the complainants' time and effort.

A short message service (SMS) (through hand phone) is cheaper for users, especially the students. It is also not surprising that many of the users opted for personal visit. This could be so, because the users believe a personal visit to the maintenance department will be accorded with more urgency. Web-based request is also another emerging means of making complaints, though it is easy on the part of the maintenance departments to monitor works, complaints and loads. But it could be difficult for students who do not own computers and worse still where Internet connection is not sufficient or strong.

Building users' gender was crossed with the preferred methods of channelling of complaints, the results of which are presented in Table 7.55. The results clearly indicate that most of those who prefer using all the methods were female building users. The interpretations of these outcomes are difficult; it was not surprising. A meaning to these outcomes is that females are more concerned with the performance of buildings in comparison with male users. However, the chi-square tests suggest none of these relationships is significant 95 % except that of Fax ($p = 0.048$). Using telephone was significant at 90 % ($p = 0.069$).

The dominant problem (32.12 %) with the maintenance organisation, according to the building users, is poor complaint reporting systems. See Table 7.56 for details. These were followed by not having enough staff (21.13 %). Not having

Table 7.55 Chi-square distribution of gender with preferred methods to make complaints

Method	Female (%)	Male (%)	Total	Chi-square results
Telephone	58	42	100	3.296, 1 d. f., $P = 0.069$
Email	53	47	100	0.142, 1 d. f., $P = 0.706$
Fax	66	34	100	3.914, 1 d. f., $P = 0.048$
Personal visit	56	44	100	0.722, 1 d. f., $P = 0.396$
Postal mail	60	40	100	1.525, 1 d. f., $P = 0.165$
Web request	57	43	100	1.937, 1 d. f., $P = 0.164$

Table 7.56 Distribution of problems with maintenance departments as rated by the building users

Problem	Yes	Percentage
Not enough staff	40	21.13
Not enough money	34	17.96
Inexperienced maintenance staff	30	15.85
Poor complaint reporting system	60.6	32.12
Inexperienced administrative maintenance staff	24.4	12.94
Total score	189	100

enough funds came third on the list of problems facing maintenance organisations. Many (15.85 %) of the building users further believed maintenance organisations were faced with the problem of inexperienced maintenance staff. In addition, 13 % of the respondents argued that maintenance organisations have administrative staff who were not experienced with the work they performed.

The frequency results contained in Table 7.57 revealed that most of the building users who claimed the maintenance organisations were facing the listed problems were users from the public universities. However, the chi-square findings further indicate that the null hypotheses cannot be rejected. These are interpreted to mean that there is no significant difference between the observed and expected frequencies.

ANOVA was conducted to test whether respondent gender, university category, student level of studies and student's residence determine the way students rate the performance of their buildings. The results of the analysis are listed in Table 7.58. It is clear from the results that only student's gender can significantly explain the rating of the student performance. The relationship was significant at 2 % ($F = 4.317$; $p = 0.002$). The results do not indicate any significant difference in the building performance with category of university ($F = 1.221$; $p = 0.301$), student levels ($F = 1.946$; $p = 0.102$) and residence of respondents ($F = 1.813$; $p = 0.125$).

Table 7.57 Summary of frequencies and chi-square of problems with the maintenance organisation

Problem	Private (%)	Public (%)	100	Chi-square results
Not enough staff	26.6	73.4	100	0.017, I d. f., $P = 0.896$
Not enough money	27.3	72.7	100	0.141, 1 d. f., $P = 0.707$
Inexperience maintenance staff	26.8	73.2	100	0.25, 1 d. f., $P = 0.874$
Poor complaint reporting system	23.6	76.4	100	2.699, 1 d. f., $P = 0.100$
Inexperience administrative maintenance staff	29.1	70.9	100	0.625, 1 d. f., $P = 0.429$

Table 7.58 ANOVA whether respondent gender, university category, student level of studies and student's residence determine the way students rate the performance of their buildings

		Sum of squares	df	Mean square	*F*	Sig.
Respondent's gender	Between groups	4.186	4	1.046	4.317	0.002
	Within groups	114.422	472	0.242		
	Total	118.608	476			
Respondent's type of university	Between groups	0.946	4	0.236	1.221	0.301
	Within groups	90.478	467	0.194		
	Total	91.424	471			
Respondent's level	Between groups	7.244	4	1.811	1.946	0.102
	Within groups	439.276	472	0.931		
	Total	446.520	476			
Respondent residence	Between groups	1.413	4	0.353	1.813	0.125
	Within groups	87.267	448	0.195		
	Total	88.680	452			

Although there is no available evidence to conclude that where building users reside will influence their perception on the performance of the buildings, theoretically, students who reside on campus will have a better understanding of the building performance.

7.3.2 Analysis of the Criteria in the User Value System

The reliability test for the 12 VALUECRITE was considered satisfactory (Table 7.59). In fact, these results were excellent; the closer to 1, the better it is. Individually, the value ranges from 0.0919 to 0.923. Therefore, this data are reliable.

Table 7.59 Reliability statistics of the criteria of the user value system as rated by the building users

Cronbach's alpha	N of items
0.926	12

Table 7.60 Scale statistics of the criteria of the user value system as rated by the users

Mean	Variance	Std. deviation	N of items
48.797	60.668	7.789	12

Table 7.60 contains the cumulative mean, variance and standard deviation score for the criteria. The information in Table 7.60 outlined that 98 % of the respondents' mean lies between 2.2 and 6.0. A convergent validity test was conducted. This means that the criteria included are adequate and represents the concept. The correlation values for the validity scores range from 0.356 to 0.628. If correlations were higher than 0.75, the criteria might have to be examined, whether they were actually the same criterion split into two. The conclusions that are drawn from both validity and reliability tests are that the "goodness of measures" was very satisfactory.

It was hypothesised that the 12 criteria collected together can significantly explain user satisfaction in buildings. In other words, it is possible to use the criteria to enhance satisfaction. In order to substantiate this hypothesis, a multiple regression analysis was conducted. The 12 criteria were regressed against meeting user expectation with building performance. The outcomes of the analysis are contained in Tables 7.61 and 7.62. The results show that the criteria can significantly explain about 50 % of meeting users' expectations.

The results further indicated that the regression findings of $F(12{,}439) = 33.576$; <0.05 are significant (Table 7.62). Therefore, there is conclusive statistical evidence to conclude that the criteria are capable of predicting meeting users' satisfaction. The criteria can actually be used as screened criteria to ensure optimum user performance.

It is gathered from the outcomes of the coefficient result (Table 7.63) that individually the most significant criteria are aesthetic appeal, dependable maintenance organisation, friendly maintenance staff and quality of maintenance service provided.

Table 7.61 Model summary of the criteria of the user value system as rated by the users

Model	R	R^2	Adjusted R^2	Std. error of the estimate
1	0.692	0.479	0.464	0.649

Table 7.62 ANOVA whether building users consider the criteria within their value system capable of explaining their satisfactions

Model		Sum of squares	df	Mean square	F	Sig.
1	Regression	169.832	12	14.153	33.576	0.000
	Residual	185.042	439	0.422		
	Total	354.874	451			

Table 7.63 Coefficients of criteria of the user value system as rated by users

	Unstandardised coefficients		Standardised coefficients	*t*	Sig.
	B	Std. error	Beta	B	Std. error
(Constant)	0.444	0.198		2.238	0.026
Optimum functional performance	0.039	0.046	0.040	0.848	0.397
Safety and health of users	−0.069	0.056	−0.063	−1.237	0.217
Time to response to complaint	0.047	0.050	0.048	0.949	0.343
Aesthetic appeal	0.191	0.049	0.186	3.898	0.000
Security of property of user	0.021	0.052	0.021	0.416	0.678
User comfort and well-being	0.046	0.057	0.043	0.798	0.425
Time to complete the maintenance works	−0.030	0.054	−0.029	−0.559	0.576
Efficient reporting systems	0.008	0.053	0.008	0.150	0.881
Reliable maintenance organisation	0.127	0.052	0.125	2.418	0.016
Convenience of appointment with user to make repair	0.026	0.053	0.026	0.496	0.620
Friendly maintenance staff	0.109	0.045	0.113	2.406	0.017
Quality of maintenance service	0.374	0.052	0.356	7.139	0.000

There are 12 criteria that are addressed to the building users for them to measure how important the criteria are. For consistent and pragmatic comparison, the same instruments and methods of data analysis were used. This has been previously explained above. Table 7.64 contains the descriptive outcomes of the survey analysis. The table indicates that minimum and maximum scores for all the criteria were 1 and 5, respectively. This means that some respondents believed part of the criteria were not important at all, while some believed it is extremely important. The minimum mode was 3, while the maximum mode score is 5. The standard deviation scores further illustrated that there is wide disparity between the respondents' opinions. All the data were negatively skewed.

Table 7.65 lists the frequency, mean scores and ranking of the criteria in the user value system. The average mean score was 4.052. Specifically, 7 of the 12 criteria have their individual mean scores more than the average mean score. The interpretation of which is that the building users are more critical of the 7 criteria. Quality of maintenance service is the most highly ranked criterion, followed by users' health and safety and then security of the user properties. The least considered criterion was building aesthetic appeal. Surprisingly, optimum building performance and dependable maintenance organisations were also poorly ranked. Less than 1 % of the respondents actually agreed that the criteria were not important at all.

Close to 3 % did not also believe that the criteria are very important. On the other hand, 24 % measured those criteria as important. While 35 % of them believed the criteria were very important, majority (37 %) of the respondents considered the criteria extremely important to them.

Table 7.64 Distribution of descriptive statistics of criteria of the user value system as rated by users

Criteria	Minimum	Maximum	Sum	Mode	Std. deviation	Skewness
Optimum functional performance of buildings and engineering services	1.00	5.00	1,799.00	4	0.90759	−0.312
Safety and health of users	1.00	5.00	2,038.00	5	0.82575	−0.948
Time to response to complaint	1.00	5.00	1,919.00	4	0.88965	−0.674
Aesthetic appeal	1.00	5.00	1,753.00	3	0.86471	−0.084
Security of property of user	1.00	5.00	1,976.00	5	0.88969	−0.830
User comfort and wellbeing	1.00	5.00	1,950.00	5	0.85232	−0.608
Time to complete the maintenance works	1.00	5.00	1,933.00	5	0.86749	−0.563
Efficient reporting systems	1.00	5.00	1,904.00	4	0.88567	−0.656
Reliable maintenance organisation	1.00	5.00	1,867.00	4	0.87294	−0.415
Convenience of appointment with user to make repair	1.00	5.00	1,908.00	4	0.88730	−0.688
Friendly maintenance staff	1.00	5.00	1,963.00	5	0.92657	−0.899
Quality of maintenance service	1.00	5.00	2,061.00	5	0.83504	−1.052

The level of importance with the quality of maintenance service spread across the five categories of measurement. Only negligible (0.4 %) of the respondents concurred that quality of maintenance service was not important at all. Two percentage of them also argued that it was not very important. However, 16 % of the respondents believed it was not important. Concurrently, while 29 % measured quality of maintenance service very important, the majority (54 %) of the respondent actually adjudged it extremely important. These outcomes were not surprising, however. It is only natural that quality of work is significant.

The respondents also ranked user health and safety important. A less than 1 % of the respondents believed that users' health and safety is not important at all. About 1 % also believed user health and safety was not very important. The majority (48.2 %) of the responding building users, however, measured user health and safety extremely important. Similarly, while only 17 % of building users agreed that user health and safety was important, 33 % opted for very important. On the one hand, while about 3 % of the respondents believed that security of properties was either not important at all or not very important, 21 % outlined that it was important. Forty-seven percentage of the survey respondents measured security of property extremely important; at the same time, 29 % considered it very important.

It is not surprising that the building users ranked user comfort and well-being next. Nearly 3 % of the respondents considered user comfort and well-being either not at all or not very important. Forty percentage of them measured it extremely important, however. Close to 60 % of the respondents further believed it was either important or very important. It is only natural that building users will want to be attended by friendly maintenance staff. Significant size (4.4 %) of the respondents either agreed that friendly maintenance staff was not at all or not very important. The majority (42 %) of the respondents considered it extremely important. More than 50 % of the responding building users considered friendly maintenance staff important or very important.

Time to complete maintenance works was ranked high. Twenty-four percentage of the respondents considered this important. Furthermore, more than 70 % of the building users measured it very important or extremely important. It was expected that procedures of reporting system is rated high. While about 4.2 % believed it was not at all or very important, 22.2 % considered this important. Furthermore, more than 70 % of them think it is very important or extremely important. The majority (38 %) of the survey respondents believed that time to respond to maintenance complaints is very important to them. This is followed closely by many (35 %) who measured it extremely important. Although 24 % of the respondents also believed it was important, 3.6 %, however, considered it not at all or very unimportant.

Less than 5 % of the respondents do not agree that convenience of making an appointment to make a repair is important. Twenty-two percentage of the respondents conceded that it was important. Sixty-four percentage of the building users further measured convenience of making appointments for maintenance very important or extremely important. Surprisingly, optimum functional performance was poorly rated. However, 71 % of the respondents considered it important or very important. And while 5 % believed it was not at all or very important, 24 % considered it extremely important. These outcomes were indeed unexpected.

However, a likely conclusion that can be drawn is that the respondents did not understand the meaning of this variable. Building users are the essence of maintenance. Nonetheless, the majority of the users considered aesthetic appeal important. Yet, 57 % of the respondents believed it was very or extremely important. However, expectedly, aesthetic appeal was the last criterion considered

Table 7.65 Distribution of frequency, mean and ranking of the criteria of the user value system as rated by users

Criteria	Level of importance					Weight–age mean	Ranking
	Not at all important	Not very important	Important	Very important	Extremely important		
Optimum functional performance	1.5	3.8	35.0	35.6	24.1	3.772	11
Users' health and safety	0.6	1.3	16.6	33.3	48.2	4.273	2
Time to respond to complaint	1.3	2.3	23.9	37.9	34.6	4.023	8
Aesthetic appeal	0.06	4.3	37.9	36.0	21.3	3.730	12
Security user property	0.8	1.7	21.4	29.1	46.9	4.195	3
User comfort and well-being	0.4	2.1	22.0	35.7	39.7	4.123	4
Time to complete maintenance work	0.6	2.1	24.2	35.8	37.3	4.070	6
Efficient reporting system	0.8	3.4	22.2	38.6	35.0	4.034	7
Reliable maintenance organisation	1.1	1.5	30.0	36.6	30.9	3.947	10
Convenience of appointment to make repair	1.0	3.6	21.8	40.5	32.9	4.008	9
Friendly maintenance staff	1.5	2.9	20.1	33.5	41.9	4.115	5
Quality of maintenance service	0.4	1.9	15.5	28.5	53.5	4.330	1
Total average score	0.84	2.58	24.24	35.12	37.22	4.052	5

important by the users. While it is good and indeed one of the plausible criteria that enhances users' satisfaction, it is the functional performance that the building users are concerned with. Criteria of the building performance include thermal comfort, noise control and sound control.

A *t* test analysis was conducted to establish relationships between genders and the criteria of the user value systems. There were no significant scores for genders and any of the criteria. However, the relationships were measured at 95 % confidence level. The results suggest that gender does not have any effect on all the criteria. Specifically, the findings indicated that whether the respondents were female or male, they will choose the same criteria. A particular or set of criteria are not unique to any gender. These findings are disturbing, however. Gender is an essential construct of building maintenance. Undoubtedly, in practice and in theory, female requirements and the requirements of the male are different. Knowing this could be an effective tool in managing maintenance services.

On the other hand, there are statistical relationships between the category of the universities and the criteria of the user value systems. See Table 7.66 for the details distributions. For instance, there is relationship between efficiency of maintenance reporting system, dependable maintenance organisation and convenience of making appointment with building users for repair. Relationships also exist between university categories with friendly maintenance staff and quality of maintenance

Table 7.66 *t* test results (group statistics) of the relationships between category of the universities and the criteria of the user value systems

Criteria	Category of university	N	Mean	Std. deviation	Std. error mean
Efficient reporting systems	Public university	344	4.0872	0.82488	0.04447
	Private university	123	3.8699	1.03194	0.09305
Reliable maintenance organisation	Public university	346	3.9827	0.87459	0.04702
	Private university	122	3.8197	0.86260	0.07810
Convenience of appointment with user to make repair	Public university	348	4.0575	0.84344	0.04521
	Private university	123	3.8618	0.97783	0.08817
Friendly maintenance staff	Public university	348	4.1897	0.90066	0.04828
	Private university	124	3.8952	0.96961	0.08707
Quality of maintenance service	Public university	348	4.3994	0.77697	0.04165
	Private university	123	4.1138	0.95988	0.08655

services. Therefore, it can be concluded from these results that university category does have an effect on only these five criteria. The results indicate that public university building users were more critical of the criteria. These are evidences from the mean scores of each of criteria and university categories.

7.3.3 Analysis of Building Defects: Users' Perspectives

Prior to the main data analysis, a Cronbacch's reliability test was performed. The results of the test are shown in Table 7.67. The results clearly indicate the criteria used for the objective were satisfactory. The reliability tests range from 0.913 to 0.921.

The results of the multiple regression analysis F (33, 374) = 1.515; $P < 0.05$ (Table 7.68) imply that the defects that were addressed to the respondents can significantly explain the major defects in the buildings.

The minimum and maximum score was 1 and 5, respectively (Table 7.69). This implied that respondents' choose from the extreme cases. The mode scores are 3, 4 and 5. A conclusion that can further be deduced from the mode score is that most of the respondents conceded that defects were important, very important or extremely important.

The standard deviation results also imply that the opinions of the respondents are not very different. With the exception of cracked floors, collapsed drains and cracks in staircases, the criteria were negatively loaded. Frequencies of the degrees of urgency building users attached to each of the defects are listed in Table 7.70.

The average mean score is 3.6114. In other words, 21 defects have their individual mean scores more than the average means. Thus, 21 defects were critical to the building users. Faulty electric circuit was the criterion that building users considered most critical. However, some 12 % of the responding users considered it not urgent at all or not very important. On the other hand, 21 % believed it is important. Furthermore, some 68 % of them measured it very important or extremely important. These findings are not surprising, however. Faulty electrical

Table 7.67 Reliability statistics of defects in building as rated by the users

Cronbach's alpha	N of items
0.966	32

Table 7.68 ANOVA on whether the concentration on the defects can enhance satisfactions as rated by the users

Model		Sum of squares	df	Mean square	*F*	Sig.
1	Regression	27.087	33	0.821	1.515	0.037
	Residual	202.667	374	0.542		
	Total	229.755	407			

Table 7.69 Distribution of descriptive statistics of defects as rated by the users

Defect	Minimum	Maximum	Sum	Mode	Std. dev.	Skewness
Damaged window	1.00	5.00	1,487.00	3	1.11825	−0.067
Blocked water closet	1.00	5.00	1,625.00	3	1.09193	−0.330
Floor tile failure	1.00	5.00	1,589.00	3	1.07652	−0.165
Wall tile failure	1.00	5.00	1,641.00	4	1.12087	−0.364
Faulty shower	1.00	5.00	1,643.00	3	1.09222	−0.221
Faulty sanitary appliances and fittings	1.00	5.00	1,736.00	3	1.07574	−0.400
Faulty fan	1.00	5.00	1,779.00	5	1.13574	−0.626
Damaged ceiling	1.00	5.00	1,771.00	4	1.09779	−0.570
Damaged door locks	1.00	5.00	1,797.00	5	1.13778	−0.509
Cracked floor	1.00	5.00	1,765.00	3	2.59935	16.222
Damaged taps	1.00	5.00	1,756.00	4	1.07896	−0.527
Damaged roof structure	1.00	5.00	1,818.00	5	1.14659	−0.685
Collapse drains	1.00	5.00	1,800.00	4	1.75000	90.646
Cracked walls	1.00	5.00	1,768.00	5	1.14121	−0.595
Faulty communication appliances	1.00	5.00	1,790.00	5	1.05843	−0.505
Failed furniture and fittings	1.00	5.00	1,697.00	4	1.08007	−0.420
Broken doors	1.00	5.00	1,757.00	4	1.12353	−0.506
Sink leakage/ blocked	1.00	5.00	1,744.00	5	1.11167	−0.512
Bad soap holders	1.00	5.00	1,530.00	3	1.14063	−0.129
Pipes leakage	1.00	5.00	1,765.00	4	1.09399	−0.544
Faulty towel rail	1.00	5.00	1,524.00	3	1.22275	−0.158
Damaged hot-water dispenser	1.00	5.00	1,603.00	3	1.17756	−0.240
Faulty bulbs	1.00	5.00	1,719.00	5	1.13475	−0.457
Faulty air-conditioning system	1.00	5.00	1,811.00	5	1.18103	−0.731
Faulty electrical circuit	1.00	5.00	1,872.00	5	1.14746	−0.870

(continued)

Table 7.69 (continued)

Defect	Minimum	Maximum	Sum	Mode	Std. dev.	Skewness
Lift failures	1.00	5.00	1,736.00	5	1.23689	−0.619
Cracked frames (beams, columns, etc.)	1.00	5.00	1,631.00	3	1.17914	−0.334
Faulty fire alarm	1.00	5.00	1,758.00	5	1.19323	−0.656
Faulty heat extractor	1.00	5.00	1,692.00	4	1.17545	−0.490
Faulty fire extinguisher	1.00	5.00	1,777.00	5	1.17260	−0.665
Faulty smoke detector	1.00	5.00	1,756.00	5	1.18764	−0.592
Cracked staircases	1.00	5.00	1,562.00	3	0.95558	0.193

circuits are life-threatening. It requires urgent attention. The building users also considered faulty air-conditioning systems very critical.

Surprisingly, however, a sizeable number of the respondents believed it is not very urgent. In fact, 5 % considered it not at all urgent, and 21 % believed it is important though. Twenty-seven percentage considered it very urgent at the same time; majority (38 %) of them saw it extremely urgent. Lift malfunction was ranked next. However, 18 % of the respondents considered lift malfunctioning not at urgent or not urgent at all. Nonetheless, while some 21 % believed it was urgent, about 60 % measured it very or extremely urgent. Like the faulty electric circuit, lift malfunctioning is also a life-threatening issue. Life becomes unbearable for the disabled if the lift fails to function. Accessibility issues for the disabled in buildings have been well documented.

Twenty-two percentage of the responding building users considered damaged roof structures as requiring urgent attention. Fifteen percentage considered it not very urgent. However, 28 % measured it very urgent and another 36 % of them considered it extremely urgent. Collapsed drain was considered not very urgent or not urgent at all by 15 % of the respondents. While 22 % measured it urgent, 34 % considered it very urgent, and 29 % measured it extremely urgent. Surprisingly, faulty communication appliance was rated high. While 27 % rated it urgent, 62 % believed it is very urgent or extremely urgent. Similarly, more than 80 % of the survey respondents considered faulty fans urgent, very urgent or extremely urgent. Fifteen percentage of the building user considered faulty or damage ceiling as not at all or not very urgent. However, 23 % think it is urgent; further, 33 % believed it is very urgent and 29 % regarded it extremely urgent.

Expectedly, faulty or malfunctioning fire extinguisher was ranked urgent. However, about 20 % of the building users regarded it not at all or not very urgent. But 23 % of them considered it urgent. Thirty percentage believe it is very urgent; majority (32 %) of the building users agreed it was extremely urgent. Pipe leakage was measured not at all urgent by 4 % of the respondents. But while 10 % think it

Table 7.70 Summary of the distribution of the degrees of urgency of the defects as rated by the users

Defect	Level of urgency					Wtg. mean	Rank
	Not at all urgent	Not very urgent	Urgent	Very urgent	Extremely urgent		
Damaged window	8.2	18.2	37.8	22.4	13.3	3.1438	32
Blocked water closet	4.7	12.4	32.6	29.8	20.4	3.4871	24
Floor tile failure	4.4	16.9	34.0	28.5	16.2	3.3523	29
Wall tile failure	5.3	13.8	29.4	30.9	20.6	3.4767	25
Faulty shower	3.8	13.2	34.9	25.7	22.3	3.4957	23
Faulty sanitary appliances and fittings	2.8	11.0	29.4	29.2	27.5	3.6780	18
Faulty fan	4.2	10.3	23.4	30.0	32.1	3.7532	7
Damaged ceiling	3.2	11.6	23.0	32.9	29.3	3.7363	8
Damaged door locks	2.7	11.6	25.9	23.2	26.5	3.7911	6
Cracked floor	3.6	11.4	31.4	27.4	25.9	3.7158	13
Damaged taps	3.8	8.4	29.1	30.8	27.8	3.7046	14
Damaged roof structure	4.0	10.1	22.1	27.7	36.1	3.8193	3
Collapse drains	4.4	10.3	21.9	34.0	29.3	3.7975	4
Cracked walls	4.4	10.7	24.0	29.9	30.9	3.7221	11
Faulty communications appliances	2.1	9.7	26.5	30.1	31.6	3.7924	5
Failed furniture and fittings	3.8	11.8	29.0	32.6	22.8	3.5877	21
Broken doors	3.8	11.6	25.8	29.4	29.4	3.6912	16
Sink leakage/ blocked	4.0	9.7	28.2	28.8	29.2	3.6949	15
Bad soap holders	7.0	20.3	30.8	27.0	15.0	3.2278	30
Pipes leakage	3.4	10.1	26.0	30.9	29.6	3.7315	10
Faulty towel rail	9.5	19.7	27.7	25.4	17.8	3.2220	31
Damaged hot-water dispenser	6.7	17.0	29.4	26.5	20.4	3.3676	28
Faulty lighting	4.2	11.5	28.0	27.6	28.7	3.6497	19
Faulty air-conditioning system	4.9	10.1	20.7	26.8	37.6	3.8207	2
Faulty electrical circuit	4.2	7.4	20.9	24.3	43.2	3.9494	1
Lift failures	7.8	10.1	23.3	27.3	31.5	3.6471	20

(continued)

Table 7.70 (continued)

Defect	Level of urgency					Wtg. mean	Rank
	Not at all urgent	Not very urgent	Urgent	Very urgent	Extremely urgent		
Cracked frames (beams, columns, etc.)	6.9	14.7	28.8	27.9	21.6	3.4265	26
Faulty fire alarm	5.9	10.1	26.6	30.8	26.8	3.7167	12
Faulty heat extractor	6.3	12.2	25.9	30.1	25.5	3.5621	22
Faulty fire extinguisher	5.5	10.1	22.5	29.6	32.4	3.7332	9
Faulty smoke detector	5.7	10.9	23.9	27.7	31.7	3.6891	17
Cracked staircases	3.5	5.2	59.3	13.9	18.2	3.3810	27
Total average mean scores	4.85	12.0	28.29	28.20	26.7	3.6114	

not very urgent, 21 % ranked it urgent. Some 60 % regarded it very urgent or extremely urgent. A cracked wall was ranked above the average mean score. However, a sizeable group (11 %) regarded it not very urgent and 4.4 % think it is not at all urgent. Twenty-four percentage agreed it is urgent, and not fewer than 61 % considered it very or extremely urgent.

Sixty percentage of the survey respondents considered faulty fire alarms not at all urgent. 10 % considered it not very urgent. But 27 % measured it urgent. Moreover, 31 % saw it very urgent and 27 % measured it with extreme urgency. Cracked floor was measured not at all or not very urgent by 15 % of the respondents. Thirty-one percentage considered it urgent. Twenty-seven percentage regarded it very urgent. Twenty-six percentage measured it extremely urgent. While 8 % of the respondents measured damaged taps not very urgent, 4 % believed it was not urgent at all. Twenty-nine percentage measured it urgent, however. About 31 % think damage tap was very urgent, yet 28 % regarded it extremely urgent. It was surprising also that sink leakage or blockage comes next. While 10 % believed it was not urgent at all, 29 % regarded it otherwise. Similarly, while 4 % considered it not urgent at all, 29 % also believed otherwise. Yet 28 % saw it requiring urgent attention.

Broken door was measured not very urgent by 12 % of the respondents and 4 % measured it not urgent at all. But 26 % considered it urgent. About 58.8 % each considered it very urgent and extremely urgent, respectively. Smoke detector was rated not urgent by 6 % respondents, and another 11 % believed it was not very urgent. However, 24 % considered it urgent. In addition, 28 % believed it was very urgent, yet 32 % affirmed it was extremely urgent. Faulty sanitary appliances and fittings were considered not at all urgent by 3 % of the building users. Eleven percentage think it was not very urgent. On the other hand, about 30 % measured it urgent in addition to another 30 % who considered it very urgent. At the same time, 28 % asserted it was extremely urgent. Faulty lighting was also highly rated. Though about 15 % considered it not at all urgent or not very urgent, 28 % saw it urgent. Furthermore, close to 60 % believed it was very urgent or extremely urgent.

Floor drain was considered requiring urgent attention by 27 % of the building users. Though the majority (31 %) of the respondents considered it very urgent, about 15 % think otherwise even though 27 % measured it extremely urgent. Failed furniture and fittings was measured not requiring urgent attention by 12 % of respondents. Whereas 29 % think it was urgent, 33 and 23 % saw it very urgent and extremely urgent, respectively. Although heat extractor was poorly rated, 26 % of the respondents considered it urgent. In addition to that, 30 % saw it very urgent and 26 % believed it was extremely urgent. Faulty shower was considered urgent by 35 % of the respondent, and at the same time, about 50 % believed it was very urgent or extremely urgent. However, many still believe it does not require very urgent attention.

Surprisingly, clogged water closet was rated poor, even though 33 % measured it urgent. Thirty percentage measured it very urgent, while 20 % asserted it was extremely urgent. Wall tile failure was not well rated, but most of the respondents measured it very important. Twenty-nine percentage believe it only required moderate attention. Whereas 14 % think it is not very urgent, 21 % saw it extremely urgent. Surprising, cracked frames were rated poorly. In fact, more than 20 % of the respondent building users believed it was not at all urgent or not very urgent. Though some considered it extremely urgent, most of them believed it only required average attention.

Sixty percentage of the responding building users believed cracked staircases require only average attention. In fact, only 14 % considered it very urgent, albeit 18 % measured it extremely urgent. Seventeen percentage of the respondents considered damaged or faulty hot-water dispenser not very urgent, and another 7 % asserted it is not even critical at all. Although 20 % of the survey respondents saw it extremely urgent, a majority (30 %) of them only measured it to require average attention. Floor tile failure was considered not urgent at all by more than 20 % of the respondents. On the other hand, 34 % believed it required urgent attention. Nonetheless, 29 % asserted it very urgent and about 17 % also considered it extremely urgent. Thirty percentage believed faulty soap holder was not requiring urgent attention, but 60 % of the respondents believed it urgent or very urgent and 15 % measured it extremely urgent. Faulty towel rail was also measured as not very urgent by 30 % respondents. Eighteen percentage considered it extremely urgent as 29 % measured it urgent. Unexpectedly, damage window was the least ranked criterion by the building users.

7.4 Mathematical Model for Building Maintenance

By now, readers could understand that much of the previous discussion centres on operation or qualitative issues in maintenance management. However, an aspect of building maintenance engineering management is concerned with is forecasting future maintenance costs or preparing maintenance estimates for future repairs whether for preventive or corrective purposes. Many criteria are considered before a decision is made as to whether to maintain or not. Due to the scarcity of resources, there is the need to prioritise maintenance—the basis of which involves both qualitative and quantitative data. Therefore, maintenance is often based on schedule or planned arrangement. In order to make a sound decision regarding what element or building (or any parts of the building) to maintain first and about which one should come later, multiple criteria require consideration. For this purpose, the AHP technique was used to prioritise maintenance works to facilitate decision-making. It is intended only to provide information of quantitative method in maintenance management.

7.4.1 Development of Hierarchy Model

The purpose of the AHP here is to develop mathematical relationships or indices for building elements, cost criteria, performance criteria and operational criteria in building maintenance. The operational factors are the broad factors distilled from the primary survey. A Pareto analysis revealed that the data could be reduced to 20 % which can be incorporated into the mathematical model. AHP model was developed for each of the indices: building elements, cost criteria, performance criteria and operational criteria. This is because independently the indices can be used to prioritise maintenance services. In practice, to maintain a building will be unlikely to require the combination of the four indices.

In Fig. 7.16 are the eigenvectors distributions for the building elements. The figure shows the contributions each of the elements makes to maintenance cost. The consistency ratio (CR) is 0.0345, which is very much smaller than the acceptable value for a 9-by-9 matrix. Therefore, there is consistency in the measurements, and thus, the results are valid and can be reported. The theory is that if the CR is not within the confined range, the worthiness of the results might not be reliable.

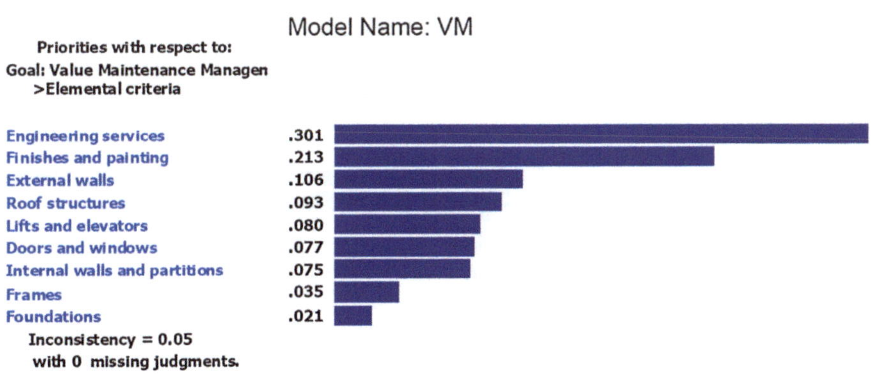

Fig. 7.16 Eigenvectors for building elements

The 0.301 means that on an elemental basis, the major element the maintenance organisations invested more money to maintain is the engineering services. 0.213 indicated that universities also spend much to maintain finishes and painting works. Further analysis of the vector for engineering services and finishes and painting revealed that the amount spent to maintain engineering services and finishes and painting is very much different. Universities also invest a substantial part of their money to maintain external walls. While doors and windows often require maintenance, foundation and roof structures often require less maintenance. However, although these findings are not very surprising, it is unexpected that roof structures are not making significant contributions because in the initial research it was found that roofs require extreme maintenance when the need arises. A plausible interpretation of this is that while roof structures seldom fail, it requires immediate attention when it fails due to its associated adverse effect on the occupants and property. The same reason is also applicable to elements like foundations.

The disparity between external walls and internal walls and partitions is relatively large. It could be expected that the external walls should contribute more because it is regarded as element providing protection from weather as well as climate modifiers controlling noise, sunlight and air that come into the buildings and thereby providing and sustaining conducive environment for learning, teaching, research and innovation. The low weighting of the internal walls and partitions may also stem from the fact that the maintenance organisation considered the internal building arrangement not very significant in influencing users' comfort and well-being. External walls together with or without internal walls are also required for image and marketing purposes. Figure 7.17 displayed the results of the operational factors in building maintenance management. The CR for the operational factors is 0.0682. This indicates a consistency in the measurements. Therefore, the results are valid and therefore can be reported.

The figure shows that the major factors that the maintenance organisations pay serious attention to are user safety and security of property. The 0.113 for financial issues clearly indicates that financial considerations are also very important factors that influence maintenance management issues. User safety and security of their properties are critical issues irrespective of the maintenance organisation structure. For instance, it does not matter whether maintenance is procured in-house or

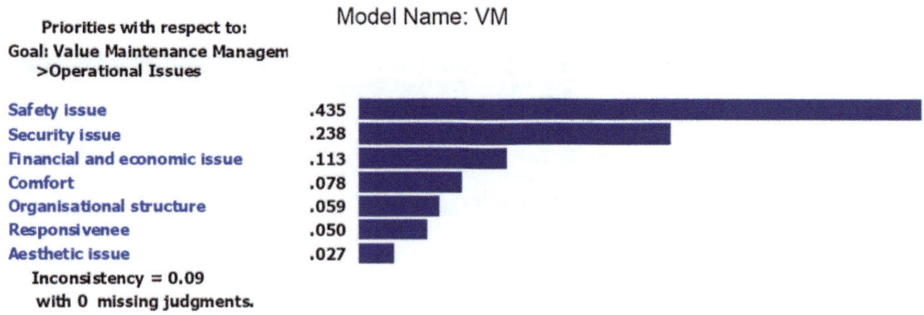

Fig. 7.17 Operational factor in building maintenance management

outsourced; buildings are attended to when it affects safety and security issues. However, maintenance organisations are not very keen in paying attention to user comfort, building appearance and responsiveness to users. These latter findings are very strange because being responsive to users' complaints and needs are very strategic issues in increasing users' satisfactions.

The results show that comfort is not highly rated. User comfort is the measure of how the building supports user activities. These findings are interpreted to mean that the buildings are adequate in terms of privacy, control noise, lighting, temperature and humidity. Responsiveness is the degree at which the maintenance organisation acts and reacts to building users' maintenance request. In this context, it includes how long maintenance requests are attended to, how users are treated by the maintenance personnel and whether repairs are made completely or often suspended after initiation, and whether the maintenance organisation always provides quality service. Surprisingly, the maintenance organisations do not consider aesthetics a contentious issue. In other words, aesthetics is not a basic requirement in maintenance service delivery; however, this is not expected. While the issue of providing conducive environment to support users' function activities is the primary requirement of building, aesthetic issue is also a major contributory factor.

In Fig. 7.18 is the list and eigenvectors for criteria that are also used to prioritise maintenance demand such as building condition; significance of the element, component or building to users; and intensity of use and the effect of lack (to suspend) of repair on the element component or building. Based on the results, the CR is 0.0555. The results revealed that the maintenance organisations consider the significance of the element or building before other criteria. The least criterion often considered by the maintenance organisation on the basis of these criteria is the intensity of use, while the highest is the significance of the element/components or building to users. So irrespective of building performance, intensity of use and effect of lack of repair or prevention on the building, the building will not be maintained unless it is very significant to the users. In general, all the criteria are making meaningful contribution —none is contributing less than 0.10. These findings are quite interesting as it shows a paradigm shift towards user-based approach. However, it is a different thing to maintain the building because it is important to the user and it is different to sustain the building in high performance. The users might be using the building because they have no choice; otherwise, they could prefer an alternative.

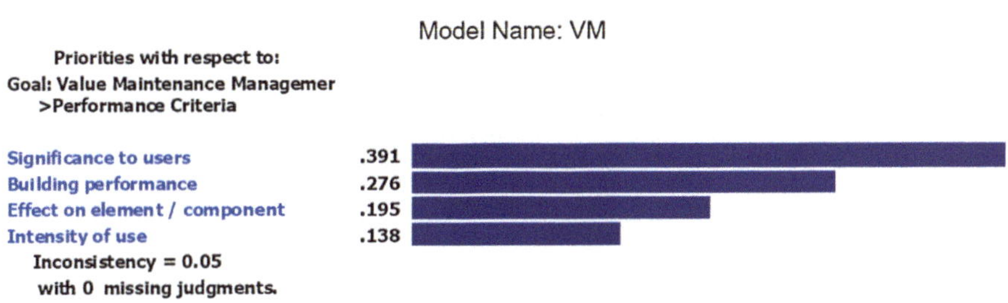

Fig. 7.18 Eigenvector of the performance criteria

Maintenance costs are also sometimes determined based on age of building, building performance/condition, gross floor area, size of user and number of buildings. In fact, most maintenance organisations actually prepare their budget based on these factors with allowance for inflation and contingency. The results of the AHP analysis on the vectors of each of the criteria are contained in Figure displayed 7.19. The inconsistency index for the five maintenance cost predictors was 0.05. This indicates the validity of the distributions based on the choices of the maintenance organisations is acceptable because the CR is 0.0446. The major predictor for maintenance cost is the built-up area and then the building performance. Specifically, the results demonstrate that 42 % of universities' annual maintenance is dependent on the gross floor area. Number of buildings is not a major determinant. The size of the building users is also making meaningful contribution to maintenance cost estimate. This is quite expected because intensity increases as the number of building occupants increases. This will ultimately lead to high maintenance costs as a result of high volume of complaints and needs.

Annual maintenance costs escalate at around 5 % subject to inflation and other economic issues. For that purpose, organisations often set aside another 1 % for capital improvement works which include replacement of ageing equipments and sustainability projects. Many organisations often relate maintenance expenditure to the asset replacement value. Organisations often consider that the total annual maintenance cost shall not exceed average of 3 % of total asset replacement including buildings, furniture and grounds. But what most organisations do is add or include 10 % of on previous expenditure to determine next allocation/ expenditure.

In general, Fig. 7.20 illustrates an AHP model for value maintenance management that can be used for prioritizing maintenance management decisions in general. It is a collection of the previous findings (Tables 7.18 and 7.19). The purpose is to illustrate major alternatives to consider in building maintenance. The results revealed that they are all comparable and they could complement each other but not necessarily compromised (Fig. 7.20). Based on the results, operational criteria (0.383) have the highest weight in maintenance management, followed by performance criteria (0.328), cost criteria (0.168) and finally elemental criteria (0.120). The CR results (0.0555) show the results of the participants are valid to be reported.

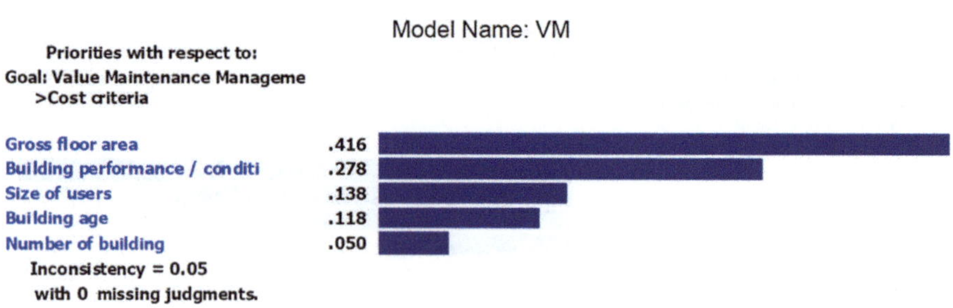

Fig. 7.19 Factor determining maintenance cost

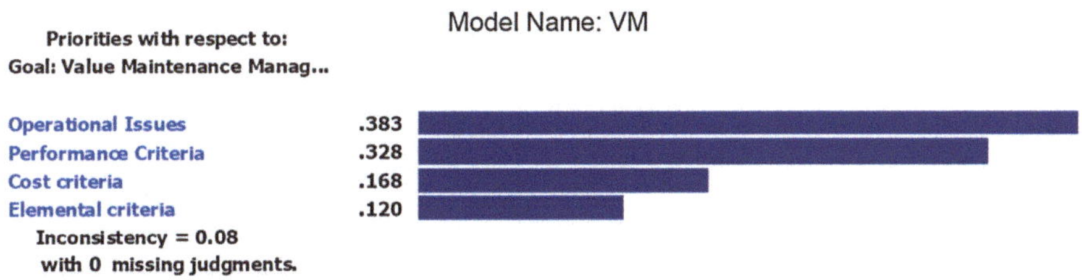

Fig. 7.20 Value maintenance management vectors

While the first three criteria, namely elemental issues, cost and performance issues, can be combined or used individually, the operational issue should always be considered, because it is the strategic issues within the entire maintenance management framework. For instance, if the purpose of prioritizing is to prepare an annual maintenance budget based on the performance criteria, the operational criteria should be factored into the decision-making process. Similarly, if decisions are to be made on what element to maintain, the operational issue should be considered. A basic reason why the elemental criterion is not making significant impact may probably be because in practice maintenance costs are not often based on maintenance basis even though it is a comprehensive approach of preparing maintenance cost. For instance, foundation seldom fails and if it fails, it normally requires the service of expert often outside the scope of the internal maintenance organisation.

In fact, most of the external contractors could not handle work relating to foundation repair like underpinning. The value in the parentheses that follows each criterion, factor or element refers to the vectors of the specific items (Fig. 7.21).

7.4.2 Model Validity

Validation is done to confirm whether the model can facilitate maintenance organisation to schedule their repair works by taking into consideration multiple criteria or factors. Usually, organisations would just address one of the repairs by intuition, but the AHP allows the organisation to explicitly consider the intermediate criteria. Therefore, each of the criteria is used to prioritise maintenance demand.

7.4.2.1 Validation 1: Elemental-Based Criteria

Data were collected from a maintenance organisation (of a university). The university intend to prioritise maintenance demand for four buildings: academic building, administrative building, hostel and auditorium. The analysis is with respect to element and criteria that affect influence prioritizing planned maintenance.

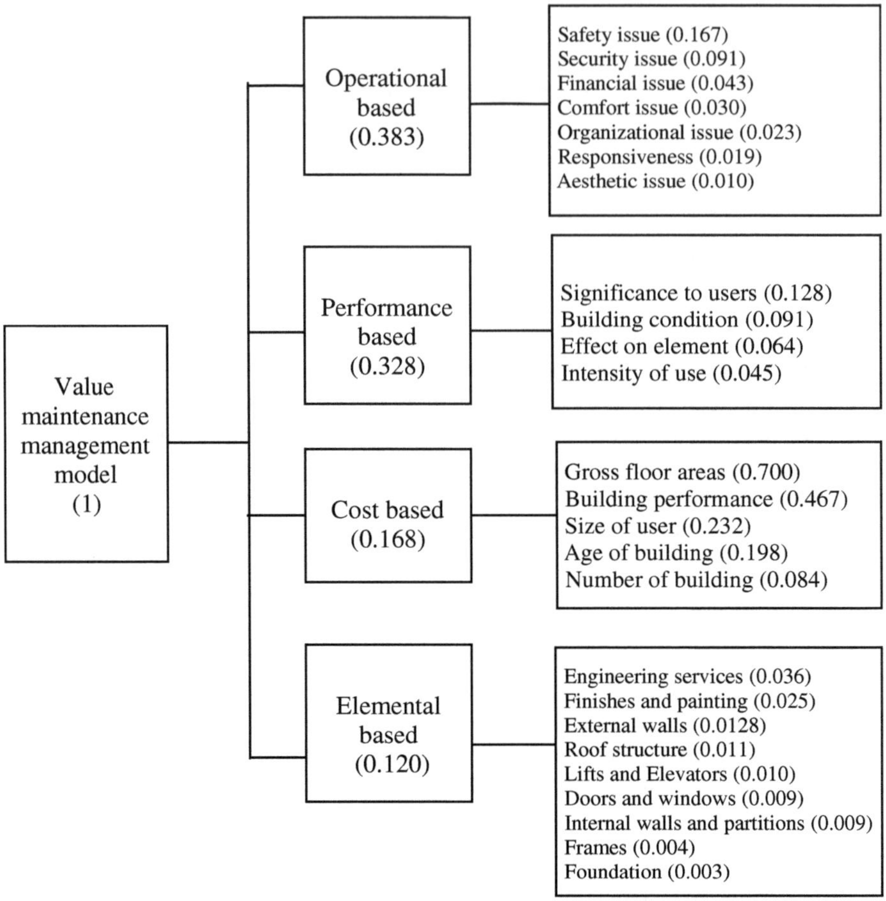

Fig. 7.21 Hierarchy model of mathematical model for value maintenance management

Initially, each of the elements is paired against one another to obtain their individual eigenvectors. Thereafter, each of the buildings is paired in terms of each of the elements. The results show that the administrative building is the most important building that requires more urgent priority in terms of scheduled maintenance before the others (Table 7.71). For instance, except for engineering services (0.206), the other elements contribute the highest each for the different types of buildings. The CR on the element basis against each building varied from 0.011 to 0.077, indicating the validity of the index. Although the CR for the doors and window is relatively high, but is acceptable, in fact, Saaty (2003) has reported the outcomes of results with a CR of 0.083.

The interpretations of the matrix is that administrative building is more important in terms of doors and windows, finishes and painting, frames, external walls, lifts and escalators and roof structures. All the buildings are equal in terms of foundation

Table 7.71 Matrix for eigenvectors of the building

Element building	Administrative	Hostel	Auditorium	Academic	CR
Doors and windows	**0.614**	0.269	0.059	0.058	0.077
Engineering services	0.206	**0.55**	0.135	0.109	0.055
Finishes and painting	**0.446**	0.072	0.277	0.205	0.055
Foundation	**0.250**	**0.250**	**0.250**	**0.250**	0.000
Frames	**0.562**	0.114	0.219	0.105	0.044
Internal walls and painting	**0.570**	0.075	0.168	0.187	0.044
External walls	**0.493**	0.078	0.190	0.239	0.055
Lifts and escalators	**0.461**	0.063	0.222	0.254	0.011
Roof structures	**0.462**	0.144	0.205	0.188	0.033

repairs, but hostels are more important to maintain in terms of engineering services. A basic reason why all the buildings are equal in terms of foundation is perhaps because it is mainly an issue of safety of users and even the passers-by. In order to determine the overall vectors for each of the buildings to select which of the building to maintain first and which to come next, the calculation involves post-multiplying the eigenvectors for the buildings and that for the elements.

The results of the calculation come out to be 0.405, 0.234, 0.187 and 0.167. These are the elements' value maintenance management vectors (VMMV) for this particular maintenance organisation. These four numbers correspond, in turn, to the relative values of administrative building, hostel building, auditorium and academic building. Specifically, the validation revealed that for the university, administrative building is almost twice (1.65) as important to maintain as compared to hostel, but hostel is not very much important compared to auditorium buildings even though the auditorium is also not very important in comparison with academic buildings. However, technically, the implication of the results is to schedule maintenance in terms of the vector as it signifies importance/criticality as this will imply transparency, consistency and being factual (Saaty 1980, 2003).

7.4.2.2 Validation 2: Performance-Based Criteria

Another university intends to prioritise its maintenance demand, and the criteria of prioritisation are based on the following performance criteria: building performance, significance of the building to users, intensity of use and effect of repair on the element itself. The eigenvectors for the performance criteria are as earlier on determined. Therefore, each of the buildings (element or component) will then be paired against each of the elements to obtain the overall eigenvectors to facilitate schedule maintenance. In line with the principle of AHP, the element with the largest eigenvector is to be maintained first followed by the element with the second largest eigenvector. Thus, the building performance vectors come out to (0.3391,

Table 7.72 Relatives' priorities for the selection criteria of the building

	BP	SU	IU	EE	Eigenvector	PR
Auditorium A	**0.380**	0.372	0.239	0.286	0.339084	1
Library (reference unit)	0.032	0.035	**0.042**	0.034	0.034943	7
Canteen	0.052	0.115	**0.323**	0.124	0.128071	3
Common lab. D	**0.155**	0.128	0.107	0.079	0.122999	4
Auditorium B	0.267	0.249	0.18	**0.378**	0.269601	2
Male common room (hostel)	0.055	0.064	**0.069**	0.049	0.059281	5
Tutorial room	**0.058**	0.035	0.039	0.051	0.04502	6
CR	**0.060**	**0.060**	**0.060**	**0.060**		

BP building performance, *SE* significance to users, *IU* intensity of use, *EE* effect on element, *PR* priority

0.034, 0.128, 0.123, 0.269, 0.060, 0.045) (Table 7.72) and could also be called performance VMMV.

The interpretation of the information presented in Table 7.72 is that the university should carry out the maintenance of Auditorium A first followed by Auditorium B. Mechanically, the implications of these results are that only three of the buildings, namely Auditorium A, Auditorium B and canteen, should be included in the next planned maintenance programme, while the others could be deferred or suspended in order not to unnecessarily put pressure on cash flow and budget. This prioritisation has provided a quick and much more reliable means of addressing urgency in maintenance without much biasness. Such prioritisation can be readily presented to management to earn their confidence, thus facilitating release of funds.

7.4.2.3 Validation 3: Operation-Based Criteria

To validate the operational issues, a case involving the prioritisation of factors is used to schedule maintenance of the request and inspect results. A different method is adopted. Additive method of prioritisation was used to prioritise the criteria. Often for simplicity and clarity, these versions are applied. Saaty and Hu (1998) further opined that different methods of prioritisation could lead to different final results for each of the alternatives and yet valid, provided the methods are explained. Therefore, to compute the priority weighting for the operational issues, it involves the multiplication of the weight of each of the factors rating scores to each of the vectors for each of the buildings.

A participant from a university was asked to rate each of the eight factors against each of the building on a scale of 1–5 where 1 denotes not critical at all, 3 denotes critical and 5 denotes extremely critical. The scales 2 and 4 are located in between. The CR for the building against each of the operational factors indicates the validity of the results because the respondents are consistent in their choices. The results revealed that using the factors as judgement criteria, laboratory should be

Table 7.73 Resulting priority values for building (part of building) by each factors

Factor building	Financial/economic	Security of property	Safety of users	Aesthetic	Organisation structure	User comfort	Responsiveness	CR	Eigenvector	Choice
Administrative building (inter-nal walls)	2	4	4	4	2	2	2	0.057	3.40	5
Canteen (in academic)	3	3	4	4	3	4	3	0.064	3.54	3
Classroom	2	2	4	3	2	4	3	0.064	3.10	7
Laboratory	2	4	5	3	3	4	4	0.064	4.12	1
Library	3	3	3	3	2	4	3	0.050	3.02	
Entrance fence	3	4	4	4	3	3	2	0.064	3.65	2
Female prayer room	3	2	4	3	3	4	3	0.064	3.28	6
Equipment and instruments room	4	4	3	2	4	3	3	0.064	3.47	4

maintained first before the entrance fence (Table 7.73). The building that will be maintained last, based on these factors, is the classroom. Analysis of the results on why the laboratory should be maintained before other buildings may be because of the contents of the room. Laboratories normally contain expensive equipment, instruments and materials that the university could not afford to lose. It is also possible for the entrance fence to require urgent maintenance because of its strategic purpose to the university's corporate image. Another reason might be security considerations. Although it might be odd to find that classroom is rated 7, as the last building to repair, the prioritisation captures all the implicit factors in the decision context.

7.5 Summary

This chapter has been able to identify, assess and rank the criteria that influence maintenance management systems of the buildings. Difference of opinion exists between what the maintenance organisations concurred to improve user satisfaction in maintenance services from what the users themselves agreed on. In another related development, there are differences in ranking of defects among the maintenance organisations and the building users. The identification and analysis of the degrees the defects contribute to directing maintenance organisations on maintenance management. This is crucial whether universities desire to succeed in modelling a systemic maintenance management system for their buildings and engineering services. Defect classification is a very strategic function of maintenance organisations. In order to maximise maintenance budget, resources should be directed to the extremely urgent defects, then the very urgent and finally to the ones that are not very urgent. Probably it is the only way that the ever inadequate fund will be spent judiciously and users' satisfaction can be achieved and maximised accordingly.

References

Chan, E., Yiu, C. Y., Baldwin, A., & Lee, G. (2009). Value of buildings with design features for healthy living: A contingent valuation approach. *Facilities, 27*(56), 239–249.

Russell, R. S., & Taylor, B. W. (2006). *Operations management: Quality and competitive in global environment* (5th ed.). USA: Wiley.

Saaty, T. L. (1980). *The Analytic Hierarchy Process*. New York: McGraw-Hill.

Saaty, L. T. (2003). Decision-making with the AHP: Why is the principal eigenvector necessary *European Journal of Operational Research,* (145), 85–91.

Saaty, T.L., & Hu, G. (1998). Ranking by the eigenvector versus other methods in the analytic hierarchy process. *Applied Mathematical Letters, 11*(4), 121–125.

Steinberg, W. J. (2008). *Statistics a life!*. California: Sage Publication, Inc.

Chapter 8
Discussing the Findings

Abstract In this chapter, the fundamental and complex issues in the management of building maintenance are discussed in detail. The common and from time to time implicit assumptions of many studies are that "user value system" in maintenance services cannot be identified let alone be measured. But, this understanding and assumption are grossly inadequate and in fact unfortunate. But if we might ask, *if you cannot measure it how then you can improve it*? In part, with this question in our minds, we revealed the services mismatch in this chapter. The variability comprises both for the value systems and in building defects. It is emphasised that to create value to the maintenance services users, the "experience of the users" is a fundamental ingredient. The service gaps are an alignment tool connecting the parent organisation with maintenance department on the one hand and the maintenance department with the service users on the other hand. Furthermore, we also emphasise here that values in "services" are created along the supply chain. In other words, value is not delivered just by addressing the compliant or defect in the building. A salient point that this chapter revealed is the changing roles of the maintenance department. Objectives and learning outcomes of this chapter (1) Understand the problems facing maintenance organisation. (2) Have a feeling of the connection between building performance and quality education. (3) Able to explain and describe solutions to maintenance management problems.

Keywords Service gaps · Priority setting · Performance of the maintenance organisation · Budgeting for maintenance · User experience

8.1 Introduction

This chapter presents the discussion of the findings of the research from which this book was developed. Research discussions are often combined in one chapter with chapter on analysis of findings. However, the choice to combine and not to combine is not a "mechanical issue". It is better to separate some research discussions from

A.L. Olanrewaju and A.-R. Abdul-Aziz, *Building Maintenance Processes and Practices*, DOI 10.1007/978-981-287-263-0_8

the data analysis to facilitate better discussions and easy understanding by the readers—this is carrying the readers along. However, it is not entirely possible that certain degrees of discussions will not be made while interpreting the research findings. If the discussion of this research is combined with that of the data analysis and findings, it could lead to fragmentation and complicate readers' understanding. This establishes the need for their separation. This chapter is structured into discussions on: respondents' profiles, criteria that influence maintenance management, criteria of the user value systems and building defects.

8.2 Discussion on the Respondent Profile and Background Information

Most of the respondents hold strategic position. Thirty-two per cent of the respondents are maintenance managers while about 19 % were facilities managers. Substantial parts of the "other" are director of development or and maintenance "executive" (this is another title or term for maintenance manager in Malaysia). Most of the respondents have more than five years' relevant industrial experience. And most of them have been working in those capacities for more than five years. These backgrounds provide the respondents with wide experience capable of providing independent opinion on information that were addressed to them.

Fifty-two per cent of the universities spent less than 10 million each on maintenance annually. Ten per cent of the universities spent about 30 million each on maintenance per annum is 10. Similarly, about 22 % of the surveyed universities occupied less than 280,000 m^2 built up area while more than 40 % occupied more than 1,400,000 m^2 built up area. Majority (42 %) of the buildings were about 15 years old while only about 10 % were between 30 and 50 years old. Therefore, it is inferred that most of the buildings are not that old per se yet a considerable amount of money is invested for maintenance. But, there are considerable complaints about buildings and their maintenance practices.

Quotes from dissatisfied building users are indicatives: "maintain efficiently and have friendly staff", "must have enough workforce and they must have knowledge on how to fix problems", "late in repairing", "have more experience maintenance staff" "alert, work faster and more friendly staff", "friendly men", "be alert and work with complaints", "staff must be friendly", "make it faster and immediately", "expert workforce", "the university should take care of the hygiene and surrounding area", improve the funding", "keep doing the maintenance according to recent problems/issue (time-to-time)" and "clean often". See Appendix for further quotations from some unhappy and frustrated building users.

Almost certainly, while there are the needs to increase maintenance budgets, the methodologies of managing the maintenance services are critical to maintenance budgets, building performance and user satisfactions. A quote from one of the maintenance organisations is an indicative "The budget allocated is not enough to carry out the proper maintenance as a result a lot of differed maintenance works occurred". Though available money that is usually allocated for maintenance is not enough, but with better maintenance practices much can be done to improve the building performance. Poor maintenance management processes will almost certainly cause work to take longer, cost more to accomplish objectives and reduces productivity. It also causes the user dissatisfaction and leads to deterioration of relationships between organisations and the maintenance department and between maintenance organisations and the building users.

It is obvious that with the ever increasing maintenance needs, the money that will be allocated will not be adequate to cater for the maintenance needs, even in future. Some of the respondents have this to say: "Please consider 5 M—man power, money, material, management and machinery to be included in your research. Manpower is the important factor to decide to have a better/efficiency maintenance of the university. Remember these phrases "World Class Buildings with 3rd Class mentality of worker", "Malaysia require law and regulation with regard to proper building maintenance" and "Require association of maintenance practice for competency purposes".

A total of 4.5 % of the maintenance organisations believed that their customers (users) are extremely satisfied with the services provided. About 5 % do not know-how satisfied their customers are even though the majority (77.3 %) believes that their customers are only satisfied with the service they have been provided with.

Most of the universities employed less than 30 employees each on full-time basis. Only 16 % of them actually employed 140–170 full-time employees even though 6.2 % of the organisations have 60–90 and 110–140 on their full-time pay rolls each. These findings are interpreted to mean that most of the maintenance organisations are in favour of outsourcing maintenance services. But, 66 % prefer to combine outsourcing with in-sourcing. The majority (46.9 %) of the organisations rate their current maintenance practices as good while 44 % rated theirs as fairly good. A minimum of 6.2 % rated theirs as very good. There is a discrepancy with these findings when compared with the fact that 77 % of the maintenance organisations claimed that their users were satisfied with the quality of services they provide to the building users.

Fifty-three per cent of the organisations have conducted users' satisfaction surveys before while 40 % have not, but the remainder does not know if they had conducted such a user satisfaction survey or not. While the majority had conducted the survey, it is clear that many have not. It is only through user satisfaction that evaluation can be made. The implications of these are that the buildings and their maintenance are not accorded commercial values. This is the case as organisations must improve their business practices to be competitive, market driven and to enhance their profitability shares. However, the maintenance department must have sufficient inbuilt flexibility to cater with every work arising from the demand. In case the maintenance organisation ran into deficit, 61.3 % apply for more money from their respective management while 32.3 % cut from money allocated for other purposes. This situation whereby sufficient allocations are not made for maintenance works is grossly inadequate. With all certainty, this could only lead to reactive and deferred maintenance. The implication is that maintenance backlogs will occur. But unfortunately, by the time the money is ready, if it ever is, the degree of the decay will increase subsequently too. Thus, the initial budget will not be sufficient to cater to the present condition of the buildings and its components.

The maintenance service users prefer to lodge their complaints through phone calls and personal visits. Those methods are perhaps easier compared to sending messages through fax and postal. Web-based method of reporting is undoubtedly not widely practised yet. For this method to succeed, there is a need for the users to be connected to the Internet. Many of the users do not even have access to Internet in their rooms and many do not even have own computers. Building users should not be expected to use cybercafes to lodge their maintenance complaints. This would be difficult for them and crude.

Therefore, maintenance organisations must look into the way maintenance complaints are lodged and the response time to complaints by the maintenance organisation. Users prefer one point of communication and not complication. They want someone to talk to person if a problem arises. In situations where time is wasted to contact the maintenance organisation, this will frustrate users the more. Users want to be able to report their complaints with ease. The contact address (phone numbers) of those concerned with specific items of maintenance should be supplied to the building's users. It will also facilitate easy reporting and communication. A free toll phone can be provided so that users will not need to pay for the complaint they made, this will go a long way towards increasing the satisfaction of the users with the service delivery. The behavioural aspects of maintenance management are critical.

The cross-tabulation conducted provides interesting facts on how maintenance organisations are arranged. Ten of fourteen responding maintenance organisations that hold a position other than those identified in the questionnaire actually had an engineering background. Interestingly, many (n = 4) of the facility managers had their professional background in facility management. Most (n = 4) of the maintenance managers had their background in engineering and facilities management. All (n = 2) of those holding the position of general managers were engineers. Many (n = 2) of those occupying other positions had their background in facility management.

The only respondent holding the position of administration manager belonged to the category of other professional background. The results were, however, difficult to interpret. Theoretically, it is expected that maintenance managers should have some background in maintenance management or, at best, in facilities management, building survey, centre management, property management and estate management. With regard to the building performance, while the duty of those with sufficient knowledge and skills in maintenance management can be described as that of a "specialist doctor", the positions of the other professionals are similar to that of the general doctor in a hospital. It would be more beneficial if the "experts" are allowed to play their roles.

Maintenance management is not yet a core course for the engineering programmes, at least at the undergraduate degree level in Malaysia. While the bulk of the work that the maintenance managers performed are concerned with building and service works, the engineers (mostly civil engineers in this case) have little knowledge and skills in those areas. The discipline of civil engineering involves public works projects such as bridge, dam and road construction. Furthermore, most of the civil engineers only have first degrees, and at the first degree level, they have only learnt about the design, installation and construction of capital projects not about their maintenance. The scope and purposes of maintenance works and capital projects are not the same. Both require different approaches and understandings.

Furthermore, the engineering discipline focuses primarily on the technical and operational aspects of project (Chang 2005) not on the managerial functions. Business skills and knowledge are not thoughts in typical engineering colleges/ university. However, maintenance management entails both the operational and strategic functions. The functions involve using theories and methodologies from varying disciplines including economic, laws, engineering, science, culture, mathematical and statistical information and knowledge. While engineers might gain some leadership and management skills through on the job-training, this is not adequate and sufficient to meet the competitiveness in modern business environments. Engineers are poor managers and perform only small leadership roles in practice. Putting all this into perspective, it is very possible that their professional competence will need to be questioned. Could the underlying reason be explained by a shortage of trained maintenance experts or the lack of adequate knowledge or simply the mindset?

Therefore, engineers that intend to occupy the position of managers of the emerging business-led maintenance organisation will need to acquire the prerequisite strategic business skills and knowledge. The implication of these findings is that those currently occupying the maintenance manager positions will need to return to college or university for a postgraduate degree in maintenance management and to continuously update their skills and knowledge through seminars and workshops. Universities should ensure that only those with strategic business acumen are employed to be maintenance managers.

It was also found that most universities prefer to outsource what they termed "non-core" to external contractors. Problems associated with outsourcing and insourcing were also considered influential. More than 70 % of the respondents considered both criteria influential, very influential or extremely influential. Though the responses do not indicate both to be very influential and extremely influential, they are recurrent issues in maintenance issue. Often organisations like universities considered building maintenance as non-core activities and as such prefer to outsource much of their services to external contractors. However, building maintenance is very critical to the survival of organisations. Students and faculty members spend most, if not all, of their productive time in or around the classrooms, laboratories, theatres and/or tutorial rooms.

Universities should also invest in training their maintenance staff as they do for their academic staff. Substantial commitment is required for continuous professional programme for the maintenance operatives. Maintenance staff must be motivated.

After all, enormous resources are committed to procuring their buildings. In fact, it is a failing part of the management to consider the management of their buildings as non-core activities. Maintenance is also a core activity of the university organisations, since without it the university would not survive. External maintenance organisations could hold their clients to ransom. It also leaves the in-house maintenance organisation staff less competent and inactive due to redundancy. It will be more profitable if the university could use in-house operative for most of their maintenance services.

Furthermore, another issue in outsourcing is trust and confidence. The outsourcing operatives do not see themselves as part of the mainstream staff leading to the perpetrations of dubious behaviour. Analysis of reports of theft cases in a university revealed that more than 60 % of the theft cases were actually carried out by operatives of the outsourcing organisations. These operatives come in during their off-duty time or in the evening to commit crimes such as stealing student laptops and other valuables. This is easy since the operatives had carried out repairs to locks, doors and cupboards. They often retain the spare key when they are supposed to return it immediately to the hostel administration. It is unfortunate, however, that about 30 % of the respondents failed to consider trust and confidence among staff as influential. In fact, only 3 % said it was extremely important. But, trust and confidence are not the only issues that are related with the outsourcing arrangement. The in-house maintenance staffers should also have confidence and trust in one another. Otherwise, service delivery could be undermined.

Further, empirical evidence has led to the conclusion that outsourcing could in actual fact be more expensive in the long run. Times have changed. It is high time that universities accepted and took care of their buildings (vis-à-vis the maintenance practices) efficiently. It is no longer acceptable for a university to invest only on improving methods of teaching and learning without improving the performance of the building assets. Outsourcing maintenance always reduces maintenance procurement to a lump sum. However, the risk and uncertainty involved with maintenance services dictate that the traditional lump sum is undoubtedly unsuitable for maintenance works. One lump sum requires the projects to be measureable, otherwise the associated risks will be very high and thus the bid will be excessive. In the opinion of this author, much of the success accredited to outsourcers are self-praise because they are actually comes from outsourcers themselves rather than from their clientele.

8.3 Discussion on the Criteria that Influence Maintenance Management

Managing maintenance activities is a complex multifaceted thought processes. Various techniques and criteria must be examined in order to make suitable decisions that facilitate service delivery, productivity and satisfactions. The inabilities of the current maintenance systems or approaches to achieve best value for money invested have been well documented. However, the current maintenance systems alone cannot guarantee best value for money invested. There is a need for organised frameworks.

Practically, the current maintenance systems (corrective, preventive, schedule and performance based) cannot be avoided altogether. For instance, concerned respondents noted that actions should be quickly taken when there are breakdowns, that is, the unfortunate reality. And even if the maintenance allocation is increased, maintenance backlogs will perhaps be reduced but the unnecessary increase in maintenance costs and poor service delivery will remain. Therefore, taking into account, the criteria that have impact on decisions that affect maintenance management will facilitate value added services. The criteria cut across technology, management, behavioural, culture and engineering aspect of maintenance issues in the buildings. The recognition of potential criteria would have a positive impact on maintenance management which will consequently influence service delivery. If maintenance organisations are unaware of the criteria, poor management is almost guaranteed, which will undoubtedly lead to an unnecessary increase in maintenance costs, delay, low productivity, conflicts, disputes and poor user satisfaction.

Although, all the criteria addressed to the respondents were considered to influence maintenance costs and maintenance service delivery, respondents rated them differently. However, it is equally acknowledged that it is not possible for each to contribute equal weight to the maintenance decision-making process. However, it is pertinent for maintenance organisations to give cognisant consideration to each of the criteria. It is the wholeness that facilitates best value service. Although, it cannot be concluded with statistical evidence that all the criteria can significantly influence maintenance service delivery and costs; however, a plausible reason for this is the sample size. The number of criteria ($n = 31$) is almost equal to the sample size ($n = 33$). Therefore, rather than relying on the test of significance, reference should be made from other information in the data. It is important we step out of the statistical area to methodological issues to fully grasp the information discussed here. Discussions of each of the criteria are presented in the followings paragraphs.

It is only natural that the quality of materials and components used for maintenance (and design) is given due consideration. Not surprisingly about 95 % of the respondents rated the criterion as influential, very influential and extremely influential. It is a common sense that the use of defective or substandard materials will soon lead to maintenance problems, if not at the commissioning stage then surely later when the facility is operational.

To reduce the impact of the problem, durable and industry-proven materials or components should be the priority. But, good quality materials can also fail to make the grade if used for the wrong purpose or in an unsuitable environment. Each material has its own peculiar characteristics and qualities that the designer must understand when specifying. As for the contractors, a maintenance bond should be secured in order for them to rectify all defects owing to the usage of unapproved or defective materials and components. Certainly, this will reduce building operation costs.

It is also not unexpected that maintenance organisations rated budget constraint as high. The reality is that it is not possible to say governments provide all the funds sufficient to cater for maintenance backlogs to the publicly owned universities. This is also true with the privately owned universities. Therefore, maintenance should be user driven rather than budget driven. Maintenance demand should be initiated in terms of its effect on the users' values systems. For instance, condition-based maintenance is not user driven, rather it is initiated because of the physical condition of the buildings. However, users are not very concerned about the physical condition of the building but rather by the building's performance. As a matter of fact, to the extent that the building is capable of allowing the users to perform their primary function, the buildings can be rated valuable by the users. However, a satisfactory way out is the effective and efficient management of the available resources.

With all built assets, irrespective of materials, components or construction methods used, the building will, over usage and time, encounter structural deterioration and systemic decline. Maintenance in terms of usage is hinged on, and very much related to, the culture of the occupants. Therefore, maintenance programmes should be introduced early, i.e. as soon as the project is completed. The personnel or department responsible for maintenance works should be encouraged to develop a maintenance inventory and keep well-organised records of all the equipment, plants and tools.

A characteristic of the construction industry's output is that it is remarkably different from the other industries is that, by and large, the products are perpetually exposed to weather conditions and the local climate. Hence, a major cause of maintenance problems in buildings is exposure to the often adverse and corrosive atmospheric effects such as that of sunlight and rain. Different building materials and components react differently to weather. Design and maintenance are not always considered together. This creates a situation whereby the left hand does not know what the right hand is doing. Therefore, there is the basic need for the involvement of the maintenance experts during design of a new development. On the Other hand, designers hardly have the opportunity to visit their previous project once completed. Where they do, perhaps is only during the defect liability period or else when invited by the clients. From the research, it reveals the importance of maintainers' involvement and suggests they have a lot to offer to the design team.

In order for maintenance organisations to actively support its core business objectives, there is need for top management to support the maintenance organisations. Maintenance issues should be discussed at the board room rather than at the technical desk. Issues receive proper attention when brought to the "big boss". The Vice Chancellors, Presidents, Rectors, Directors or as the case may be should have direct access to maintenance personnel. It is also recommended that maintenance mangers, executives or directors and general managers should be active members of the governing council.

For effectiveness, the structure of the maintenance organisation is very important. This will define and clarify the line of communication. In fact, while about 90 % of the respondents measured organisation structure as influential, very influential and extremely influential close to 90 % also believed that communication is influential, very urgent and extremely influential.

Lack of effective communication could lead to problems, reworks, waste of time and contribute to poor service delivery. If organisation is not well defined and effective communication is lacking, it will affect the scope, quality of maintenance and at the same time, maintenance objectives will be ambiguously defined. With all of these, there is no way that the user's expectations and perceptions can be met and/or achieved. However, while only 7 % of the universities believed that to meet user expectation was extremely influential, a sizeable group (16 %) believed it was not influential. However, we are sceptical to interpret this last outcome. User expectation should constitute the foundation of the maintenance service delivery.

However, a likely interpretation of this result is that it means that the respondents did not quite understand what this implied or they believed, and there is no way user expectations can be met. Alternatively, it could mean respondents were much more concerned about buildings not users. However, the first justification is very unlikely. The second justifications were the likely probable reasons. However, it is very unfortunate if the service providers believed the users' expectations cannot be met. This could be so where maintenance systems are not holistic and systemic. However, user expectations must in actual fact be exceeded. This cannot, however, be attained under the classical maintenance management philosophy.

This is the only plausible interpretation. Given that 50 % of the respondents measured having information on user performance satisfaction as influential, while many (25 %) opined that it is not even very influential. Whereas it is only through user satisfaction surveys that the criteria of the value system can be revealed. It is therefore, very unfortunate that maintenance organisations do not appreciate this instrument. Surprising, however, is the fact that many of the respondents claimed that they had conducted users' satisfaction surveys though they claimed that the outcome of the survey was only satisfactory.

It is interesting to find that many of the respondents failed to understand that having information on building performance is critical. In fact, 25 % of them also failed to appreciate the important of a building register. Information on the building performance and previous maintenance records should be well documented in a safe register. Where, this is not the case, a lot of time and effort will be wasted while locating necessary information when needed. Many of the respondents did not understand the areas where the competencies of maintenance mangers or executives are influential. It is well known that a competent manager will perform better that those that are not competent.

However, it was good that many of them considered skilled personnel as influential and very influential at same time that more than 90 % of the respondents measured staff training and development as influential and very influential. Labour is very important in maintenance operations. In fact, about 70 % of maintenance costs are attributable to labour cost. The maintenance personnel must display a high level of competency and be able to work in sync. It will not be difficult to postulate that even with high-quality materials and components, buildings will eventually fail if the workers that put the components together were not competent.

Poor workmanship could be the result of lack of know-how, training and skill or just plain negligence. Competent managers and operatives will undoubtedly save costs and at the same time improve users' satisfactions. Supporting issue of competency of maintenance organisations, many of the building users groups strongly suggested employment of company operatives.

Quoting from concerned users reinforced this: "improve staff skill", "efficiency of the workers", "hire experience maintenance staff" and "improve your maintenance management and don't be lazy".

Problems associated with outsourcing and in-sourcing were also considered influential, both criteria rated down the scale. However, more than 70 % of the respondents considered them influential, very influential or extremely influential. Interestingly, however, most of the respondents claimed to prefer outsourcing their services with only a small number preferring to in-source all their service. While most (66 %) of the organisations preferred to combine in-sourcing and outsourcing, a considerable number (22 %) preferred to outsource their service to external specialists. Fewer than 10 % preferred to in-source their maintenance services.

These findings are in tandem, with the outcome on the number of full-time employees. These two findings only pointed to the fact that most of the universities prefer to outsource their maintenance services. Often, universities have failed to realise that maintenance is also part of their core services. Building maintenance is very critical to the survival of universities. Students and faculty members spend most if not all their productive time in or around the classrooms, laboratories, or theatres or and tutorial rooms.

Outsourcing maintenance function reduces maintenance to corrective maintenance. There are also some organisational political-proprietary data that the clients would not like to expose to external parties. Tsang (2002) warned against organisations outsourcing facilitating functions such as maintenance. He also cautioned organisations not to consider outsourcing activities that are critical to their core activities. No doubt building maintenance is very critical to the continuous existence of all organisations and particular the "building-based type". However, decisions on what to outsource should be based on objective and quantitative facts even though a situation where a organisation outsources substantial parts of its services is not entirely healthy for any business unit.

According to Sherwin (2000), it is still too early to say outsourcing maintenance services is an alternative approach to maintenance service delivery. However, times have moved on. It is high time, organisations accept and take care of their buildings (vis-à-vis the maintenance practices) efficiently. It is no longer acceptable for a university to invest only on improving methods of teaching and learning without improving other assets.

Outsourcing if not properly managed often leads to maintenance backlogs and unnecessary increases in maintenance costs. It is argued here that organisations should not outsource the management of their maintenance services, albeit part of the maintenance implementation could be outsourced where extremely necessary. However, it will be more profitable if universities could use in-house staff for most of their maintenance services. External maintenance organisations could hold their clients to ransom. It also leaves the in-house maintenance organisation staff less competent and inactive due to redundancy.

In order for users to measure the value of the services being rendered, whatever results they get from maintainers will inevitably be compared with the service they received from retail and manufacturing industries. Hence, the construction industry players must embrace the culture of value added best practices as being observed in many other different sectors of the economy. In order to reduce maintenance costs through user approaches and behaviours, users can be educated and a maintenance manual would also assist greatly in this. Although a sizeable number of the respondents did not think a maintenance manual had very much influence on building management, about 80 % of them thought otherwise.

Other aspects that a maintenance organisation must look into are the way maintenance complaints are lodged and the response time to complaints by the maintenance organisation. In fact the users commented that: "poor reporting system", "improve complaint system", "give a very good and faster feedback for any report or problem exist" and "the reporting system should be more efficient". These comments clearly suggest there is the need for an alternative reporting system.

Often, the building users were confused about who should be notified regarding building repairs, such as leaky faucets, non-functioning windows or broken furniture. Many users would report the same problems to many different departments, individual or units or as the case may be. Further, suggesting that repairs are not being made promptly. However, usually the procedure is that problems would be reported to department office managers who would report the problems to a coordinator in the dean's office. The coordinator would in turn inform the maintenance staff. As such users will have to contact their office managers when they find a maintenance problem. Repairs could then begin to be made more promptly and there would be less confusion about how to report problems. However, awareness of the new procedures seems to be fading.

8.4 Discussion on Criteria of User Value System

This section discusses the criteria of the user value systems. This subsection was designed to identify, evaluate and prioritise criteria of user value systems that create value for the services that maintenance organisations provide. These criteria are a major source of satisfaction or dissatisfaction of the users with the service delivery. Therefore, in this section, the rating of the criteria by the maintenance organisation will be compared and contrasted by the rating of the same criteria by the users. It is a service providers and consumers analysis-service gap. This is done in order to make obvious where values are mismatched. Building users were asked to indicate further improvement that should be made to the management of buildings. They were also asked to indicate the problems they observed with their maintenance organisations. About 100 indications were extracted by the respondents opinions (refer to Appendix).

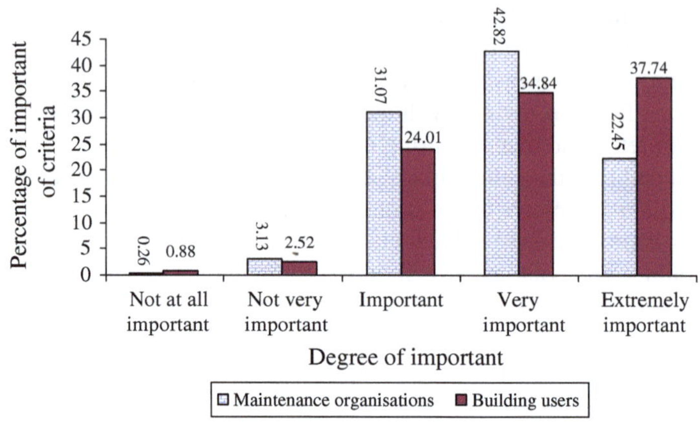

Fig. 8.1 Comparing the analyses of the criteria of the user value system

The question which naturally emerges is if there is any difference with the measurement of maintenance organisations and users and the question will thus be answered in this section. The next paragraphs will provide a summary of the users' profile. Figure 8.1 displays the cumulative compound histogram of the degrees of the importance of the criteria to both the maintenance providers and building users.

Analysis of the data indicated that of the 477 respondents, about 46 % were male and the remaining 54 % were female. In terms of the respondent category, 74 % were from public universities and 26 % came from the private universities. Most (80 %) of the respondents have spent more than a year at the universities. Above 70 % of the responding students resided on the campus. More than 90 % of the respondents were not very pleased with the condition and performance of the buildings. Similarly, almost 90 % of the respondents were not very pleased with the current maintenance practices. The analysis indicated that 71 % stressed that their opinions were very influential if meaningful maintenance is significant.

Most of those that lived on campus were females (52 %). Females constituted the majority (58 %) that live outside the campus. Most of those that rated the condition and performance of the buildings very good (52 %) and good (62 %) were females. On the other hand, those that rated the variable fairly (51 %) and not good (74 %) were males. The majority (58 %) of users that rated the building condition and performance very good were males. But most, (61 %), of the users that believed that the building condition and performance were not good were females.

The three most preferred methods that building users wanted to channel their maintenance complaints through were telephone (22 %), email (20.5 %) and personal visit (19.13 %). Based on building users' experiences, the three frequent problems facing maintenance organisations are poor complaints reporting systems (32.12 %), not enough staff (21.13 %) and not enough money (17.96 %). The largest part of the respondents that preferred telephone (58 %), fax (66 %) and personal visit (56 %) were females. Those that believed that the maintenance organisations were faced with the above problems were from public universities.

Table 8.1 Variability on criteria of user value system

No	Criteria	Users	Providers	Variability	*f*-value	Sig. level
1	Optimum functional performance	88.01	71.52	−16.49	24.33	0.000
2	User's health and safety	85.45	79.40	−6.05	27.93	0.000
3	Time to respond to complaint	80.46	81.21	0.75	35.93	0.000
4	Aesthetic appeal	74.51	64.85	−9.66	39.32	0.000
5	Security of user property	83.82	75.15	−8.67	34.26	0.000
6	User's comfort and well-being	82.31	75.76	−6.55	41.36	0.000
7	Time to complete maintenance work	81.30	79.40	−1.90	42.84	0.000
8	Efficient reporting system	80.67	75.1	−5.57	36.05	0.000
9	Reliable maintenance organisation	78.95	80.61	1.66	47.86	0.000
10	Convenience of making appointment repair works	80.17	73.61	−6.56	38.60	0.000
11	Friendly maintenance staff	82.31	73.93	−8.38	35.31	0.000
12	Quality of maintenance service	86.63	84.24	−2.39	75.97	0.000

User measurement of building conditions and performance did not depend on their gender, university category, their level of studies or whether they lived inside or outside the campus.

Table 8.1 contains the variability scores between the providers and building users with respect to the measurement of the criteria. The variability was derived by subtracting the scores of the maintenance organisation from that of the users. The Table presents a bleak outlook on the performance of maintenance organisations. The overall score for the maintenance organisations is 914.78, whereas that for the building users is 984.59. Variability has a negative score of −69.81 meaning that the overall requirements of users with regard to maintenance service delivery fall below expectations. The overall variability scores were found to be significantly different at 99 % confidence level. The *t*-test was performed to see if the variability was different for both the maintenance organisations and building users for each of the criteria. The test confirmed the variability.

Surprisingly, the size of the users (24 %) that measured the criteria important was lower compared to the size of the maintainers (32 %). However, most of the users (37 %) actually measured the criteria as extremely important. Only 23 % of the maintainers believed in that order. The percentage of the providers that believed the criteria were very important is 40. Conversely, only 35 % of the users believed in that order. The average mean scores rated by the users was 4.0516 while the providers rated it 3.8169. These were interpreted to mean that the users paid more attentions to the criteria compared to providers.

The maintenance organisations measured 6 of the 12 criteria critical to the users. From the providers' perspective, quality of maintenance was the most critical criterion. Surprisingly, the users also rated 6 of the 12 criteria critical. At the same time, they rated quality of maintenance service the single most critical criterion. Aesthetic appeal was the lowest criterion in the users' rating list. The maintenance organisations also believed that the users were not critical of the aesthetic appeal. While about 60 % of the providers considered quality of maintenance service very important or extremely important, more than 70 % of the users measured the criteria very important or extremely so. This is not surprising, however. It is only that the buildings users accorded more importance to the quality of services. However, the 60 % of the providers also suggested that providers were cognisant of their customer needs. But the major question is if the providers are meeting up to the expectations. Because it is one thing to know what is right and it entirely different to provide what is right. The right thing might not be provided due to certain constraints, however.

Although, maintenance organisations ranked time to respond to complaints as the second criterion, the building users actually desired issues of health and safety as second most critical criterion to them. Maintenance organisations actually ranked issues of health and safety as the fifth criterion. Instead, the building users ranked friendly maintenance staff as fifth criterion. Security of properties was ranked next by the users. But the maintenance organisation thought that dependable maintenance organisation is what users preferred. Whereas issues of comfort and well-being was ranked high by building users, the maintenance organisation thought time to complete maintenance work was next in the list of user values system. Building users actually ranked time to complete maintenance as the last in the list of most critical criteria.

Even though both the providers and consumers rated six criteria most critical, the criteria were ranked differently. Therefore, if the satisfaction of the building users was critical, a re-alignment was required. It was the users' requirements that should lead the way to create user-based services. Nonetheless, the other criterion needed to be evaluated. Building users are also critical of the efficient methods of reporting complaints. This is also corroborated by the ranking of the same criterion by the service providers. In fact, about 70 % of the maintenance organisations measured the criterion as very important or extremely important. This is a very important criterion that needs great consideration.

From the analysis, it is obvious that users preferred to report their complaints through telephones, emails and personal visits to the maintenance organisation. The users did not want any complications in reporting their complaints. Indeed, telephone was perceived to be the quickest and most convenient means of communications these days. Users could make use of the telephone anywhere and anytime. This could be a good method to employ as nearly everybody owns a handphone these days. Much of the comments received from the building users clearly indicated that users also preferred an online reporting system.

Therefore, maintenance organisations should CMMS online reporting system. This will go a long way in information management. The enormous quantity of information received by the maintenance organisation would mean that if a maintenance organisation does not own one, it might be difficult to manage information efficiently. However, while a number of the maintenance organisations did own the CMMS, poor usage of the software was prevalent. The software was not used to its best advantage.

The building users will require the telephone numbers of the maintenance staff. Universities need to publish the contacts of those responsible for maintenance, while some universities do this, for the users to have constant and reliable access to those in care of building managements is another issue. The maintenance organisations could provide free access staff for building users to be able to report maintenance complaints at all times. It was interesting to find that users also preferred to lodge complaints through emails. This is also a cheap and convenient method to report complaints. However, it might sometime prove expensive for users that do not own computers or and where that Internet is not effective and adequately available.

Users also preferred personal visits; apparently because of the unique consideration attached to face-to-face contact. It is only natural that complaints will be given consideration if both the provider and consumers comes in contact physically. To this, it is not surprising that methods such as fax, postal delivery and Web request were not the preferred method of communication. People want to talk to people, not machines. Apparently, users cannot bear the unnecessary time and complexity associated with the low-ranked methods.

The maintenance organisations did not measure security of user property as very critical even though users thought otherwise. However, in reality, most of the maintenance organisations considered it very important. In fact, more than 60 % believed it was very important or extremely important. However, based on these findings, it was pertinent that the maintenance organisations provided more security to user properties. It was only natural that users would consider the security of their properties crucial. Maintainers cannot appreciate this fact like occupants themselves can. Unwanted guest must be kept at arm's length. Cases of theft and rape are becoming more rampant these days, particularly among the students.

Both the maintenance organisations and building users actually ranked convenience of making appointments for repair as the ninth criterion. This is not surprising, however. Students are critical of times. Except in very few cases, students have tight schedules, Monday through Friday and morning till night. Building users that commented on the improvement required in the building maintenance management noted: "Improve the reporting system to make easy for the complainants and be quick in responding to the complaints". It is also natural that users would want to be around when repairs are ongoing to look after their property. This calls for concern particularly with the high rate of crime in the hostels these days. This is also to make sure that the maintenance work is properly done to the required standard; otherwise, they would have to face the hurdle of re-complaining again. However, this might not be the case if say perhaps the repair was required to common facilities such as toilets, bathrooms and kitchens.

Unexpectedly, however, both the providers and users ranked optimum performance of buildings and service low. In fact, it was ranked second to the last by both parties. This is indeed strange. As a matter of fact, only about 50 % of the providers believed it was very important or extremely important. These outcomes are difficult to interpret. But what this implies is that optimum building performance is not deemed critical. Or else the outcomes could be the result of errors of measurement. However, if the building users could not understand the construction, it was thought that the providers could. A plausible interpretation of the findings was that the providers were more concerned about the condition of the building rather the functional performance. If this is the case, then it is quite unfortunate.

Building occupants must be seriously engaged in order to make their value systems known. If this is not done, the service will not be appreciated by the occupants. For better satisfaction, when planning for maintenance services, one has to consider the interests of those the works are planned for. Concerned users of buildings have seriously commented on how they are often neglected when maintenance service is to be provided. The following quotations are indicators.

> "maintenance department should manage good links or connections with the students so that the problem can be solved very fast and easily. Department of maintenance needs to do their work properly and effectively to solve the problem", "Don't try to take advantage of a student facility that it is just for students", "Good co-orperation between users and staff", "Should provide accommodation preferable to the students", "Should hear the problems of Malaysia regarding the management of buildings in Malaysia and respond to complaints that have been made", and "Give the best facilities to students so that it is in line with the fees paid for each semester. The most important is hygienic"

The condition-based maintenance has various shortcomings. It is the functional performance of buildings that users desire, not the condition. To the extent that the building allows the users to perform the required functions optimally, then the building is a source of value creation. In fact, this can be further supported by the ranking of the aesthetic appeals. Both the providers and building users ranked aesthetics as the last criterion in the user value systems.

8.5 Discussion on Defects in the Buildings

The average means score for the defects according to the maintenance organisation was 3.42. On the other hand, the users measured this as 3.61. The interpretation of these is that, users are more critical of the defect compared to the maintenance organisation. However, it is only natural that users are concerned about the defects. The defects affect the users' productivity, comfort and total satisfaction. Although only 2 % of the maintenance organisation believed that all the defects did not require urgency at all, it was only 5 % of the users that thought the same. However, while 55 % of the users measured the defects very urgent or extremely urgent, it was only 44 % of the maintainers that believed that defects were not very urgent or extremely urgent.

Although only 3 % of the maintenance organisations believed that all the defects did not require urgency at all, it was only 5 % of the users that had similar thoughts (Fig. 8.2). However, while about 60 % of the users measured the defects very

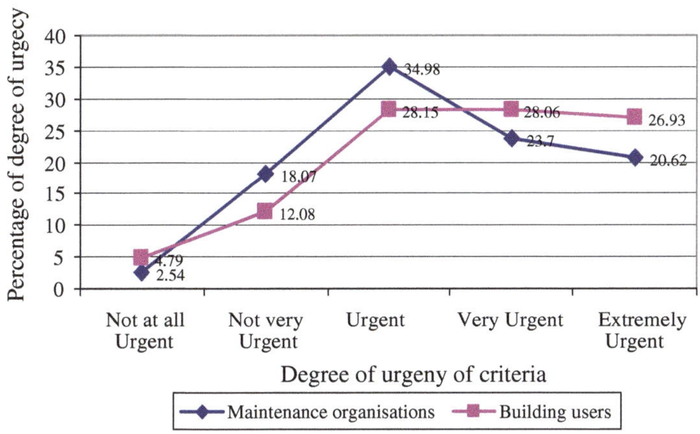

Fig. 8.2 Measuring the variability in defect rating

urgent or extremely urgent, it was only 44 % of the maintainers that believed that defects were not very urgent or extremely urgent.

Table 8.2 contained the urgency index, variability and the *t*-test value for the defect. Twenty-one defects were ranked very critical according to the users, while the maintenance organisations only ranked 16 defects as very critical. The columns on the variability contain figures with both negative and positive signs. The negative numbers mean the maintenance organisations were providing more than required on these defects. On the other hand, the positive sign indicated that the maintenance organisations were not meeting up to the requirements or expectations on effecting repairs or preventing them from occurring. As can be seen, there is wide variability on service provisions.

Twenty-one defects were ranked very critical according to the users, while the maintenance organisation only ranked 16 defects as very critical. However, the maintainers ranked faulty electrical circuits as the second most critical criterion. These outcomes were, however, not surprising, both defects are life threatening issues that require considerable attention. This is expected, as it indicates that both the providers and the building users could not afford to lose lives. While maintenance organisations believed lift mal-functioning was the criterion that required urgent attention, the users believed it was faulty electrical appliances. The users measured damage windows as the least-ranked criterion in contrast to this the maintainers believed bad soap holders to be the least critical. In fact, the maintainers believed a damaged window was very critical.

The third most critical defect according to the maintainers was damaged or faulty roofing. Coincidentally, the users also ranked damage roofing in that order. This is not surprising, however, as roof leakage is a very complex building defect with all its associated damage. It allows rain into the building and it affects the entire climatic condition within the building. Rain penetration causes further damage to the roofing members and other parts of the buildings and also damages the

Table 8.2 Variability of defects

	Defect	Users	Organisations	Variability	*t*-test	
					f-value	Sig. level
1	Damaged windows	63.73	67.88	−4.15	7.345	0.000
2	Clogged water closets	69.69	81.21	−11.52	9.509	0.000
3	Damaged floor finishes	67.00	53.33	13.67	6.704	0.000
4	Damaged wall finishes	72.16	51.52	20.64	10.827	0.000
5	Faulty shower	69.77	60.61	9.16	10.861	0.000
6	Faulty fan	73.42	61.82	11.6	12.008	0.000
7	Damaged ceiling	75.05	61.82	13.23	11.156	0.000
8	Faulty door locks	74.72	75.76	−1.04	5.953	0.000
9	Faulty towel rail	53.25	39.40	13.88	7.667	0.000
10	Cracked floor	72.21	59.40	12.81	8.753	0.000
11	Faulty hot water system	74.05	54.55	19.5	13.134	0.000
12	Faulty lighting	76.35	70.30	6.05	12.545	0.000
13	Cracked walls	74.68	60.00	14.68	7.551	0.000
14	Faulty air conditioning system	74.38	81.82	−7.44	6.719	0.000
15	Faulty electrical circuit	75.85	87.27	−11.42	7.162	0.000
16	Lifts failures	71.95	88.49	−16.54	3.721	0.005
17	Faulty taps/faucet	73.80	69.10	4.70	10.979	0.000
18	Damaged roof structure	73.88	84.85	−10.97	7.941	0.000
19	Collapse floor drains	64.57	70.30	−5.73	9.882	0.000
20	Faulty IT package/communications appliances	74.55	77.58	−3.03	14.505	0.000
21	Failed furniture and fittings	62.56	61.21	1.35	6.953	0.000
22	Broken door	67.34	70.30	−2.96	7.788	0.000
23	Sink leakage/blocked	70.53	74.55	−4.02	4.800	0.001
24	Bad soap holders	76.35	65.46	10.89	3.242	0.012
25	Pipe leakage	78.87	76.36	2.51	8.717	0.000
26	Faulty sanitary appliances and fittings	72.91	73.33	−0.42	9.789	0.000
27	Cracked frames	68.51	59.40	9.11	7.374	0.000
28	Faulty fire alarm	74.34	81.82	−7.48	3.702	0.006
29	Faulty heat extractor	71.20	65.46	5.74	9.453	0.000
30	Faulty fire extinguisher	74.63	75.15	−0.52	8.118	0.000
31	Faulty smoke detector	73.75	79.40	−5.65	8.845	0.000
32	Cracked staircases	67.38	60.00	7.38	3.814	0.005

occupants' belongings. All these can indeed be grave concerns with the Malaysian tropical weather. For instance, dampness is a serious problem in the tropical climate albeit buildings are designed to prevent and protect against dampness, the effects of dampness in buildings cannot be over stressed, however. The tropical climate also encourages the growth of plants including algae, fungi, mosses, figs and ferns.

Next in the list of users' concerns with defect management were collapsed drains. Contrarily, the maintainers ranked faulty fire alarms in that position. Collapsed drains were actually ranked sixteenth criterion by the maintainers. This ranking actually requires further consideration by the maintainers. The needs of the users should be dominant. The maintainers should investigate why the users believed collapsed drains as so critical to the users.

Certainly, collapsed drains are associated with a lot of damage to the users. When the toilets, kitchen and/or bathrooms drains collapse, it leaves the occupants uncomfortable. Collapsed drains are also associated with offensive odour particularly in the case of toilets and bathrooms. Considering these issues, it is not surprising that users ranked it in that order.

A faulty communication appliance was ranked next by the building users, but the maintainers ranked faulty air conditioning systems as the eighth criterion. Instead, faulty air conditioning occupied the fifth position in the ranking of the maintainers. While differences exist between the parties, they are not that great. No doubt air conditioning systems are important but may not be as demanding as the communications systems. It could be the fact that air conditioners were only in classrooms or workshops, and hostels lacked those facilities. Ceiling fans were the most common means of cooling the rooms in the hostels. In fact, many of the universities did not have air conditioning systems in the classroom. The roles of communication appliances such as the telephone cannot be emphasised enough. The faculty members and students greatly depend on these facilities for their day to day activities. Booking and keeping appointments with lecturers and supervisors these days are by means of telephone.

While maintainers ranked faulty detectors as sixth criteria, building users identified damaged door locks instead. In the view of the building users, smoke detectors occupied seventeenth position. This is a wide disparity. While both criteria are security issues naturally, door locks are critical to the building users. The views of the users are empirical, the effect of which is immediate to them. The users know the implication of not having functioning doors. This is not surprising and calls for concern because of the high rate of crime on campuses. Much of the user group commented on the need to improve on the security in the Malaysian university campuses. Quotations from university building users suffice: "security 24 h because many problems at night such as blackout", "security system at hostel of university very poor. I hope university will take serious about this problem", "safe building and comfortable to study" and "the hostel for students' safety. Building for faculty is not important but build the hostel is important. We are scare to live at the hostel. Think about it!!!"

Damaged ceilings were ranked high by the users as compared to the service providers' ranking. The service provider actually ranked ceiling damage as the twenty-first criterion. In the opinions of the maintainers, damage is not critical. This is surprising, however. It was the researcher's expectation, considering the impact of damaged ceilings on the building occupants, that the maintainers would rank it high. Damaged roofing exposes the ceiling to direct contact with the weather elements. These outcomes are difficult to argue. It was understood that providers ranked damage ceilings as critical, requiring very urgent or extremely urgent attention. Pipe leakage was ranked ninth by the maintenance organisations, whereas users preferred faulty fire extinguishers. No doubt, pipe leakage calls for serious attention yet fire outbreak is very serious indeed. With the recent issue of fire outbreak in Malaysia, it will not be difficult to argue why the user ranked it so high. Fire outbreaks destroy lives and properties, whereas pipe leakages in buildings do not have such corresponding devastation.

The implication of pipe leakage is alarming, and it allows water to penetrate the walls and floors and ceilings. It also triggers offensive odours which make users uncomfortable. It was interesting that the users also ranked piped leakage as tenth. This was interpreted to mean that pipe leakage was quite critical to the view of both parties. Issues related to pipe leakage are severe in Malaysia as many of the user groups commented on the bad state of toilets on Malaysian campuses. Apparently, pipe leakage was common in the toilet area compared to other places such as living rooms, classrooms and halls. In fact, one of the respondents commented on how pipe leakage had damage some walls on the campus.

Cracked walls in the view of the users were very urgent. The maintainers did not see them as so critical, instead they measured blocked sinks or sink leakage in that position. These were interesting findings. Walls are required for protection (against human intervention and the weather) sinks are not. However, sink leakage and blocked sinks could render the facilities or the entire toilets or bathroom areas useless. Damage can be caused to users' belonging as well as they could produce odour. However, if the wall damage is structural, it would call for very urgent consideration.

Maintainers ranked cracked floor as the thirteenth criterion, whereas users ranked faulty sanitary appliances and fittings in that order. Users' ranked damaged taps next to the sanitary appliances and fittings but providers believed faulty lighting should come next. These findings are unexpected, and it is difficult to reconcile both perceptions. Nonetheless, service providers ranked damaged taps or faucets as the seventeenth criterion. However, interpretations of these outcomes are that users were concerned about their comfort and well-being. Users may not see reasons why damaged floors are critical as these may not be structural. On one hand, damage to floors may not be so obvious to most buildings users except among those few who study in construction-related disciplines. Conversely, such defects are obvious to maintenance organisations.

While maintainers ranked faulty fans as twenty-first, users considered failed furniture and fittings in that position. Next in the list of user ranking was heat extractors. Users ranked faulty fans as very critical. A faulty fan is critical as nearly all of the hostels rooms had this facility. It is strange that maintainers considered this so low. Fans were used to control temperatures in the hostels and canteens. This opinion will further be strengthened with the fact that many of the user groups commented negatively about the poor performance of air conditioning units in their lecture room and theatre halls. It was strange, however, that users ranked clogged water closets so low. Or does that mean that this problem does not occur often, otherwise it a very critical issue if it does happen. Apart from the fact that the equipment cannot be used, it pollutes the surrounding environment and causes other dissatisfaction to the building users.

In the view of both the maintainers and building users, floor finish damage was not urgent. Incidentally, they did not consider damaged wall finishes urgent. The users were satisfied if the floor and wall finishes were in moderate condition in as much as they can perform their tasks in the rooms, workshops and laboratories. In fact, about 20 % of the building users believed wall and floor finishes were not urgent at all or not very urgent to maintain. Less attention was paid to these finishes. These outcomes are not very surprising, however. The maintainers need to allocate much more resources to these criteria beyond normal.

Building users did not measured bad soap holders, faulty towel rails and damaged windows as critical. However, the maintainers considered damaged windows as critical. But like the building users, faulty towel rails and bad soap holders do not require urgent consideration. In other words, they do not believe, in that order; users will be dissatisfied if these criteria were not urgently addressed. However, it was surprising that building users ranked damaged windows so low. Albeit, it might not be a structural problem, it is a critical element in a building. It provides ventilation and protection for occupants. A damaged window provides easy access to dangerous insects. Apart from that it protects occupants against unwanted guests (rapists, thieves, burglars, etc.). Although a considerable number of the maintenance organisations believed it does not require very urgent attention none of the respondents measured damaged windows as not at all urgent. While about 50 % of them measured it urgent about 40 % considered it very urgent or extremely urgent.

From the *t*-test results conducted between criteria of the users' value system and university's categories, there is an active relationship and significance. However, the interpretations of the results are difficult and ambiguous, albeit the outcomes were unexpected. It is impractical to classify maintenance services in terms of class. In reality, in most or all cases, students in the different degrees are accommodated in the same building. Only in some of the universities are postgraduates and undergraduates separated or else married couples separated from the others. This is not often the case though but where this exists, the number involved is small, however. Managing maintenance in this order will undoubtedly lead to fragmentation. From the organisation's perspective, this will lead to an increase in maintenance costs without a corresponding increase in efficient service delivery.

However, the *t*-test results conducted indicate that there is an active relationship and significance between gender with criteria of user value systems and defects. These outcomes are not surprising and in fact, they were expected. It might be surprising if the research found no relationships between the constructs. The fact that male and female hostels are separated coupled with the fact that gender plays considerations in the area of needs and wants. The female students have different and perhaps more discerning tastes than most males. Females are particular about their pleasures and creature comforts. This is a very critical construct in Malaysia. This is taking into account that Islam is the official religion in Malaysia. In Islam, males and females are uniquely separated.

8.6 Discussion on AHP Model

Different approaches to prioritise maintenance demands have been investigated, and models developed for each of them, and subsequently validated. The results show that by using the AHP technique, it is possible to speed up the prioritisation setting procedure with a high degree of simplicity and accuracy and thereby reduce the intuitive approach that normally characterises schedule maintenance. Comparing the results of both, the general survey and AHP survey imply that the important prioritisation criteria have common premise that the criteria, factors and elements are crucial towards value-based maintenance management.

The impact of safety and operational issues in maintenance management decision-making is high. Regardless of method of procurement and the structure of maintenance organisation, safety and security issues will always prevail. However, the elimination of safety and security from the model resulted in distribution of the vectors of 0.395, 0.242, 0.177, 0.134 and 0.052 for financial, comfort, organisational responsiveness and aesthetic issues, respectively. But the results do not affect the almost 40 % attributable to the combined operational issue. So, even if the safety and security issues are not included in the model, operational issues still account for nearly 40 %.

If safety and security are not considered, in terms of priority for maintenance, the equipment and instrument room ought to be maintained first (3.61) while least is the administrative building (2.1). However, safety and securing issues are the issues in maintenance standards and maintenance standards are operational aspects.

However, sometime, two or more criteria, factors or elements could have the same weighting. Where such happened, the stakeholders could revisit the rating or consider pertinent criteria to make final judgement. For instance, other user requirements should be considered along with how the repair influences the corporate missions and visions. Also, the impact of the repairs on social, economic and environmental issues could be taken into account. Some obvious factors that could also be used are operational issues.

The consideration of including the operational issues at each stage is to provide long-term value for money invested. For instance, many parties contribute to the decision-making process which governs the design, construction and the maintenance of buildings. Therefore, any maintenance policy that does not consider the users' functional requirements in the formulation of a maintenance management system is inadequate as a value adding strategy.

8.7 Summary

Greater responsibilities are placed on the maintenance department. The building and its management need to be considered as an investment rather than as liabilities. Maintenance management requires a multi-disciplinary approach: from the engineering, technological, commercial, economic and social perspectives. Building maintenance management is a complex and multi-faceted thought process that involves planning, directing, controlling and organising organisations' resources for the sustenance of the building performance. In order to successfully manage maintenance, various issues require cogent considerations.

The maintenance department should be seen as a commercial unit in any organisation. Its significant impact can be illustrated by reviewing it in terms of the characteristics, complexities and roles. It should be strengthened by including it in the strategic business units. Therefore, the department should not be seen as engineering-based unit, but entrepreneurial so that it can be creative, innovative and competitive. Thus, there is the need for the development of the maintenance department with forethought. The lack of this will mean that the department is not considered as one of the university strategic business units. Organisation theory has described the importance of organisation structure has on productivity and customer satisfaction.

While the buildings must be seen as direct income and profit generators, the objectives of maintenance must be properly linked with the mission of the university. However, a unique role towards bringing positive changes in an organisation is to have competent leaders. The competences of the most senior person in the maintenance department cannot be over emphasised. As an illustration, the degree and scope to which the manager can pull together the members in organisation to accomplish tasks and reach their goal as a team depends on the competencies of the manager.

The maintenance manager must be an effective leader, able to contribute with inspired ideas as well as to inspire and encourage staff and colleagues. The manager should possess strong business acumen, a strong vision, and the ability to convince others that the vision will be successful. As an entrepreneuring-led organisation, the organisation would be able to generate ideas, develop the ideas, implement them, operate them and understand the risks involved in managing the maintenance processes. The organisation constantly seeks new and original ways of doing something.

There is no compelling argument against using in-house maintenance organisation to provide the maintenance service. After reading this book, readers will be convinced on where we stand as far as using either outsource or in-source for maintenance service. In this regard, there are also the needs for capable maintenance organisations. Instead of outsourcing maintenance services, the department should engage the services of consultants to train the staff from time to time in order to keep the department abreast with the latest best practices. The maintenance department should be able to provide information on the allocation and expenditure in relation to the customers'/users' experiences (expectations on building performance). This will go a long way towards better service delivery. However, the maintenance department must have sufficient inbuilt flexibility to cater with every work project arising from the demand. Following the different facets, it is possible to develop a systemic maintenance management paradigm that is value based.

References

Chang, M. C. (2005). *Engineering management: challenges in the new millennium*. Upper Saddle River, NJ: Pearson Education, Inc.

Sherwin, D. (2000). A review of overall models for maintenance management. *Journal of Quality in Maintenance Engineering, 6*(3), 138–164.

Tsang, A. H. C. (2002). Strategic dimension of maintenance management. *Journal of Quality in Maintenance Engineering, 8*(1), 7–39. ISSN: 1355-2511.

Chapter 9
Elements of the Proposed Value Maintenance Management Model—"Harmonising It All Altogether"!

Abstract Having presented the results of the data analysis and discussions on the findings in the last two chapters, the value maintenance management model (VMMM) is presented here. To recap on what were explained on the methodological issues (Chap. 6), the development of the model is based on grounded theory. Cohen et al. (Research methods in education. RoutledgeFalmer, Oxon, 2006) contain useful information on grounded theory. For this book, all the bits and details of information and knowledge gained from literature, observations, survey, discussions and interview were brought together into a unified quantity of information and knowledge. Following are the objectives and learning outcomes of this chapter: (1) appreciate the duties and functions of all the stakeholder in maintenance organisations, (2) understand the validation process, (3) understand the purpose of the functions of each of the elements in the model, (4) understand the process and procedures of applying the model and (5) harmonise the knowledge and information gained in the previous chapters.

Keywords Value maintenance management model · Validations · Performance metrics · Maintenance objectives · Organisation structure

9.1 Introduction

This chapter presents the value maintenance management model (herein after referred to as VMMM) by bringing together the major issues involved in planning, directing, controlling, organising, and implementing building maintenance services. The model contains the steps in the maintenance management processes, procedures, and techniques. However, the model dwells partly on the validation which is achieved by the interview. The VMMM comprises of a main graphical component, namely the building maintenance value chain, which in turn is composed of the maintenance demand model and the maintenance performance model. The VMMM is the entire series of organisational processes that add "value" at each step beginning with the process of initiation to the finished phase. The VMMM presents

A.L. Olanrewaju and A.-R. Abdul-Aziz, *Building Maintenance Processes and Practices*, DOI 10.1007/978-981-287-263-0_9

a methodology for undertaking building maintenance processes. However, as the elements of the model have also been drawn from various types of buildings or constructed facilities, the model can as well be applied to other types of construction projects. Simply explained, the VMMM is developed to be generic in nature; as such, any organisation would be able to adapt the model to meet their specific organisation performance requirements.

The model creates and recognises opportunities for achieving value in maintenance service. A pertinent issue in the model is the creation of value culture by all stakeholders. Value is created over time as a result of continuous strategic and operating decisions. In an organisation, value is created by the stakeholders through opportunity and execution. Resources get directed to their most productive use and handled by a well-informed workforce. In this way, the productivity of those resources is maximised which results in better services, optimum user satisfaction and protection of other stakeholders' interests. A key consideration in the VMMM is that the maintenance demand and implementation be users focused. However, many of the functions often overlap as the function most of the time falls into more than one category. For instance, organisation function sometimes overlaps with the planning function. It is worth mentioning that some detailed information will not be provided in this book. Readers that require more detailed information could refer to the thesis from which this book was developed.

9.2 Design Strategy for the Validation

For the purpose of the validation, a total of 10 universities, comprising both public and private founded universities, were selected as "respondents" here in after referred to as participants. Seven of these universities are located within the Klang Valley, as about 70 % of universities in Malaysia are located in this region.

Another deciding factor to whether to include a university in the list is the SETARA ranking published by the MOHE in 2010. SETARA ranking is the local version of international ranking systems for universities like the QS world university rankings and TIMES ranking system. The essence of the various ranking systems is to "guide" students, academics, investors, captains of industry, politicians and policy makers, on a general outlook of universities.

Location, accessibility, and historical factors are also considered. Some of the universities were very "cool" to grant the author's request to have an audience with their maintenance organisations. Some of the reasons cited by many of them are that they cannot disclose any information that involves the maintenance process as that it exposes their corporate image to an "outsider". Financial constraint and time constraints also preclude the extension of the validation to regions beyond the Klang Valley. On another note, there is no simple rule of thumb regarding the number of participants to interview and the number of interviews to conduct with the participants. However, the underlying fact is that as many participants as necessary should be involved in order to gain the required information and knowledge.

In as much as *sufficient* numbers of participants are included, the validation is described as adequate in terms of response rate. Thus, 10 out of 50 universities are considered sufficient for the purpose of this validation.

It is instructive to stress that the data collection and data analysis stages were not kept separate from each other. This is because the two are intermingled, ongoing, and a complementary process (Mei 2002). Combining the two stages together is normally applicable and appropriate to a study like this, which is a grounded theory type of research. However, it could be argued that the best method to validate a model like this is through action research. However, this will be practically impossible for pure research as opposed to applied research. In applied research, the researcher(s) is (are) part of the main organisation that commissions the work. Thus, the findings can be administered by the sponsor to confirm whether or not their service products have improved and to what degree. However, for research in engineering or technology management like this, a conceptual model can be validated through techniques including focus group, interview, nominal group technique, videoconferencing, panel and Delphi technique. However, the interview approach is the most consistent among the techniques with this research.

9.3 Model Development Process

On the basis of the empirical findings and discussion, a conceptual model of the VMMM was developed. There are no hard or fast rules in the developing model; *inter alia,* it should be developed based on the peculiar requirements of those that have a stake in the operation, application, effectiveness and efficiency of the model. Figure 9.1 displayed the validation process.

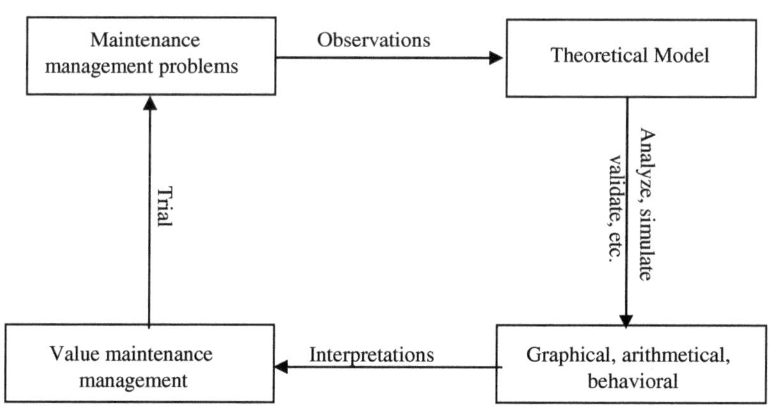

Fig. 9.1 Model development process

9.4 Overview of the Demography of Participants

The validation was conducted in two stages, namely survey questionnaire and interview. With respect to the first stage, the conceptual model was emailed or posted to some forty (40) participants that were responsible for the management of buildings. By end of the cut-off date, only three of them returned their completed questionnaires. However, after several telephone calls to those that were yet to return their survey, six of them returned theirs within a week, with apologies for the delay. Altogether, nine participants returned completed a validation report/form after a period of almost 2 months. Four of the respondents were from the public universities, while the other five were from the privately owned universities. From the outcomes, the total number of buildings in the portfolio of the nine respondents is 675, while the size of the floor area totalled 1,850,000 m^2. However, it is quite unusual to conduct a validation exercise using questionnaire survey. This is because it is more common to meet face-to-face with the participants where clarification may be provided to the participants; thus, the first stage of the validation was considered as a *pilot* validation.

Thus, arising from the comments and suggestions obtained through the pilot validation, some aspects of the model were slightly modified. However, the overall assessments on the model were quite encouraging and positive. The entire group of participants concurred with high confidence that the model can significantly lead to better service delivery. However, one of the nine participants was sceptical about some aspects of the model, and the participants believed that many of the parameters in the model are on the higher side for maintenance organisation. To this respondent, the model is suitable to hotel buildings, but not for university buildings. In the understanding of this participant, maintenance of university buildings requires low standards in terms of maintenance. Apparently, his concern was related to maintenance standard and funding.

With respect to the second stage of the validation—the face-to-face interview—a total of 10 universities comprising both private and public owned were selected based on the criteria previously set. The experts were met within their respective offices at the agreed time. The participants had between 10 and 27 years of experience with the asset, facilities, or maintenance organisations of their respective universities. Their positions in the organisation include the following: Maintenance Executive, Head of Facilities and Maintenance Division, Senior Manager, Department of Maintenance, Director of Development Division, Facilities Engineer and Head of Electrical and Electronic Division. The entire participants with one exception were male. The only female was working with a private university with close to 10,000 students.

On the average, about 65 min was spent with each of the participants. The maintenance index of about half of the universities is between 1 and 2 %. However, in many of the universities, it is between 4 and 8 %. Maintenance index is the fraction of annual maintenance expenditure with the total building value of a university's fixed assets. From the outcomes, the total number of buildings in the portfolio of the nine respondents is 1,050, while the size of the floor area totalled 3,900,000 m^2.

9.5 The VMMM

The value-based maintenance management model consists of a maintenance value chain, a performance matrix, a reporting procedure and criteria of the user value system. The model is a maintenance operation manual for the maintenance organisation. Different methods of problem-solving techniques are applied in each of the blocks. A detailed explanation on how each of the techniques is applied is outside the scope of this book. Interested readers could consult some of the books referenced in this book. These problem-solving techniques include the following: functional analysis, cost benefit analysis, SWOT analysis, stakeholder analysis, need analysis, maintenance driver analysis, building-through analysis, REDReSS and value versus cost analysis. The value maintenance management models borrow much from the value engineering and management methodologies. For instance, a prime purpose of the model is to introduce a systemic approach in defining maintenance objectives vis-à-vis building performance. The model connects the maintenance objective with the corporate mission and vision. The model allows ideas to be logically and systematically manoeuvered in the decision-making process. Thus, the model can as well be described as a maintenance management job plan or MMJP.

The building maintenance value chain consists of the five major partitions of planning, control, organisation, directing and implementation. The elements of the building maintenance value chain are shown in Fig. 9.2. These elements must form parts of the organisation maintenance philosophy and principle. The diagram consists of five main consecutive but overlapping functions. This is to provide a step-by-step methodology for maintenance demand and implementation. With this sequential arrangement, maintenance is strategically and consistently initiated.

Sequentially following the chain will drive maintenance objectives as profit generating functions. In other words, maintenance is considered an investment that contributes to organisation corporate business objectives. The chain stressed the need for decision takers. Involving decision makers in the decision-making is not sufficient. Decision takers must be seriously involved if meaningful outcomes are significant. Maintenance should be discussed at the board meeting. The section that follows provides an overview of various activities involved within each of the blocks (Fig. 9.2).

While some of the functions (sub-functions) are process or financial related, the others are functional or operational oriented. These multidimensional perspectives are to provide a robust, inclusive and comprehensive template for the maintenance service.

To facilitate on the data collection and data analysis with regard to the BMVC, the functions are tabulated. Furthermore, to measure the degree to which each of the functions can contribute to the provision of value added services, the degrees were measured on a five points continuum of not agree at all (1), not very agree (2), agree (3), very agree (4) and extremely agree (5). Quantitatively, what these imply are contain in Table 9.1.

The interpretations in Table 9.1 were disclosed to the participants during the validation. The results of the validation are discussed against each of the functions.

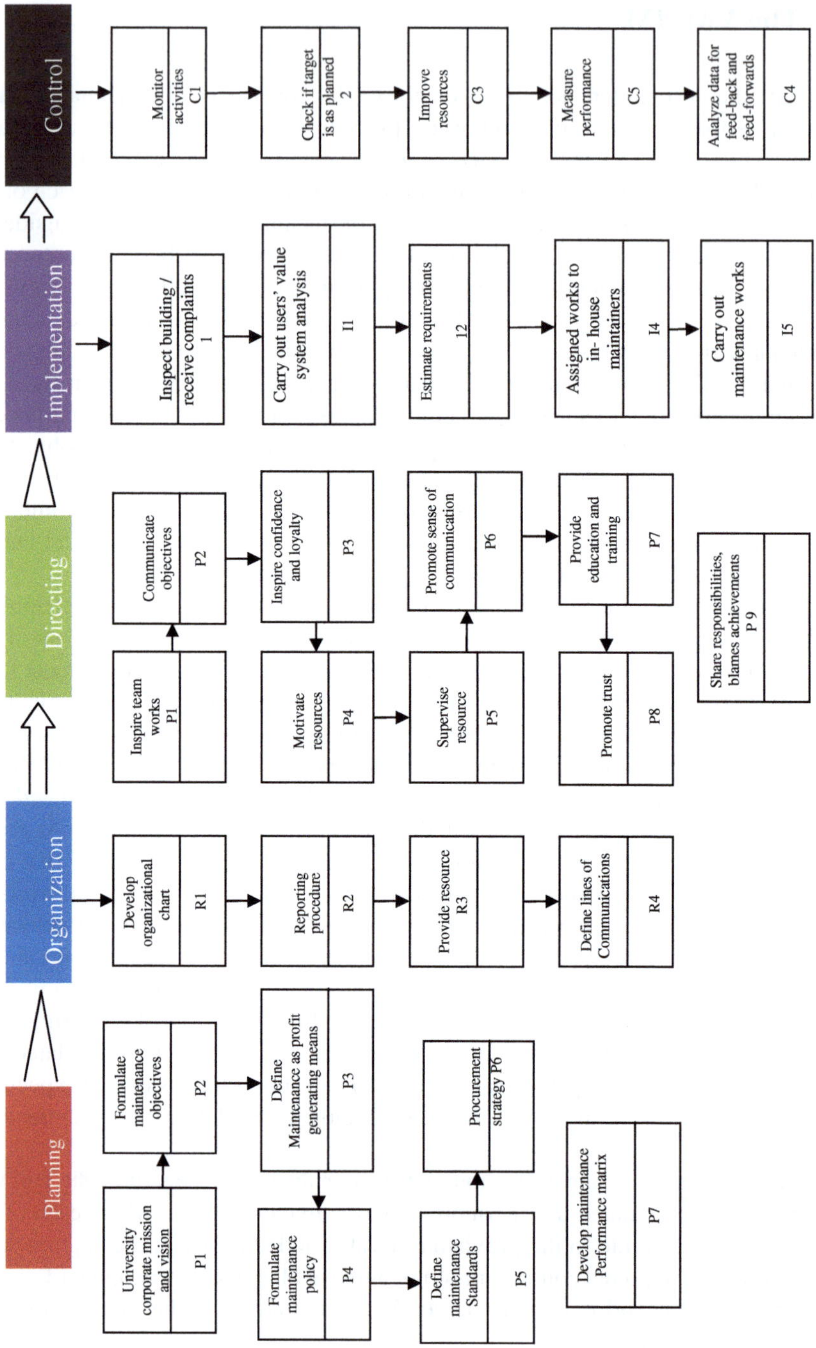

Fig. 9.2 Schematic of conceptual of the building maintenance value chain “matching it all together”

Table 9.1 Scale interpretation for validation

No.	Scale	Interpretation (%)
1	Not agree at all	20
2	Not very agree	40
3	Agree	60
4	Very agree	80
5	Extremely agree	100

Table 9.2 contains the distributions of the validation of the BMCV. The results revealed that virtually, all the participants agreed very much that the BMCV can move maintenance management towards value added initiatives. This is indicated by the weight–age mean score of 4.88. This is an excellent score on a scale of 1–5, where 5 is the maximum value.

On the one hand, less than 3 % (actual: 2.59 %) of the participants moderately agreed that the BMVC can contribute to increasing user satisfaction, reduce maintenance backlogs and reduce unnecessary maintenance costs. On the other hand, about 97 % of the participants extremely agreed that BMVC can lead maintenance to provide added services. To assign work to in-house maintenance organisation has a mean score of 3.8, which though, in reality is still considered very good. Other than that the mean scores of the others exceeded four points.

9.5.1 Maintenance Management of Planning

Maintenance planning is the first stage of any successful maintenance management system. At this stage, a university sets a specific maintenance target (standard and quality) and decides on how to achieve the target. In other words, it is about determining its value system and what has to be done prior to carrying out the technology and engineering aspect of the maintenance services. It also involves who does it and when a task has to be done. Maintenance planning involves the developing of maintenance performance indicators and all other policy strategies. Decisions are also made on whether to outsource or to use in-house or combine both methods.

Brief explanations and descriptions of the various sub-functions in the maintenance planning process are provided next.

P1: Organisation's corporate mission and vision

This is a very critical function of the maintenance organisation. The corporate mission and vision of the maintenance organisation must be linked and intertwined with the corporate business of the organisation. However, the purpose and function of the buildings must be connected with and to the business objectives. Where this is not done, there will be value mismatches. This would lead to a waste of resources and poor service delivery. Universities provide platforms for the transfer and

Table 9.2 Distribution of statistics of validation

Function	Frequency					With mean
	1	2	3	4	5	
Linked maintenance with corporate missions and visions	0	0	1	2	7	4.6
Formulate maintenance objectives	0	0	0	3	7	4.7
Define maintenance as profit generating functions (see maintenance as investments)	0	0	0	4	6	4.6
Formulate maintenance policy	0	0	0	2	8	4.8
Define maintenance standards	0	0	0	4	6	4.6
Formulate procurement strategy	0	0	1	3	6	4.5
Develop maintenance performance matrix	0	0	1	2	7	4.6
Develop maintenance reporting procedure	0	0	0	2	7	4.8
Develop organisation chart	0	0	0	2	8	4.8
Provide resource and ensure (i.e. components, materials, technology…)	0	0	0	3	7	4.7
Define line of communication (e.g. with management)	0	0	0	3	7	4.7
Inspire teamwork	0	0	0	0	10	5.0
Communicate maintenance objectives	0	0	0	1	9	4.9
Inspire confidence and responsibilities	0	0	0	1	9	4.9
Staff motivation	0	0	0	1	9	4.9
Promote sense of communication	0	0	0	3	7	4.7
Supervise works	0	0	0	0	10	5.0
Provide education and training	0	0	0	3	7	4.7
Promote trust and loyalty	0	0	1	2	7	4.6
Share responsibilities, blames and achievements	0	0	0	2	8	4.8
Inspect building and process complaints	0	0	0	2	8	4.8
Identify problem	0	0	0	1	9	4.9
Analysis user value system	0	0	0	3	7	4.7
Analyse and estimate resources	0	0	0	4	6	4.6
Assigned works to in-house organisation (at least 75 % of all work)	0	1	3	3	3	3.8
Implement work: (e.g. repair, replace)	0	0	1	4	5	4.4
Monitor works, progress and systems	0	0	0	3	7	4.7
Check if target is achieved or not and why (plan versus actual)	0	0	0	2	8	4.8
Improve service delivery, resources and system	0	0	0	0	10	5.0
Analyse data for feed word and feedback	0	0	0	1	9	4.9
Measure performance	0	0	0	1	9	4.9
Weightage average (%)	0	0.32	2.59	21.68	75.41	4.88

development of knowledge to societal benefits. The essence of buildings is to provide and sustain an enabling environment to support and stimulate learning, teaching, innovation and research. The need for maintenance must therefore be tied to this purpose if value added service is to be provided.

Based on the validation, the mean score of this function is 4.6 on scale of five, which naturally is very high and encouraging. However, most of the participants are finding it very difficult to relate maintenance with business objective. Even the managements have failed to link this unique role with the corporate business (Validation: 15 December 2010 and 04 January 2011 (Validation (15 December 2010) Expert validation)). However, there is no way value added services could be provided if this linkage is not made. It is a failure of this linkage that makes some of those concerned with the building maintenance to think that the model is too high for organisation.

Therefore, the business mission should be at the back of the minds of all in the maintenance organisation. University administrators do not even want to think about maintenance, believing it is not part of the main services (Validation: (13 December 2010) Expert validation). However, not connecting the maintenance objective with business objective is a recipe for poor services. Administrators will need to consider the maintenance of their buildings as an extension of their core services and as such should be shown the same amount of care and attention. Building maintenance is a core issue for university. The failure of universities to accept maintenance as a core service is a serious failure to a sector that prides itself as a vehicle for scientific and technological advancement.

P2: Formulate maintenance objectives

The establishment and subsequent consideration and inclusion of the user value criteria into the maintenance management system cannot be stressed enough. It precedes maintenance organisations making decisions on maintenance strategy. In fact, it also forms the basis of all maintenance budgets and controls (Kelly 2006). Needless to say, formulating maintenance objectives involves active participations from the administration and the users and other stakeholders (i.e. government and promoters) that signify their expectations and perceptions from the buildings. Therefore, maintenance management stems from clear and realistic maintenance objectives (Validation (21 December 2010a) Expert validation). It is useful to state that maintenance objectives perform the same function as maintenance performance indicators. It is only when maintenance objectives are properly articulated that maintenance demands can contribute to value added initiatives. From the foregoing, the following are maintenance objectives for buildings (Table 9.3).

The outcome of the validation with regard to the formulation of maintenance objective is encouraging with the mean score of 4.7. The participants did not have much problem with this function except that they were not actually developing clear objectives for maintenance. What this means is that they were still stuck with the traditional objective of maintenance, which is, to carry out repair works.

Table 9.3 Objectives of initiating maintenance service in building fabrics, structure and engineering service

No.	Objective
1	To ensure that buildings are fit for purposes through design
2	To ensure that the condition of the buildings meet all statutory and legal obligations
3	To sustain the value of the buildings
4	To ensure dependability through safety, availability and reliability
5	To promote sustainable development
6	To meet the service delivery and expectations reflected in the standards to which buildings are to be maintained
7	To minimise whole-life cost of the buildings
8	That the building is conducive for learning, teaching and research

P3: Define maintenance as profit generating functions

Traditionally, maintenance is considered a burden. But times have moved on. Universities are beginning to set an income budget in terms of the functional performance of spaces. In many countries, and including Malaysia, higher education is now a business. The business considerations of the university sector entail viewing the university as an entity or organization that depends on the connections of the other economics activities. For instance, a RM10 million spent on education will yield a fourfold return in economic activity (Nordin 2010). Therefore, it is not difficult to infer that universities can focus on maintenance management to reduce operating costs and dispose of excess of property as well as increasing productivity. Functional performance of space will also increase the learning, teaching, researching and innovating which the university is primarily established for.

The prime objective of maintenance is to ensure as far as practicable that the building continues to perform at its peak throughout its design life/service life. University stakeholders may not attach the same importance to the building as in case of the hotel sector, though. Maintenance seeks to ensure that a building is operating at its peak efficiency and thereby creating value for investment. It is the only way that the design life/service of a building can be attained.

Students, parents and faculty members might pay much attention to the availability of the latest and relevant reading, learning and research materials. However, times have moved on, and there are greater concerns and pressures on the organisations to ensure optimum performance of the buildings in order to be competitive, attractive and innovative. A greater understanding of the building performance is shifting a lot of pressure to the maintenance department. In fact, there are continuing realisations that the maintenance department should be considered as a university strategic business unit or USBUs. This is the only way the maintenance department can provide value added services. Managing the building is a huge business. It is established in Chap. 1 of this thesis that the government has spent about RM13 billon for the maintenance of its universities alone for the past 5 years, with buildings consisting of about 30 % of the university's capital base.

However, there is the need to broaden the definition of maintenance process beyond technical and engineering issues. This is to facilitate holistic understanding by the top management. The VMMM depicts maintenance as part of the corporate mission. This implies a linkage to the business plan. Consequently, maintenance will be managed as a strategic business function rather than regarding it as technical and engineering issues. The lack of understanding by top management about the business of maintenance management is the result of an inadequately defined linkage to the broader mission and vision.

Although the mean score of this function is 4.6, which is excellent enough, almost all of the participants have problems with this function, however. They just cannot understand how maintenance can be translated to investment. Much effort is spent here to explain to the participants how maintenance contributes to profit generating functions. However, at the end of the explanation, they all agreed with the researcher's view and explanations. Initially, most of them submitted that maintenance is a problem and liability which managements do not even want to think about (Validation (14 December 2010) Expert validation). *A necessary evil costing what it must cost no matter what you do.*

However, without maintenance, the full service life of buildings cannot be attained. It is obvious that if the building is not well maintained, it affects the corporate image and which consequently affects the profits and turnover. It has already been identified elsewhere in this thesis that universities use the condition and performance of their buildings to attract students.

P4: Formulate maintenance policy

The mean score for this function is 4.8. This is an excellent assessment. In other words, the entire participants believed it is very significant to formulate policy for maintenance. As a matter of fact, 8 of the 10 participants extremely agreed that this function can contribute to value added initiatives. The other two very much agreed that it is equally important. A critical element in building management processes is the maintenance policy. The maintenance department should provide its specific processes and practices to enable the personnel responsible for building maintenance to proactively manage the maintenance services. The policy should also address service to outsource and in-source and provide briefs on maintenance standards. The policy should also provide information on maintenance strategy for different elements of the buildings. For instance, part of the building requiring corrective or preventive work must be identified and managed accordingly. Maintenance policy also contains decisions on the maintenance procurement strategy for all aspects of the building and its associated services.

P5: Define maintenance standard

Organisations must specify the level at which the buildings should be maintained. A performance rating for each of the buildings, building components and services must be documented with regard to buildings' performance, condition, functional purposes, intensity of use, future workloads, and detailed description and explanation on the rating criteria must be equally documented. Different buildings or components have different maintenance standards. This affects the building

Table 9.4 Maintenance Standard

Rating	Description	Maintenance standards
1	Not important	Carry our only essential maintenance
2	Low importance	Defer non-essential maintenance where possible
3	Fair importance	Carry out maintenance based on risk assessment
4	Important	Maintain to the best standard that resources allow
5	Very important	Maintain to a very high standard

maintenance strategy. The mission statement or MS (the underlying purposes and values) of the maintenance department should be to enhance user satisfaction firstly and then to improve productivity. The other purpose includes safety and reliability. The MS must be translated into SMART (specific, measurable, achievable, realistic and time bound) objectives. The maintenance department should be able to provide information on the allocation and expenditure in relation to the customers/users experiences (expectations on building performance).

Defining and formulating maintenance standards is also considered extremely critical in providing value added services. This is explained with a high mean score of 4.6. However, the problem with most of the organisations is having a clear and consistent maintenance standard. They often seem not to have clear maintenance standards or not ones concisely committed on paper. Although a few had maintenance standard, but it was not often adhered to. Only 6 out of the 10 participants extremely agreed that it could contribute to value added service. In the opinion of many (4 of 10) of them, they are only about 80 % confident that to have a maintenance standard can lead to better services. Table 9.4 contains some information on maintenance standards. Each building's fabric, structures or/and services are allocated an importance severity rating of 1–5.

Maintenance standards vary depending on the importance of the facility. There will always be some buildings and facilities, which because of their importance, their prominence, aesthetic value or historical significance, must be maintained at a higher standards than the others. Maintenance could further be classified as emergency, urgent and normal. Emergency maintenance concerns maintenance works that relate to the safety of the building users and the general public, whereas urgent relates to repairs that concern security of properties (Validation (15 December 2010) Expert validation). Normal works is related to neither safety nor security. They could be repairs that affect users' and clients' comfort. While emergency works require immediate attention, urgent work might be delayed to a few hours or even days or as the case may be. Normal works could be settled within few days or weeks. In some cases, it could take a month.

P6: Procurement delivery methods

Maintenance organisation should be part of management core functions. This is necessary because when organisation owners (stakeholders) are separated from its management, an agency problem often arises. The outsourcing organisations often make decisions that benefit themselves at the detriment of the building stakeholders' interest and particularly the users' interest. This is business, it is a means to maximise profits. The building provides values not only to the organisation, but also to the students, faculty members, parents and other users and stakeholders. Thus, they require better attention than what they are presently receiving. However, this will require the development of capable maintenance organisation.

The structure of the department should grow as the organisation grows in size, age and complexities. Certainly, at the initial stage of the business, it is just natural (assuming the company starts operation with new buildings), and the structure of the maintenance department will be small. A business with old buildings will require a larger maintenance department. Over the years, the maintenance department expands in relation to demand. Due to the complexities and significance of maintenance works, there is the need for a proper maintenance department. From the organisation point of view, the ways an organisation is structured and operates its business have significant effects on its outputs which consequently influence the environments it exists in.

The outcomes of the validation revealed the procurement function score 4.5, while 6 of the participants are 100 % confident that this function is required at the management level to drive maintenance to value added service, three are 80 % sure and one is only 60 % sure. Even though this is important, many universities do still consider this as a strategic function.

P7: Develop maintenance performance matrix (MPM)

Performance measurement is described to entail the arrangement and making use of performance criteria—e.g. cost, quality, satisfactions, time and productivity—as the basis for measuring effectiveness and efficiency. Performance measurement guides organisations towards achieving their goal effectively and efficiently. An extensive literature review contains that performance measurement should include the following characteristics (Gleich et al. 2009):

1. Provide past- and future-oriented management control information
2. Reflect the demands of both internal and external stakeholders
3. Provide management control information at all performance levels
4. Corporate group to business segments to employees
5. Allow improvement considerations on a short-term and long-term basis
6. Corporate and divisional levels
7. Contain financial KPIs which can be extended by parameters which influence
8. Long-term financial performance capabilities of a company
9. Contain not only quantitative (hard facts) but also qualitative (soft facts) information.

The above characteristics could be described as comprehensive characteristics of the modern performance indicators. However, it might not readily correspond to a maintenance organisation. The characteristics require structural contextualisation to fit in. Within the context of organisation's maintenance organisation, performance matrix is the criterion that facilitates the stimulation of user satisfaction and increase productivity. In that respect, the criteria greatly influence maintenance organisation productivity, user satisfaction, maintenance backlogs and building performance. There is sufficient literature on the development of performance measurement for new built (Jones and Sharp 2007; Wadugodapitiya et al. 2010; Lordsleem and Rabbani 2010) and maintenance services, (Shohet 2010), but the same cannot be claimed for maintenance.

The theory behind performance measurement is that completed building or maintenance service should be measured based on outcomes. In other words, the fundamental issue is not about the input per se rather with meeting the clients'/users' functional requirements. This theory emerges based on the philosophy and understanding that buildings are capital goods. Therefore, maintenance services should be initiated based on the same theory and philosophy. However, there are evidences that the current performance measurements are not conclusive. They do not in any way link maintenance expenditure with business performance and most importantly with users' satisfaction.

However, there is a shortage of literature on performance measurement for the maintenance organisations. A review of literature would suggest the existing performance measurements are not specific for the maintenance organisations. Also, the existing indicators are concerned with **maintenance technology** but not the management aspect. Simply put, the existing models do not indicate performance in terms of users' satisfactions. However, for the indicators to be useful, it must be developed with the active participation of those who use and learn from the indicators. The MPM relies on the users' satisfaction for initiating maintenance and improvement in services deliveries. This is crucial because customers' satisfaction is a significant performance measurement metric.

Table 9.5 contains a proposed generic MPM for a maintenance organisation. The proposed MPM consists of 11 metrics. Based on the targeted value, a weekly, monthly or quarterly achievement will be computed. Simple descriptive and illustrative statistics are sufficient for easy understanding and meaningful comparison. A five continuum scale of measurement is adopted. Where one denotes not satisfied at all and five denotes extremely satisfied. For instance, the percentage of the users that are satisfied, very satisfied or extremely satisfied with the performance of the building and services can be determined through simple mean technique. On the basis of the outcomes of periodic achievement computation, assessment is made whether the target is achieved or not. This is inserted in the remarks column.

These metrics must form parts of the maintenance organisation processes and procedures.

Maximum complaints to be received from users for each of the building category must be set, albeit this could be based on the age of building, budget, nature of users and quality of previous maintenance works. This is to guide against an unnecessary increase in maintenance backlogs. However, the fact that less maintenance complaints are received does not simply imply the maintenance management is value based. There should be continuous monitoring. Users are the direct beneficiaries of the service. They are the only people who can define the quality of service they are receiving. Getting them involved will not only result in commitment to efficient

Table 9.5 Maintenance performance metric or MPM

No.	Metric	Location	Target value	Monthly achievement	Remark
1	Customer satisfaction survey	Administrative, academic and residential	Rating of minimum of 4 (on scale of 5)		
2	Customers complaints	Academic and administrative	Maximum of 10 complaints/building		
		Residential	Maximum of 100 complaints/building		
3	Response time to complaints	Academic and administrative	85 % of customer complaints responded within 30 min of complaint received		
		Residential	80 % of customer complaints responded within 30 min of complaint received		
4	Response time to complaints	Academic and administrative	100 % of customer complaints responded within 48 h of complaint received		
5	Turnaround time to resolve complaint	Academic and administrative	80 % of customer complaints resolved within same day		
		Residential	40 % of customer complaints resolved within same day		
6	Recurring complaint	Administrative, academic and residential	Maximum of 5 % of total W.O		
7	Engineering system stability	Administrative, academic and residential	Minimum of 90 % PPM executed		
8	Efficiency of work order execution	Academic and administrative	Minimum of 90 % closed		
		Residential	Minimum of 85 % closed		

(continued)

Table 9.5 (continued)

No.	Metric	Location	Target value	Monthly achievement	Remark
9	Poorly executed work	Administrative, academic and residential	Maximum of 5 % W.O rated as poor		
10	Efficiency of work order execution within 60 days	Academic and administrative	100 % closed		
		Residential	100 % closed		
11	Procurement of maintenance works	In-source	Not less than 80 %		
		Outsource	Not more than 20 %		

services but also influence the service delivery, and apart from the fact that users create their value system, they have the ability to bring their requirements to notice and to take action if their requirements are not fulfilled (see Soderholm et al. 2007). However, the maintenance organisation seldom carries out user satisfaction surveys.

It is also in the best interest of the building maintainers and users if times to respond to complaints are set. This is necessary as it can contribute to best value delivery. Where this is monitored periodically by the maintainers, it would contribute to the enhancement of users' satisfactions. Times should be set as per when all complaints must be addressed. This will not only reduce maintenance backlogs but provide a proper check for labour management. While most of the work orders must be efficiently closed within 1 month, all work orders must be finalised within 6 months. But, the order should not entail those services that could unnecessarily be a discomfort to the user or those that will disturb or disrupt business operation.

While recurring maintenance must be checked and kept to a minimum, say 5 %, corrective maintenance should also be kept to minimum. Much of the works should be based on properly planned preventive maintenance or PMM measure. This will reduce unnecessary increases in maintenance costs and disruptions to the barest minimum. Provision should also be made to poorly executed work. The size of work to outsource or in-source should be addressed. The university must critically consider what aspect of the works to outsource. Unfortunately, however, universities often consider building maintenance as non-core area. As such, most of the work is outsourced.

How else can organisation claim a resource that constitutes about 30 % of their capital base as non-core? This is unfortunate. Without buildings, university cannot operate. Even the virtual university requires some minimum amount of buildings to carry out their business. Perhaps, it might not be possible for all work to be 100 % effective and efficient, however. So, it is therefore necessary that the universities change their mindset and accept the reality that the management of the buildings is their core service. The degree of ineffective or inefficient works should be near negligible. In no circumstance should the poorly manage works exceed 5 % of the total works.

The remarks column should indicate whether the periodical target is achieved or not. To avoid complications, a yes or no answer is sufficient. No provision should be made for may be, neutral or do not know. However, this stage is very critical. The maintenance organisation should provide sufficient evidences as to why the target is not achieved. This is very useful to a person performing the control management functions. It is also good for the maintenance managers. The financial controllers also require the information to obtain more funds from the government (in case of public university). On the other hand, the simple reason that the target is achieved does not warrant the maintenance organisation not searching for innovative approaches to improve service delivery. The organisation will need to identify whether or not the target is achieved by accident or it is as a result of good maintenance management philosophy. The MPM should be primarily used by the maintenance organisation to drive in value and drive out unnecessary maintenance costs.

The maintenance organisation needs to know how much they are doing in order to know whether they are doing well or not. This is imperative as this will provide opportunity for improvement. The organisation should provide support and measurable or quantitative facts to show that service delivery is optimum. On the other hand, it is imperative for the maintenance organisation to understand the commercial benefits of maintenance. The MPM is a benchmark to measure the before and after service. Otherwise, it will be only a mere issue of talking the talk against walking the talk. With VMMM, there is a potential that better service could be delivered with optimum users' satisfaction.

As is expected, the mean score of the participants on this function is 4.6, which is an excellent representation of its significance. While seven of the participants were 100 % certainly, it can contribute to value added services, and the other three were only 80 or 60 % confident. The results of the validation on the criteria within the MPM are contained in Table 9.6. In seven of the cases, all the participants agreed with the metrics. In relation to size of complaints emanating from each building, only eight participants agreed, the remaining two were not very sure. However, this is not surprising, because complaints depend on age, use and types of buildings. This metric is quite subjective in nature, however. Discussions with some of the participants on whether size of maintenance could be related to size of buildings, number of buildings or even the users proved very difficult. It should be noted that the "no" does not actually indicate rejection.

The entire participants approved the affirmative that a minimum of four on a continuum scale of five is acceptable or reasonable enough on the buildings (building fabrics, structure or/services). The implication of the findings is that good service should be provided to the extent that users will not be satisfied only to the level of less than four point. Should there be a case where the organisation is rated less than 4 or 80 % on the total service satisfaction level, systemic effort is thus required.

Similarly, most of the participants concurred that a maximum of 10 complaints (of defects) is good enough. Although one of them believed it should in fact be

Table 9.6 Validation of MPM

No.	Metric	Location	Target value	Frequency	
				Yes	"No"
1	Customer satisfaction survey	Administrative, academic and residential	Minimum of 4 (on a scale of 5)	10	
2	Customers complaints	Academic and administrative	Maximum of 10 complaints/building	8	2
		Residential	Maximum of 100 complaints/building	8	2
3	Response time to complaints	Academic and administrative	85 % of customer complaints responded within 30 min of complaint received	9	1
		Residential	80 % of customer complaints responded within 30 min of complaint received	9	1
4	Response time to complaints	Academic and administrative	100 % of customer complaints responded within 48 h of complaint received	6	4
5	Turnaround time to resolve complaint	Academic and administrative	80 % of customer complaints resolved within same day	10	0
		Residential	40 % of customer complaints resolved within same day	10	0
6	Recurring complaint	Administrative, academic and residential	Maximum of 5 % of total work order	10	0
7	Mechanical and electrical engineering	Administrative, academic and residential	Minimum of 90 % based on planned preventive maintenance	10	0
8	Efficiency of work order execution	Academic and administrative	Minimum of 90 % closed	10	0
		Residential	Minimum of 85 % closed	10	0
9	Poorly executed work	Administrative, academic and residential	Maximum of 5 % work order rated as poor	10	0
10	Efficiency of work order execution within 60 days	Academic and administrative	100 % closed	10	0
		Residential	100 % closed	10	0
11	Procurement of maintenance works	In-source	Not less than 80 %	5	5
		Outsource	Not more than 20 %	5	5

reduced to less than five complaints per month in buildings. However, it is interesting to also find that a participant did not agree with this respondent. The participant, who disagreed, requested that 10 complaints should be the minimum because it affects their KPI (Validation (17 December 2010) Expert validation). One respondent also indicated that the "number of complaints" did not reflect actual performance, but rather "time for solving" the complaint should be of critical importance.

Though the nature of complaints should be taken into account, the building users sometimes complain even where it might not be necessary. This observation could sometimes be the case with specific reference to female students. Female students are more particular about their surroundings compared with their male counterparts in regard to the condition and performance of their buildings. However, with respect to the building maintenance, what matters to the users is not even the quality of service provided, and rather it is their perceived value in terms of their experience, perception and an expectation that matters most. Hence, a complex reason the maintenance management process is complicated and sophisticated.

However, with regard to the "ceiling level", we set for the maximum complaints in a month per building for the residences the pattern of response is not very much different from the one relating to the administrative and academic buildings. As an illustration, while some agreed that the 100 complaints we set are realistic, some believed the 100 was too much, but at the same time, one of the respondents argued vehemently that it was not realistic to achieve that target. The respondent believed there should not be a maximum limit because it might put the maintenance organisation under high pressure. However, based on a recent study, it was discovered that about 20 complaints were taken up by a maintenance worker.

All of the participants agreed that the response time to complaint (RTC) was very critical in service delivery. This was indicated in the response to this aspect of the model. Some believed the 85 % was on the higher scale, but some argued it was achievable enough because of the criticality of maintenance. In fact, all of those that made their interests known indicated that at least 70 % of the complaints must be responded within 30 min. However, in the VMMM, it is required that 85 % of complaints for the academic and administrative buildings must be responded within 30 min, and for the residence (i.e. student hostels), 80 % of the complaints must be responded within the same timescale.

In another regard, the participants also agreed that all complaints had to be responded within 48 h. However, a participant argued that the 48 h to respond to all complaints was not realistic; however, the participant failed to provide a convincing case to the total objection. But some of the organisations, lump building maintenance with work such as landscaping (Validation (21 December, 2010b) Expert validation).

The turnaround time to resolve complaints was found to be reasonable. In fact, some of the respondents believed that the 80 % we set for the turnaround time to resolve complaints in academic and administrative buildings within the same day should be 100 %. The participants held that all complaints must be resolved on the same day. The participants also believed that the 40 % we set for residence was too low and that it should be closed to the same scale with that of academic and administrative buildings. However, a participant suggested that in general, the turnaround to resolve complaints "shall be priorities based on urgency and plan works to avoid unnecessary maintenance cost" (Validation (15 December 2010) Expert validation). This concern was taken into account in the model development we assured him.

On another aspect, the entire number of participants concurred with high confidence that the maximum number of recurrent complaints should not be more than 5 % of the total work executed in the entire buildings category. Similarly, all the participants concurred that more than 90 % of engineering service should be based on planned preventive maintenance. These results are not however unexpected. In fact, our undisclosed hypothesis was that all mechanical and electrical faults would be based on planned maintenance. This is to avoid necessary disruption of work. While some maintenance works cannot be prevented, a lot can however be minimised through the preventive maintenance.

With regard to the efficiency of work, our target values of 90 % for academic and administrative and 85 % for residence were realistic enough. Yet a participant opined that it should be subject to various factors including skilled workforce, materials and availability of fund. That notwithstanding, some of the participants further believed that the figure should be 100 % in actual reality. While it might not be possible to achieve 100 % efficiency with all works executed, these figures should be kept at the barest minimum. To this, all the participants concurred that it should not exceed 5 % of the total work executed. At least, 95 % of the works must be efficient. Though as expected, all the participants held that all repairs must be efficiently maintained within 60 days. Five per cent (or less) of the repairs must be rectified within 60 days. Though the 60 days seems a long period, the reality is that many of the maintenance organisations outsource their maintenance, and sometimes, some materials and components might not be available locally.

With respect to whether universities should outsource or in house most of their maintenance service, most seem to prefer to outsource maintenance services. We had set the target at 80 % of in-sourcing, while the remainder of 20 to be outsourced. These targets did not go down well with many of the organisations. As an illustration, a participant wrote: "Dependent on what type of business you are doing. For Education Sector shall maintain less staff to cut cost on salary and fringe benefits etc. Lab Unit shall be in house to build their own strength on Equipment and Experts. Out—sourced more to reduce manpower overhead cost. Further we don't keep unskilled workers".

The closest target value the participants agreed with was that 60 % of works could be outsourced. In fact, some of participants revised the target values that we set. Meaning to say, 40 % of the maintenance service should be outsourced. This aspect of the data is however the most interesting, though it not unexpected. While there is sufficient literature on the shortcomings of outsourcing, universities have failed to see this. However, many of the participants were of the opinion that though all the maintenance could be executed in-house, shortage of experienced staff was an albatross issue vis-à-vis the government regulation on employment. But, the failure of the top management to appreciate their investment in maintenance to the corporate business objectives is a fundamental issue that needs to be addressed first (Validation: 14 December 2010 and 04 January 2011 (Validation (14 December 2010) Expert validation)).

The most interesting metric of the MPM is the procurement strategy. Based on the initial findings and analysis, it was identified that for the maintenance organisations to be adequate, there is the need for organised maintenance organisations to execute about 80 % of the works in house. Even though all the participants agreed with the benefits of this approach, most of them did not see this would happen because of the way management perceives maintenance. The government has a role to play however. For example, the government will need to relax the rule of maintenance organisation (Validation: 15 December 2010 and 04 January 2011 (Validation (15 December 2010) Expert validation)). The government needs to allow the university (public) maintenance organisations to recruit more competent staff if the maintenance is to be procured in house.

9.5.2 Maintenance Organisation

The meaning of maintenance organisation is used in different context here. Here, maintenance organisation is defined as a process and activities not as a unit or department dedicated to maintenance tasks. Organising is the process of procuring and allocating resources to accomplish the planned target. Simply put, it is the maintenance management function of assembling and coordinating human, physical, and financial information and other resources required to achieve the planned target. It also involves staffing the required workforce to carry out the maintenance as well as making decisions on how resources can be maximised. This includes delegating duties and/or authorities to others (subordinate). This is usually displayed on the organisation chart. The competency of the person that leads and performs various duties is very important towards achieving the organisation's corporate objectives. The relationship between the maintenance operative and the higher level executive is very important. To achieve user objectives, resources are required and need to be assigned to respective functions in the organisation. For instance, efficient organisation of labour is crucial in maintenance management since maintenance is labour intensive in established universities as compared to those with new buildings. Therefore, great potential cost saving in executing maintenance service can be obtained with efficient labour management.

The section that follows provides an overview of the various functional activities in each of the sub-blocks under the maintenance organisation function.

G1: Define lines of communication

Value and its delivery mode must be clearly defined and communicated across the organisation as well as to external parties. Top management in the parent organisation must have direct access to the maintenance organisation. The reverse must also exist. The internal organisation structures must permit easy communication among management and operatives. A unique component in the defining communication link is the maintenance reporting procedure. This entails a clear route regarding how building users channel their complaints to the maintenance organisation. The people in the organisation that have direct contacts with customers/users often see the need for changes in the organisation before even the senior management. The users must also have direct access to the top officers in the maintenance organisation. This is very necessary in order to enhance users' satisfaction. A participant reiterated that complaints received utmost attention when brought to the attention of the big boss (Validation (04 January 2011) Expert validation).

G2: Provide resources

The provisions of materials, components, technology and other resources received high mean scores of 4.7 and with seven of the participants extremely agreed and the remaining three also very much in agreement. It is the responsibility of organisation to provide the necessary funds required in maintaining the buildings. For public universities, larger amounts (more than 90 %) of the maintenance cost expenditures are born by the government, while the remaining sum is often provided by the university itself through internally generated revenue or donations or both. In case of the privately owned university, all the money required for maintenance is supplied by the user (i.e. students) directly or otherwise. Apart from providing funds and materials, the role of the university is to provide competent human resources for the maintenance organisations. On many occasions, a university also provides some resources to the outsourcing organisation.

Even though the participants did not have a problem agreeing with this role, a major problem was noticed; however, the time required for resources to be provided. Sometimes, resources are not provided on time. In many of the cases, materials and components are not procured until when there are problems. This will make a mess of the repairs, and since maintenance is contagious, a delay in the supply of materials will increase maintenance cost and affect total service delivery (Validation (22 December 2010) Expert validation). The university should as a matter of strategic planning identify the most common repair complaints and stock the relevant resources.

University organisations and ministries (finance and higher education) must allocate sufficient funding for the maintenance budget to enable the buildings in their portfolio to be maintained to the optimum performance standards ratings specified and documented in the maintenance policy. Maintenance organisation must formulate an annual maintenance budget which is a realistic calculation of the quantum of funding required to address the maintenance demand. This relies upon reliable data extracted from the strategic maintenance plan, the maintenance assessment reports, condition and age of the building and services, assessment of maintenance demand and maintenance backlog. However, currently, only about 1 % of allocation to education is allocated to building maintenance. This is very inadequate. Fundamentally, not less than 5 % of the total amount allocated to high education should be allocated for building maintenance in the case of public university. With respect to private universities, 5 % of the present value of the building could be a sensible guide.

G4: Develop organisation chart

Eight of the ten participants were 100 % confident that to develop an organisation chart for the maintenance department was critical. At the same time, the other two participants were 80 % certain. However, the problem did not lie with the development of the chart alone; rather, the contents of the chart was equally significant. The chart should be clear and concise enough as to indicate who does what. It also involve, identifying the maintenance organisation, the current buildings and the size of them and expected future expansion and also to outline maintenance organisation's responsibilities, functions and scopes. This is mainly to organise activities and to facilitate cooperation and avoid misconceptions among the maintenance team and to provide information as to who is responsible for particular maintenance tasks.

However, among the universities contacted, there is no proper linkage between the maintenance organisation and the management. Thus, management does not accord proper attention to maintenance. On another related note, if the organisation chart is not concise and precise as to who does what, users often channelled their complaints to the wrong personnel (Validation (21 December 2010b) Expert validation). An obvious implication of this is the unnecessary delay in processing the maintenance work.

G5: Develop maintenance demand procedure

Maintenance organisation must look into the way maintenance complaints are lodged and the RTC by the maintenance organisation. Users prefer one point of communication and not complication. They want someone to talk to if a problem arises. In situations where time is wasted in order to contact the maintenance organisation, this can only further frustrate users. Users want to be able to report their complaint conveniently. The contact address (phone numbers) of those concerned with specific items of maintenance should be supplied to the building's users. It will also facilitate easy reporting and communication, and a free toll phone can be provided so that users will not need to pay for the complaint they need to make. This will go a long way to increasing the satisfaction of the users with the service delivery.

Though some universities have developed online reporting, complainants want to "talk" to people but not a machine. What about the issue of access to a computer and Internet connection? How many of the students had access to Internet in the hostel? How will students make emergency complaints? What about the distance from the computer lab to the residence? Are the computer labs in operation 24 h a day? And again, during the off hours, the person at the desk cannot be contacted. The behavioural aspects of maintenance management are critical. The need to develop maintenance a reporting strategy obtained a high mark, with a mean score of 4.8. Seven of the participants believed the development of the reporting procedure would extremely contribute to value added services and the remaining very much agreed to the same fact too.

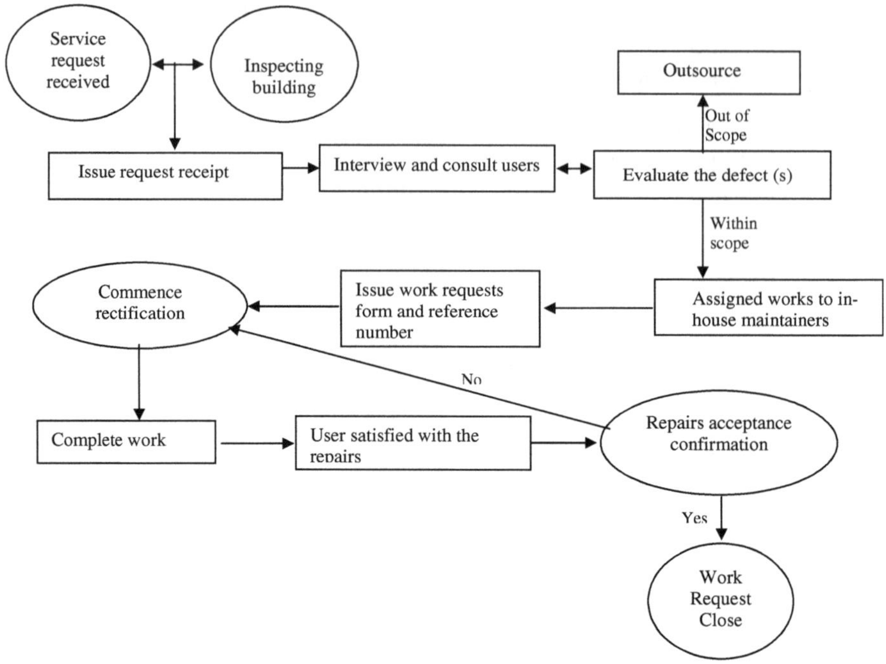

Fig. 9.3 Proposed maintenance demand model (MDM)

Figure 9.3 displayed the maintenance reporting developed and validated. Although some of the universities were already using the online complaint reporting system, the entire group of participants agreed with the model developed. However, none of the universities was actually practicing the developed model. For instance, what the participants' universities did was to receive user confirmation on the completion of repairs. But one of the participant organisations allowed the users to assess their satisfaction with the service provided on completion of works; however, no criteria were used except to measure the level of satisfaction. Certainly, this cannot lead to better services. We proposed a five point scale that should be used to obtain the level of user's satisfaction with the service provided. This was necessary, and otherwise, how would we know the level of user satisfaction? Meanwhile, all the 10 participants, except one, were pleased with issue of consulting and talking with the users before or during repairs. While some of the participant organisations interview the user before repair commences, the consultation is casual and inconsistent. However, if users are not consulted before or during repairs, there are often problems with the works whereby the users will keep on complaining about the quality of works or even sometimes that works they requested were not intended by them. However, the users' satisfaction is a good avenue to reduce vandalism by the students (Validation (14 December 2010) Expert validation).

9.5.3 Maintenance Control

This is another critical function in value maintenance management processes. Here, the organisation determines if resources are used as planned and if the maintenance organisation is meeting its set target. For instance, the maintenance management function of controlling involves monitoring performance, making corrections and making continuous improvement. It also involves the determination of whether organisation is achieving the level of quality set as well as if maintenance activities are being completed on schedule, with high reliability and safety and ultimately meeting the users' satisfaction. A key function here is the monitoring of the maintenance performance indicators.

In the sections that follow, we have outlined the overview of activities required in each of the blocks under the maintenance control function.

C1: Monitor activities
A mechanism must be put in place to monitor the maintenance process. This is to ensure improvement when necessary. There is the need for the maintenance organisation to continue to benchmark its activities against the best in the system. Resources must be monitored to see if they are used as planned and if not the causes of the disparity must be established. To monitor works, progress and system scored 4.7. Specifically, while three of the participants were 80 % confident, the remaining seven were 100 % confident that the maintenance organisation needs to monitor their works, progress and system in order to provide better services. This will provide an indication on the position of the organisation.

C2: Check if target is as planned
As part of the above function, decisions will be arrived at as to whether the actual achievement is in consonance with the planned target. Where the target is not achieved, an explanation is provided as to why it is not achieved. The maintenance organisation also needs to put mechanisms in place to see whether targets are achieved or not. Much of this can be obtained through maintenance activities monitoring. It is instructive to mention again that the fact that the target is achieved does not imply maintenance is done in the most effective and efficient way. Therefore, even if a target is achieved, effort should be made to establish why the target it achieved was at that level. It could be a coincidence. Thus, this model is both problem-solving and problem seeking. The significance of this function is indicated with the high mean score of 4.8. The entire group of participants agreed extremely (100 %) or very much agreed (80 %) that it was pertinent.

C3: Improve resources
This is very necessary where a target is not achieved. However, even if a target is achieved, alternative ways of putting the resources into optimum satisfaction must be recognised as continuous processes. The mean score for this function was overwhelming. All the 10 participants were 100 % confident that there was the need to improve service delivery, resources and system continuously. This was indicated by the mean score of five on a scale of 1–5 with the maximum score of five.

C4: Analyse data for feedback and feed-forward
All information related to maintenance, renovation, refurbishment and other similar initiatives should be well documented in a concise format or medium as intellectual property. Fundamentally, maintenance organisations should have information relating to safety, the building portfolio, the condition and performance of the buildings as well as the maintenance expenditure. This will allow the trend to be monitored and appraised. With this, informed decisions can easily be reached as if the maintenance, refurbishment, alteration or reconstruction are required. The results of the validation indicated that the participants considered analysing data for feedback and feed-forward very important. This is indicated with the mean score of 4.9. Specifically, nine of the participants were confident that the function was extremely significant. The remaining one also very much agreed however.

C5: Measure performance
Performance of the building and maintenance management philosophy must be measured. This is in order to provide a feedback to the organisation on the state of the building performance and user satisfaction. Tools and techniques applicable for this exercise include the following: customer relationship management, customer service management and customer demand management. The mean score of the function is also 4.9 indicating its high level of significance.

9.5.4 Maintenance Directing

This is sometimes called the management function of leading or mobilising. This is a very important function of the maintenance managers because after making plans, creating a structure and employing the right personnel to do the various assigned tasks, someone must direct or lead the organisation in order to achieve the organisation goals. The main function that the manager is expected to perform here is to communicate the organisation's goal to the employee clearly. Also the employees have to be motivated to perform their tasks highly as well as encouraging them to assume responsibilities in solving problems.

The sections that follow provide an overview of the various activities in each of the blocks under the maintenance directing function.

D1: Inspire teamwork
Each of the maintenance groups or teams must be provided with conducive and supportive environments that facilitate team building. The team members receive instructions about the challenges and tasks to be performed. From there, a relationship is created. Members of the team are educated on the need for collective responsibility. They must be informed of the principle of total output, which is greater than individual contributions. A section of the team must be trained to look out for ways to increase the team productivity. Able, willing, talented and attractive people should be employed. People with positive drive will make a good team. This will go a long way in increasing productivity and user satisfaction. Quotes on the essence of

teamwork are indicatives: “Talent wins games, but teamwork and intelligence wins championships”—Michael Jordan. “No matter what accomplishments you make, somebody helps you”—Althea Gibson. The entire participants were 100 % confident that there was need to inspire teamwork, particularly among the subordinates. The most senior person—maintenance manager—must do all things possible to inspire his staff to work. He/she must keep the momentum going at all times.

A unique role towards bringing positive changes in an organisation is through competent leaders. The competences of the most senior person in the maintenance department cannot be over stressed enough. As an illustration, the degree and scope to which the manager can pull together the members in the organisation to accomplish tasks and reach goals as a team depend on the competencies of the manager. A manager is someone who organises the members of a workforce and delegates responsibility to those who can do the job. The maintainers must be the people-oriented type, who will always be sensitive to and aware of what is happening around their workplace.

A change in attitudes, behaviours, communication guide and styles, and a scope of other verbal and non-verbal cues can be readily picked up by managers who are sensitive to the various nuances. Regardless of whether it is concerned with finance, accounting, management, marketing or administrative matters, and regardless of the sophistication of the machines and the Internet, in the ultimate analysis, it is the people who achieve the goals and make things happen.

D2: Communicate objectives

People can only do things better, if they are clear of its objectives. The maintenance objectives must be clearly communicated to everybody in the organisation. Certainly, the operatives cannot be blamed if they failed to achieve the objectives if they do not know them. A situation where only those at the top management have access to the objectives of maintenance is crude. In fact, the objectives of the maintenance organisation must clearly be written and hung on walls so that everybody can see it at all time. Various pictorial tools can as well be employed to make the messages informative. Apart from that, they should be constantly reminded on the set objectives. Considering the significance of this function, nine of the participants were 100 % confident that the function was required. However, most of the universities were of the opinion that they never thought it was necessary to communicate maintenance objectives to the staff. To them, the staff ought to naturally understand. But in reality, this is not the case. However, such issues cannot be based on assumptions.

D3: Inspire confidence and loyalty

Most of the participants found it difficult to consider inspiring confidence and loyalty among the staff as necessary. They undermined the essence of this function. However, maintenance organisations must earn the confidence and loyalty of their operatives in order to increase service delivery and increase productivity. This will also go a long way to increase users' satisfaction. However, it is equally imperative for the operatives to be loyal to the organisation and to create confidence among colleagues too. Operatives must learn to be obedient and patient in educating the participants on the essence of these roles, and they cannot but agreed that it is critical for service delivery. Their understanding of this function is a proof of the high mean score of 4.9 earned by the function. The entire number of participants either agreed extremely or very much agreed.

D4: Motivate resource
Also, the mean score of motivating the staff members is 4.9 with nine of the participants having 100 % confidence that there is a need to motivate staff. However, most of the organisations never thought it was necessary until the author took time to explain them how this could lead to better service delivery. In any organisation, the staffs require encouragement to exert high levels of effort to attain organisational objectives and conditioned by the effort to satisfy some individual and personal needs; needs here are the internal state that makes certain outcomes appear attractive. Although salary of the operatives contributes significantly to operatives loyalty, but the operatives want their work to be challenging. Above that they want a clear career path as a source of motivation and loyalty.

Unfortunately though, the maintenance personnel were seldom promoted. This also applied to all staff including those occupying the management position. The implication of this is that on reaching a certain age, they tended to quit their jobs. The unfortunate consequence of this is that the time they quit also always coincides with the time they have garnered sufficient experience to significantly contribute to the organisation's value creation chain. In another area, staffs have less confidence in the work they do as a result of the lack of promotion. Similarly, the staff often felt discouraged doing the same type of work over the years. In fact, in many cases, some had being in the same position for a period of up to 10 years.

D5: Supervise resources
Money, materials, components, labour and information require adequate supervision. This also brings to fore the importance of employing a competent workforce. It is very critical that the maintenance manager possesses the necessary qualifications required to lead the maintenance organisation. Having just a degree in engineering or architecture is not sufficient. Only very few universities actually include maintenance management in the curriculum. And in fact, those that do only provide outline discussions on maintenance strategies. Times have moved on, and experts from backgrounds including facility management, property management, maintenance management and asset management should lead the maintenance organisation. There should be practise of division of labour in the organisation for better service delivery. With respect to the validation, all the participants were 100 % confident that work supervision was highly essential. The acceptability of the participants with this was indicated by the mean score of 5 on a scale with maximum value of 5.

D6: Promote sense of communication
Everybody in the organisation should be encouraged to communicate ideas that will increase organisational productivity and at the same time increase user satisfaction. Primarily, the corporate mission and vision of the organisation and specifically the

maintenance organisation must be communicated to all members of the maintenance organisation. In particular, emphasis is also the maintenance objective. The fundamental objective of providing maintenance service must be communicated to all and sundry working in the organisation. This function earned a mean score of 4.7. Seven of the participants argued that the function can 100 % contribute to better service. Therefore, there is the need to promote a sense of communication among the staff members.

D7: Provide education and training

The VMMM can only work best where the employee understands and accepts the fundamental premise of the model so that they are able to implement it in their daily operation. Therefore, sound education and training are absolutely essential to the success of the model. The VMMM is about transforming behaviours, and for the VMMM process to be successful, it is vital that the employees understand what they are being asked to do, why it is necessary and how their own personal well-being will be affected. Systemic maintenance management systems must have staffs that are willing, ready and able to take advantage of business opportunities, albeit that can only be achieved through informed and educated employees. The more training and experience employees have, the less direct supervision they need. For example, managers who have well-educated and experienced employees will have high productivity and with less span of control.

In fact, the competencies of those concerned with the management of maintenance have been questioned. Tapsir (2005) opined that the management organisation of high rise buildings and stratified properties in Malaysia are not qualified to manage maintenance works. This she argued is contributing to poor service delivery and maintenance practices. Thus, in order to maximise the benefits of the VMMM, it requires that employees develop an adequate degree of skill and continuously refresh the workforce. This will enable the employees to be equipped with updated knowledge and relevant skills. Learning must continue.

Training contributes in no small measures to staff development. On-the-job training operates to ensure the importance of training to the needs of employees. As a matter of fact, the maintenance organisation needs a strong corporate team to stand behind the manager to run the business and to grow it. Many of the maintenance managers recognise this as important, though. As hypothesised, most of the participants agreed that this function was highly essential; this is indicated with a mean score of 4.7.

D8: Promote trust

Most of the participants had some trouble in relating trust among the staff with service delivery. However, the effectiveness of any manager depends on the ability of the leadership to gain the trust of their subordinates. In fact, the mean score is 4.6, but one of participant, even after an explanation of this function, still considered it moderately important. For a better working environment, integrity, competency, loyalty and openness must be maintained. This is in order to establish trust by being honest, open and forthright about the challenges and rewards of organisation success.

D9: Share responsibilities, blames and achievements
Providing an enabling environment will encourage employees to think and act like the owner of organisation as well as the beneficiary of the service they provide. By establishing and reinforcing a "we" spirit, an organisation will recognise that it takes the coordinated effort of all the employees to achieve an objective. The contribution of each employee's efforts must be reinforced and guaranteed. More than in most services, maintenance service depends on the teamwork of its workforce for success.

In the VMMM, staffs have neither the time nor the interest in grabbing the limelight or diverting the blame. Everyone is empowered, responsible, accountable and motivated to use their initiative and to take whatever steps are necessary to fix problems when they arise. This attitude is directly opposite to the traditional case in an organisation that supported a blame culture; when anything went wrong, no one would own up to the error. Some of the participants were initially sceptical of this function. Those participants did not want to be responsible for the failure of their staff member, yet they wanted to be associated with the success of the staff. However, they all agreed that it was in actual fact a very important factor and that if put into consideration, it would generate a better working relation.

9.5.5 Maintenance Management of Implementation

Usually, all maintenance services are grouped under this function. Often, this is a result of poor methodological issues or due to poor understanding of the maintenance management process. The maintenance department is often considered a technical-driven unit. It is in this stage of the maintenance management that all efforts are focused under the classical system. Traditionally, the loop is open and closed in this stage. This stage is very critical, because if adequate and sufficient efforts are not used in the implementation phase, much of the "soft" saving in the preceded stages will be channelled into the drain. Hence, it should be handled with outmost care. This stage entails the technical requirements of maintenance. It involves investigating the cause and effects of defects, the diagnosis of defects, condition assessments and making decisions on alternative solutions, materials and components and supervises corrective and curative works. The cause of defects must be properly analysed. Before any corrective maintenance work is proposed, the primary causes of defect must be identified and analysed. If this is not done, it is almost certainly, and the correct solution cannot be obtained.

The phase is dominated by the third-level decision makers and operatives. The operatives must be well trained in human relations. The lower gang leader or any member of the team should be encouraged to the take responsibility for follow up with the works. Human relations and diplomacy are necessary here so that the service recipients/users appreciate the service. The operatives must be responsive, respective friendly and courteous to the building users. Prior to this stage, a suitable maintenance procurement strategy should be addressed. In the following paragraphs, the major roles within this function are briefly provided.

I.1: Inspect building and engineering system
Surveys, tests and investigations must be carried out to ascertain the nature, kind and quality of materials required in sustaining the building to an optimum operable functional performance. Stakeholder and need analyses are required in order to ensure that the actual value system of the building user is factored into the maintenance process. Providing a quality building with the optimum cost alone is not sufficient to obtain the satisfaction of the users. Quality that is just a criterion in the user value system is objective. The qualitative aspect which is the primary concern of the user can only be known by asking the users themselves.

I.2: Analysis and estimate of resources
Most of the participants were convinced that analysis and estimating the required materials and components were highly necessary before the intended repairs commenced. This is, among other things, to avoid unnecessary delay and frustration. The mean score of this function is 4.6, however. While six of the participants were 100 % confident that it was a critical function, the remaining four were only 80 % positive.

I.3: Analyse user value system
These criteria included both qualitative and quantitative factors that influenced user satisfaction with the service. The following includes criteria of the user value system: comfort, convenience, quality, friendlessness, reliability, safety, aesthetics and time. However, the major task here was to determine the user requirements for the buildings and identify the worker and numbers required for the maintenance work. The people that understand buildings best are the people that use them every day. Therefore, there is the need for the maintenance organisation to learn from their building users whether the building is performing or not. The mean score of this function is 4.7. While three of the participants were 80 % confident of this, the remaining seven of them were 100 % sure it was very important and unique; however, most of the respondents did not think it was actually significant.

A user's value system is the most important element in building maintenance management. There is no one who can understand the problems of the building more than the occupants themselves, because the person who wears the shoe knows where it pinches. While a number of criteria will have to be considered when taking decisions on building maintenance, the dominant criteria should always be that of the worthiness of the building to the users.

I.4: In-house and outsourced work
The decision whether to operate on planned or unplanned maintenance is addressed here. The fundamental of this stage is that the outcomes are physical; they can be seen, felt and in fact smelled. Unlike with the top management functions that are not visible in actual reality, the implementation stage is a characteristic with physical outputs. Repair of doors, painting of walls and repairs to defective pipes are examples of outcomes of the implementation stage. This is another stage where the

author expends much effort in trying to convince the participants that it is difficult to have a capable maintenance organisation that can handle all major work in-house. The difficulty encountered is in convincing them which is the obvious mean score of this function. In fact, it is the least-rated criteria, with a mean score of 3.8. Further discussions will not be provided here as many of the issues have been presented earlier in this book.

However, many universities still prefer to outsource their building maintenance to external contractors. In spite of this, many will prefer to carry out the routine repair of the facility like the general lubrication, cleaning, sweeping, servicing it themselves and contract the major preventive and corrective aspects to other organisations. The maintenance management planning should not be outsourced, however, under whatever condition. This is because maintenance management must be tied with the organisation's cooperative objective. Maintenance management cannot be strategically achieved by an external agent. The maintenance contractor would not understand the in-house politics that are often necessary for the organisation to be in operation. These politics are usually concealed to the external parties.

Yet, many universities' top management do not see why they should concentrate on building management when the universities' function is to provide education. This is a fallacy, however. They should create better maintenance organisation to cater for improved building performance. Further discussion on this topic is not to be provided here, as much has already been provided in this book. However, instead of outsourcing maintenance services, the department should engage the services of consultants to train the staff from time to time in order to keep the department abreast with the latest best practices.

I.5: Implement work

Whether the work is to be carried out by the in-house or external contractor, the physical repair or prevention works is required to be made. This phase in the value maintenance management process is the tactical aspect of the framework. Strategic decisions are not fundamental at this stage. This is where plans are translated into physical action. Much of the roles and functions at this stage are corrective and preventive rather than managerial and systemic. In other words, it is not the main stream of management functions. Rather, it is the engineering aspect. This phase is where the operation of the work is carried out. The mean score of this function is 4.4. However, the main reason on why the score is low can be explained by the fact that most of the universities were at that time more comfortable with outsourcing.

It is pertinent to disclose that participants were allowed to include other functions they believed were necessary to the list provided. However, none of the participants believed to the contrary. Simply put, the content of the validation is sufficient to provide value added service. Due to the limitation on the size of the participants, it was not useful or even lead to conducting any inferential statistics. As a result, in order to determine the relative significant of the validation, participants were asked to rate each of the model in terms of five point confidence levels: 0–20 %; 21–40 %; 41–60 %; 61–80 %; and 81–100 %. The results of which is contained in Table 9.7.

Table 9.7 Distribution of rating of the model

No.	Maintenance performance metrics	Building maintenance value chain	Maintenance demand model
1	61–80	81–100	61–80
2	61–80	81–100	61–80
3	61–80	61–80	61–80
4	81–100	81–100	81–100
5	81–100	81–100	81–100
6	81–100	61–80	41–60
7	81–100	81–100	81–100
8	61–80	61–80	61–80
9	81–100	81–100	61–80
10	81–100	81–100	81–100
Average (%)	82.5	84.5	76.5

Overall, the participants were more than 80 % confident that the model would positively improve their service delivery. Seven of the participants were 100 % positive that the building maintenance value chain was sufficient, and some of the participants were only 80 % confident with model, yet none of them believed the contents of the model was not sufficient. In the like manner, six of the participants were 100 % sure that the maintenance performance metric could improve the maintenance organisation's service delivery. With respect to this, one participant asserted that: "your proposed MPM—seem acceptable and reasonable. At (…) we set a target of 80 % and most of the time we achieve more that 90 %". Furthermore, "At (…) we do carry out preventive maintenance and it helped to reduce the number of complaints".

However, with respect to the maintenance demand model, a participant rated it only about 50 % (41–60 %) positive. In relation to this, the participant submitted that: "it is clear for the owner, but makes it clear and simple to customers to understand the flow chart whenever they make complaint" (Validation (17 December 2010) Expert validation). But when probed further on how it could be improved as he suggested, the participant continued thus: "make the flow chart above drawn vertically". However, other participants do not see the need to make it vertical after all it is not for the buildings users per se. In general, the participants overtly appreciated the practicality and simplicity of the model, particularly as it was devoid of complex mathematical or software models.

9.6 Summary

A VMMM has been developed and extensively validated. The model aims to provide a step-by-step methodology in making decisions on maintenance management. The model provides a logical, systematic and yet simple model to guide businesses in building maintenance management. The model is widely acceptable by the maintenance organisation as a standard operating manual. It further encourages improved service delivery. Identifying and meeting shifting users' expectations have become an essential part of an effective system, particularly as users are growing ever smarter and savvy.

Within the context of maintenance organisations, performance matrix is the criterion that facilitates the stimulation of user satisfaction and increased productivity. The maintenance unit or department is the most neglected unit among most organisations. This can be explained as a result of the missing link between the building performance and organisation's mission and vision.

References

Cohen, L., Manion, L., & Marrison, K. (2006). *Research methods in education* (5th ed.). Oxon: RoutledgeFalmer.

Gleich, R., Henke, M., Quitt, A., & Sommer, L. (2009). New approaches in performance measurement: Methods for specification and operationalisation within the context of supply management. *International Journal of Business Excellence, 2*(2), 105–123.

Jones, K., & Sharp, M. (2007). A new performance based process model for built asset maintenance. *Facilities, 25*(13/14), 525–535.

Kelly, A. (2006). *Strategic maintenance management: As essential guide for manages and professionals in engineering and related fields*. UK: Butterworth Heinemann.

Lordsleem, A. & Rabbani, E. (2010). Maintenance management of building projects work through performance indicators. In P. Barrett, D. Amaratunga, R. Haigh, K. Keraminiyage & C. Pathirage (Eds.) *Proceedings of CIB 2010 World Congress*. UK: The Lowry, Salford Quays, 10–13 May 2010.

Mei, A. T. (2002). *Malaysian private higher education: Globalization, privatisation, transformation and market places*. UK: Asean Academic Press.

Nordin, M. K. (2010). *Private-finance initiative can save government RM500 million*. The Star, 5 May 2010. N4.

Sekaran, U. (2004). *Research methods for business: A skill building approach* (4th ed.). London: Wiley.

Shohet, I. M. (2010). Performance-based-maintenance of public facilities: Principles and implementation. In P. Barrett, D. Amaratunga, R. Haigh, K. Keraminiyage & C. Pathirage (Eds.) *Proceedings of CIB 2010 World Congress*. UK: The Lowry, Salford Quays, 10–13 May 2010.

Soderholm, P., Holmgren, M., & Klefsjo, B. (2007). A process view of maintenance and its stakeholders. *Journal of Quality in Maintenance Engineering, 13*(1), 19–32.

Wadugodapitiya, R. R. M. M. K., Sandanayake, Y.G. & Thurairajah, N. (2010). Building project performance evaluation model. In P. Barrett, D. Amaratunga, R. Haigh, K. Keraminiyage & C. Pathirage (Eds.) *Proceedings of CIB 2010 World Congress*. UK: The Lowry, Salford Quays, 13 May 2010.

Tapsir, H. S. (2005). Strategy for sustainability in affordable housing_a challenge to Malaysia construction industry. I*n Proceedings of the 2005 World Sustainable Buildings Conference* (pp. 2886–2891).Tokyo, 27–29 September 2005 (SB05 Tokyo).

Chapter 10
Sustainability Maintenance Initiatives

Abstract Maintenance plays strategic major functions in improving building performance for users and the environment. In fact, the major aim of the maintenance is entirely about sustainability of the constructed facilities. For instance, primarily, what building maintenance seeks to emphasise is that it is more valuable to the users and environment to ensure that the existing buildings are in optimum performance and thereby discourage the construction of new buildings. This makes a lot of sense from the sustainability requirements' point of view. The duties and responsibilities of modern maintenance organisations are primarily concerned with the creation of sustainable future. On completion of studying this chapter, one will (1) understand the roles of maintenance in meeting the sustainability agenda, (2) understand how maintenance organisations in Malaysia are approaching the greening initiatives, (3) understand the roles various stakeholders could achieve in sustainable development through maintenance and (4) discover organisations that are yet to come to terms with the issue of green initiatives.

Keywords Buildings · Maintenance management · Green initiatives · Sustainability · Residential buildings · Rethinking sustainable maintenance management · Malaysia

10.1 Introduction

This is one of the papers which the first author has authored on how maintenance contributes to building sustainability. It is included here to demonstrate through literature and survey the understanding and perceptions of maintenance organisations on the sustainability concept. A good text on (construction/building) maintenance must make some references to the contributions of maintenance to sustainable development. The impact of the construction industry on the natural environment is huge. To construct or build, a lot of resources are required and a lot of wastes are generated in the construction stages as well as during operation processes of the built assets. The land, materials, water, energy and other resources

A.L. Olanrewaju and A.-R. Abdul-Aziz, *Building Maintenance Processes and Practices*, DOI 10.1007/978-981-287-263-0_10

that are required to have the constructed facilities in place are provided by the natural environment. It is also the environment that will "receive" or "store" the wastes that are generated in production and operation of the buildings. It thereby means that the building sector and the construction industry have big roles to play in order to achieve sustainable development. Specifically, the construction industry could not deliver sustainable construction if the maintenance processes and procedures that the industry has are not sustainable.

Like other strategies such as refurbishment, maintenance, conversion, reconstruction and retrofitting, maintenance is a tool that contributes to the sustainability drives. A unique feature of green building is durability and reliability. By this, the building is expected to last a long time to delay or avoid reconstruction. Sustainable maintenance will make certain the continuous functioning of the building, energy cost will reduce, and reworks will minimise. However, buildings cannot remain new forever, and for that purpose, maintenance is required to enable the engineering service to perform optimally. However, there have been various regulations, rules and codes adopted in some countries or by some organisations to improve or achieve sustainable buildings with the understanding that such rules, regulations and codes will lead to sustainable buildings. But in reality, in many of the cases, this assumption has been proved to be wrong after all. There are many buildings designed or constructed by renowned architects/contractors, yet such building used close to 10 times it designed energy and produced more waste than was designed for. Therefore, concerted efforts should be directed towards actual performance rather than designed performance.

> I BRING TO YOU THE VOICE OF MALAYSIANS WHO ARE OF NO DOUBT THAT CLIMATE CHANGE, AND THE CATACLYSMIC CONSEQUENCES ARISING FROM IT ARE INDEED REAL. DESPITE THE MANY PROBLEMS AND DIFFICULTIES WE SEE ON THE LONG ROAD AHEAD TO ADDRESS CLIMATE CHANGE, MALAYSIA IS WILLING TO CONTRIBUTE TO THE GLOBAL EFFORTS…

Dato' Sri Mohd. Najib Tun Abdul Razak COP15, United Nations Framework Convention on Climate Change (17 December 2009).

The quotation above was made by the Prime Minister regarding Malaysians' commitments towards sustainable development. This commitment is to actually re-emphasise government, because there are being commitments in terms of both policy formulations and implementations also by the previous governments. Therefore, there are consistencies in the government policies in this direction.

Case #1 Contributions of the housing sector towards sustainable development.

10.2 Sustainable Housing Maintenance Management Model

The drive towards sustainable development is achievable through the maintenance of the existing housing stock. But the major problem that could impede maintenance from contributing to the sustainable development drives is the lack of clear platform that provides the required steps and evidences that maintenance has significant roles to play. This lack of connection can be explained by the poor perception and understanding of the scope of maintenance. Based on synthesis of literature and discussions with experts, we proposed a conceptual sustainable housing maintenance management model that would allow sustainable development to be set as the objective/motivation for maintenance was presented. The model does not only show that sustainable development can be set as the maintenance objective, but also both sustainable requirements and maintenance requirements can be assimilated into the management theory and practices. The key variables in the sustainable housing maintenance management can be categorised into planning, controlling, organising, directing and implementing functions. The study though at the initial stage represents a potential contribution to sustainable maintenance management for use in the built environment. The model that we propose here is an adaptation of the VMMM model.

10.2.1 Introduction

Research on maintenance management is not lacking, though it carries a different meaning. However, the addition of the term sustainable to the "maintenance management" concept changes it meaning and scopes. This new concept/definition means that the scope of maintenance management must not only focus on increasing the building performance but should also consider maximising the users' satisfactions, productivity and issues in sustainable development. There is a fundamental need to consider sustainable development in maintenance, because many of the processes and procedures involved in maintenance management have great impact on social, economic and environmental issues. Maintenance managers/organisations and users should be cognisant of the social, economic and environmental consequences of their actions and reactions. The role of maintenance organisations is becoming more complex and complicated with the maintenance organisations taking over unfinished construction or construction that do not meet with sustainability requirements yet mandated to provide sustainable maintenance. It has been reported that more than 90 % of existing stock does not comply with the sustainable requirements, but even if all buildings meet with the requirements of sustainable development, building cannot remain new throughout their lifespan. Furthermore, while standards and regulations are becoming available for sustainable construction, the same is not available for sustainable maintenance. Although new "sustainable housing" is encouraging, in existing cities, it must occur alongside the transformation of the current housing stock and infrastructure that their construction does not comply with the requirements of sustainable development.

There are significant researches on sustainable construction, but research that focused on sustainable maintenance still remains little and very little if at all actually dwells on sustainable maintenance management. This could be for a purpose.

For one, sustainable maintenance is construed to mean the same as sustainable maintenance management. This definition or understanding needs rethinking. Management and planning are two terms that are often considered as being synonymous. For example, one finds expression like "a tool to manage and control your work management and material management processes" and "manage and plan of maintenance works" and "to manage and control". But they are not the same, either as a term or as concept. With specific reference to maintenance, the scope of maintenance management has been consistently limited to planning and implementation. Even at that, the planning aspect is not even given full thought in practice. Both in theory and in practice, there is no linkage between the various management functions in maintenance management.

A question that we seek to provide answer to here is that what significance does management bring to sustainable maintenance that each of the functions if consider separately will not offer? Our answer is clear. The value that management has on sustainable maintenance includes the following: to determine maintenance objectives, establish strategic objectives, set maintenance standards, determine maintenance performance indicators, identify user and their value systems, set priorities, reduce "re-maintenance", deliver better service and maximise productivity and profits. If one considers each of the functions of management separately, best value cannot be achieved. But to achieve the above objectives, there is a need to correctly articulate sustainable maintenance into a management theory and practice. Because focusing on one aspect without systemic consideration of the others would not add "value to money". Some authors have addressed maintenance planning (see for instance Mills 1994; Chanter and Swallow 2007; Wood 2009). With specific reference to sustainable maintenance, Jones and Cooper (2007) have sort to know the extent that sustainability issues are considered in the development of maintenance plan. But such studies do not provide the systemic/holistic link the planning function has with the other management functions.

There is a need to consider all the functions consecutively. Each of the functional stages should serve as functional "gate" facilitating the connection with the next stage. If all functions are not consistently considered, the process chain will be disjointed and feedback loop cannot be maintained; thus, it is punitive and regressive to sustainable maintenance. It will be difficult to measure the "actual" importance towards sustainable development functions that are not considered. It is imperative that maintenance organisations and developers realise the importance to improve their sustainable performance in order to grow their businesses. Therefore, to contribute to knowledge in this space, this section reports a study that provides that sustainability could be set as the objective/motivation/need for housing maintenance management. Some of complexities were observed in using maintenance as means towards sustainable development. However, to facilitate using maintenance as a platform to manage sustainable criteria in the maintenance of existing housing stock, a sustainable housing maintenance management model SHM3® was developed.

10.2.2 Management: An Overview

Management is a generic term. However, management principles are applicable to organisation, business or process. Management involves process of forming a strategic vision, avoiding mismatch, setting objectives, formulating a strategy, and then implementing and executing the strategy. It includes recording and storing facts and information for future use. Essentially, it aims to deliver results in line with recipients' expectations, enhance productivities and maximise profits. The functions or duties of management vary and depend on the organisation to which it is applied. However, the more specialised or complex the field of applications, the more the required skill set away from the fundamental/pure management theory and services. For instance, construction managers are expected to be well grounded in construction service in addition to having knowledge on business and commercial perspectives.

10.2.3 Maintenance: An Overview

Before examining maintenance theory and practice, it is pertinent to define the scope of maintenance because maintenance is subject to misinterpretation. For instance, the objective of building maintenance is still considered mainly as building repairs. However, this study challenges this understanding and practices and goes ahead to justify the need for alternative definition. Pitt (1997) defined maintenance as a task of providing a repair service that protects the fabric and the use of the building as a standard that represents value for money. Best and de Valence (2002) labelled maintenance as control engaged to make sure that assets remain functional. According to BS3811: 1964, "maintenance is a combination of any actions carried out to retain an item in or restore it to an acceptable condition". Booty (2006) extends the definition a bit further, as the technical and administrative actions that are required in order to retain buildings in a state in which it serves its required functions.

However, these definitions are restrictive, subjective and regressive for the building/construction performance and user satisfactions (see also Jones and Sharp 2007; Wood 2009). The purpose of the building/construction has been compromised if these definitions are to be meaningful. Buildings are to perform the functions as required by the occupants/clients. From these definitions, maintenance is concerned with the buildings, thereby putting the physical conditions of the building at the centre of the maintenance services and not the building users. These definitions are obstacles towards allowing maintenance to contribute to sustainable development, since maintenance is not initiated based on user value systems except the building conditions. With these definitions, some authors will consider that maintenance has no relevance as far as sustainable development is concerned (see Killip 2006).

Thus, maintenance is defined as the processes and services required to preserve, protect, improve and care for a building's fabric, structure and engineering services after completion, refurbishment or replacement to current standards to enable it to serve its intended functions throughout its entire lifespan without drastically upsetting its basic (physical) forms and use. The definition seems to be long (Wood 2009), but meaningful. It portends the reality, because in practice, maintenance involves the optimisation of the users/clients' value system. From this definition, it is seen that the essence of maintenance stems from the user value system. In other words, maintenance demand is a function of the client/user value system. To this end, maintenance need/demand is considered in terms of three requirements: physical, functional and technological requirements. Therefore, if sustainability is the major criterion within the user value system, it will be set as the main maintenance objective. This definition considers the building as a systemic input of the users/clients' activities. In this context, maintenance involves any work on existing construction whether to meet the latest technological improvement or changing users' requirements is maintenance provided the basic features, forms and function/use of the construction is not altered.

10.2.4 Housing as a Sustainable Development's Driver

Sustainable construction is defined as construction that strives to improve the capital outlay, the operational performance and cost and not harmful to the environment and communities and acceptable by the users and other stakeholders. Construction operation involves resources such as energy, water and raw materials. It also generates wastes and emits potentially harmful atmospheric emissions. While all these are contributing to the climatic changes, climate change is a major challenge to sustainable development.

Following Abraham Maslow's theory of need, housing is a basic need of human. It is immaterial whether one owned their building or rent the buildings one lives in. Adequate, affordable and functional housing has significant/competing consequences on productivity and well-being of the occupants. Housing is one of the fundamental aspects of sustainable development, because the impact of housing construction and its operation on sustainability issues is huge (Organ et al. 2013; Swan and Ruddock 2013; Miller and Buys 2013). For example, the building sector consumed more than 40 % of the world energy, release 1/3 of CO_2, release 2/5 of compounds that caused acidic rain, consumed about 1/6 of freshwater, use 1/4 of harvested woods, release about 1/2 of fluorocarbons that destroy ozone layer, produce about 40 % of landfill materials, use 45 % of energy in operations, responsible for about 40 % of greenhouse gas emissions (GHGs) and at the same time make use of about 15 % of world's usable water (see Wood 2006; Killip 2006; Sherwin 2000). Furthermore, about 40 % of world's solid waste is generated through building construction and operation (see Wood 2006; Killip 2006; Sherwin 2000). Building interiors were also estimated to contain about five times more pollutants than the air outdoor. The major implication of the statistics presented is that the way buildings are built and operated is part of the major problems or threat towards achieving sustainable society.

10.2.5 Towards Sustainable Housing: Malaysia's Case

Since the 1980, when the report of "Our Common Future" (Brundtland 1987) was realised, the concern on climate changes is intensifying globally. The Malaysian government also recognised the adverse effect of climate change. Due to this recognition, developmental goals were set up based on five principles: (1) development on a sustainable path; (2) conservation of environment and natural resources; (3) coordinated implementation; (4) effective participation; and (5) common but differentiated responsibilities and respective capabilities (Department of Statistics Malaysia 2013). The developmental goals and the principle show that the government also recognises the roles of the various sectors and players. Deductively, the climate changes require that the construction industry should respond to the requirements of climate change. Malaysia commitments towards sustainable development are impressive, and it is a signatory to the Kyoto protocol. However, climate change is a major threat to sustainable development. Malaysia has outlined various initiatives towards sustainable development (Government of Malaysia 2010). Table 10.1 contains some initiatives by the government towards achieving sustainable development.

In Malaysia, there are a number of sustainable projects. Very few have been completed, but many are still at infancy. Some examples of these are now cited: Tanarimba sustainable housing projects, Demonstration and Documentation Centre for Sustainable Urban Household Energy Usage, Heriot-Watt University, Putrajaya, Setia City Mall, Amanjaya Specialist Centre Sdn Bhd, Bangunan Suruhanjaya Tenaga (The Energy Commission Diamond Building), Green Energy Office (GEO) building and Low Energy Office (LEO) Building. As of 2013, about 140 projects have been certified based on the Green Building Index tools. Various agencies and ministries including the Centre for Environment, Technology and Development, Malaysia and Ministry of Energy, Green Technology and Water, Malaysian Green Technology Corporation and Department of Statistics are playing leading role towards creating viable platforms for sustainable development. Though there are many dimensions of sustainable development (Enquist et al. 2007), for simplicity, the three most often cited dimensions are considered here. They are economic, social and environment.

In the context of housing maintenance, the economic aspect relates to financial commitments required to maintain the stock throughout its lifespan, taking into account various stages in its management. With some discount tools, this is measurable. The social aspect would mean the impact that the maintenance or the lack of it would have on the communities within which the building is located. In other words, this dimension measures the value the maintenance on building brings to

Table 10.1 Some commitments by governments of Malaysian towards sustainable society

S. No.	Commitment
1	An amendment to the Uniform Building By-Laws 1984—this requires newly built detached, semi, and bungalow buildings with a roof area of 100 sqm and above to equipped with system of the reuse of rain water
2	Total of certified emission reduction units from Clean Development Mechanism projects are 673,857 CO_2
3	Strategy and action plan on Environmentally Hazardous Substances (EHS) Management developed in 2009
4	500 sewerage treatment plants were rectified and upgraded to suit the effluent quality requirement standard (Environmental Quality Act 1972)
5	Aim achieve to 40 % reduction in carbon emission
6	RM1.5 billion green technology founding scheme made
7	RM20 million under 2010 budget to address sustainability issues
8	Reduces water consumptions by 30 %
9	PM Office complex upgraded to meet gold standard rating
10	Central Forest Spine project covering 4.3 million hectares
11	Establishment of a Renewable Energy Fund and Renewable energy to expand from <1 % in 2009 to 5.5 %
12	Formation of national green technology council
13	Reduce non-revenue water (NRW) percentage
14	Increase coverage and efficiency of sewerage services
15	Increase urban public transport network coverage and facilities
16	Implementation of EURO 2 M standard
17	Development of integrated action plan to prevent pollution and improve water quality for Sg. Linggi, Sg. Sepetang, Sg. Merbok, Sg. Kuantan and Sg. Kinabatangan
18	The development of strategy and action plan on environmentally hazardous substance (EHS) management
19	Installing and laying 14,988 km of new pipes
20	95 % compliance to national standard for drinking water quality set by Ministry of Health
21	Malaysia's total electricity generated by 2015
22	Improving solid waste management
23	Adoption of national climate change policy and national green technology policy
24	Promoting energy efficient to encourage productive use of energy through various initiatives
25	Develop Clean Air Action Plan
26	EURO 2 M standards implemented in 2007
27	Reduce dependence on petroleum products by increasing the use of alternative fuels
28	Increase use of renewable energy (RE) for power generation and by industries
29	Intensify energy efficiency (EE) initiatives in the industrial, transport and commercial sectors as well as in government buildings

Source Government of Malaysia (2010) and others

communities, and this is pertinent since building does not exist in isolation. The social aspect is also concerned with the cultural constraints placed on some buildings that make us to require some standard of maintenance. It has been suggested that the lack of maintenance could set up inner psychological tension in the users, if the defect is not rectified (Sommerville and McCosh 2006). The environment dimension comprises air, water, forest and lands. Environmental dimension denotes the impact that maintenance works and process and the lack of it has on the environment. To maintain, resources including water, energy, materials and components are required. The processing, harvesting, storage, transporting and installation have influence on climate changes. Similarly, if the stock is not maintained, the environment will be affected. The aspect is largely concerned with the effect that maintenance or the lack of it has on water, air, land and forest due to waste, pollutions and carbon emission.

For simplicity, it could be surmised that housing that cost more than it "real cost" or cost overruns, generate more waste, consume high electricity and water, produces more CO_2 emission is not a sustainable housing. Insufficiency in the housing delivery could also be considered as unsustainable housing. In Malaysia, there are 4,655,067 housing units in Malaysia (NAPIC 2012) but only about 100,000 homes are added annually, representing about 2 %. Therefore, to replace existing housing that does not comply with the sustainable development will take about 50 years. The implication of this is that Malaysia is experiencing housing crisis. The crisis will lead to poor standard of living, unaffordable rental price, abandonment, dilapidation of the existing stock, among others. One implication of this is that, house rent will take position of other more desire need like foods, clothing and health. While the existing building stock is insufficient, the existing stock is ageing and this coincides with the time the governments' expenditures on housing supply are not inconsistent and grossly inadequate.

10.2.6 Sustainable Housing Maintenance Management

Housing maintenance can contribute to global sustainability. Maintenance is a key to housing performance and occupants' productivity, and satisfactions and the performance of the housing and occupant's satisfactions have far-reaching impact on sustainable development and national prosperity. Building maintenance constantly affects everyone's life because people's comfort and productivity are relative to the performance of the building they live, learn, conduct research, work and the impact it have on social fabric and economic growth (Rendeau et al. 2006). A building that do not meet the required maintenance standards will certainly produce more waste and pollutants (e.g. toxic build-up, carbon dioxide emission) and affect users' well-being as well as consume more energy, water and other resources (Olanrewaju and Kafayah 2008). It is arguable that perhaps more than the design team, the maintenance management aspect has more roles to play to achieve sustainable housing.

To understand the importance of maintenance towards building performance and to the sustainability targets, internationally recognised systems for assessing building performance including Leadership in Energy and Environmental Design (LEED), Building Research Establishment Environmental Assessment Method (BREEAM), Greenstar (Australia), Sustainable Building Certificate (Germany), Haute Quality Environment (France) and Green Building Index (Malaysian) consider maintenance issues in their assessment criteria. But, like government regulations and laws, these third-party tools are more appropriate to new project but not for maintenance.

Sustainable building provides energy saving, water saving, healthier indoor environment and better connectivity to social facilities and amenities (public transport) embracing recycle initiative (see also Wee 2009; Ervin 2011). While new building can be designed in accordance with the sustainable agenda, building will not remain new throughout its lifespan. They require maintenance in order to attain its design life. The need for maintenance in buildings will be intensifies only as the value of the building, and the associated engineering services must be preserved and sustained for the building to be meaningful to its users since. Basically, there is no way existing housing can comply with the sustainable agenda without maintenance. The correlation between maintenance and sustainability is positive. The current approaches to sustainable maintenance are product-based. A variable that is common to both sustainable development and maintenance is that both are future-based. Sustainable development and maintenance are both tactical or technical issues, therefore, the prime purpose of the management aspect is to blend the requirements in both to deliver "value added services/functions" or develop new set of objectives and purpose (Fig. 10.1).

While study of the actual impact of maintenance on sustainable development (Miller and Buys 2013) is not conclusive, a number of researches have proved that maintenance has study impact of maintenance on sustainable development (Jones 2002; Jones and Sharps 2007; Wood 2006). The researchers have identified various sustainable factors in maintenance practice. These factors include life cycle approach, the 3R of material and components, use of green technologies, multi-disciplinary approach and use of local materials. With specific reference to housing maintenance, it was found that it is possible to reduce maintenance cost by taking into account factors that affect operational costs (El-Haram and Horner 2002; Stoy and Kytzia 2008). But the authors only consider maintenance cost at design and construction stages. There are also studies on technological aspect of sustainable

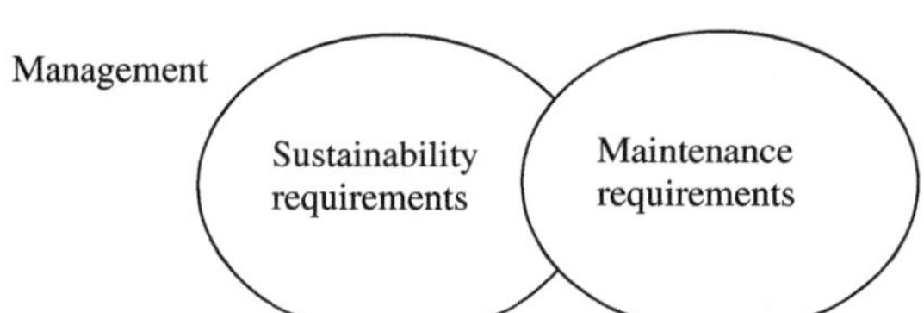

Fig. 10.1 Simplified sustainable maintenance management framework

issues in the housing stock (see also Zhu and Lin 2004; Swan et al. 2013). But the major problem in housing maintenance is not about the technology but rather about its management. For instance, a significant aspect of the housing users understanding of the sustainable maintenance cost is as a result of how the users understand the scope and benefits of sustainable maintenance. Most users and in fact the facilities/maintenance managers limit maintenance to the operational aspect and specifically to repair/corrective maintenance.

But such poor perception of the maintenance will remain if the maintenance demand is based on the building rather than the users. But its maintenance and maintenance management are not redefined; if all the various functions in management are consistent in process chain, the importance of maintenance in sustainable development would not be clear and optimised. There is the need to correctly, adequately and clearly spell out the various functions in sustainable maintenance management. For sustainable maintenance to be seen as value added service, the requirements of both theories and practices should be matched with the management framework. Much focus has been on maintainability rather than on maintenance management. This deficiency is prevalent because sustainable maintenance is not in position in management theory and practices but rather limited in scope to one aspect/function of management. Against the problems illustrated above, this study situates sustainable maintenance into the management theory and practice.

The strategies and initiatives in sustainable maintenance management refer to adopting a sustainable mindset in the maintenance processes, methods and materials, in order to minimise pollution, reduce waste, offer sound financial prudence and enhance community and social integration and meet the overall objective of sustainable development. If buildings are not well maintained, then the future generations would, among others, face very serious financial burden. Too often during maintenance works, the contractors (and consultants) are faced with problems of shortages of materials and components due to incompatibility or outdated items.

Based on a comprehensive review of literature and a brief discussion with selected members of CIOB Malaysia, recently, on the relationship between sustainability development and housing maintenance, a sustainable housing maintenance management model SHM3® was proposed for inculcating sustainable issues in the management of housing, as shown in Fig. 10.2.

The SHM3® consists of five major partitions of planning, control, organisation, directing and implementing. These elements must form parts of the maintenance organisations' philosophy and principle. This is to provide a step-by-step methodology for maintenance demand and implementation. With this sequential arrangement, maintenance is strategically and consistently planned, organised, directed, controlled and executed. Sequentially following the chain will drive sustainable development as maintenance objectives. In other words, maintenance is considered an investment that contributes to corporate business objectives. A critical stage in the model is the "maintenance objectives-mission statement", and at this

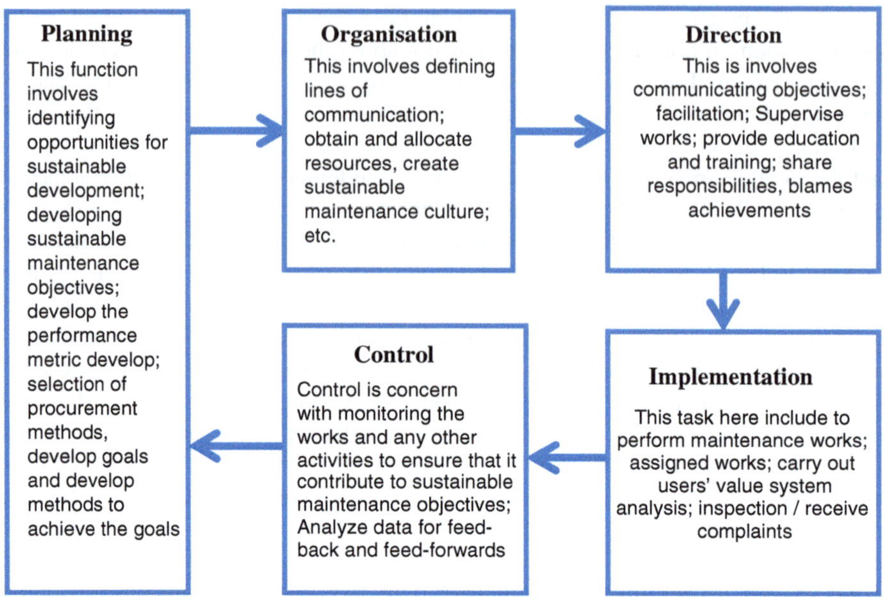

Fig. 10.2 A simplified sustainable housing maintenance management mode—"harmonising it all altogether"—SHM3®

stage, sustainable development will be set as the maintenance objective. It is on this basis that other sub-functions in the planning stage will stem from. This subsequently forms the basis for other functions in the management theory and practice.

As can be seen under this model, implementation is not the "beginning" and planning is not to be treated in isolation. During the discussion with the CIOB members, it was confirmed that since the clients/occupants are those who benefit more in the long run with sustainable maintenance management, it thereby becomes imperative that the clients/users should demand for it and be willing and ready to pay for it. Furthermore, it was observed that the maintenance manager/facilities' managers have poor understanding on scope and meaning of maintenance, maintenance management and sustainable development. But, the maintenance organisations are not doing enough to promote sustainable maintenance management. The government should enforce in the way of regulations, and laws enforce the maintenance organisations to SMMM.

10.2.7 Conclusions

This paper seeks to prompt active discussion about what are important to the operators/managers of sustainable buildings. This paper has drawn attention to some of the practical difficulties that exist within the definition and scope of maintenance, thereby limiting it to serve as a platform to drive sustainable development, but this is due to poor perception not reality. The original research question for this study revolved around whether considering one aspect of management is sufficient to define the scope and objective of sustainable maintenance. It is surmised that this is not sufficient, anyway. There is a need to consider all the various aspects of management in sustainable maintenance to obtain sustainable maintenance and consequently sustainable development. Maintenance and maintenance management if properly defined is a major key towards sustainable development. The owners of the housing stocks should demand for sustainable maintenance management in efforts to derive best value on their investment. The SHM3® considers sustainability requirement as the motivation for maintenance demand. SHM3® is a concept, whereby organisations integrate the principle of sustainable development in the management of housing maintenance. The prevailing justifications for this have been mentioned earlier. But to reiterate, it involves setting sustainable development as the target for maintenance, setting strategic objectives, identifying housing owner value system, providing consistency link with other functions, using both objectives and subjective criteria in housing maintenance decision practices, evaluating performance, enhancing communication, user-focused decision-making and continuous improvements, encouraging team work and complying with other guidelines on sustainable development.

Case #2 Contribution of the universities towards sustainable maintenance. The original version of this article is reported in Olanrewaju (2011)

10.3 Green Campus

10.3.1 Introduction

To go green or greening is a principle, philosophy and mindset. The term means different things to different people, however. The word is often used interchangeably with words including sustainability and eco-friendly. In another respect, the word *green* has being prefixed with numerous words such as economy, tourism, building, banking, education, technology and information. It is common today to hear or see expressions such as green economy, ecotourism, sustainable education and green technology. However, irrespective of the words or terms used, the fundamental intent is the same. Greening is a blend of energy optimisation, durability, waste minimisation, social impacts, good indoor environment, pollution control, life cycle cost, user-friendliness, user comfort and satisfaction. Strictly speaking, the greening concept is about tackling change in the world's climate, however. Climate change is caused as a result of an increase in temperature. Meaning to say, the temperatures close or around the Earth's surface are warmer than normal. However, from the other hand, it has been shown that the major force behind the change in temperature is human activities.

This paper aims to examine building in the context of the green initiatives. While there is sufficient literature on how new buildings can contribute to meeting the green initiatives, literature on how maintenance management could contribute to the green initiative is grossly inadequate. Therefore, this paper adds to the scanty existing body of knowledge. Data obtained on approaches taken by the maintenance organisations in going green were analysed. The paper concludes by arguing that building maintenance is a strategic issue for academic institution. The failure of universities to accept maintenance as a core service is serious failure to a sector that prides itself as vehicle for scientific and technological advancement.

Advantage of complying with the sustainability target

1. Tax benefits and incentive for sustainable building increase the saving and value of buildings
2. Reduced maintenance cost
3. Reduced operational cost
4. Increase annual energy saving and thereby creating higher value for building.

10.3.2 Theoretical Framework and Background

Terms including eco-building, green building, sustainable building and energy-efficient building are synonymous with green building, and they are sometimes used interchangeably. The green initiative is a vague and ambiguous term. The concept places much emphasis on striking balance between the natural and the built environment, while taking into account the impact of man's activities on the environment in meeting the need of the present generation without inflicting harm to the future generation. The greening concept is akin to refraining from passing our debts to our yet-to-be-born grandchildren. In this book, green building can be defined as building that strives to preserve the economic, ecological, spiritual, social and cultural requirements for societal well-being and at the same time preserve the natural environment for the future users both during and after its construction and specified service life (Olanrewaju and Kafayah 2008). Building construction and operation involve resources such as energy, water and raw materials. It also generates wastes and emits potentially harmful atmospheric emissions (Freed 2008). All these contribute to the climatic changes directly and indirectly. From the conceptualisation phase of a building to its disposal phase, it impacts greatly on the environment.

Building construction and operation are also claimed to account for close to half of the water used globally. The building sector makes use of one-third of the world's resources and is responsible for about 40 % of GHG emission and at the same time makes use of about 15 % of world's usable water. Furthermore, about 40 % of world's solid waste is generated through building construction and operation. The building envelope is also argued to contain about 5 times more pollutants than the air outdoors. The quantity of energy used and CO_2 emission in buildings during construction is very much smaller as compared to the amount used and emitted during operation and maintenance (Doran et al. 2009).

A building serves to enclose a space and modify weather conditions. Further, it usually serves to protect and maintain privacy. In short, it is a facilitator of human's activities. Buildings consist of several types of structures such as private buildings, health facilities, public buildings, public housings, town halls, educational buildings and offices. A building is an expensive item to own and to operate. Buildings, generally once completed, are expected to perform certain functions for a certain period. However, buildings may not perform satisfactorily as a result of inadequate design, poor workmanship, defective materials and components, wrong installations and applications and the failure to provide the required maintenance.

However, assuming that the initial design was adequate for the intended use, the quality of workmanship was high, the materials and components selected were of high quality and installed properly, and any inconsistencies from the predicted service life can be attributed to maintenance. In the same vein, new buildings are often beautiful, meet the required energy standards and are functional for the occupants and are of state of the art and stakeholders are happy, but for them, to achieve their design performance, they must be well maintained (Schrag et al. 2007). Maintenance, therefore, is essentially required to delay defects to ensure that buildings perform optimally throughout their life cycle so that they represent value to the users. If defects are allowed to occur, the performance of the building is undermined. In another words, the essence of building maintenance is to increase the life expectancy of a building and to conserve energy. A building requires maintenance in order to attain its design life. Thus, the need for maintenance in buildings will only intensify and the value of the building and the associated engineering services must be preserved and sustained for the building to be meaningful. About 75 % of the total expenditure on the life cycle cost of a building is attributed to maintenance (Booty 2006). More than 90 % of the life cycle of building projects requires maintenance.

Maintenance has been defined as "the processes and services taken to preserve, repair, protect and care for a building's fabric and engineering services after completion, repair, refurbishment or replacement to current standards to enable it to serve its intended functions throughout its entire life span without drastically upsetting its basic features and use" (Olanrewaju 2009). From this definition, maintenance is not necessarily about the building itself per se rather it is about the occupants of the buildings. "User care" is the focus of maintenance! This definition is based on the understanding that a building is procured for the sake of the services (i.e. comfort, protection, accommodation, security and esteem) they offer to their users. It is the correct functioning of the building that the users desire and not the physical condition of the building. To the extent that the building is capable of allowing the users to perform their functions, the building is a source of value creation to the functional service of accommodating, learning, teaching and doing research with specific reference to the university buildings.

While new buildings should be designed and operated to incorporate green initiatives, the maintenance of existing buildings must be made to address green criteria, if saving and protecting the mother earth is critical. Designing and constructing green buildings is not enough, and the maintenance objectives must include green criteria; otherwise, a long-lasting solution will not be achieved. To underscore this, internationally recognised systems for assessing buildings' environmental performances including LEED, Green Building Challenge (GBC) and BREEAM, High Environmental Quality (HQE), Comprehensive Assessment System for Building Environmental Efficiency (CASBEE) and VERDE include building and service maintenance in their assessment criteria.

In Malaysia, the Green Building Index (GBI) has clearly articulated maintenance as a critical component towards meeting the sustainable development goal. Apparently, to provide holistic framework for sustainable construction, maintenance consideration is significant. A sustainable building must be easy to maintain anyway! Building maintenance constantly affects everyone's life because people's comfort and productivity are relative to the performance of the building they live, learn, conduct research and work in and the impact it has on social fabric and economic growth (Rendeau et al. 2006).

The recent amendment to the building regulations that compel sustainability issues be considered during any building maintenance work (Wood 2006) is a point that maintenance drives towards the sustainability agenda. The amendments stipulate that all repair and replacement objectives must go down well with the ideas of sustainability. If a building is not maintained in accordance with the green initiative, certainly, it will affect the users' quality and productivity. It will generate more waste, and more funds will be required to keep such a building operational. Buildings that fail in their efficiency criterion or do not meet the required maintenance standards will certainly produce more waste and pollutants (e.g. toxic build-up, carbon dioxide emission), affect users' well-being, as well as consume more energy, water and other resources (Olanrewaju and Kafayah 2008). The strategy and initiative in the green construction drive is green maintenance, which refers to adopting a green mindset in our maintenance processes, methods and materials, in order to minimise pollution to our environment, sound financial prudence and community and social integration and meeting the overall objective of the green environment.

10.3.3 Green Design and Construction Without Green Maintenance?

Maintenance organisations must not only be proactive but systemic in their maintenance philosophies. Because whether or not the building is designed or constructed sustainably, the maintenance organisation has the control and liberty over the maintenance processes and procedures to make certain a systemic stance is taken with respect to be environmentally and socially sustainable. While it is imperative that the building is a sustainably designed and built, the maintenance organisations can still perform maintenance sustainably. This is perhaps the stark reality maintenance organisations must face. While the exact figure of the size of buildings that were designed and built sustainably globally is not available, it is probably not arguable that more than 90 % of these buildings are not sustainable. Therefore, if the maintenance organisations are not systemic, the organisations will have to wait for all buildings to be designed and constructed sustainably. Green maintenance becomes reactive if it is prescribed only because the building is built sustainably or because of the requirements that are contained in the service-level agreement. This is the traditional approach to green maintenance. However, these mindsets need to change, and though this is changing, its speed is slow. Maintenance organisations should be more systemic towards maintenance service delivery. Concerned employers are now asking prospective employees whether or not they hold any of the certifications in any of the sustainability promoter organisations such as the LEED, CASBEE, BREEAM or VERDE before they are considered for employment.

10.3.4 Existing Buildings and Green Maintenance Initiatives

In this section, our aim is to represent the complexity of the body of literature capturing the range of conflicting assumptions and understandings about what theories and practices of maintenance of existing and other initiatives such as refurbishment, reconstruction and retrofitting are there. Before proceeding further, however, it is important to acknowledge that although we attempt to offer a balanced portrait of opposing views, our opinions and biases will come through whether we want them to or not. Although we are more comfortable with the usual impersonal academic writing style, we believe that it will help readers to differentiate what we believe from what others believe if we are honest and explicit about where we stand on some of these issues.

We do this here and again wherever we view it is necessary. This is a kind of discussion of the preference and an opinion of the author. Simply put, the author's personal ideas and thoughts cannot entirely be differentiated from the content of this paper. This kind of investigation is particularly important in maintenance issues, in which so many divergent assumptions are often left unsaid or asserted as truth. While some could argue that some issues are better left unsaid, it is felt, it will not be in any one's interest to continue to pretend that everything is right, and thus, it fails to present our side of the case. At least, this could well serve as impetus to some writers and commentators.

Published literature revealed a wide range of opinion which tends to polarise towards new buildings, maintenance or other initiatives. However, maintenance is only justifiable to reconstruction or refurbishment if the building remains fit functionally since the object of maintenance is much related to the use value as compared to other criteria such as the physical condition. While initiatives including refurbishment, alteration, conversion and reconstruction are one-offs, maintenance is a must in the building life cycle. Clients may decide not to alter, refurbish or convert their buildings, but this is not possible in the case of maintenance. Almost certainly, buildings that will stand for more than 5 years or even less require maintenance to remain useful. A building that would not require maintenance throughout its lifespan will compromise its lifespan and the functional requirements.

Even if all buildings are to be refurbished, converted or improved, they still require some considerable amount of maintenance to avoid deterioration and unnecessary wastage of the investment. Too often during maintenance works, the contractors (and consultants) are faced with problems like shortages of specific materials and components due to incompatibility or outdated items. Naturally, such materials and components should be repaired or replaced with new and improved ones that are designed to meet existing standards, codes and regulations, as well as being helpful to the sustainability agenda. Strictly speaking, there is no way all existing buildings can comply with the sustainable agenda without going through the painstaking process of maintenance. It is practically impossible to replace or rebuild all organisation or national buildings at one time. Even a nation, no matter how well endowed it is, cannot embark on such an unachievable mission. This is an illustration. The replacement cost of the 1960s' buildings in English universities alone is estimated to cost £11 billion (Rawlinson and Brett 2009). Summarily, through efficient maintenance management, energy usage can be minimised, toxic materials will not be used, safe and durable materials will be preferred, and living conditions will be improved.

10.3.5 Green Maintenance of University Buildings

Recent studies (Housley 1997; Fleming and Storr 1999; Amaratunga and Baldry 2000; Price et al. 2003; Green and Turrell 2005; Leung and Fung 2005; Wong et al. 2007; Fianchini 2007; Lavy and Bilbo 2009) have affirmed the positive correlations between performance of educational buildings and quality of education. University buildings (and services) are procured to create suitable, conducive and adequate environments to support, stimulate and encourage learning, teaching, innovation and research. Building in this context includes the building's fabrics, structures and the engineering services. University buildings include offices, lounges, reception areas, conference rooms, storage, treatment rooms, workrooms for equipment (photocopiers, fax machines, printers and mailbox) classrooms, teaching and research laboratories, libraries, residence halls, cafeteria, mosques, tea rooms and sport facilities. University buildings constitute the most difficult collection of large buildings to maintain due to their complex engineering services and their heterogeneous nature.

Buildings provide values not only to the organisation, but also to the students, faculty members, parents and other users and stakeholders. Although there is no available and comparable numerical data on the state of disrepair, decay, deterioration and unfitness of the buildings, it is possible that they suffer from a similar degree of care and neglect like other institutional buildings. University buildings in Malaysia are characterised by ceramic sanitary appliances and fittings. The administrative and academic buildings are with split air conditioners, while hostels fitted with three-bladed fans. The walls are plastered and rendered in cement and sand in preparation for emulsion paints. The maintenance managers of universities manage an extensive array of buildings such as residential (hostels and staff apartments), administrative, academic and religious (e.g. mosques) buildings. Malaysia is in a hot tropical zone. Hot tropical regions are located within the tropical belts (23.5°N and 23.5°S). The region is characterised with high temperature, heavy rainfall, bright sunlight and high humidity. The temperature is humid with rainfall in almost all the months in the year.

In the maintenance of university buildings, the roles of the maintenance organisation are prominent in order to achieve the greenery or green scorecard. The following are some of the obvious strategies or initiatives that are applicable or can be taken. Plants and equipment should be maintained at optimum standards. Defects will lead to deficiency which will account for an increase in energy consumption and release of greenhouse gases to the earth's surface. These gases are carbon dioxide, methane, nitrous oxide, water vapour, halocarbon and halogenated components. Repair maintenance should be carried out with local materials, components, instruments and labour. Maintenance should be systemic, holistic and proactive; this is in order to delay refurbishment and reconstruction. This is because initiatives such as refurbishment and reconstruction consume more energy and water and at the same time generate more wastes as compared to maintenance.

The criteria of the user value system should in addition to the environmental, social and economic consideration form part of the basis for maintenance demand. If this is done, often the building will have to be maintained again, even where the environmental, social and economic issues are taken into account. In other words, poor user satisfaction will lead to maintenance. Repair maintenance should ensure the applications of natural or green materials and other resources. In other words, when the need for maintenance arises, materials and components that their productions and operation require complex high-technology they should be discouraged and at same time, durable materials and components should be given high preference. A unique feature of maintenance is that it is contagious, meaning to say, (lack or poor) maintenance leads to maintenance.

As part of an ongoing research on the validation of the systemic building maintenance management model that was developed for Malaysian university organisations, a question was addressed to the participants. The model was validated through face-to-face interview technique. The participants were from both the public and private universities. The participants were senior officers in the university facilities, asset or maintenance organisation or as the case may be. The 12 participants were asked about their strategies towards greening or attaining the sustainability status and how they were approaching the green or sustainable maintenance. Out of the 12 participants interviewed, only two of the participants had some good ideas about what it actually represented. However, most of the participants said that their universities had planned to inculcate sustainability in their design and construction in future. A participant said that to them, in the maintenance department, the issue of greening was confusing and complicated. He also argued that the agenda was to them a purely academic issue. While he was making reference to a building constructed on the campus that was dabbed "green building", but to him, he could not see anything sustainable or green about the building. Actually, the building was pilot project. Meaning to say, in practice, sustainability had not being clearly understood.

However, among the two participants that had some idea about the concept, they had made it a matter of principle to replace all defective parts and components with those that were certified sustainable. Also, measures were put in place to reduce energy usage. For example, previously, all chillers were in full operation throughout the year, but now the operations of chillers were constrained. Also, the quantities of water supply to all the hostels have been reduced and users provided with tips on water usage. In another note, the efficiency of plants and equipment had been improved and constantly monitored with the building automation systems. This university has since last year, gradually being replacing major mechanical and electrical installations with those that meet with the sustainable status. The other universities though they did not have a laudable programme towards greening also have some encouraging initiatives. For instance, it had invested into water harvesting. This is whereby rain water is collected and stored for use. Water recycling was also another measure that the university was currently embarking on.

With respect to the third university, the only measure it had towards green was with regard to the building operation but not with the building maintenance. A device was installed into part of the academic and all of administrative buildings to control energy usage. This device will work based on the temperature in the buildings. The device allowed all the lighting installations in the building to switch automatically at 7 pm on the working days. But if the devices indicated the presence of people in the building, the light would switch on automatically. However, the installations were only in parts of the buildings on the campus. In another words, half of the university's buildings did not have such devices. Furthermore, there was a serious campaign at the university mosque on how to use water and power wisely. Stickers with some quotations were placed in the entrances to toilets and ablution (wudhu) areas.

In general, however, the common approaches were mainly related to switching off the lighting and air conditioning systems when not in use. Since most of the lighting and air conditioning systems were centrally controlled, a lot of energy could be saved this way. However, all the universities surveyed had serious campaigns on recycling used items such as paper, plastic and glass. In fact, three categories of bin, one each for paper, plastic and glass, were placed in a strategic location for this purpose. Universities also do not engage in the indiscriminate disposal of electronic gadgets such as computers. These items are returned to the manufacturers for reuse if the gadgets are no longer useful. There is the need for the maintenance, facilities or the asset management unit of the university to work together with the academic unit, so that the lighting and air conditioning system in the classes, workshops and laboratories that will not be used could be identified to be switched off.

This working together would also enable the maintenance organisation could advise the academic unit on alternative classrooms to use. An example of this may be cited. The lecturers may never think of the need to swift a class to a smaller classroom facility that could comfortably serve the required purpose yet save energy used for the air conditioning system. Often, one finds less than 10 students in a room that is meant to accommodate 50 or more students. Certainly, this class will consume more energy compared to a small classroom. Furthermore, maintenance management should be introduced to engineering disciplines in order to produce engineering knowledge and skills in maintenance management functions.

The chapter concludes by advocating for professional roles for maintenance management if delivering best value is critical. There is the need for the development of the maintenance department from forethought. The lack of this will mean that the department is not considered as one of the USBU. The structure of the department should grow as the university grows in size, age and complexities. Certainly, at the initial stage of a university's life, is it just natural (assuming the university starts operation with new buildings), and the structure of the maintenance department will be small. A university with old buildings will require a larger maintenance department. Over the years, the maintenance department expands in relation to demand.

Organisation theory has described the importance organisation structure has on productivity and customer satisfactions. Due to the complexities and significance of maintenance works, there is the need for a proper maintenance department. From the organisation point of view, the ways organisation is structured and operates its business have significant effects on its outputs which consequently influence the environments in which it exists. Put in other words, a university maintenance department has greater effect on the university business. The maintenance department must have sufficient inbuilt flexibility to cater with every works arising from the demand. Universities are now realising that greater saving is possible with a well-functioning maintenance department. Emphases are now places on the way maintenance is procured and the competencies and experience of the members.

10.3.6 Conclusion

Green initiatives for university organisations have alighted. While there are no hard or fast rules towards meeting the greening initiatives, a number of positive principles can facilitate meeting waste reduction, energy maximisation, water reduction and other aspects of the greening agenda. However, there are not concrete steps towards the greening initiatives among the universities in Malaysia. Fortunately, much research on the benefits and applications of sustainability is coming from the universities. It is recommended that synergies be formed among the researchers in the sustainability disciplines with the maintenance organisations of universities so that those concerned with the building management will appreciate the benefits of the approaches. In this way, practical problems and other application issues can be addressed and jointly worked on the researchers and practitioners.

References

Amaratunga, D., & Baldry, D. (2000). Assessment of facilities management performance in higher education properties. *Facilities, 18*(7/8), 293–301.

Best, R., & De Valence, G. (eds.). (2002). *Design and Construction: Building in Value*. Woburn: Butterworth-Heinemann.

Booty, F. (Ed.). (2006). *Facilities management* (3rd ed.). Oxford: Butterworth-Heinemann.

Brundtland, G. H. (1987). 'Report of the world commission on environment and development: Our common future', United Nations.

Chanter, B., & Swallow, P. (2007). *Building maintenance management*. Oxford: Blackwell Publishing.

Dato' Sri Mohd. Najib Tun Abdul Razak, COP15, Quotation by the Dato' Sri Mohd. Najib Tun Abdul Razak at United Nations Framework Convention on Climate Change, Government of Malaysia (2009), Ten Malaysian Plan 20011–2015. Putrajaya: Economic Planning Unit Prime Minister's Department.

Department of Statistics Malaysia. (2013). *Compendium of Environment Statistics Malaysia.*

Doran, D., Douglas, S., & Pratley, R. (2009). *Refurbishment and repair in construction*. UK: CRC Press.

El-Haram, M. A., & Horner, M. W. (2002). Factors affecting housing maintenance cost. *Journal of Quality in Maintenance Engineering*, 8(2), 115–123.

Enquist, B., Edvardsson, B., & Sabhatu, P. S. (2007). Value-based service quality for sustainable business. *Managing Service Quality, 17*(4), 385–403.

Ervin, C. (2011). Market transformation: the green building story. In G. S. Mcnall, C. J. Hershauer & G. Basile (Eds.) *The business of sustainability: trends, policies, practices and stories of success* (Chap. 8, pp. 107–128). Volume 3: The road to sustainability. Santa Barbara: Praeger (an imprint of ABC-CLIO, LLC).

Freed, E. C. C. (2008). *Green building and remodelling for dummies*. Hoboken: Wiley Publishing Inc.

Fianchini, M. (2007). A performance evaluation methodology for the management of university buildings. *Facilities, 25*(3/4), 137–146.

Fleming, D., & Storr, J. (1999). The impact of lecture theatre design on learning experience. *Facilities, 17*(7/8), 231–236.

Government of Malaysia (2010). Ten Malaysian Plan 2011–2015. Putrajaya: Economic Planning Unit Prime Minister's Department.

Green, D., & Turrell, P. (2005). Schools buildings investment and impact on pupil performance. *Facilities, 23*(5/6), 253–261.

Housley, J. (1997). Managing the estate in higher education establishments. *Facilities, 15*(3/4), 72–83.

Jones, K. (2002). Sustainable building maintenance. In J. Kelly, R. Morledge, & S. Male (Eds.), *Building in value in construction*. Oxford: Blackwell Publishing.

Jones, K., & Cooper, J. (2007). The role of routine maintenance in improving the sustainability of existing social housing. In *Proceedings of the European Network for Housing Research*

W07—Housing Regeneration and Maintenance. Sustainable Urban Areas, 25–23 June 2007 [Rotterdam 2007].
Jones, K., & Sharp, M. (2007). A new performance based process model for built asset maintenance. *Facilities, 25*(13/14), 525–535.
Killip, G. (2006). The housing maintenance and sustainability debate, In *Proceedings of the annual research conference of the royal institution of chartered surveyors*, London.
Lavy, S., & Bilbo, D. L. (2009). Facilities maintenance management in large public schools, Texas. *Facilities, 27*(1/2), 5–20.
Leung, M., & Fung, I. (2005). Enhancement of classroom facilities of primary schools and its impact on learning behaviors of students. *Facilities, 23*(13/14), 585–594.
Miller, W., & Buys, L. (2013). Factors influencing sustainability outcomes of housing in subtropical Australia. *Smart and Sustainable Built Environment, 2*(1), 60–83.
Mills, E. D. (Ed.) (1994). *Building maintenance and preservation; a guide to design and management* (2nd edn). Oxford: Butterworth-Heinemann.
National Property Information Centre (NAPIC) (2012). Summary of existing stock for residential, commercial, leisure and industrial property sub-sector as at Q1–Q4.
Olanrewaju, A. (2009). Building maintenance management in Malaysia. *Journal of Building Appraisal*, 4(3), 207–214.
Olanrewaju, A (2011). Green maintenance management initiatives for university buildings. *Built Environment Journal*, *8*(1), 17–24.
Olanrewaju, A. A., & Kafayah, S. (2008). The need to maintain our buildings: sustainable development. In *Proceedings of PSIS enviro (The 1st national seminar on environment, development and sustainabilit*y). Selangor Malaysia: Politeknik Sultan Idris Shah(PSIS).
Price, I. F., Matzdorf, F., Smith, L., & Agahi, H. (2003). The impact of facilities on the student choice of university. *Facilities, 21*(10), 212–222.
Pitt, T. J. (1997). Data requirements for the prioritization of the predictive building maintenance. *Facilities*, *15*(3/4), 97–104.
Rawlinson, S., & Brett, L. (2009). Cost model universities. Building Magazine. Available at http://www.building.co.uk/data/cost-model-universities/3143057.article. Accessed on 15 June 2010.
Rendeau, E. P., Brown, R. K., & Lapides, P. D. (2006). *Facility management* (2nd ed.). New Jersey: Wiley.
Schrag, S., Smith, K., & Stollenwerk, B. (2007). Designing for the care and maintenance of buildings. *Buildings, 101*(11), 34–35.
Sherwin, D. (2000). A review of overall models for maintenance management. *Journal of Quality Maintenance Engineering, 6*(3), 138–164.
Sommerville, J., & MoCosh, J. (2006). Defects in homes: an analysis of data on 1,696 new UK houses. *Journal of Structural Survey*, *24*(1), 6–21.
Stoy, C., & Kytzia, S. (2008). Utility cost: a survey of Swiss office building, Journal of Facilities Management. *6*(2), 120–131.
Swan, W., Ruddock, L., Smith, L., & Fitton, R (2013). "Adoption of sustainable retrofit in UK social housing". *Structural Survey*, *31*(3), 181–193.
Wood, R. B. (2006). The role of existing buildings in the sustainability agenda. *Facilities, 24*(1/2), 61–67.
Wood, B. R. (2009). *Building maintenance*. Oxford: Blackwell Publishing Limited.
Zhu, Y., & Lin, B. (2004). Sustainable housing and urban construction in China. *Energy and Buildings, 36*, 1287–1297.

References

Chapter 11
Conclusions and Recommendations

Abstract In this chapter, the major issues in previous chapters are brought together. The findings, discussion and implications of each of the objectives of the study are summarised here. While the primary data are concerned with university buildings in Malaysia, the principles, process, strategies and decision-making comprise a core set of techniques and methods, which are applicable to most building types anywhere in the world.

Keywords Value added service · Maintenance management · Business objective · Maintenance users · Service delivery

11.1 Introduction

This chapter presents the general inferences from the previous chapters. It draws together the major themes in the study. Despite the benefits of existing maintenance practices that organisations use for building maintenance management, their ontology (the what), epistemology (the how) and axiology (the why) have been the focus of continuous criticisms for those that have a stake in maintenance service delivery. The precise targets of the attacks have been the lack of a systemic view of maintenance, which by definition and application excluded the consideration of user value system, taking maintenance as a liability rather than as an investment and a failure to connect maintenance with business objectives. This has prompted the need for a value based model for building maintenance management. Such a model must be practical, simple and easy to use. From the classical perspective, maintenance is concerned with fixing broken or failed building components. However, such a view only confines maintenance to corrective, reactive or a replacement process. In other words, these are the technical aspects in the maintenance framework, but which in reality is not even the major aspect in the whole maintenance management processes. The scope of maintenance is far beyond that perspective. There is no way, the current practices can move towards value-added service, if maintenance is not looked at from different perspectives.

A.L. Olanrewaju and A.-R. Abdul-Aziz, *Building Maintenance Processes and Practices*, DOI 10.1007/978-981-287-263-0_11

By taken a systemic view of maintenance, maintenance entails much broader decisions that shape the scope of maintenance services. Decisions of maintenance processes must be proactive, systematic, holistic and consistent. This broader perspective of the maintenance scope allows maintenance to be viewed as value-added services and as a factor of production. Therefore, the scope of maintenance management must cover every value-added stage in maintenance processes in the building life cycle and in the management process. While the book is focused on the academic buildings, many public and private sector institutions face similar maintenance management problems. Therefore, the discussions presented here have broader applications as the findings of this research are applicable to other types of buildings and not only in Malaysia but also elsewhere.

The common understanding is that the perspectives of the users are not important and are not realistic and thus cannot be provided. However, it is very possible to combine the perceptions of the two groups. Maintenance organisations can improve the degree of users' satisfaction by paying attention to the highly important criteria on the user value systems. It is more beneficial that the maintainers carry the building users along while making decisions that directly or indirectly affect the users. It is difficult to assume the criteria that increase satisfactions, however. The maintainers need to do more in this regard if users' satisfaction is significant. Users should be carried along. Providing quality service is not enough. The users' experiences, expectations and perceptions are serious issues in the service industry, and which maintenance is part of. Though, this might involve a little increase in maintenance expenditure, the trouble is worthwhile.

While what we do is critical, how we do it is very critical to long time organisational benefits. Maintenance management is about how, what, when and who does it and who we are doing it for. The lack of a comprehensive maintenance management model can significantly impede a maintenance organisation from meeting its core objectives. Consequently, maintenance departments are under considerable pressure to cut maintenance costs, to provide satisfactory services to users and to support the organisation's corporate mission and vision. Poor planning, poor organisation, poor directing, control and implementation all cause maintenance work to take longer time and cost more as well as leading to a lack of users' satisfaction.

The VMMM, within its general philosophy and applications, might not be new per se. In fact, many maintenance strategies and initiatives have considered some of the attributes of the VMMM. However, the methodologies, comprehensiveness, holistic, logic and consistency in the VMMM are fundamental. Traditionally however, the user value systems are not factored into the maintenance services. Major stakeholders' interests are also not exploited. VMMM is a business process and procedure with the understanding purpose of achieving the best use of buildings and maximising the maintenance expenditure.

However, the VMMM alone cannot be proclaimed as a single source of all these superior performances, but it is generally accepted as a major criteria; this is because it generates a greater clarity of purpose in the organisation. In practice and theory, VMMM is a conceptual term that coordinates a variety of tactics and systems to achieve its prime goals. There is a need for continuous improvement. Measurement is a very important aspect of the VMMM. VMMM preserves and protects the performance of buildings and everyone must work together for common goals. Applying a model for buildings lock, stock and barrel cannot provide the needed improvements. The requirements of university buildings are very critical and versatile. The users are more highly informed and enlightened than the users of other types of buildings.

11.2 Features of the VMMM

The VMMM is a philosophy that facilitates maintenance organisation to deliver best value service to buildings' users, reduce maintenance backlogs, reduce maintenance costs and increase the maintenance organisation's productivity. The VMMM provides natural steps for dealing with the multi-perspectives of providers, service users and other stakeholders (e.g. government) in building maintenance services. It provides both an intuitive and attractive interface by which maintainers can model highly interacting sets of variables as well as a data structure that lends itself naturally to the design of efficient decision-making under complex algorithms. The model is simple and easy to apply yet effective and efficient. The VMMM is both a problem-solving and problem-seeking approach. As a problem-seeking tool, it proactively and remotely anticipates a problem before it happens. Thus, systemically it develops solutions in anticipation of the problems.

VMMM is both product and process oriented. It incorporates both the hardware (technology) and software (management, qualitative issues) into account, though the software is the most important. This research significantly contributes to the advancement of the knowledge and discipline of the maintenance service delivery with respect to user value systems. Currently, the object of maintenance is to extend the service of building without necessarily relating it to building performance and user satisfaction. However, a building is not procured for its own sake; rather, it is procured because of the services it provides to the users. In order words, buildings are capital goods. Buildings only have value if they continue to provide the intended services adequately, failure of which will render the existence of the building insignificant if not even useless.

11.3 Recommendations

Arising from the findings of the research from which this book was developed, the following recommendations were developed:

An empirical study is required to determine the condition and performance of university buildings in Malaysia. The government of Malaysia should carry out an audit of the condition and performance of public university buildings. While the results of the audits might not be static, they will provide framework for maintenance allocation. The need for possible reconstruction might also be obvious.

Many of the participants claimed that a basic reason for outsourcing maintenance is due to inabilities to stock the necessary materials and components. Therefore, study is required to develop a maintenance costs model. By this, organisations could keep in stock the materials and components to facilitate in-house procurement. However, because of the nature and class of the defects of the buildings as identified in this research, this claim may not be substantiated.

The competency levels of those currently occupying the positions of maintenance managers require investigation. It is very likely that the investigations could indicate those occupying the position on of maintenance managers are doing so without adequate knowledge and they may need to return to college or university to obtain the prerequisite degrees in maintenance management or even facilities management, property management or asset management.

A professional role for maintenance management is required if delivering best value is critical. Furthermore, maintenance management should be introduced into engineering disciplines in order to produce engineers that have some knowledge and skills in maintenance management functions. The VMMM should be empirically tested; even though the model has gone through several scrutinisations by those concerned with building maintenance.

11.4 General Summary

The management of building maintenance has become more challenging to building based organisations. Organisations are now operating within highly competitive and turbulent environments. However, this is happening when governments are cutting down budgets to their agencies and departments which are also coinciding with the time the buildings are ageing. This is again happening when stakeholders are becoming more demanding and concerned about the building performance. Private organisations are also cutting down their expenses to the maintenance departments to meet other needs. Thus, the only way today's organisation can survive and progress is to change their business model as it relate to building performance.

To support the rapidly changing requirements of maintenance organisations, conceptual models, which constitute the foundation of maintenance management, should be sufficiently flexible to be able to incorporate changes easily and smoothly. There are valuable criteria that influence building maintenance, and the performance indicator criteria of the user value system can be used as indirect measures of increasing quality of maintenance services through VMMM. The VMMM illustrates a linear process. The step by step processes may not be prescriptive; they are simply intended to illustrate how the conceptual model translates into value-added services. This linearity exists most dominantly at the upper level of the abstraction, however.

Appendix

Quoting verbatisms from selected dissatisfied building users

1. Workforce must try to do the work
2. Be professional
3. Always survey the buildings
4. Provide better facilities
5. Upgrade the quality of management and staff
6. Always do checking
7. Hopefully to upgrade the maintenance system in university to give a comfortable situation for students
8. Take fast action and make it easier for student to study
9. Make online maintenance report
10. Security 24 h because many problem at night such as blackout
11. Upgrade, spend more to student. Students have more priority than others
12. The maintenance staff should the attitude
13. I think the air-condition usually not work properly
14. Do work systematic
15. Do the best among the best
16. They ignore our complaints
17. Do the maintenance asap
18. Be punctual
19. Fast in action
20. Improve the maintenance system to become quickly
21. The staff must be working quickly after have the report of the maintenance problems
22. Take an experience person and depend on their profession
23. I would like to propose that the management should upgrade and keep it up
24. Make sure all maintenance is in good condition
25. Staff did not do their works
26. University maintenance department should manage good link or connection with the student so that the problem can be solve very fast and easy. Department of maintenance need do their work properly and effective to solve the problem

A.L. Olanrewaju and A.-R. Abdul-Aziz, *Building Maintenance Processes and Practices*, DOI 10.1007/978-981-287-263-0

27. Ignore complaints
28. Build a system in internet and always give feedback to the problem as faster as possible
29. The management system should be more effective to avoid any of problems
30. The management should take action quickly and faster
31. Fast and quick maintain the facility
32. Paint beautifully
33. Repair system water in hostel
34. Good feedback from management to solve any report
35. Maintenance must be check weekly for make comfortable and good for us
36. Reduce the bureaucracy when student help
37. The facilities should be upgraded
38. Try to make it more friendly
39. Fast response on maintenance system
40. Don't try to take advantage of student facility that just for student
41. Provide roof for walkway pedestrians
42. Be alert to all complaints
43. Should be more efficient
44. Improve the condition of cleaning area
45. Do the maintenance on time
46. Upgrade the system and always check the system
47. The damage should take quickly action for the damages
48. The maintenance system have to improve the defect urgently and give action to them
49. Be a very good maintenance system
50. Always take fast action if there is wiring maintenance system
51. Please make the maintenance activity by following schedule
52. Security system at hostel of university very poor. I hope university will take serious about this problem
53. Unfriendly staffs, reporting very slow when we make a report
54. The response to complaint very slow
55. Take urgent responsibility
56. I always make spot check

Nomenclatures: Concepts and Definitions

A number of terms are used widely in this book, and it is deemed essential to define (and explain) these terms; the terms shown in bold are those used in this book. Where alternative terms are commonly used in the construction industry, they are shown in italic. Following are the nomenclatures used in this thesis:

Benchmarking. Is the activity whereby a measure of a service, process, or product that organization or individual provides or obtains is compared with a similar service, operation, process, or product that another individual or organization provides or obtains. Primarily, benchmarking aims to learn and improve performance and productivity. Benchmarking involves planning, evaluating, analyzing, adapting, and reviewing. In this book, this benchmarking is closely is related to variability.

Benefit. This is the achievement, 'profit', reward, or pleasure obtained from the choice we made. The benefits that service user obtained from the service they received depend on the user's preference. For instance, the benefit obtained by maintenance service user of university building is higher learning or teaching as a result of conducive or stimulating environment in or around the buildings. In economics, money is the unit of comparison. Therefore, benefit is primarily an economic issue.

Building services. *Engineering services* The term for the services or utilities required to operate in the interior environment and functioning of a building—heating, ventilation, air conditioning, acoustic and sound control, fire, safety, plumbing, draining, lift, escalator, and electrical installations.

Business-to-user services or B2U. Are the services that organizations provide to individuals, an example being the service the maintenance department provides to users.

Building condition. The physical state of a building.

Customer. This is a complex term because it means different things to different people. 'Customer' represents individuals, organizations, and communities of that organization or individual deliver service. Customer can be internal or external depending on whether the service is provided within the organization or to external customer.

A.L. Olanrewaju and A.-R. Abdul-Aziz, *Building Maintenance Processes and Practices*, DOI 10.1007/978-981-287-263-0

Customer experience. Is the customer's direct and personal interpretation of, and response to, their interaction with participation in the process, and its outputs, involving their journey through a series of touch points/steps. Customer experience is personal, no two individuals have the same experience.

Decay. Decomposition of building or/and building service often caused by fungi.

Defect. Nonfulfillment of an intended requirement and expectation.

Deterioration. Physical manifestation of failure caused by environmental or natural autogenous influence on the building and/or building services. The loss of the original condition of a building and/or building service over time due to natural, man-made, or human activity.

Develop. Formulate, design, and construct model.

Efficiency. This is productive goods and services provided at the lowest cost and at a quantity that delivers the greatest benefits. Factors of production This is the productive resource used to produce other goods and/or service. Factor of production is also called investment goods and services. In general, these productive resources comprise land, capital, labor, and entrepreneur. Building and other built assets are subsets of capital. The income that capital earns is interest.

Functional use requirement. *Performance requirement* The user practices or procedures that the building and/or its services must perform to fulfill the user needs.

High building performance. Buildings that are energy efficient, comfortable, conducive, and easy and efficient to maintain.

Labor. This is work time and work effort that operatives offer to produce goods and services.

Maintenance management model. A framework containing the processes, activities, and tasks involved in the development, operation, and maintenance of a building, spanning the life of the system from the definition of its requirements to the termination of the use.

Maintenance manager. *Maintenance Executive or Building Manager* The person or individual that performs technical, administration, direction, and control roles to individual or groups performing tasks or activities within the maintenance department. Everyone working on maintenance is tasked by the maintenance manager.

Maintenance. The required processes and services taken to preserve, repair, protect, and care for building's fabric and engineering services after completion, repair, refurbishment, or replacement to current standards to enable it to serve its intended functions throughout its life span without drastically upsetting its basic features and use.

Non-residential. Buildings other than for dwellings.

Not-for-profit service. Includes the type of service provided by non-governmental organization, voluntary organization, or not-for profit organization. Maintenance service whether provided by public organization or private organization is not an example for not-for-profit services.

Model. The word "model" in the research is synonymous with decision-making framework or guideline for managing building maintenance. Thus building maintenance management model means decision-making framework for the maintenance of buildings. In this context, model consists of personnel, procedures, activities, requirements, and tasks.

Performance. A qualitative and subjective measure characterizing a building or engineering service relating to the initiation and implementation of maintenance. Performance criteria include readiness in service delivery, quality of service, friendliness of provider, and timeless in service delivery.

Preference. Is the explanation of what users like or dislike.

Procurement. The entire processes that are involved in acquiring construction projects. The process could be for new works or for maintenance, refurbishment, conversion, or extension.

Rational choice. This is the choice that users made when they compared cost of the service with the benefits they obtained from the service. As a result, users will aspire for higher benefit when they compare the benefit of a service with the money they paid in return. The money they exchange for the service is the denominator. However, on the choice of maintenance service, users are not rational, but a rational or irrational because the behavior of the users extends beyond rational choice and equilibrium understanding. Psychological and sociological issues need to be matched with economic issues. The rational choice is limited to a *certainty* situation but the decisions of service users are made under *uncertainty* situations. In most cases, the maintenance services are not even aware of how much they paid for the maintenance service they received. In general, people display emotion, preference, and sentiment when deciding on the choice they make on the quality of service they obtain.

Repair. The replacement or the correction of decomposed, deteriorated, or damaged building (or its part) and building service.

University buildings. Building that provide conducive environment for learning, teaching, innovation, and for conducting research.

User. An individual, group, or organization that uses the building and/or its services to perform a specific function, or the person or group that specifies functional requirements and specifically they refer to students and staff (academic and nonacademic).

User focus model. A model which has as its prime intention or focus the consideration of the interests or needs of the individual and/or groups who will use the building and its services. The focus of a building user is the capability and ability of the building to facilitate, sustain, and encourage learning, studying, and doing research.

User participation and involvement. Is the behaviors, perceptions, expectations, and responsibilities that users perform contributing to the building maintenance, especially during the planning and implementation phases.

User requirements and expectations. Include conformance, reliability, dependability, durability, appearance, serviceability, user-friendliness, safety, and eco-friendliness.

Utility. This is the benefit that service users obtain from the service they received from the maintenance organization.

Services are outputs produced to users and typically consist of changes in the conditions of the consuming units realized by the activities of producers at the demand of the consumers; by the time their production is completed they must have been provided to the consumers. Service is an activity, a process, or procedure to establish steps that involve the treatment of a user or customer themselves or items owned by the users.

SETARA. This is a Rating System for Malaysian Higher Education Institutions (Sistem Penarafan Institusi Pengajian Tinggi Malaysia SETARA) introduced in 2007 to enhance quality and promote the best practices among universities. The performance rating is intended to produce a uniform and objective assessment to deliver greater transparency and enhanced pressure on performance to raise the overall quality of the education system. Funding to a public university is explicitly linked to SETARA's rating. Initially, it was directed to the public university when first established in 2007 but now covers both privately and publicly owned universities.

Systemic management. Is synonymous with value management.

Value. A measure of desirability of the building and/or its services. Providing the required service to the user at the required time and at the correct price.

Value-based maintenance management. The systemic representation of a system and processes, expressing the content of that system and processes in terms of the criteria that explain maintenance service delivery. It is also a process and methodology, which enhances the performance values of a building and engineering by the maintenance services from initiation, through implementation, through systemic analysis of decisions against maintainer productivity, and user satisfaction.

Variability. As we have explained above, this is related to benchmarking. The purpose of variability analysis is to understand the differences that exist between what a producer produces and the service that one organization deliver with what another organization obtain from the product or service. In general, variability analysis aims to reconcile demand and supply side. Variability analysis or benchmarking is performed based on sound design in order to offer strategic information to maintenance organization to understand the service mismatch in a service delivery.

Wear and tear. The defect that takes place in the operation stage of building. It is caused by use, weather, and age of the building.

Wudhu. Ablution, this is ritual washing of body parts (i.e., legs, hands, face, and head limbs) by Muslims prior to prayers.

Index

A.L. Olanrewaju and A.-R. Abdul-Aziz, *Building Maintenance Processes and Practices*, DOI 10.1007/978-981-287-263-0

FSC
www.fsc.org
MIX
Papier aus verantwortungsvollen Quellen
Paper from responsible sources
FSC® C105338

If you have any concerns about our products, you can contact us on
ProductSafety@springernature.com

In case Publisher is established outside the EU, the EU authorized representative is:
Springer Nature Customer Service Center GmbH
Europaplatz 3, 69115 Heidelberg, Germany

Printed by Libri Plureos GmbH
in Hamburg, Germany

Building Maintenance Processes and Practices

IntechOpen

Supporting open minds since 2005

Ecosystem and Biodiversity of Amazonia
http://dx.doi.org/10.5772/intechopen.91089
Edited by Heimo Juhani Mikkola

Contributors
Gabriela N. Nicoleta Tenea, Pablo Jarrin-V, Lucia Yepez, Paola González Carvajal, Klaus Hilbert, Jens Soentgen, Natali Gomes Bordon, Niwton Leal Filho, Tony Vizcarra Bentos, Gelson Dias Dias Florentino, Lucieta Guerreiro G. Martorano, José Reinaldo Da Silva Cabral De Moraes, Sandro Augusto Lima dos Santos, Ires Paula de Andrade Miranda, Maria De Lourdes Pinheiro Ruivo, Hermes R. Luz, Thiago F. Martins, Sebastián Muñoz-Leal, Francisco B. Costa, Sérgio L. Gianizella, João Luiz H. Faccini, Marcelo B. Labruna, Alexandre Guida Navarro, Pedro Pablo Cardoso Castro, Niriva Ravena, Zoran Stiperski, Tomica Hruška, Anna Kanele, Claude François Béguin, Heimo Juhani Mikkola

First published in London, United Kingdom, 2021 by IntechOpen
IntechOpen is the global imprint of INTECHOPEN LIMITED, registered in England and Wales, registration number: 11086078, 5 Princes Gate Court, London, SW7 2QJ, United Kingdom
Printed in Croatia

British Library Cataloguing-in-Publication Data
A catalogue record for this book is available from the British Library

Additional hard and PDF copies can be obtained from orders@intechopen.com

Ecosystem and Biodiversity of Amazonia
Edited by Heimo Juhani Mikkola
p. cm.
Print ISBN 978-1-83962-812-2
Online ISBN 978-1-83962-813-9
eBook (PDF) ISBN 978-1-83962-814-6

For EU product safety concerns:
IN TECH d.o.o., Prolaz Marije Krucifikse Kozulić 3, 51000 Rijeka, Croatia, info@intechopen.com or visit our website at intechopen.com.

Meet the editor

Heimo Mikkola, Ph.D., has more than 35 years of experience in sustainable development through working with growing responsibilities in international organizations, in particular the African Development Bank, European Union, United Nations System and the World Bank. Dr. Mikkola received his Ph.D. from the University of Kuopio, Finland. He has visited and worked in 133 countries (46 of those in Africa). In South America, he worked with the Food and Agriculture Organization of the United Nations in Colombia and Uruguay visiting also Brazil, Chile, Costa Rica, Cuba, Curaçao, Guadeloupe, Panama and Venezuela. Since 1988 Dr. Mikkola has been an adjunct professor at Eastern Finland University. Between 2010 and 2018, he served as a visiting professor at three different universities in Almaty, Kazakhstan and one in Bishkek, Kyrgyzstan. He has published more than 660 zoological papers and texts and has edited seven books.

Contents

Preface

"If humans continue to cut down the Amazon, the forest may start to die off in the 2030s, destabilising South America and the world."

David Attenborough in his book *A Life on Our Planet* [1].

The Amazon River is the largest in the world in terms of volume and second only to the Nile in length. The climate is very warm and humid resulting in the largest tropical rainforest, which represents more than half of the world's remaining rainforests [2]. It includes parts of eight South American countries: Bolivia, Brazil, Colombia, Ecuador, French Guiana, Guyana, Suriname and Venezuela. The Amazon basin covers roughly 634 million hectares, but only 529 million hectares are classified as primary forest [3]. The basin is named after the river, known in Spanish and Portuguese as 'Rio Amazonas'. That name was derived from a tribe of mighty women warriors, 'Amazons', in ancient Greek mythology [4] because back then several tribes in the region practiced headhunting for trophies [5].

The first humans migrated from Central America to the Amazon region some 13,000 years ago [6]. Francisco de Orellana was the first European to travel the length of the Amazon River [7] and found there a complex civilization in the 1540s. It is suggested that up to 10 million indigenous people lived in fortified settlements, creating ceremonial artworks and growing food in small fields. Unfortunately, that civilization was later devastated by the spread of diseases originating from Europe such as measles, smallpox and typhus.

The Amazon rainforest is an ecosystem with vertical layers differing in plants and wildlife. Trees form the forest canopy with the dense top layer shading the trees below. The next layer includes shade-tolerant shrubs that cover the lower herbaceous plants. All large mammals, many small mammals, and a few birds live at ground level, but most of the birds and all of the bats live among the upper layers. Various insects inhabit all layers, but most species inhabit the top layers. Many of the species and much of the ecology of this ecosystem remain unknown [8].

In this introduction, I attempt to answer several questions [cf. 3 and 9] about the Amazon region.

Are the Amazon rainforests Earth's lungs?

The Amazon rainforest is often referred to as the "lungs of the planet." It produces oxygen through photosynthesis in the daytime but absorbs a lot of oxygen at night via respiration. Therefore, the rainforest is not a real net producer of oxygen. It does, however, absorb carbon dioxide, which is a greenhouse gas; hence why the word 'lungs' is often used. Through transpiration, the rainforest releases large amounts of moisture into the atmosphere. This helps stabilize the global climate, generating rainfall at local, regional, and trans-continental scales [3]. Ricardo Galvão, former head of Brazil's National Institute for Space Research (INPE), rightly concluded: "If the Amazon is destroyed, it will be impossible to control global warming" [6].

What causes fires in the Amazon?

Fires result from either natural lightning or intentional setting by people. At least 75 percent of fires are due to human activities like farming and logging, which allow fires to spread from dry areas into the rainforest.

Shifting agriculture and cattle ranches are responsible for 80–85 percent of deforestation in the Amazon [10]. Deforestation increases the vulnerability of rainforests to fire as it dries out the forest interior.

Climate change is largely contributing to this worsening situation, as it leads to increasing average temperatures and larger and more frequent fires during the dry season. If farmers, ranchers, and land speculators start fires, these fires can easily move to the rainforest. A vicious circle is taking place as fires also reinforce global warming [11].

According to INPE, 93,485 fires have been recorded in the Brazilian rainforest in 2020 prior to November, which is 25 percent more than in the same period in 2019 [12].

Why is the Amazon rainforest in danger?

Cattle ranching and soy plantations in massive landholdings of thousands of acres are the dominant drivers of deforestation in the Amazon. New threats emerge from palm oil plantations, illegal mining, and hydropower development [13]. Forest degradation due to unsustainable selective logging practices may not be the largest problem. It is more likely that accelerating deforestation is a real concern because it may affect the global climate. If climate warming continues, a large part of the Amazon could change from rainforest to savanna. This shift would mean huge economic and ecological consequences in the form of rainfall losses. Normally rain falls for 200 days each year, and total rainfall exceeds 80 inches per year [2]. Current rains feed the region that produces two-thirds of South America's GDP and their absence would trigger large carbon emissions from forest die-off. These emissions will further worsen climate change [3].

How much of the Amazon rainforest has been destroyed?

Brazil has the power to save the greatest rainforest – or destroy it [14]. It seems that South America's natural wonder may be perilously close to the tipping point beyond which its gradual transformation into something closer to the steppe cannot be stopped or reversed, even if people lay down their axes. An area equivalent in size to Turkey and bigger than that of Texas has been lost to logging, farming, mining, roads, dams, and other forms of "development." Between August 2017 and July 2018, Brazil lost 7900 km^2 of Amazon forest and nearly a billion trees [15]. During the following reporting period of the INPE, the forest loss was 9205 km^2 from August 2018 to July 2019. Due to soaring levels of illegal deforestation, the July 2020 figures from the Brazilian Amazon alone showed a 28 percent increase in the number of fires over the same period in the year before. The INPE recently reported that between August 2019 and July 2020 a total of 11,088 km^2 of rainforest were lost in Brazil; this being the greatest deforestation rate since 2008 [16]. Now the World Wildlife Fund (WWF) calls for immediate emergency measures to be implemented to ban deforestation in the Amazon for five years [11].

How do we stop the global and Amazon deforestation?

The restoration of forested land at a global scale is likely to be the most effective strategy to limit the rise of atmospheric carbon and mitigate climate change [17]. It is promising that some organizations around the world are planting trees at an unprecedented rate in a bid to combat global heating. For example, TreeSisters is planting 2.2 million trees each year across Africa as well as Brazil. The International Tree Foundation (ITF) is supporting local charities to work with local women to establish small nurseries of native trees at the forest fringe. These native trees are then planted directly into deforested areas. Women are the primary caretakers and can drive massive change; planting trees will make their lives better [18]. Although there is no doubt that the restoration of trees is a good strategy for climate change mitigation and wildlife conservation [17], it is important to decide the kinds of trees as well as how and where they are grown. In some contexts it has been shown that increasing tree cover can elevate fire risk if reforestation is achieved with single-species tree plantations, which decrease water supplies and can cause crop damage by wildlife. We must view restoration as a system and concentrate on restoring native forest ecosystems [19]. Avoiding forest fires if of utmost importance in the future, as smoke from burning forests in the Amazon has intensified glacier melt in Zongo Glacier in Bolivia fueling concern about a water crisis in South America [20]. Since the 1970s, the Peruvian Andes have lost nearly half their glacier-ice surface area. This impacts hundreds of thousands of people who use glacier water for agriculture and hydropower. In addition, the Patagonian ice fields between Chile and Argentina are melting at a very fast rate [6].

Who owns the Amazon rainforest? Why can't we just buy it?

Within the eight countries covering parts of the rainforest, the land is privately owned or held by indigenous people in legally formed territories. Some parts are owned by collectives or are government-controlled public lands and protected national parks. Attempts to buy the entire Amazon are not viable, but sovereign nations of the region could be financially assisted in setting aside some lands for conservation. It is shown that the protected status of indigenous lands is an effective measure against deforestation [21].

I want to thank Author Service Manager Sandra Maljavac at IntechOpen for all her efforts during the preparation and publication of this book. Without her active but friendly attitude, many important chapters would not have come to be.

Heimo Mikkola
University of Eastern Finland,
Kuopio, Finland

References

[1] Attenborough D. A Life on Our Planet: My Witness Statement and a Vision for the Future. London: Ebury Publishing; 2020. 272 p

[2] Keen C. Amazon Basin. 1997. Available from: https://www.greatestplaces.org/notes/amazon.html [Accessed: October 19, 2020]

[3] Butler RA. The Amazon Rainforest: The World's Largest Rainforest. 2020. Available from: https://rainforests.mongabay.com/amazon/ [Accessed: October 19, 2020]

[4] Taylor I. Names and Their Histories: A Handbook of Historical Geography and Topographical Nomenclature. London: Rivingtons; 1898. ISBN 978-0-559-29668-0

[5] Anonymous. Amazon's head hunters and body shrinkers. The Week. 2012. Available from: https://theweek.com/articles/478804/amazons-head-hunters-body-shrinkers [Accessed: January 20, 2012]

[6] Sandy M. The Tipping Point. Time Magazine. 2019;**23**(Sept.):67-79

[7] Smith A. Explorers of the Amazon. Chicago: University of Chicago Press; 1994. ISBN 978-0-226-76337-8

[8] Dinesen C. The Ecosystem of the Amazon Rainforest. 2017. Available from: https://sciencing.com/ecosystem-amazon-rainforest-6495612.html [Accessed: September 15, 2020]

[9] Phillips D, Camargos D. Scorched earth. How do we stop the Amazon burning? The Guardian. 2020;**15**(May):28-29

[10] Salisbury DS, Schmink M. Cows versus rubber: Changing livelihoods among Amazonian extractivist. Geoforum. 2007;**38**:1233-1249. DOI: 10.1016/j.geoforum.2007.03.005

[11] World Wildlife Fund. 2020 Forest Fires Globally Could be Worse Than 2019, WWF Warns. WWF Press Release, August 27. 2020

[12] BBC 2020. Amazon fires: Year-on-year numbers doubled in October. BBC News Online November 2, 2020. Available from: https://www.bbc.com/news/world-latin-america-54779877

[13] Yale University 2020. The Amazon Basin Forest. Available from: https://globalforestatlas.yale.edu/region/amazon [Accessed: October 19, 2020]

[14] Anononymous. Deathwatch. The Economist. 2019;**3**(August):9

[15] Nova Xavantina & Santarém. Briefing The Amazon. On the brink. The Economist. 2019;**3**(August):16-18

[16] BBC 2020. Brazil's Amazon: Deforestation 'surges to 12-year high. BBC News Online December 1, 2020. Available from: https://www.bbc.com/news/world-latin-america-55130304

[17] Bastin J-F, Finegold Y, Garcia C, Mollicone D, Rezende M, Routh D, et al. The global tree restoration potential. Science. 2019;**365**:76-79

[18] Barkham P. Can trees save the planet? The Guardian Weekly. 2019;**2019**(June):21-25

[19] Chazdon R, Brancalion P. Restoring forests as a means to many ends. Science. 2019;**365**:24-25

[20] Gill V. Amazon fires intensify Andes glacier melt. BBC Reel. 2019

[21] Valqui M, Feather C, Espinosa Lanos R. Revealing the Hidden-Indigenous Perspectives on Deforestation in the Peruvian Amazon – The Causes and the Solutions. Lima, Peru: AIDESEP (Inter-Ethnic Association for the Development of the Peruvian Amazon) and Moreton-in-Marsh: FPP (Forest Peoples Programme); 2015. 127 p

Section 1

Animal and Plant Diversity

Chapter 1

Diversity of the Owl Species in the Amazon Region

Heimo Juhani Mikkola

Abstract

The Amazon basin is the largest tropical rainforest in the world covering almost 40 per cent of the South American continent. For centuries, its vastness and inaccessibility have been protecting this area. The general belief is that the Amazonia region has unparalleled biodiversity which means that one in five of all bird species known in the globe lives in the Amazon Jungle. The author is testing this claim against the known occurrence of the world owl species. There are fewer owl species in Amazonia than expected namely 34 (against 54 expected). Two possible reasons are that our knowledge of the Amazon region species is still incomplete as may be indicated that new owl species have been found recently. The other reason could be that the rainforest is not after all the most wanted biotope for many of the owl species which often depend on the prey available in the bottom tier of the forest which is flooded during the wet season. Whatever the reason is, more detailed research is needed on all species of the Amazon region as there must be many new owls and other animal species out there waiting to be discovered. We have no idea how many of those new species have been or will be lost due to the rainforest destruction before they can be discovered. Therefore, it is necessary to stop deforestation in the Amazonia, be it due to forest logging, uncontrolled fires (often deliberately set) agriculture expansion or industrial development. Deforestation is not the only reason that the Amazon basin is changing. The rainforest suffers also from the global climate change when the higher temperatures reduce the rainfall in the tropical Atlantic region, causing drought and increasing the fire susceptibility of the rainforest. Luckily the owls are not so sensitive to surrounding forest cover as many other tropical forest-dependent bird species, like the large-bodied avian frugivores. Impacts of forest fragmentation on owls will need additional research.

Keywords: Amazonia, conservation, owl species, number of wildlife species, climate change

1. Introduction

The history of geology tells us that at one time Amazon River flowed westward when it was still inside the Gondwana continent and part of the proto-Congo River system [1]. Later the African and South American continents separated and the Atlantic Ocean widened enough to provide a moist and warm climate to the Amazon region. The rainforest formed during the Eocene era between 56 and 34 million years ago [2]. The Andes have formed 15 million years ago and some 5 million years later the Amazon River began to flow eastward [1].

Now the Amazon River is the largest of the world's rivers in terms of water volume discharged into the sea, carrying more than five times the volume of the Congo River or twelve times that of the Mississippi River [1]. The Amazon River basin covers almost 40 per cent of the South American continent and the main river is some 4080 miles long, second only to the River Nile in length [3]. The Amazon river and its 1100 tributaries have a drainage area covering 2.7 million square miles belonging to nine nations: (the percentage share of each nation is shown in brackets) Bolivia (7.7%), Brazil (58.4%), Colombia (7.1%), Ecuador (1%), French Guiana (1.4%), Guyana (3.1%), Peru (12.8%), Surinam (2.5%) and Venezuela (6.1%) [4]. Even in this book, the percentages vary as the perceived extent of the Amazon depends on the definition. Often people as well include areas outside the basin and that biogeographic Amazon ranges 3–3.2 million sq. mi, of which just over 80 per cent is forested [1]. Whatever definition is used the Amazon basin has the world's largest rainforest, bigger than the next two largest rainforests – in the Congo River basin and Indonesia – combined. The moist broadleaf forest comprises an estimated 390 billion individual trees representing some 16,000 species [5]. Billions of trees pump daily huge quantities of water vapour into the atmosphere. A large part of this water rains down locally, but part of this moisture is carried in the air to other areas including the agricultural heartland of South America. It has been estimated that two-thirds of Brazil's national product comes from the areas that receive Amazon rainforest generated rainfall [1].

It was for a long time a general belief that the Amazon rainforest has been always only sparsely populated by the humans due to the poor soils. Recent archaeological findings, however, suggest that the Amazon was earlier densely populated and that up to 5 million people were living in the region in 1500 AD. However, by 1900, the population had fallen to one million and by the early 1980s, it was estimated to be less than 200,000. Indigenous civilizations were devastated by the spread of new diseases like smallpox and typhus after the arrival of the first Europeans, in 1542 and after [6]. The UK broadcaster BBC 4 presented evidence that the Amazon rainforest has been shaped by humans for at least 11,000 years through activities like forest gardening and'terra preta' indigenous soil management, instead of being only a pristine wilderness [7]. In recent decades the human population has accelerated and incoming people have introduced mechanised agriculture and integrated the region into the global economy. Large quantities of the Amazon-produced commodities like cattle beef and leather, timber, soy, oil and gas, and minerals are sold to China, Europe, Russia and the U.S. [1].

Number of plants/and animal group species	Butler/Mongabay [1]	Wikipedia [2]	WWF [9]	Ascent of the Amazon [10]
Plant species	40,000	At least 40,000	40,000	Na
Mammals	430+	427	427	More than 1400
Birds	1300	1294	1300	1500
Fishes	3000	2200	ca. 3000	2200
Reptiles	400+	378	378	Na
Amphibians	1000+	428	Over 400	More than 1000
Invertebrates	Na	96,660 – 128,843 Brazil alone	ca. 100,000	Na
Insects	Na	2.5 million	Na	90% of all animal species are insects

Na = Not available.

Table 1.
Biodiversity of Amazonia plant and wildlife [1, 2, 9, 10].

The wet forests in Amazon have consistently higher species diversity than any comparable forests in Africa and Asia [8]. The Amazonia Region is home to a large collection of living plants (40,000 or more) and animal species in the world. However, **Table 1** above shows considerable discrepancies in the stated species diversity.

The largest discrepancies seem to be in the numbers of mammals and amphibians whilst the number of birds and reptiles is relatively uniform. It is possible that the highest numbers represent the number of South American species in its entirety and not just in Amazonia.

It has been said that one in ten known species in the world occurs in the Amazon region and one in five of all bird species are found in the Amazon rainforest [11]. The author knows best "Owls of the World" [12] and compares here how these statements hold for the world owls.

2. Material and methods

The history of owls in the Amazon region is far less known than that of human history. Fossil records of the Amazon owls are very few if any [13] so we do not know what kind of owls lived in the region before the present species. Even today, Amazonian owl species are not well known, and few studies have focused on the species from the region (see [14–24]. In the nine countries sharing the Amazon basin the number of owl species is on average 22 species ranging from 13 to 30 per country (**Table 2**).

However, it is important to note that only part of the above-listed owl species lives in the Amazon basin of that country. In the following, the scientific names are in the same form as in the "Owls of the World" [34] to avoid the complicated taxonomic discussions. In Bolivia, the tropical lowland evergreen forests are found in all the Amazonian ecoregions of the country. They represent the habitat with the second highest diversity of owls, with the following species: *Tyto furcata*, *Mecascops choliba*, *M. watsonii*, *M. hoyi*, *M. guatemalae*, *Pulsatrix perspicillata*, *Bubo virginianus*, *Strix chacoensis*, *S. virgata*, *Glaucidium brasilianum* and *Aegolius harrisii* [25]. In Brazil, there are five owl species, which occur exclusively in this biome: *Megascops watsonii*, *M. usta*, *M. roraimae*, *Glaucidium hardyi* and *Lophostrix cristata* [26]. From Colombian 28 owl species 10 are found in the Amazon area of the country. This corresponds well with the fact that in the Amazon region Colombia represents 30% of the national territory. Amazon species include *Tyto furcata*, *Megascops choliba*, *M.*

Country	Number of owl species	Source of information
Bolivia	25	[24]
Brazil	23	[25]
Colombia	28	[26]
Ecuador	28	[27]
French Guiana	13	[28]
Guyana	16	[29]
Peru	30	[30]
Suriname	15	[31]
Venezuela	22	[32]
Average 22.		

Table 2.
A total number of owl species listed in the 9 countries sharing the Amazon basin. Sources of information [25–33].

watsonii, *Lophostrix cristata*, *Pulsatrix perspicillata*, *Pulsatrix melanota*, *Strix virgata*, *Strix huhula*, *Glaucidium brasilianum*, and *Athene cunicularia* [27]. In Ecuador, the rather homogenous forests of the Amazon lowlands are home to nine species mostly distributed throughout the entire region including *Tyto furcata*, *Megascops choliba*, *M. watsonii*, *Lophostrix cristata*, *Pulsatrix perspicillata*, *Strix virgata*, *Strix huhula*, *Glaucidium brasilianum*, and *Asio clamator* [28].

Unfortunately, there is no available specific study on owls in Guyana, Peru and Venezuela. Also in the French Guiana and Surinam the total of 13 and 15 owl species sounds very small and it is equally not well defined which owls live in the Amazon basin of the country.

The occurrence of different owl species in the comparison of this chapter is based entirely on the latest known distribution maps [12, 13] and no new field research was undertaken in the region. Therefore, especially in the Andean region, it was not so easy to decide which "Pacific side" owls may or may not belong to the Amazonian fauna. As all these owl distribution maps are still in a state of flux so one can expect that the total numbers may hold up very well as any mistakes have most likely been made in both directions.

3. Results

There are some 268 owl species in the world [12] and as **Table 3.** shows that a new Pygmy Owl (yet to be validated) has been discovered by Luis Fabio Silveira, Curator of Birds at the Museum of Zoology, the University of São Paulo by its call during an ongoing expedition to the Neblina National Park area, in Brazil. Provisionally this

1. American Barn Owl *Tyto furcata* LC Stable
2. Tropical Screech Owl *Megascops choliba* LC Stable
3. Maria Koepcke's Screech Owl (Koepcke's Screech Owl) *Megascops koepckeae* LC Stable
4. Peruvian Screech Owl *Megascops roboratus* LC Stable
5. Rufescent Screech Owl *Megascops ingens* LC Decreasing
6. Cinnamon Screech Owl *Megascops petersoni* LC Stable
7. Cloud-forest Screech Owl *Megascops marshalli* NT Stable but in Bolivia classified as vulnerable [24]
8. Northern Tawny-bellied Screech Owl *Megascops watsonii* LC Stable
9. Southern Tawny-bellied Screech Owl *Megascops usta* Na
10. Roraima Screech Owl (Foothill Screech Owl) *Megascops roraimae* Na
11. Rio Napo Screech Owl *Megascops napensis* Na
12. White-throated Screech Owl *Megascops albogularis* LC Stable
13. Great Horned Owl *Bubo virginianus* LC Stable
14. Magellanic Horned Owl (Magellan Horned Owl or Lesser Horned Owl) *Bubo magellanicus* Na
15. Spectacled Owl *Pulsatrix perspicillata* LC Stable
16. Band-bellied Owl *Pulsatrix melanota* LC Stable
17. Mottled Owl *Strix virgata* LC Decreasing
18. Rufous-banded Owl *Strix albitarsis* LC Stable
19. Black-and-White Owl *Strix nigrolineata* LC Stable
20. Black-banded Owl *Strix huhula* LC Decreasing
21. Crested Owl *Lophostrix cristata* LC Stable

22. Amazonian Pygmy Owl *Glaucidium hardyi* LC Stable
23. Ferruginous Pygmy Owl *Glaucidium brasilianum* LC Decreasing
24. 'Neblina Pygmy Owl' *Glaucidium* spp. Nov. [35] Na
25. Subtropical Pygmy Owl *Glaucidium parkeri* LC Stable
26. Andean Pygmy Owl *Glaucidium jardinii* LC Stable
27. Yungas Pygmy Owl *Glaucidium bolivianum* LC Decreasing
28. Chaco Pygmy Owl (Tucuman Pygmy Owl) *Glaucidium tucumanum* Na
29. Long-whiskered Owl (Long-whiskered Owlet) *Xenoglaux loweryi* EN
30. Burrowing Owl *Athene cunicularia* LC Decreasing
31. Buff-fronted Owl *Aegolius harrisii* LC Stable but in Ecuador ranked as vulnerable [27]
32. Stygian Owl *Asio stygius* LC Decreasing
33. Striped Owl *Asio clamator* LC Decreasing
34. Short-eared Owl *Asio flammeus* LC Decreasing

Note that all listed 'species' are not, yet, fully approved by the taxonomists. Known conservation status is indicated for those species listed to be in danger [36]. CR = Critical; EN = Endangered; LC = Least concern;Na = No conservation status given; NT = Near-threatened and VU = Vulnerable.

Table 3.
List and scientific names of Amazon region owl species [12, 13] (+one new species proposed after 2012 and referenced separately in the table).

Figure 1.
Amazonian pygmy owl Glaucidium hardyi distribution map [from 34] with a scan of an excellent photo of Kurazo Okada from Macapá, Brazil.

new owl has been named the 'Neblina Pygmy Owl' [35]. That brings the total of the world owl species to 269, so 34 species listed for Amazonia represent 12.6 per cent. Similarly, 88 owl species of the world lives in the Neotropical zone of the world. This is 32.7 per cent of the world species. Amazon Region owls make 38.6 per cent of the Neotropical owls, respectively [12, 13]. There is only one owl species that has been nominated due to the region as Amazon Pygmy Owl *Glaucidium hardyi.* Its distribution map shows very well how difficult it is to know the exact limits of some species in the Amazon region (**Figure 1**). This tiny owl is difficult to observe in the rainforest canopy [12]. The latest book on Colombian owls [37] does not include this species, so the presented map extends a little too much to the west.

4. Discussion

Based on the assumption that one in five, i.e. 20 per cent of the world owl species should live in the Amazon Region we can see that the rainforest has twenty species less (34 *vs.* 54) than expected. There can be at least two very distinct reasons for this. Firstly, we may not know, yet, all of the owl species living in the Amazon region. The practical limitations of calculating and identifying birds, especially night-time, mean that data collection is often very difficult if not impossible. This explanation is supported by the owl study difficulties summarised below:

Compared with other avian groups, owls are difficult to study	[13, 38, 39]
Studying them is a challenge because......... Owls are typically not covered by land bird monitoring programmes because of	[13] [38]
• they are uncommon • their low densities	[13] [38]
• elusive behaviour	[38]
• nocturnal habits	[13, 38]
• usually go unnoticed • incomplete detectability or false absences	[13] [39]
• and live in places difficult to access	[13]

The second reason could be that the rainforest may not be the most ideal biotope for many of the owl species which often depend on the prey available in the bottom tier of the forest which is flooded during the wet season. The wet rainforest prey availability and catch-ability can be very different from the dry forest environment. In Ecuador, it has been noted that owl diversity is higher in forested ecosystems than in open areas and in humid forests than in dry forests. However, even in Ecuador the Pacific lowlands harbour more species than the Amazon lowlands [28] somewhat supporting my second reason. It gains further support at the end of the conclusions.

5. Conclusions

More detailed research is needed on all species of the Amazon region. Only a small percentage of the basin's plant life has been documented or studied for their potential medicinal value [40]. Much of the ecology of this ecosystem remains unknown, and one can suspect that there must be many new owls and other animal species out there waiting to be discovered. It may be the only area in the world where you can still hope to find a new owl species, as the recently found and not, yet, officially described new *Glaucidium* spp. Nov. owl shows (**Table 3**). Also, we will have no idea how many of those new species were or will be lost through rainforest destruction before they can be discovered. Therefore, stating the obvious, it is necessary to stop deforestation in the Amazon. South America's natural wonder may be perilously close to the tipping point beyond which its gradual transformation into something closer to steppe cannot be stopped or reversed, even if people lay down their axes. An area equivalent to the size of Turkey, and bigger than that of Texas has been lost to logging, farming, mining, roads, dams, and other forms of 'development'. Between only August 2017 and July 2018, Brazil lost 7900 km^2 of Amazon forest – nearly a billion trees – the highest rate of deforestation for a decade [41]. And in the following year in Brazil, alone, deforestation rose more than 88 per cent in June 2019 compared with the same month in 2018 [42]. Since then deforestation has again been steadily rising in the Brazilian Amazon, with deforestation alerts from August 2019 to July 2020 and is 33 per cent higher than in the same period of the previous year [43]. Unfortunately, it is not Brazil alone burning and deforesting the Amazon region. After two years in Colombia, I sadly concluded with my biologist wife that people in that country have very little respect for the forests and wildlife [44]. Since that time deforestation has increased especially in the Amazon region of the country. In 2017 alone, deforestation destroyed 60,300 hectares in Caquetá (one Amazon department) [45]. If current trends in the Amazon region continue, millions of extra tonnes of carbon dioxide will be released in the atmosphere with devastating long-term consequences. There will be immediate impacts like decimating biodiversity, and vital ecosystem destruction. The impact on regional economies could also be substantial due to the dramatic implications for regional rainfall, leading to drier conditions across South America's breadbasket and major urban areas [1]. Deforestation is not the only reason that the Amazon region is changing. Global climate change is having an important impact because the higher temperatures reduce the rainfall in the tropical Atlantic. This is causing regular drought periods over years thus increasing the susceptibility of the rainforest to fire [1]. This widespread devastation in the natural habitats is likely to have deleterious implications for the Amazon basin owl populations due to the sensitivity of many owl species to habitat disturbance [46]. A recent study in Ecuadorian owls found no relationship between species richness and the amount of surrounding forest cover [47] which tends to support my second hypothesis that the rainforest environment may not be the first choice of the owls. However, another concurrent study in Ecuador noted that the large-bodied frugivores birds in the same fragmented forests showed that frugivore richness corresponded strongly and positively to surrounding forest cover [48]. Clear discrepancies between different bird groups suggest that additional research is needed to evaluate how the impacts of forest fragmentation may vary among the bird guilds.

Author details

Heimo Juhani Mikkola
University of Eastern Finland, Kuopio, Finland

*Address all correspondence to: heimomikkola@yahoo.co.uk

References

[1] Butler, R.A. 2020. The Amazon Rainforest: The World's Largest Rainforest. https://rainforests.mongabay.com/amazon/ Accessed 18/10/2020

[2] Wikipedia 2020. Amazon rainforest. https://en.wikipedia.org/wiki/Amazon_Rainforest. Accessed 15/07/2020

[3] Keen, C. 1997. Amazon Basin. In Greatest Places Physical Geography. July 1997. https://www.greatestplaces.org/notes/amazon.html Accessed 18/10/2020.

[4] Coca-Castro, A., Raymondin, L., Bellfield, H. & Hyman, G. 2013. Land use Status and Trends in Amazonia. https://web.archive.org/web/20160319140993l/http://segamazonia.org/sites/default/files/press_releases/land_use_status_and_trends_in_amazonia.pdf Accessed 15/07/2020

[5] Field Museum 2013. Field Museum scientist estimate 16,000 tree species in the Amazon. http://www.eurekalert.org/pub_releases/2013-10/fm-fms101413.php Accessed 15/07/2020

[6] Smith, A. 1994. Explorers of the Amazon. University of Chicago Press, Chicago. ISBN 978-0-226-76337-8.

[7] BBC Four. 2011. Unnatural Histories – Amazon. http://www.bbc.co.uk/programmes/b0122njp.

[8] Turner, I.M. 2001. *The ecology of trees in the tropical forest.* Cambridge University Press, Cambridge. ISBN 0-521-801183-4

[9] WWF 2020. From the boa to leafcutter ant, and back to red piranha, Amazon wildlife comes in all shapes and sizes. https://wwf.panda.org/knowledge_hub/where_we_work/amazon/about_ wildlife_amazon/ Accessed 15/07/2020

[10] Ascent of the Amazon 2020. Amazon Animals A-Z. https://www.ascentpftheamazon.com/learning-resources/amazon-animals-a-to-z/ Accessed 15/07/2020

[11] World Wide Fund for Nature 2008. Amazon Rainforest, Amazon Plants, Amazon River Animals. http://www.worldwildlife.org/wildplaces/amazon/index.cfm Accessed 15/07/2020

[12] Mikkola, H. 2013. Owls of the World. A Photographic Guide. 2nd Revised Edition. – Bloomsbury/Christopher Helm, London. 528 p.(ISBN: 978-1-4729-0593-2).(Also, in Canada and USA in 2014 by Firefly Books in Ontario and New York, ISBN-13: 978-1-77085-274-7).

[13] Enriquez, P.L. (Ed.) 2017. Neotropical Owls Diversity and Conservation. Springer International Publishing, AG. Cham, Switzerland. 670 p. (ISBN 978-3-319-57107-2). DOI 10.1007/978-3-319-57108-9.

[14] Barros, F.M. & Cintra, R. 2009. The effects of forest structure on occurrence and abundance of three owl species (Aves: Strigidae) in the Central Amazon forest. Fortschr. Zool. 26(1): 85-96.

[15] Borges, S.H., Henriques, L.M. & Carvalhaes, A. 2004. Density and habitat use by owls in two Amazonian forest types. J. Field. Ornithol. 75: 176-182.

[16] Mikkola, H. 2018. Again, a New Owl Species from Colombia – Santa Marta Screech Owl. Tyto March 2018: 6-10.

[17] Muñoz-Garcia, J., Castaño Vargas, C.C., Reyes-Bueno, S. & Rodríguez-Villamil, D.R. 2019. Registros de reproducción y notas del cuidado parental de Ciccaba huhula en Colombia. Spizaetus 28: 15-20.

[18] Padilla, O. 2019. Descripcíon de un evento de anidacíon del búho real (*Bubo virginianus nacurutu* Vieillot, 1817) y del búho moteado (*Ciccaba virgata virgata* Cassin, 1849) en Huila, Colombia. Intropica 14(1):1-8 DOI : 10.21676/23897864.2696.

[19] Penagos, A.P., Martínez, A. & Rodríguez-Bolaños 2018. Nueve registro y ampliación de distribución del búho bicolor (*Aegolius harrisii*) en Colombia. Biota Colombiana 19(2): 140-146. DOI: 10.21068/c2018.v19n02a13.

[20] Restrepo Cardona, J.S., López Betancur, A. & Córdoba-Córdoba, S. 2013. El Búho de Anteojos (*Pulsatrix perspicillata*) en Fragmentos de Bosque Seco del Valle de Río Cauca en Neira, Caldas, Colombia. Spizaetus 15: 19-24.

[21] Restrepo Cardona, J.S., López Betancur, A. & Cano Castaño, N. 2015. Abundancia y nuevos registros de búhos simpátricos en Manizales y Villamaría (Caldas, Colombia). Bol. Cient. Mus. Hist. Nat. U. De Caldas 19(2): 220-229. DOI: 10.17151/bccm.2015.19.2.13

[22] Restrepo Cardona, J.S., Sáenz-Jiménez, F., Echeverry-Galvis, M.A., Marín-C, D. & Poveda, J. 2019. Diet of the Great Horned Owl (*Bubo virginianus*) during the breeding season in the paramo of Laguna Corazón, Tolima, Colombia. Ornitologia Colombiana 17: 1-5 http://associacioncolombianadeornitologia.org/ Accessed 15/02/2020

[23] Rodríguez-Villamil, D.R. 2018. Extensión de la distribución de Pulsatrix melanota (Strigidae) en el piedemonte llanero colombiano. Ornitología Colombiana 16: 1-6 http://associacioncolombianadeornitologia.org/ Accessed 15/02/2020

[24] Seminario Moscol, L.E. & Infante Castro, Y.I. 2018. Primer registro documentado de Búho Blanco y Negro (*Ciccaba nigrolineata*) en el bosque de Suyo, Ayabaca, departamento de Piura, Perú. Boletín UNOP 13(1): 22-25.

[25] Méndez, D.R. 2017. The Owls of Bolivia. Chapter 5: 75-95. In: Enriquez, P.L. (Ed.) 2017. Neotropical Owls Diversity and Conservation. Springer International Publishing, AG. Cham, Switzerland. 670 p.

[26] Motta-Junior, J.C., Rocha Braga, A.C. & Monteiro Granzinolli, M.A. 2017. The Owls of Brazil. Chapter 6: 97-158. In: Enriquez, P.L. (Ed.) 2017. Neotropical Owls Diversity and Conservation. Springer International Publishing, AG. Cham, Switzerland. 670 p.

[27] Chaparro-Herrera, S., Córdoba-Córdoba, S., López-Ordoñez, J.P., Restrepo-Cardona, J.S. & Cortes-Herrera, O. 2017. The Owls of Colombia. Chapter 9: 317-371. In: Enriquez, P.L. (Ed.) 2017. Neotropical Owls Diversity and Conservation. Springer International Publishing, AG. Cham, Switzerland. 670 p.

[28] Freile, J.F., Guevara, E.A., Pacheco, C. & Santander, T. 2017. The Owls of Ecuador. Chapter 10: 373-395. In: Enriquez, P.L. (Ed.) 2017. Neotropical Owls Diversity and Conservation. Springer International Publishing, AG. Cham, Switzerland. 670 p.

[29] Claessens, O., de Pracontal, N. & Ingels, J. 2017. The Owls of French Guiana. Chapter 12: 419-445. In: Enriquez, P.L. (Ed.) 2017. Neotropical Owls Diversity and Conservation. Springer International Publishing, AG. Cham, Switzerland. 670 p.

[30] Wikipedia 2020. List of birds of Guyana https://en.wikipedia.org/wiki/List_of_birds_of_Guyana Accessed 20/10/2020

[31] Wikipedia 2020. List of birds of Peru https://en.wikipedia.org/wiki/List_of_birds_of_Peru Accessed 20/10/2020

[32] Ramcharan, S. & Ottema, O. 2017. The Owls of Surinam. Chapter 19: 633-644. In: Enriquez, P.L. (Ed.) 2017. Neotropical Owls Diversity and Conservation. Springer International Publishing, AG. Cham, Switzerland. 670 p.

[33] Wikipedia 2020. List of birds of Venezuela https://en.wikipedia. org/ wiki/List_of_ birds_of_Venezuela. Accessed 20/10/2020

[34] König, C., Weick, F. & Becking, J-H. 2011. Owls of the World. 2nd Reprinted Edition. – A&C Black/ Christopher Helm, London.528 p. (ISBN: 978-0-7136-6548-2).

[35] BBC News 2018. A New Pygmy Owl species from Amazon. Amazon Discoveries 06/04/2018

[36] IUCN 2020. The IUCN Red List of Threatened Species. https://www. iucnredlist.org Accessed 20/10/2020

[37] Chaparro-Herrera, S., Enríquez, P.L. & Lopera-Salazar, A. 2020. Búhos de Colombia Guía Ilustrada y Sonora. Puntoaparte Ltda, Bogota. 107 p.

[38] Fuller, M.A. & Mosher, J.A. 1987. Raptor survey techniques. Pp. 37-65. In: Giron Pendleton, B.A., Millsap, B.A., Cline, K.W. & Bird, D.M. (Eds.) 1987. Raptor management techniques manual. National Wildlife Federation, Washington, D.C., USA.

[39] Wintle, B.A., Kavanagh, R.P., McCarthy, M.A. & Burgman, M.A. 2005. Estimating and dealing with detectability in occupancy surveys for forest owls and arboreal marsupials. Journal of Wildlife Management 69: 905-917.

[40] Dinesen Rogers, C. 2017. The Ecosystem of the Amazon Rainforest. https://sciencing.com/ecosystem-amazon-rainforest-6495612.html Accessed 15/07/2020

[41] Anonymous 2019. Deathwatch for the Amazon. The threat of runaway deforestation. – The Economist, August 3rd–9th 2019.

[42] Deutsche Welle 2019. Brazil registers huge spike in Amazon deforestation. July 3, 2019. https://www.dw.com/en/ brazil-registers-huge-spike-in-amazon-deforestation/a-494627773 Accessed 15/08/2020

[43] WWF 2020. 2020 Forest Fires Globally Could be Worse Than 2019, WWF Warns. Press Release 10/19/2020. Accessed 20/10/2020

[44] Mikkola, H. & Mikkola, K. 1978. Tragedy of Colombian nature. Suomen Luonto 37: 14-16 and 47 (In Finnish with summary in English).

[45] MINAMBIENTE 2017. Resultados. Monitoreo de la deforestación 2017. Gobierno de Colombia. http:// documentacion.ideam.gov.co/ openbiblio/bvirtual/02835/Resultados_ Monitoreo_Deforestacion_2017.pdf Accessed 15/02/2018

[46] Stotz, D., Fitzpatrick, J.W., Parker, T.A. & Moskovits, D.K. (Eds.) 1996. Neotropical birds: ecology and conservation. The University of Chicago Press, Illinois, USA.

[47] Walter, S.T., Browne, L., Freile, J., Gonzáles, N., Loor, J., Darkes, M., Gillespie, T.W. & Karubian, J. 2017. Nocturnal bird diversity in forest fragments in north-west Ecuador. Journal of Tropical Ecology 33: 357-364.

[48] Walter, S.T., Browne, L., Freile, J., Olivo, J., Gonzáles, M. & Karubian, J. 2017. Landscape-level tree cover predicts species richness of large-bodied frugivorous birds in forest fragments. Biotropica 49: 838-847.

Chapter 2

Microbiota of Wild Fruits from the Amazon Region of Ecuador: Linking Diversity and Functional Potential of Lactic Acid Bacteria with Their Origin

Gabriela N. Tenea, Pablo Jarrin-V and Lucia Yepez

Abstract

Subtropical wild fruits are a reservoir of microbial diversity and represent a potential source of beneficial microorganisms. Wild fruits provide essential nutrients, minerals, and antioxidants that contribute to human health. Many of these wild fruits are used by indigenous peoples for medicine and food, but there is yet an unexplored potential in the study of their properties and benefits. Wild fruits from the Amazon region and their associated active substances or bacterial communities can prevent disease, provide appropriate nutrition, contribute to new sources of income, and improve lives. Despite its condition as a megabiodiverse country, Ecuador suffers from limited access to its genetic resources, and particularly for research. A total of 41 isolates were obtained from six wild Amazonian fruit species and were molecularly classified into the genera *Lactiplantibacillus* (31 isolates), *Lactococcus* (3 isolates), *Weissella* (3 isolates), and *Enterococcus* (1 isolate). Three isolates showed large divergence in sequence variability and were not identified by the taxonomic assignment algorithm. Inferred phylogenies on the 16S rRNA gene explained the relationship between lineages and their origin. Carbohydrate metabolism and antimicrobial profiles were evaluated, and the isolates were classified from a functional perspective. Antimicrobial profiles showed a wide-range spectrum against several Gram-positive and Gram-negative bacteria. To our knowledge, this is the first study assessing the diversity of LAB in native tropical fruits from the Amazon region of Ecuador and their promising functional properties. The obtained isolates and their assessed properties are valuable genetic resources to be further investigated for industrial and pharmaceutical applications.

Keywords: microbial diversity, Amazon region, lactic acid bacteria, fruit origin

1. Introduction

Lactic acid bacteria (LAB) are a versatile group of microorganisms and with a long history of use in fermented foods. LAB are distributed in diverse environments and are of considerable economic interest due to their "Food Grade Status" [1].

LAB have been extensively used as starter cultures, as probiotics, and in the production of inhibitory compounds. The latter has been considered as a new generation of antibiotics [2].

The growing number of people with lactose intolerance, allergy to cow's milk protein, or high cholesterol, which cannot consume dairy products, is affected by the lack of access to beneficial probiotic bacteria. Thus, fermented fruits or vegetables, which are an alternative source of probiotic bacteria, have received considerable attention from the consumers. The increasing trend for veganism or vegetarianism has promoted an increase in consumer interest for functional foods that contain beneficial bacteria [3].

When evaluating aspects of safety, taxonomy, and potential to produce pathogenic toxins and resistance to antibiotics, LAB have been recognized by the Food and Drug Administration (FDA) and the European Food Safety Authority (EFSA) as "generally recognized as safe" (GRAS) [4]. Although resistance genes in bacteria, including Lactobacillus, are located in transposable elements that can be transferred to other species, the US legislation has no guidelines that contemplate the potential development of resistance for microorganisms used in foods [5]. Alternatively, the European Food Safety Authority (EFSA) established guidelines that define the safety standards for microorganisms used "from farm to fork" [6]. The probiotics belonging to the species included in the category "qualified presumption of safety" (QPS) by EFSA show excellent safety records [4]. Undoubtedly, a full safety assessment of newly identified strains begins with proper identification and an in vitro evaluation of the potential risks [7].

Numerous research studies have attempted to design fermented foods based on fruits or vegetables with probiotic bacteria. LAB can display a remarkable degree of phenotypic and genotypic diversity, allowing them to survive in a variety of habitats and stress conditions. Most species in the LAB group are found colonizing the human or animal intestine but also in several fermented foods, vegetables, and fruits [8, 9, 10].

The capacity to adapt to environmental changes depends on the genetic repertoire and the competency to use micronutrients along with the ability to counteract and overcome externally exerted physical-chemical challenges [11]. Despite an immeasurable arsenal of microbes, each plant or fruit harbors a distinctive microbiota that represents a remarkable niche to several LAB species [12]. However, the microbial composition in these environments is fluctuating and depends on intrinsic (physical and nutritional conditions) and extrinsic (environmental and harvesting conditions) parameters of the plant matrix [13]. Depending on the plant or fruit origin, the LAB species differ; thus, the *Leuconostoc, Lactobacillus, Weissella, Enterococcus,* and *Pediococcus* genera are those most frequently identified as epiphytes within the microbiota [14]. Although several species have been identified and characterized, the selection of new strains with valuable biotechnological properties remains a topic of interest [2].

Ecuador is known as a megabiodiverse country, where access to genetic resources for research is still deficient. Studies on the microbial diversity of unexplored niches and environments are of interest as these might lead to the identification of an endless number of species with unique characteristics and properties. It has been claimed that the microbial population present in raw plant material that has been originated from unexplored niches, differs among samples, as the plant matrix might carry a particular microbiota in a specific geographical region and at a specific time-point [15].

Recently, we proposed to investigate the microbiota associated with wild fruits; especially, to select and characterize strains of beneficial lactic acid bacteria [16]. Tropical forest fruits provide essential nutrients, minerals, and antioxidants that keep a healthy body and provide resilience to disease [17]. The physical and nutritional properties of fruits, together with the environmental and harvesting conditions of the plant matrix, may influence the final microbial community composition [2]. However, we hypothesize that the wild fruits from the Amazon region differ in their native availability of nutrients and physicochemical conditions, especially when compared to domesticated fruits from commercial crops. These substantial differences might influence the range of potential microbiota.

Wild tropical fruits are used for nutrition and medicine by the local inhabitants; yet, the lack of research makes this natural resource an yet unknown source of benefits to the food industry [18]. The prevention of diseases through appropriate nutrition with such fruits or the active substances extracted from such fruits or the products from the associated bacterial community can save lives and income [19]. The role of LAB isolated from fruits is not clearly understood; nonetheless, as they are naturally present on the surface of fruits, they produce antimicrobial compounds that might be used as a biological agent to control the growth of spoilage bacteria, and without changing the sensory properties of foods [20, 21]. Thus, we expected to find microorganisms with unique characteristics unlike those of the strains found in ordinary fermented materials, such as fermented milk or vegetables. Nonetheless, to use such strains as part of probiotic foods implies the application of numerous functional analyses to determine their susceptibility or resistance to antibiotics, capacity to prevent the growth of harmful bacteria by competitive exclusion, and production of organic compounds.

The resistance of LAB to antibiotics is a relevant biosafety issue. Some strains with intrinsic antibiotic resistance could be useful to restore the antibiotic resistance of the gut microbiota after antibiotic treatments [22]. Specific antibiotic resistance genes are carried by transposable elements and constitute a reservoir of resistance for potential food or gut pathogens, thus representing a major biosafety issue [22, 23]. In this regard, special attention should be given to the presence of antibiotic resistance determinants and their potential mobility. Lactobacilli are usually sensitive to the cell wall-targeting penicillin and β-lactamase, but more resistant to cephalosporins. Also, some species are susceptible to low concentrations of several inhibitors of protein synthesis, such as chloramphenicol and tetracycline [22]. Some LABs are considered the most promising natural food preservatives, as they secrete antimicrobial substances (i.e. bacteriocins) that, when applied as crude-extracts to food, precipitated peptides that protect the products from deterioration by microbial activity [24]. LAB can also be used in the pharmaceutical industry to produce compounds such as esters, through the reduction of fructose, as the most abundant sugar present in fruits [25].

This research aimed to investigate the lactic acid bacteria diversity associated with the microecosystem of wild fruits from the Amazon region of Ecuador. We sought to find unique characteristics, unlike those of the strains found in ordinary fermented materials such as milk or vegetables. As a first step to select the most promising strains with superior functional properties, we present broad insights into this microenvironment. To our knowledge, this is the first study that assessed the diversity and functional properties of LAB in native tropical fruits from the Amazon region of Ecuador.

2. Materials and methods

2.1 Sampling, isolation, and purification of lactic acid bacteria

The source wild fruit samples were collected during the rainy season (April–July, 2014-2015) from the tropical forest of the Sucumbíos Province in the Amazon region of Ecuador (**Figure 1**). Several units of fruits from *Chrysophyllum oliviforme,*

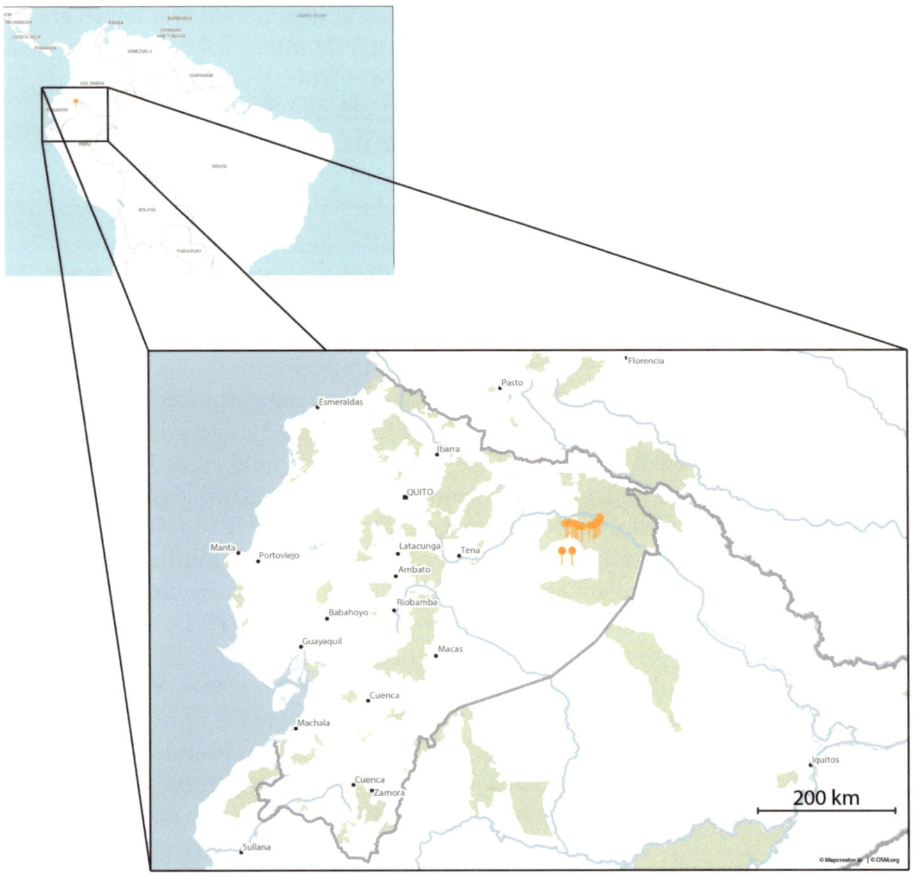

Figure 1.
Location of samples obtained for this study in the Amazon region of Ecuador. Colored areas represent national parks. Most of the samples were obtained in the areas adjacent to the Cuyabeno National Park.

Malus sp., *Costus* sp., *Solanum stramofolium, Cheilocostus specious*, and *Theobroma grandiflorum* were collected aseptically in sterile bags and stored at 4°C for transport to the laboratory. The fruit samples were washed with a 5% bleach solution for 5 min, then twice with distillate water, smashed in small pieces, and transferred into Erlenmeyer flasks (500 ml) that contained peptone water (0.1%, bacteriological peptone, Difco Detroit, MI, USA). Samples were incubated statically for up to 5 days at room temperature. MRS agar [26] plates containing 5 g/l $CaCO_3$ were inoculated and incubated under anaerobic conditions (37°C for 72 hours), and individual colonies were randomly selected and purified. The purified colonies (<75 colonies/each fruit) were Gram-stained, characterized by microscopic morphology, and tested for mobility, indole and catalase production, spore formation, and gas production from glucose [16]. The selected colonies were screened for their capacity to inhibit four bacterial pathogens (Section 2.4), resistance to bile (1%), acidity (pH 2.5-3.0), and temperature range (15-45°C). A total of 41 isolates complying with these criteria were further selected and identified. Each isolate was assigned a collection code (UTN) and was stored at −80°C in MRS medium containing 20% glycerol.

2.2 Classification of isolates using 16S rRNA gene sequence analysis

16S rRNA gene sequencing was used for taxonomical assignment (Macrogen Inc., Korea, custom service). The primers 27F 5′ (AGA GTT TGA TCM TGG CTC

AG) 3′ and 1492R 5′ (TAC GGY TAC CTTGTT ACG ACT T) 3′ were used for the PCR amplification. The PCR reaction was performed with 20 ng of genomic DNA in a 30 μl reaction mixture with EF-Taq (SolGent, Korea) and through the following protocol: activation of Taq polymerase at 95°C for 2 minutes, 35 cycles of 95°C for 1 minute, at 55°C and 72°C for 1 minute, finishing with 10 minutes at 72°C. The amplification products were purified with a multiscreen filter plate (Millipore Corp., Bedford, MA, USA). The sequencing reaction was performed using a PRISM BigDye Terminator v3.1 cycle sequencing kit. The DNA samples containing the extension products were added to Hi-Di formamide (Applied Biosystems, Foster City, CA). The mixture was incubated at 95°C for 5 minutes, followed by 5 minutes on ice and then analyzed by an ABI Prism 3730XL DNA analyzer (Applied Biosystems, Foster City, CA).

A preliminary fast search of the sequences was conducted using the megablast algorithm on the 16S ribosomal RNA database at NCBI, as implemented in Geneious Prime 2020.2.3 [27]. This first search was used to obtain a maximum of 100 hits and associated search quality parameters that provided an initial reference for taxonomic classification. A final taxonomic assignment was made by the RDP Bayesian classifier algorithm [28] with 100 bootstrap replicates and a K-mer of size 8, as implemented in the function "accurate, high-resolution sample inference from amplicon sequencing data" (assign Taxonomy) of the DADA2 package [29] in R [30]. The Genome Taxonomy Database [31] was used as the reference for the Bayesian classifier algorithm. A threshold for bootstrap values equal or over 80% was used to filter the taxonomic correspondences. The resulting final table of taxonomic assignments was used for downstream statistical analyses.

We used reference samples to root the inferred phylogenetic hypotheses and provide a comparison to the experimental samples used in this study. This reference samples were: KJ660075.1: *Brachyspira_hampsonii*; AJ276460.1: *Enterococcus_faecalis;* NR_114312.1: *Weissella oryzae*; NR_104573.1: *Lactobacillus*; LC063164.1: *Weissella confusa;* NR_029133.1: *Lactobacillus.*

LAB represent a small fraction of the autochthonous microbiota in the fruit and plant phyllosphere. However, due to the inherent limitations of in vitro cultures, the most frequent isolates often belong to the genera *Lactobacillus, Leuconostoc, Weissella, Enterococcus*, and *Pediococcus* [14]. More recently, based on conserved pairwise average amino acid identity, core genome phylogeny, physiological criteria, clade-specific signature genes, and ecology, *Lactobacillaceae*, and *Leuconostocaceae* were merged into *Lactobacillaceae* [32]. The genus *Lactobacillus* was split into 25 genera including *Lactiplantibacillus, Fructiplantibacillus, Lactobacillus*, among others. Thus, we allocate the study samples according to this new classification.

2.3 Metabolic profile of the selected isolates

The BBL Crystal anaerobe identification system (cat # 245010, BD Company, US) is a miniaturized identification method for tests of fermentation, oxidation, degradation, and hydrolysis of diverse substrates. It also includes chromogen and fluorogenic linked substrates to detect enzymes that microbes metabolize. The results obtained after incubation of strips for 24 hours at 37°C were analyzed using the BBL Crystal analysis software (according to the manufacturer instructions). The results were qualitatively defined as negative (−) or positive (+) according to the detected color. As a reference for the metabolic profile test, we used *Lactobacillus plantarum* ATCC8014 and *Lactobacillus fermentum* CNCM 1-2998 (Lacc), which were recovered from the commercial probiotic Lacteol Fort (*Lactobacillus* LB, Axcan Pharma, France).

2.4 Antibiotic susceptibility testing

Susceptibility to several antibiotics was determined by commercial discs of Amoxicillin (AMX: 25 µg), Ampicillin (AM: 10 µg) Gentamicin (CN: 10 µg), Kanamycin (K: 30 µg), Amoxicillin/Clavulanic Acid (AMC: 20/10 µg), Tetracycline (TE: 30 µg), and Cefuroxime (CXM: 30 µg). For the disk diffusion assay, we used concentrations recommended by the Scientific Committee on Animal Nutrition (discs provided by Merck) [12]. We used *Escherichia coli* ATCC 25922 as quality control. The microbiological breakpoints reported by the FEEDAP standards were used to categorize lactobacilli as susceptible or resistant [4]. *Lactobacillus plantarum* ATCC8014 and *L. fermentum* CNCM 1-2998 (Lacc) were used as a reference to the test. Results were qualified as R (resistance), I (intermediate), and S (susceptible).

2.5 Antimicrobial spectrum of the selected isolates

The antimicrobial spectrum of the selected isolates was determined against Gram-positive and Gram-negative bacteria, as previously described by Garzon et al. [33]. The indicator bacteria were: *Salmonella enterica subsp. enterica* (Kauffmann and Edwards) Le Minor and Popoff ATCC 51741, *Salmonella enterica subsp. enterica* serovar Abaetetuba ATCC35640, *Escherichia coli* ATCC25922, *E. coli* ATCC10536, *Shigella sonnei* ATCC25931, *Streptococcus thermophilus* ATCC19258, *Enterobacter aerogenes* UTNEag1 (laboratory collection), *Salmonella* UTNSm2 (laboratory collection), *Shigella* UTNShg1 (laboratory collection), and *E. coli* UTNEc1 (laboratory collection). LAB was grown in MRS broth at 34°C for 24–27 hours and the supernatants were collected by centrifugation at 13,000 x g for 20 minutes at 4°C. The crude extract (CE) was recovered and filtered with a 0.22 µm porosity syringe filter. The indicator strains (100 µl) were grown in broth medium (7 log CFU/ml) and mixed with 3.5 ml of soft MRS agar (0.75%). It was then overlaid on nutrient agar plates and incubated at 37°C for 2 hours. The CE of each strain (100 µl) was transferred onto the reaction wells (6 mm) on overlaid agar, incubated at 37°C and subsequently examined for the presence of an inhibition zone at 48 hours. To rule out the possible inhibitory activity of organic acids, the CE was heated at 80°C for 10 minutes, the pH adjusted at 6.0 and the activity was determined. *Lactobacillus plantarum* ATCC8014 and *L. fermentum* CNCM 1-2998 (Lacc) were used as reference strains. MRS broth was used as a negative control. The experiments were run in triplicate and the mean value of the inhibition zone was determined. A numeric scale from zero to ten was included in the statistical analysis and the results were also qualitatively defined as narrow (inhibit less than 5 indicator strains) or broad activity (inhibit more than 5 indicator strains).

2.6 Statistical and phylogenetic analyses

The interpretation of the antibiogram results was assisted by the package AMR [34], which provided corresponding frequencies on the qualitative responses. The distances (Bray-Curtis) among samples were then projected in canonical space through a non-metric multidimensional scaling. Either putative genera of bacteria, assigned through the RDP Bayesian classifier algorithm [28], or the host plants were included as the grouping variable.

The metabolic profiling resulted in a matrix that could be interpreted in binary form, and from which it was possible to determine a set of distances (binary Bray-Curtis) for classifying samples through a cluster analysis (unweighted pair group method with arithmetic averages, UPGMA). Ordination methods were carried out through vegan [35] and ggdendro [36] in R [30].

Sequences were aligned trough Clustal Omega, as implemented in Geneious Prime 2020.2.3 [27]. A proper substitution model was obtained through jModeltest v. 2.1.10 [37] and selected by a Bayesian information criterion. A phylogenic hypothesis was inferred by Bayesian inference with Mr. Bayes 3.2.6 [38]. The selected HKY85 model included a proportion of invariable sites and varying rates across sites with a discretized gamma distribution (HKY85 + I + G). The Bayesian analysis included 1.1 million generations and four chains per run. Hypotheses were sampled every 200 generations and the first 10% of these samples were discarded. The remaining 90% of the trees and parameters were respectively summarized in a 50% majority-rule consensus tree.

3. Results and discussion

3.1 Wild fruits: A microenvironment of diverse lactic acid bacteria

Out of 41 isolates, the most frequently observed genus was *Lactiplantibacillus* (31 isolates), followed by *Lactococcus* (3 isolates), *Weissella* (3 isolates), and *Enterococcus* (1 isolate). Three isolates showed large divergence and were not identified by the taxonomic assignment algorithm (i.e. UTN39, UTN41, and UTN88) (**Figure 2a**). The former three isolates may represent unreported lineages or species. The presence of *Lactococcus* in plants or fruit is rare: thus, we only found few isolates belonging to such genus.

Isolates showed a remarkable distance to the outgroup reference samples, and most were included within a clade formed by *Lactiplantibacillus,* where they showed relatively small distances (i.e. small branch lengths). However, within this inclusive group of *Lactiplantibacillus* there was an ingroup with strong support (Bayesian posterior probabilities = 1) and formed by *Weissella, Enterococcus,* and *Lactococcus*. This paraphyletic ingroup, within a more inclusive group formed mostly by *Lactiplantibacillus* should be eventually resolved by the inclusion of additional molecular markers (**Figure 2a**). The paraphyletic ingroup contains samples that belong to the plant species *Costus* sp., which occurs exclusively to this clade (**Figure 2b**).

3.2 The metabolic profile reveals the divergent properties of selected LAB

LAB strains may present specific metabolic traits as a result of their microenvironmental origin (i.e. different species of fruits) and possess a unique portfolio of enzymes that allow them to metabolize various compounds found in the host plant or fruit matrices. We present a metabolic profile together with other properties that were analyzed in the obtained isolates (**Figure 3**). Within *Lactiplantibacillus*, the isolates showed a variable capacity to ferment sugars and hydrolyze esculin. These features were strain-dependent. Among the available isolates, UTN39 and UTN76 were the only two samples that metabolized ARG, while UTN37 and UTN39 hydrolyzed urea. The latter is a relevant characteristic for selecting probiotic strains [2]. Of particular relevance were the locations of UTN76 and UTN39 on the previously presented phylogenetic hypotheses (**Figure 2**); these two isolates showed large genetic distances relative to the other samples, or unique positions in clades that diagnose them as different or remarkable lineages. The observed patterns in the metabolic profile are rather complex. UTN37, UTN39, and UTN76 show noteworthy properties; they, however, differ in their response to other substrates (**Figure 3**); thus, the corresponding dendrogram, which results from clustering the observed metabolic profile, hints to four main groups of isolates (**Figure 4**).

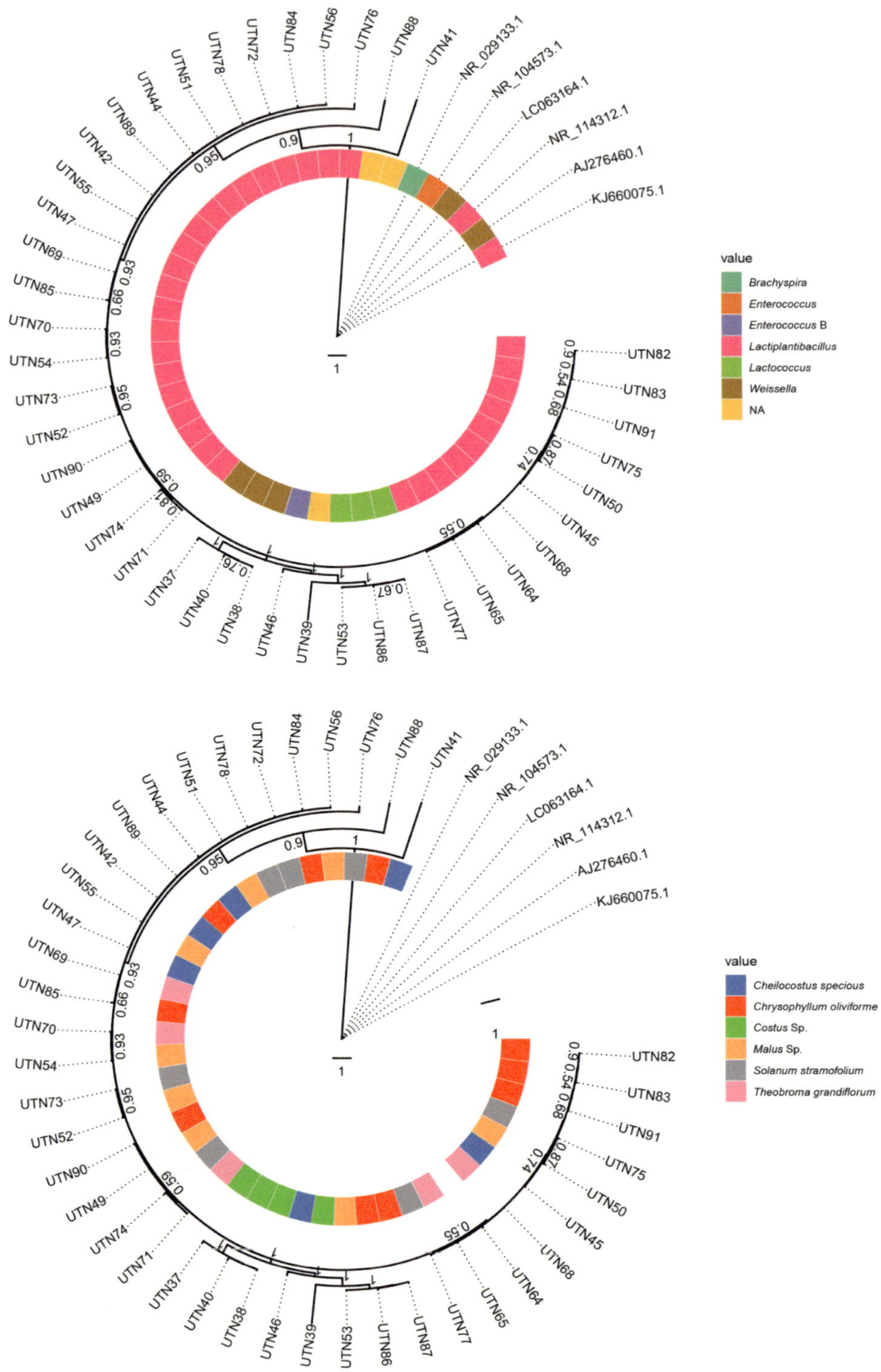

Figure 2.
(a) Phylogenetic hypothesis of the isolated samples and corresponding genera of bacteria. (b) The hypothesis with the corresponding host genera of plants. Branch support values are posterior probabilities from the applied Bayesian inference.

The ability to utilize the α-galacto-oligosaccharides-family (αGOS), d-melibiose [α-Gal-(1 → 6)-Glu], as well as the raffinose-family oligosaccharide (RFO) d-raffinose, seems to be a common feature among all the tested isolates. On the

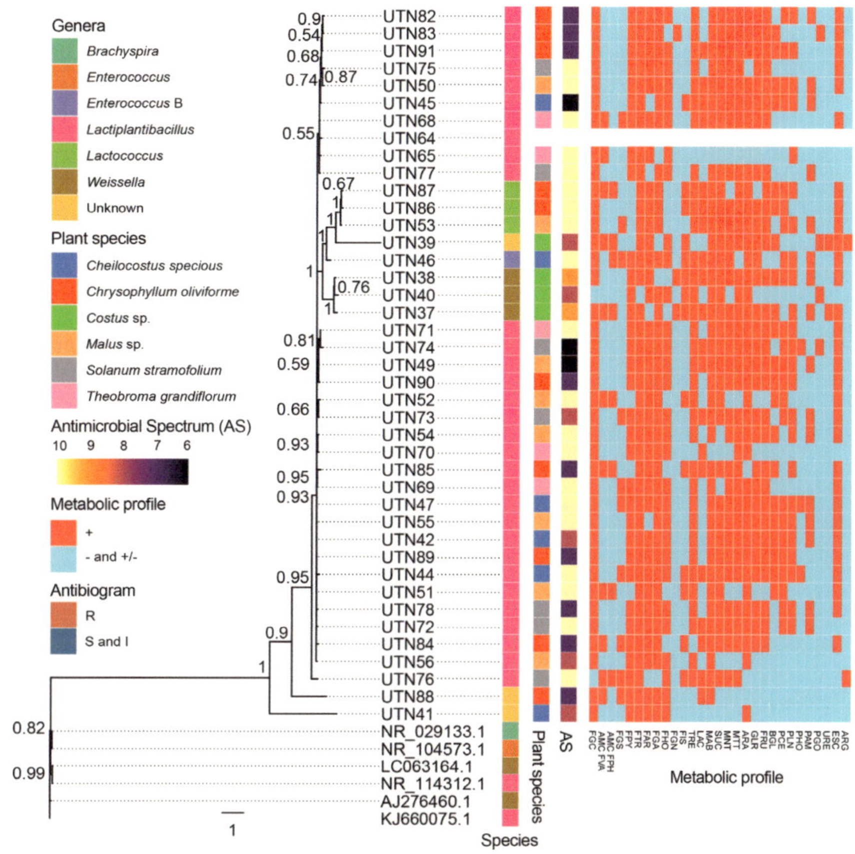

Figure 3.
Annotated phylogeny and heatmap showing the metabolic profiles of the LAB isolates. The phylogeny is an alternative depiction of the one shown in ***Figure 2*** *and serves as a reference to the profiles. The antimicrobial spectrum (AS) is also presented for each isolate.*

other hand, the metabolisms of the trisaccharide d-melezitose (α-Glu-(1 → 3)-β-Fru-(2 → 1)-α-Glu) and the disaccharides d-trehalose [α-Glu-(1 → 1)-α-Glu] and d-turanose (α-Glu-(1 → 3)-α-Fru) were strain-specific and restricted mainly to the *Lactoplantibacilaceae.* Although not classified, in a distinct lineage near the base of the tree, and with relatively large genetic distances, the isolates UTN41 and UTN88 did not metabolize most of the oligosaccharide substrates. The dissimilarity of individual isolates was supported by the different use of glycerol, trehalose, and sucrose. UTN40 was in the same lineage group as UTN37 and UTN38; however, it showed a distinct metabolic and antibiotic pattern, being the only one with resistance to AM10 and CN10. The isolate UTN76 showed a similar metabolic pattern to UTN88, both were the only isolates that did not metabolize p-n-p-β-D-cellobioside, p-n-p-phosphate, and proline. The *Lactococcus* species (UTN53, UTN86, and UTN87) were grouped in the same clade but were different in both the metabolic and antibiotic profiles.

The results revealed the production of β-galactosidase, α-glucosidase, β-glucosidase, and p-n-p-α-β-galactoside in some isolates. Although all isolates were originated from plants, there were differences in the utilization of mannitol and fructose. Previous studies have indicated that strains of *L. plantarum*, which were isolated from plant environments, often were able to metabolize these carbohydrates, and which has been considered as a fructophilic property [39].

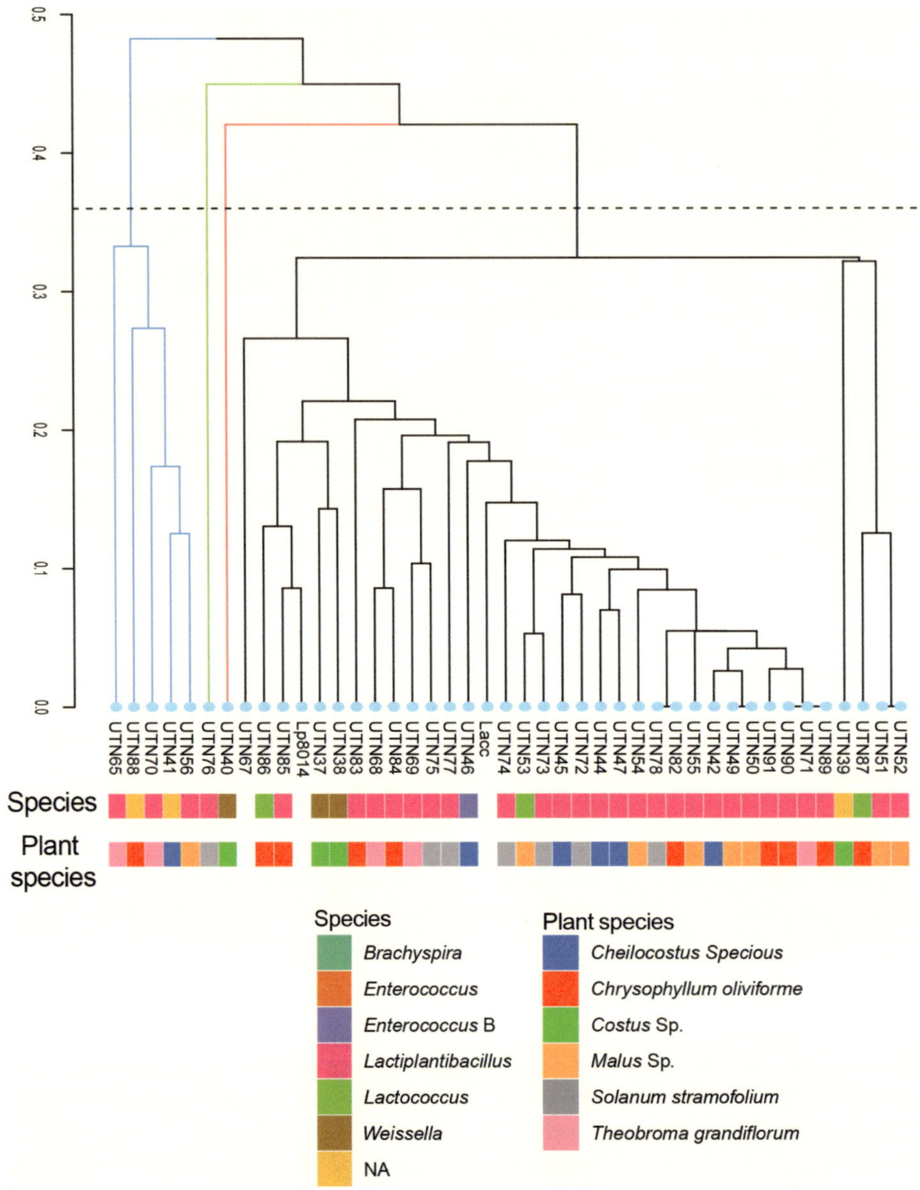

Figure 4.
Dendrogram from a cluster analysis on the metabolic profile for the obtained isolates. Four groups are possible to define at the established distance (dotted vertical line). The assigned species and the host plant species are represented as color codes.

3.3 The antimicrobial profile reveals that inhibitory activity is related to microenvironmental origin

AMX25 was the only antibiotic for which all isolates showed susceptibility, while all isolates showed innate resistance to MET5. Previous studies have indicated that Lactobacilli have a high natural resistance to metronidazole, as well as antibiotics that inhibit the synthesis of proteins such as chloramphenicol, erythromycin, clindamycin, and tetracyclines [40]. UTN41 was susceptible to almost all antibiotics. Among the *Lactiplantibacillus* genus, the isolates UTN76 and UTN88 were resistant to AMC30, and UTN84 and UTN89 were resistant to TE30. However, this relevant pattern did not show consistency with the fruit host, as the plant species

Chrysophyllum oliviforme, which was the origin of the resistant samples, was also the origin of other isolates in *Lactiplantibacillus*, but which did not coincide in their resistance pattern (**Figure 5**). Another relevant aspect was that of UTN68 (*Lactiplantibacillus*), which originated from *Theobroma grandifolium* and UTN40 (*Weissella*) from *Costus* sp., which were the only isolates showing resistance to AM10 (**Figure 5**).

Although resistance to gentamicin and kanamycin is considered a health concern, the isolates that showed resistance to both antibiotics were control reference strains that were isolated from dairy and human intestine. In general, patterns in the antibiogram were broad and yet inconclusive, as the position of isolates and plant hosts in canonical space show that further evidence is necessary to establish definitive patterns of association (**Figure 6**). However, this is also evidence of large variability across samples, with no conclusive patterns in isolated species or their host plants.

Although antibiotic resistance is a criterium to be considered from a biosafety perspective, it has been shown that the probiotic strains of starter culture strains

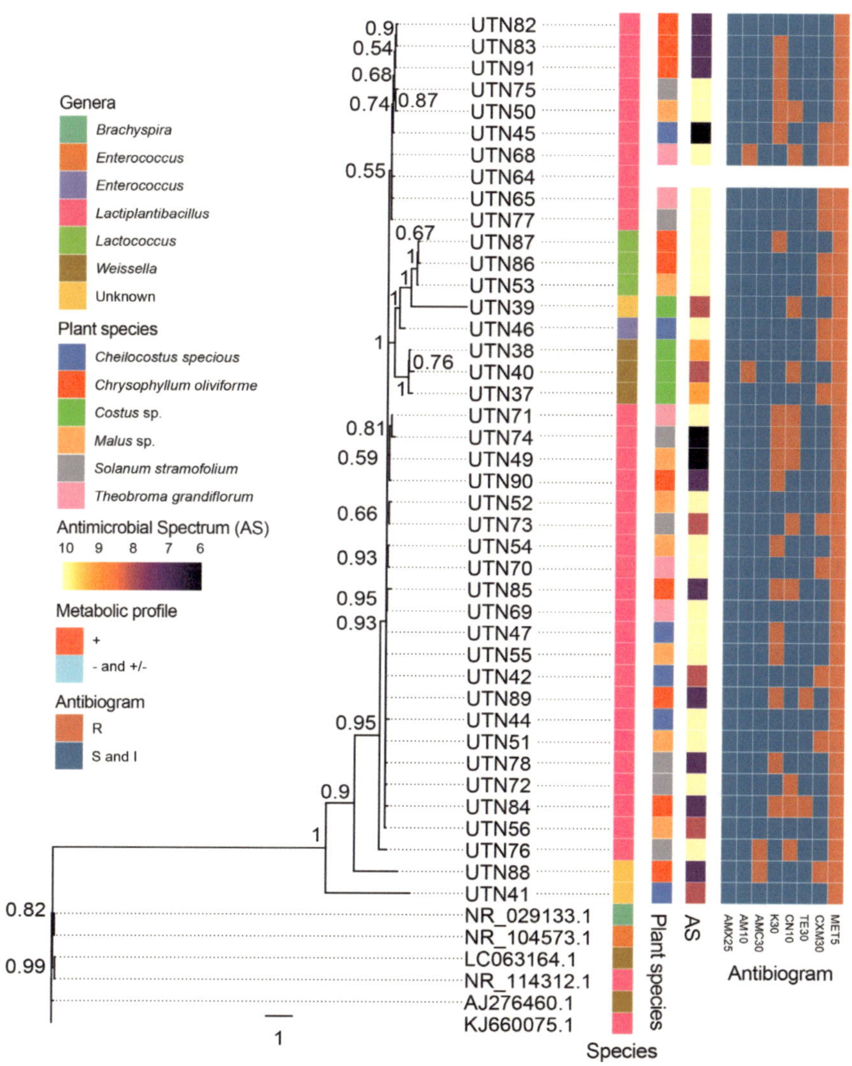

Figure 5.
Annotated phylogeny and heatmap showing antimicrobial profiles of the LAB isolates.

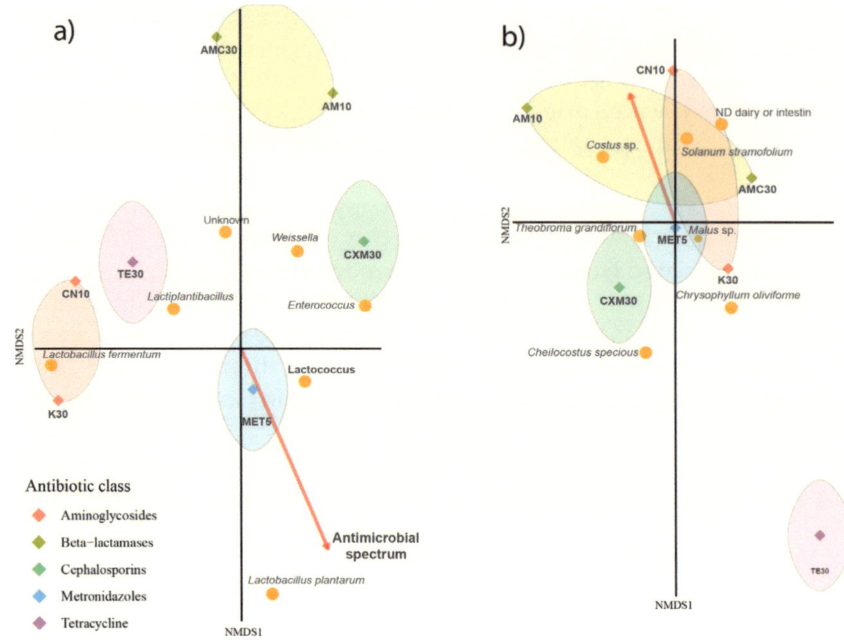

Figure 6.
Non-metric multidimensional scaling (NMDS) that represents, in two dimensions, the distances among isolates and host plants according to the observed pattern of antibiotic resistance. a) Tile for the spatial relationships in NMDS space between antibiotics and the studied bacterial isolates. b) Tile for spatial relationships in NMDS space between antibiotics and the host plants of the studied bacterial isolates. Short distances among these objects in the canonical space represent strong associations of either resistance or susceptibility. The antimicrobial spectrum is shown here as a cofactor showing a directional trend. AMX25 was removed from this analysis as it was not informative (i.e. susceptibility as a zero-response constant). The colored areas facilitate the interpretation of the antibiotics position. L. fermentum *and* L. plantarum *were used as a reference in the tests.*

are resistant to ampicillin, vancomycin, gentamicin, kanamycin, streptomycin, erythromycin, clindamycin, tetracyclines, and chloramphenicol [4, 7]. Another study has reported that Lactobacilli isolated from fermented olives were resistant to cephalosporins, streptomycin, and kanamycin [41]. Overall, the observed pattern is for high variability or diversity in the response to antibiotics as there is considerable dispersion in the response within and among bacterial species or plant hosts (**Figure 6**). However, there were also isolates with unique resistance patterns, which will require a further inquiry into their molecular and physiological properties. Undoubtedly, a full safety assessment with a robust identification of the strains and an in vitro evaluation of the potential risks is needed; particularly if these are intended to be used as additives in food products.

It is known that some LAB strains produce a wide variety of anti-pathogenic compounds, like bacteriocins, ethanol, organic acids, diacetyl, acetaldehydes, hydrogen peroxide (H_2O_2), and peptides [42, 43]. When we analyzed the antimicrobial spectrum against ten Gram-negative and Gram-positive bacteria, including closely related species and pathogens such as *Salmonella enterica, Shigella sonnei, Escherichia coli, Enterobacter, Staphylococcus aureus,* we observed that the isolates showed high inhibitory potential, as none had values below six and were defined as broad-spectrum (**Figures 3** and **4**). The inhibitory effect of LAB strains may result from a combination of competition for metabolic substrates, growth suppression by organic acids, and bacteriocin secretion. Recently, we showed that some of the Lactobacilli strains inhibited *Salmonella enterica subsp. enterica* ATCC51741 and *E. coli* ATCC25922 at both the early and logarithmic stages of bacterial growth *in vitro* and *ex vitro* [44, 45, 46]. Also, we showed that one selected LAB strain from the

Lactococcus genera harbored interesting functional properties to be used in starter culture formulations for dairy-based fermented food products [46].

4. Conclusions

The Amazon rainforest is a sizeable reservoir of plants, animals, and bacterial diversity. For Ecuador, the Amazon region could be a significant source of new bioproducts, based on the transformation of biodiversity [47]. Subtropical wild fruits have a relevant ethnobotanical significance, as they are mostly consumed by indigenous people as food or natural medicine; however, the bacterial microbiota of those fruits has not been assessed. In this research, we investigated the lactic acid bacteria diversity associated with several wild fruits collected from the Amazon region of Ecuador. Their remarkable inhibitory potential towards Gram-negative bacteria might be related to their capacity to produce various antimicrobial substances, that when applied to food products might prevent the growth of undesirable microorganisms. A better understanding of the metabolic capacity of these microorganisms will further complement our knowledge about the development of a novel starter or preservative culture for fruit- and vegetable-based foods. The prospective comparative exploration of the genomes of LAB strains from various plant or fruit origins would be of particular interest to provide information on their adaptations to different food-matrices and to further explore biotechnological applications.

Genotype-functional correlation studies contribute to the discovery of new biotechnological properties for several species. The results from the present study supported our hypothesis that LAB strains from wild fruits of the Amazon Region of Ecuador carry noteworthy characteristics that could be inherent to their ecological niches or environmental origin and that could be developed for biotechnological applications. Several strains were found capable of producing antimicrobials with high inhibitory potential against commensal and spoilage bacteria and are promissory natural food preservatives.

Acknowledgements

This research was financed by the Technical University of the North, Centre of Research (CUICYT)-Grant no. 01388/2014 and Grant 0179/2016. The authors gratefully acknowledge the generous technical support of Ulcuango M, Torres J., and Benavidez A. GNT was supported in part by the Prometeo Project of the Secretary for Higher Education, Science, Technology and Innovation (SENESCYT, 2014-2016).

Conflict of interest

The authors report no conflicts of interest. The authors are responsible for the content and writing of this article.

Author details

Gabriela N. Tenea[1*], Pablo Jarrin-V[2] and Lucia Yepez[1]

1 Biofood and Nutraceutics Research and Development Group, Technical University of the North, Ibarra, Ecuador, Av. 17 de Julio s-21 y José María Córdova, Barrio El Olivo, 100150, Ibarra, Ecuador

2 Population and Environment Research Group, Universidad Regional Amazónica Ikiam, Tena, Ecuador

*Address all correspondence to: gntenea@utn.edu.ec; gtenea@hotmail.com

References

[1] Garcia EF, Luciano WA, Xavier DE, da Costa WC, de Sousa Oliveira K, Franco OL. et al. Identification of lactic acid bacteria in fruit pulp processing byproducts and potential probiotic properties of selected *Lactobacillus* strains. Frontiers in Microbiology. 2016;7: 1371. DOI: 10.3389/fmicb.2016.01371

[2] Ruiz Rodríguez, LG, Aller K, Bru E, De Vuyst L, Hebert EM, Mozzi F. Enhanced mannitol biosynthesis by the fruit origin strain *Fructobacillus tropaeoli* CRL 2034. Applied Microbiology and Biotechnology. 2017;101:6165-6177. DOI: 10.1007/s00253-017-8395-1

[3] Szutowska J. Functional properties of lactic acid bacteria in fermented fruit and vegetable juices: a systematic literature review. European Food Research and Technology. 2020;246:357-372. DOI: 10.1007/s00217-019-03425-7

[4] European Food Safety authority (EFSA). Guidance on the assessment of bacterial susceptibility to antimicrobials of human and veterinary importance. EFSA J. 2012: 10:2740-9.16. 732:1-15. DOI: 10.2903/j.efsa.2008.732

[5] Mattia A, Merker R. Regulation of probiotic substances as ingredients in foods: Premarket approval or "generally recognized as safe" notification. Clinical Infectious Diseases. 2008; 46(supplement_2):S115-S118. DOI: 10.1086/523329

[6] Laulund S, Wind A, Derkx PM, Zuliani V. Regulatory and safety requirements for food cultures. Microorganisms. 2017;5:E28. DOI: 10.3390/microorganisms5020028

[7] Fraqueza MJ. Antibiotic resistance of lactic acid bacteria isolated from dry-fermented sausages. International Journal of Food Microbiology. 2015, 212, 76-88. DOI: 10.1016/j.ijfoodmicro.2015.04.035

[8] Chao SH, Huang HY, Kang YH, Watanabe K, Tsai YC. The diversity of lactic acid bacteria in a traditional Taiwanese millet alcoholic beverage during fermentation. LWT - Food Sci and Technology. 2013;51:135-142. DOI: 10.1016/j.lwt. 2012.09.015

[9] Emerenini E, Afolabi O, Okolie P, Akintokun A. Isolation and molecular characterization of lactic acid bacteria isolated from fresh fruits and vegetables using nested PCR analysis. Brazilian Microbiology Research Journal. 2013;3:368-377. DOI: 10.9734/BMRJ/2013/2520

[10] Mustopa AZ, Fatimah F. Diversity of Lactic acid bacteria Isolated from Indonesian traditional fermented foods. Microbiology Indonesia. 2014;8(2):48-57. DOI: 10.5454/mi.8.2.2

[11] Barraclough T, Balbi K, Ellis R. Evolving concepts of bacterial species. Evolutionary Biology. 2012;39:148-157. DOI: 10.1007/s11692-012-9181-8

[12] Naeem M, Ilyas M, Haider S, Baig S, Saleem, M. Isolation characterization and identification of lactic acid bacteria from fruit juices and their efficacy against antibiotics. Pakistan Journal of Botany. 2012;44:323-328

[13] Di Cagno R, Filannino P, Gobbetti M. Vegetable and fruit fermentation by lactic acid bacteria. In: Mozzi F, Raya RR, Vignolo GM, editors. Biotechnology of Lactic Acid Bacteria: Novel Applications. Chichester:John Wiley & Sons; 2015. p. 216-230. DOI: 10.1002/9781118868386.ch14

[14] Filannino P, Di Cagno R, and Gobbetti M. Metabolic and functional paths of lactic acid bacteria in plant foods: get out of the labyrinth. Current Opinion in Biotechnology. 2018;49: 64-72. DOI: 10.1016/j.copbio.2017.07.016

[15] Samuni-Blank M, Izhaki I, Laviad S, Bar-Massada A, Gerchman Y, Halpern M. The role of abiotic environmental conditions and herbivory in shaping bacterial community composition in floral nectar. PLoS ONE. 2014;9:e99107. DOI: 10.1371/journal.pone.0099107

[16] Benavidez A, Ulcuango M, Yépez L, Tenea, GN. Assessment of the in vitro bioactive properties of lactic acid bacteria isolated from native ecological niches of Ecuador. Revista Argentina de Microbiologia. 2016;48:236-244. DOI: 10.1016/j.ram.2016.05.003

[17] Xing J, Wang G, Zhang Q, Liu X, Gu Z, Zhang H, Chen YQ, Chen W. Determining antioxidant activities of lactobacilli cell-free supernatants by cellular antioxidant assay: a comparison with traditional methods. PLoS ONE. 2015;10(3):e0119058. DOI: 10.1371/journal.pone.0119058

[18] Khoo HE, Azlan A, Kong KW, Ismail A. Phytochemicals and medicinal properties of indigenous tropical fruits with potential for commercial development. Evidence-Based Complementary and Alternative Medicine. 2016;ID 7591951. DOI: 10.1155/2016/7591951

[19] Gul K, Singh AK, Jabeen R. Nutraceuticals and functional foods: The foods for the future world. Critical Reviews in Food Science and Nutrition. 2016;56(16):2617-2627. DOI: 10.1080/10408398.2014.903384

[20] Tenea GN, Hurtado P, Ortega C. Inhibitory effect of substances produced by native *Lactococcus lactis* strains of tropical fruits towards food pathogens. Preventive Nutrition and Food Science. 2018;23(3):260-268. DOI: 10.3746/pnf.2018.23.3.260

[21] Tenea GN, Pozo Delgado T. Antimicrobial peptides from *Lactobacillus plantarum* UTNGt2 prevent harmful bacteria growth on fresh tomatoes. Journal of Microbiology and Biotechnology. 2019;29:1553-1560, DOI: 10.4014/jmb.1904.04063

[22] Gueimonde M, Sánchez B, de Los Reyes-Gavilán GC, Margolles A. Antibiotic resistance in probiotic bacteria. Frontiers in Microbiology. 2013;4:202. DOI: 10.3389/fmicb.2013.00202

[23] Ceapa C, Lambertb J, van Limptb K, Welse M, Smokvinaa T, Knol J, Kleerebezemd M. Phenotype profiling links genotype to carbohydrate utilization signatures in the species *Lactobacillus rhamnosus*. Applied Environmental Microbiology. 2015; DOI: 10.1128/AEM.00851-15

[24] Bali V, Panesar PS, Bera MB, Kennedy JF. Bacteriocins: Recent trends and potential applications. Critical Review in Food Science and Nutrition. 2016;56(5):817-34. DOI: 10.1080/10408398.2012.729231

[25] Saha BC. A low-cost medium for mannitol production by *Lactobacillus intermedius* NRRL B-3693. Applied Microbiology and Biotechnology. 2006;72:676-680. DOI: 10.1007/s00253-006-0364-z

[26] De Man JC, Rogosa M, Sharpe E. A medium for the cultivation of lactobacilli. Journal of Applied Microbiology. 1960;23:130-155. DOI: 10.1111/j.1365-2672.1960.tb00188.x

[27] Kearse M, Moir R, Wilson A, Stones-Havas S, Cheung M, Sturrock S, Buxton S, Cooper A, Markowitz A, Duran D, Thierer, T, Ashton B, Meintjes P, Drummond A. Geneious Basic: an integrated and extendable desktop software platform for the organization and analysis of sequence data. Bioinformatics. 2012;28(12):1647-1649. DOI: 10.1093/bioinformatics/bts199

[28] Wang Q, Garrity G, Tiedje J, Cole J. Naïve bayesian classifier for rapid assignment of rRNA Sequences into the new bacterial taxonomy. Applied and Environmental Microbiology. 2007;73(16):5261-5267. DOI: 10.1128/aem.00062-07

[29] Callahan BJ, McMurdie PJ, Rosen MJ, Han AW, Johnson AJA, Holmes SP. DADA2: High-resolution sample inference from Illumina amplicon data. Nature Methods. 2020; 13:581-583. DOI: 10.1038/nmeth.3869

[30] R Core Team R: A language and environment for statistical computing. R Foundation for Statistical Computing. Vienna, Austria. 2020. URL: https://www.R-project.org/

[31] Alishum A. DADA2 formatted 16S rRNA gene sequences for both bacteria & archaea (Version 3). Zenodo. 2020; DOI: 10.5281/zenodo.3951383

[32] Zheng J, Wittouck S, Salvetti E, Franz C, Harris H, Mattarelli P, O'Toole PW, Pot B, Vandamme P, Walter J., Watanabe K, Wuyts S, Felis GE, Gänzle MG, Lebeer S. A taxonomic note on the genus *Lactobacillus*: Description of 23 novel genera, emended description of the genus *Lactobacillus* Beijerinck 1901, and union of *Lactobacillaceae* and *Leuconostocaceae*. International journal of systematic and evolutionary microbiology. 2020;70(4):2782-2858. DOI: 10.1099/ijsem.0.004107

[33] Garzón K, Ortega C, Tenea GN. Characterization of bacteriocin-producing lactic acid bacteria isolated from native fruits of Ecuadorian Amazon. Polish Journal of Microbiology. 2017; 66(4): 473-481. DOI: 10.5604/01.3001.0010.7037

[34] Berends MS, Luz CF et al. AMR - An R Package for working with antimicrobial resistance data. bioRxiv. 2019. DOI: 10.1101/810622

[35] Oksanen JF, Blanchet G, Friendly M, Kindt R, Legendre P, McGlinn D, Minchin PR, O'Hara RB, Simpson GL, Solymos P, Stevens HH, Szoecs E, Wagner H. Vegan: community ecology package. R package version 2.5-6. 2019. https://CRAN.R-project.org/package=vegan

[36] de Vries A, Ripley BD. ggdendro: Create dendrograms and tree diagrams using 'ggplot2'. R package version 0.1.21. 2020. https://CRAN.R-project.org/package=ggdendro

[37] Darriba D, Taboada GL, Doallo R Posada D. jModelTest 2: more models, new heuristics and parallel computing. Nature Methods. 2012;9(8):772

[38] Ronquist F, Huelsenbeck JP. MRBAYES 3: Bayesian phylogenetic inference under mixed models. Bioinformatics. 2003;19:1572-1574

[39] Gustaw, K., Michalak, M., Polak-Berecka, M. et al. Isolation and characterization of a new fructophilic Lactobacillus plantarum FPL strain from honeydew. Annals of Microbiology. 2018;68:459-470. DOI: 10.1007/s13213-018-1350-2

[40] Abriouel H, Casado Muñoz MD, Lerma LL, Perez Montoro B, Bockelman W, Pichner R, et al. New insights in antibiotic resistance of *Lactobacillus* species from fermented foods. Food Research International. 2015;78:465-481. DOI: 10.1016/j.foodres.2015.09.016

[41] Casado Muñoz MC, Benomar N, Lerma LL, Gálvez A, Abriouel H. Antibiotic resistance of *Lactobacillus pentosus* and *Leuconostoc pseudomesenteroides* isolated from naturally-fermented Aloreña table olives throughout fermentation process. International Journal of Food Microbiology. 2014;172:110-118. DOI: 10.1016/j.ijfoodmicro.2013.11.025

[42] Venegas-Ortega MG, Flores-Gallegos AC, Martínez-Hernández JL, Aguilar CN, Nevárez-Moorillón GV. Production of bioactive peptides from lactic acid bacteria: a sustainable approach for healthier foods. Comprehensive Review in Food Sciences and Food Safety. 2019;18:1039-1051. DOI: 10.1111/1541-4337.12455

[43] Simons A, Alanout K, Duval RE. Bacteriocins, antimicrobial peptides from bacterial origin: overview of their biology and their impact against multidrug-resistant bacteria. Microorganisms. 2020;8:639. DOI: 10.3390/microorganisms8050639

[44] Tenea GN, Lara MI. Antimicrobial compounds produced by *Weissella confusa* Cys2-2 strain inhibit Gram-negative bacteria growth. CyTA – Journal of Food. 2019;17(1):105-111. DOI: 10.1080/19476337.2018.1561520

[45] Tenea GN, Olmedo D, Ortega C. Peptide-based formulation from lactic acid bacteria Impairs the pathogen growth in *Ananas comosus* (Pineapple). Coatings. 2020;10: 457. DOI: 10.3390/coatings10050457

[46] Tenea GN, Suárez J. Probiotic Potential and Technological Properties of Bacteriocinogenic *Lactococcus lactis* subsp. *lactis* UTNGt28 from a native Amazonian fruit as a yogurt starter culture. Microorganisms. 2020;8:733. DOI: 10.3390/microorganisms8050733

[47] Jarrín Valladares PS, Carrillo LT, Zamora G. The internal colony as a current issue: transformation of the human territory in the Amazonian region of Ecuador. Letras Verdes. 2016;20:22-43. DOI: 10.17141/letrasverdes.20.2016.2063

Chapter 3

Ticks from the Brazilian Amazon: Species, Distribution and Host-Relations

Hermes R. Luz, Thiago F. Martins, Sebastián Muñoz-Leal, Francisco B. Costa, Sérgio L. Gianizella, João Luiz H. Faccini and Marcelo B. Labruna

Abstract

Ticks are important ectoparasites and can transmit a wide variety of pathogens to animals and humans worldwide. These ectoparasites are the most important vectors of diseases causing pathogens in domestic and wild animals, and the second for man. In Brazil, Spotted Fever is the only disease transmitted to humans by ticks, while for animals we can highlight babesiosis, ehrlichiosis and anaplasmosis. Although ticks are found in all of Brazil's biomes, little is known about its diversity and host-relations in the Amazon biome. The existing gaps and the lack of research indicate that the diversity of ticks and their possible pathogens are underestimated in the Brazilian Amazon. Therefore, in order to guide the next studies in the Amazon biome, we present in this chapter a compilation of the records of hard ticks and soft ticks parasitizing wild and domestic animals, and humans. We present the general list of ticks for this biome, their distribution, hosts and importance for public health and veterinary. Finally, the ixodofauna found in the Brazilian Amazon and presented here does not match the vertebrate diversity of this biome, which is one of the largest on the planet. Therefore, more acarologists and epidemiologists are needed in this region.

Keywords: Ixodida, domestic animals, wild animals, humans, Amazon, Brazil

1. Introduction

The Amazon or Amazon Rainforest is the largest remnant of tropical forest in the world, occupying a region of approximately 6.7 million/km^2, covering nine countries in South America: Ecuador (≅ 2%), Suriname (≅ 2%), Bolivia (≅ 4%), Venezuela (≅ 4%), Guyana (≅ 3%), French Guiana (≅ 2%), Colombia (≅ 10%), Peru (≅ 13%), and Brazil (≅ 60%) (**Figure 1**) [1, 2]. In Brazil, the Amazon biome, also known as "Legal Amazon" occupies approximately 49% of its territory, covering the states of Acre, Amapá, Amazonas, Mato Grosso, Pará, Rondônia, Roraima, Tocantins and Maranhão (**Figure 1**). The Brazilian Amazon is known for its high richness of landscapes composed of 23 ecoregions, whose main domain is the humid rainforest (≅ 78%). Due to this heterogeneity, the Amazon has an immeasurable amount of essential habitats for the maintenance of flora and fauna, represented

IntechOpen

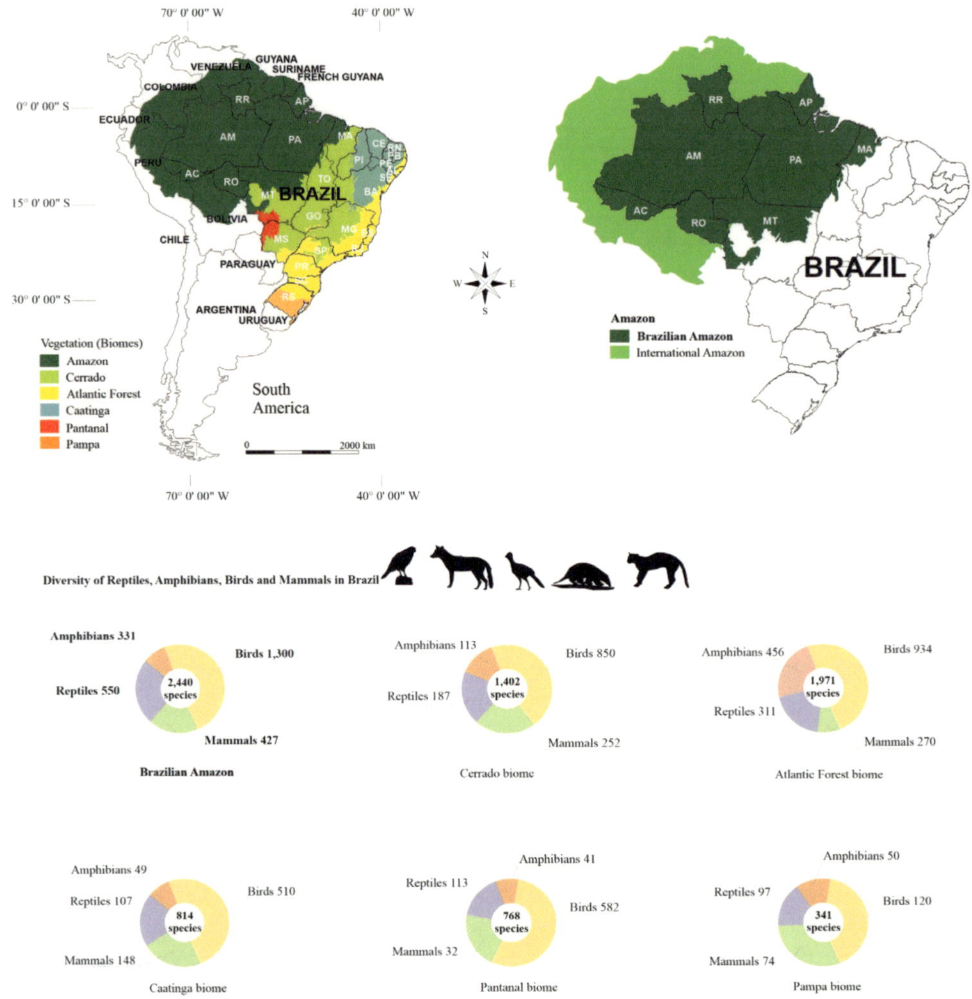

Figure 1.
Diversity of reptiles, amphibians, birds and mammals in Brazil, including the Brazilian Amazon.

with high biological diversity, which is regarded as the largest in the world [3, 4] The Amazon has approximately 45,000 species of flora (39,474 species) and fauna (5,526 species) [2, 5]. However, even with this expressive diversity and the numerous faunal studies carried out in the region, there are still many gaps in the biological diversity of the Amazon. Due to its enormous extension and high degree of preservation (e.g., unexplored areas), new bioecological associations and new species are discovered every year. In the past 20 years, more than 1,200 new species have been described in the Amazon region from which we can highlight 16 birds, 39 mammals, 55 reptiles and ≅ 100 amphibians [2]. However, anthropic action has negatively impacted the animal-forest relationships for decades, resulting in the extinction of ecologically demanding species and, at the same time, the appearance of opportunistic and/or generalist species [6].

Many domestic and wild animals are responsible for the maintenance and dispersion of ticks in nature. In addition, vertebrates act as amplifiers and/or reservoirs for viruses, protozoa and bacteria transmitted by these ectoparasites. The antropic action affects the population dynamics of both ticks and their wild hosts directly or indirectly, and consequently, the epidemiology of tick-borne diseases once restricted to wild fauna can reach domestic animal and humans interfaces [7]. Ticks are responsible for more than 100,000 cases of diseases in humans and animals worldwide [8], therefore a concern for public health professional. In Brazil, ticks are vectors of diseases such as babesiosis, ehrlichiosis, anaplasmosis and rickettsioses, including Brazilian Spotted Fever (BSF) [9, 10].

Ticks belong to the Sub-Class Acari, Super-Order Parasitiformes, Order Ixodida, and four families: Ixodidae, Argasidae, Nuttallielidae and Deinocrotonidae [11, 12], the latter extinct. Of these, only the Ixodidae and Argasidae families occur in Brazil, with nine genera and 75 species [11, 13–16]. Ixodidae family is the most diverse with 51 species and five genera: *Amblyomma* (33 species), *Ixodes* (12 species), *Rhipicephalus* (two species), *Haemaphysalis* (three species) and *Dermacentor* (one species). Argasidae family is represented by 24 species and four genera: *Ornithodoros* (18 species), *Antricola* (three species), *Argas* (one species) and *Nothoaspis* (two species). In general, the immature stages (larvae and nymphs) of two or three host ticks feed on small mammals and birds, while adults exploit medium to large-sized hosts [17]. One-host species *Dermacentor nitens* and *Rhipicephalus microplus*, complete the life cycle mainly on large animals such as horses and cattle, respectively. Second Esser et al. [18] the diversity of ticks increases with the increase of diversity of hosts, inasmuch as these hematophagous arthropods rely on the hosts to complete their life cycles. Therefore, if we take into account the high diversity of hosts living in this biome and the amount of unexplored ecoregions, it is reasonable to infer that the diversity of ticks in the Amazon biome is insufficiently addressed. Thus, a review on the subject is necessary to point out the existing gaps and encourage new studies on ticks in the Brazilian Amazon.

2. Hard ticks associated with amphibians and reptiles

The class Amphibia includes the orders Anura, Caudata, Gymnophiona, whereas the class Reptilia includes the orders Squamata, Testudines and Crocodylia. There are 331 amphibian and 550 reptile species in the Brazilian Amazon (**Figure 1**), although that faunal records are far from complete [2, 19]. Brazil has witnessed an increasing number of reports on tick parasitism of amphibians and reptiles over the past few years [20–24]. However, knowledge of this tick fauna as their hosts remains incomplete. To date, major tick-amphibian associations reported in Brazil are the ixodids *Amblyomma dissimile*, *Amblyomma rotundatum*, *Amblyomma humerale*, *Amblyomma fuscum* and *Amblyomma goeldii* [20–24], and the argasids *Ornithodoros saraivai* and *Ornithodoros faccinii* associated with species *Cycloramphus boraceiensis* and *Thoropa miliaris*, respectively [25, 26]. Occasional records on reptiles have also been reported for *Ornithodoros mimon* and *Ornithodoros rietcorreai* [22]. However, none of these argasids have been reported in the Amazon biome. Although the Brazilian Amazon has a high diversity of amphibians and reptiles, studies on ticks in association with these hosts are still rare, with a dominance of species *A. dissimile*, *A. rotundatum* and *A. humerale* [21–23, 27, 28].

The tick *A. dissimile* is constantly misidentified with the morphologically similar *A. rotundatum*. Recently, the distribution of *A. dissimile* in the country was reorganized, showing that this tick is restricted to the Pantanal and Amazon biomes [28]. Therefore, reports outside these biomes are considered misidentifications and must be viewed with caution. Most publications on this species are simple records of occurrence with some authors including data on prevalence and intensity of infestation both in the Amazon biome and elsewhere. According to Luz et al. [29] 12 (54.5%) out of 22 *Rhinella marina* toads captured in Amapá state, were parasitized by a total of 97 ticks (6 males, 39 females, 31 nymphs, 21 larvae) and mean intensity of 8.1 ticks per infested toad. In the Amazonian biome, *A. dissimile* is common on *R. marina* (Anura) and Boidae (Squamata) (**Table 1**). The experimental life cycle

Host	**Specie**	***A. dissimile***			***A. rotundatum***			
		L	**N**	**A**	**L**	**N**	**A**	
Amphibian (Anura)								
Bufonidae	*Rhinella marina*		37	12M; 16F		164	133F	[23, 29–33]
	Rhinella major			2F		15	2F	[33–35]
	Rhaebo guttatus					1	1F	[23]
	Rhinella margaritifera					3	1F	[23]
	Rhinella gildae					1		[36]
Leptodactylidae	*Leptodactylus pentadactylus*						1F	[23]
Reptile								
Boidae	*Boa constrictor*		1703	723M; 1346F		12	1M; 77F	[23, 30, 32–34, 36–40]
	Corallus hortulanus		3	1M				[31, 34]
	Eunectes murinus		305	21M; 3F				[31, 40]
Viperidae	*Bothrops atrox*	4	11	24M; 14F			1M; 2F	[32, 33, 36, 37]
	Crotalus durissus			1M; 1F				[34]
	Lachesis muta		1					[34]
Elapidae	*Micrurus averyi*			3F				[31]
	Micrurus lemniscatus			12M; 9F				[31]
Colubridade	*Leptophis ahaetulla*					3		[33]
	Chironius multiventris					3		[33]
	Chironius scurrulus			1M; 1F				[34]
	Chironius laevicollis		12	15M; 6F				[31]
	Mastigodryas boddaerti					1		[33]
	Hydrodynastes gigas					1	7F	[30]
	Helicops polylepis		1					[34]
	Leptodeira annulata		1					[34]
	Phimophis guerini			1F				[34]
	Erythrolamprus reginae		1					[34]
	Spilotes pullatus		72	42M; 9F				[31]
	Erythrolamprus reginae semilineatus		1					[34]
Dipsadidae	*Xenodon severus*			8				[40]
Testudinidae	*Chelonoidis denticulatus*			3M		3	2M; 42F	[31, 41]
	Chelonoidis carbonaria		1	1M; 1F		11		[32, 39]
Podocnemididae	*Podocnemis expansa*						1F	[39]
	Podocnemis unifilis					1		[39]
	Trachemys dorbigni					2	12F	[39]
Kinosternidae	*Kinosternon scorpioides*						2F	[42]

Host	**Specie**	***A. dissimile***			***A. rotundatum***			
		L	**N**	**A**	**L**	**N**	**A**	
Chelidae	*Platemys platycephala*						5F	[32]
Tropiduridade	*Uranoscodon superciliosus*					7	6F	[33]
	Tropidurus sp.						1M	[38]
Teiidae	*Ameiva ameiva*		4				1F	[31, 34]
Iguanidae	*Iguana iguana*		31	118M; 69F			2F	[31, 36, 39, 40]
Dactyoloidae	*Norops auratus*		1					[34]
Gekkonidae	*Hemidactylus mabouia*		1					[34]
Alligatoridae	*Paleosuchus trigonatus*			9M; 15F		4	12F	[31, 32]
	Caiman crocodilus		3					[31, 39]
Total		4	2189	983M; 1496F		221	307F	

Table 1.
Hosts of Amblyomma dissimile *and* Amblyomma rotundatum *in the Brazilian Amazon.*

including pre-attachment periods for each parasitic stage lasts approximately 350 days, as reported by Schumaker et al. [43] who started a colony from one engorged female collected from the Amazonian biome. Ogrzewalska et al. [37] reported *Rickettsia bellii* and '*Candidatus* Rickettsia colombianensi' in ticks collected from *Bothrops atrox* from Pará state. Luz et al. [29] reported '*Ca.* R. colombianensi' in ticks collected on *R. marina* from Amapá state.

The tick *A. rotundatum* is an obligate parthenogenetic species, although there are six reports of adult males, two males in the laboratory [44] and six in the Amazon region parasitizing *Tropidurus* sp., *Boa constrictor*, *Iguana iguana*, *Chelonoidis denticulatus* and *B. atrox* [30, 36, 44–47]. Most publications on *A. rotundatum* are simple records of its occurrence with some additional data on prevalence and intensity of infestation in the Cerrado, Atlantic forest and Amazon biomes [22, 23, 31, 34, 48]. In the Brazilian Amazon, the families Bufonidae (Amphibian) and Boidae (Reptilia) are the most frequently parasitized by *A. rotundatum*, but there are reports on other species of Amphibian and Reptiles (**Table 1**). Recently, Gianizella et al. [31] reported *A. rotundatum* in different municipalities of the Amazonas state, on *B. constrictor* and several unknown hosts. The bioecology of this species under quasi-natural environment demonstrated a peak of larvae and nymphs in the dry season and females in the rainy season; however, there is a hypothesis of the absence of seasonality in this tick [48, 49]. Although *A. rotundatum* is a three-host tick on amphibians, it can behave as a two-host tick when feeding on snakes [50]. The complete experimental life cycle, including pre-attachment periods for each parasitic stage, ranged from 126 to 228 days on toads [51] or 56 to 163 days on snakes [50] and ulcerative lesions and hemorrhages after *A. rotundatum* feeding [48] were reported for ticks collected on toads from the Cerrado biome. Transmission of the hemogregarine *Hemolivia stellata* by *R. marina* collected in Belém, state of Pará was reported by Petit et al. [52]. This tick has also been found infected with *R. bellii* in the state of Amazon in the municipalities of Cacaulândia and Monte Negro, in the state of Rondônia; in the municipalities of Amapá (Ilha de Maracá) and Santana, in the state of Amapá; and in the municipality of Rio Branco, in the state of Acre [29, 53, 54].

Most reports of *A. humerale* mention the adult stage parasitizing species of tortoises, namely the "yellow-footed tortoise" *C. denticulatus* and the "red-footed tortoise" *Chelonoidis carbonarius* [27, 55]. There is also a record of *A. humerale* on *Rhinoclemmys punctularia* in central Amazon [56]. Labruna et al. [27] collected 215 adult ticks from six *C. denticulatus* and nine *Chelonoidis* sp. from state of Rondônia, with mean infestation of 14.3 ± 12.0 ticks per tortoise. In addition, seven engorged nymphs were collected on lizards. Morais et al. [57] collected 120 adult ticks from 18 (75%) out of 24 *C. denticulatus* captured in a transitional area between the biomes Cerrado and southwestern Amazon rainforest, in the state of Mato Grosso. The mean intensity of infestation was 6.7 ticks/tortoise. In both surveys, male ticks were mostly attached in clusters on the carapace whereas females were found attached to the tortoise skin. The sex ratio (males:females) for *A. humerale* were different in both surveys, 10.3:1 [27] and 1.1:1 [57]. There is one additional record of two nymphs on *Paleosuchus trigonatus* (Crocodylia) [32]. The life-cycle in the laboratory, including pre-feeding periods for each of the parasitic stages, could be completed in an average period of ca. 200 days [58]. *Rickettsia bellii* and *Rickettsia amblyommatis* have been detected in *A. humerale* [42, 53]. The records of *A. fuscum* in association with amphibians and reptiles are rare in Brazil. Dantas-Torres et al. [59] collected one male tick in one out of 490 caimans (*Caiman latirostris* and *Paleosuchus palpebrosus*) trapped in the Atlantic rainforest biome in Pernambuco state, north-eastern Brazil. *Amblyomma goeldii* has been recorded only in the Amazonas state. Martins et al. [60] reported two males collected on *B. constrictor* and recovery of ≈100 (20%) engorged larvae out of ≈500 unfed larvae experimentally infested on a *B. constrictor.*

3. Hard ticks associated with wild birds

The importance of birds to maintain biodiversity and ecological balance of nature is notorious [61]. Due to migration, wild birds are of concern to human and animal health worldwide [62] because they can carry infected ticks over long distances, directly influencing the epidemiology of tick-borne diseases in animals and humans. In addition, wild birds themselves can be reservoirs of *Borrelia burgdorferi* sensu lato, and potentially to *Anaplasma phagocytophilum* and *Rickettsia* spp. [62, 63]. Wild birds play an important role in maintaining and dispersing immatures (larvae and nymphs) of several tick species into new locations [61, 64].

Of the total genera of ticks described in Brazil, five have at least one species recorded in association with wild birds. The most common are the hard ticks of the genera *Amblyomma*, *Haemaphysalis* and *Ixodes* [61, 64]. There are also occasional reports of the genera *Rhipicephalus* and *Ornithodoros*. Ticks of the genus *Amblyomma* are the most common on wild birds in the Brazil including the Amazon biome, exclusively for the larvae and nymphs [61, 64]. Adult ticks are only occasionally found on wild birds, with the exception of *Ixodes paranaensis* and *Ixodes auritulus*, which have the entire cycle synchronized with birds [65]. In Brazil, there are no reports of wild birds as a source of pathogens transmitted by ticks to humans, but they can serve as disperser hosts for vectors of Brazilian Spotted Fever (BSF) as *Amblyomma sculptum*, *Amblyomma aureolatum* and *Amblyomma ovale*, in the larvae and nymph stages [64]. Therefore, wild birds act indirectly in the epidemiology of BSF by dispersing and maintaining their vectors in nature.

Over more than 1,900 birds recorded in Brazil, approximately 1,300 reside in the Brazilian Amazon, with a 20% of endemism [66] (**Figure 1**). Of these, approximately 7% are migrants from the northern hemisphere and southern South America, including migrations from other Brazilian biomes [67]. To date, 86 bird species of Brazilian Amazonian have been recorded in association with at least one tick species (**Table 2**). This is equivalent to approximately only 7% of bird species found in this biome and 5% of the total birds recorded in Brazil (**Table 2**). Similar to other studies regarding tick-bird associations in Brazil [73–75], Passeriformes birds were the most parasitized in the Amazon biome, including 14 families and 72 species (**Figure 2** and **Table 2**). In this group, the greatest diversity of parasitized birds was Thamnophilidae (20 species) followed by Dendrocolaptidae (16 species) and Tyrannidade (10 species) (**Figure 2**). The least parasitized families were Conopophagidae, Furnariidae, Xenopidae, Tityridae, Cardinalidae, Columbidae, Cuculidae, Momotidae, Capitonidae, Ramphastidae, Psittacidae, Accipitridae and Falconidae with only one species of parasitized bird each (**Figure 2** and **Table 2**). Non-Passerines were represented by 10 different orders and 11 families, with

Hosts			**Ticks**			**References**
			Species	**Stages**		
				LL	NN	
Passeriformes	Thamnophilidae	*Thamnomanes schistogynus*	*A. humerale* *A. nodosum*		1 1	[68]
		Thamnophilus doliatus	*A. nodosum*		2	[68]
		Thamnophilus schistaceus	*A. nodosum* *Amblyomma* sp.	2	2	[68]
		Thamnophilus aethiops	*A. calcaratum* *A. longirostre*		5 1	[68]
		Myrmelastes hyperythrus	*A. nodosum*		9	[68]
		Sciaphylax hemimelaena	*A. nodosum*		1	[68]
		Schistocichla leucostigma	*A. humerale* *Amblyomma* sp.	4 5		[69]
		Phlegopsis nigromaculata	*A.coelebs* *A. longirostre* *A. nodosum* *Amblyomma* sp.	1 1 24	5	[68, 69]
		Hypocnemis subflava	*Amblyomma* sp.	3		[70]
		Hylophylax naevius	*H. juxtakochi* *A. humerale* *A. longirostre* *A. calcaratum* *Amblyomma* sp.	1 3 1 8	1 2 1	[69]
		Hylophylax punctulatus	*Amblyomma* sp.	1		[69]
		Thamnomanes caesius	*A. geayi* *A. longirostre* *A. humerale* *Amblyomma* sp	1 3 4	2 1	[69, 71]
		Hylophylax poecilinotus	*A. geayi* *A. humerale* *A. longirostre* *Amblyomma* sp. *H. juxtakochi*	1 1 3 22	1 1 1 1	[69]
		Myrmornis torquata	*Amblyomma* sp.	3		[69]

Hosts			Ticks			References
			Species	Stages		
		Myrmotherula hauxwelli	*Amblyomma* sp.	1		[69]
		Myrmotherula longipennis	*A. longirostre*	3		[69]
			Amblyomma sp.	3		
		Pyriglena leuconota	*A. coelebs*	1	1	[69]
			A. humerale	1		
			A. longirostre	9		
			Amblyomma sp.	4		
		Myrmoborus myotherinus	*A. longirostre*		1	[69]
		Epinecrophylla leucophthalma	*A. longirostre*		1	[71]
		Willisornis poecilinotus	*A. calcaratum*		1	[71]
	Conopophagidae	*Conopophaga aurita*	*Amblyomma* sp.		1	[69]
	Scleruridae	*Sclerurus cf. rufigularis*	*A. humerale*		1	[68]
		Sclerurus caudacutus	*A. humerale*		1	[69]
	Furnariidae	*Automolus paraensis*	*Amblyomma* sp.		1	[69]
	Dendrocolaptidae	*Dendrocincla merula*	*A. longirostre*	7	1	[68, 69, 71]
			Amblyomma sp.	34		
		Dendrocincla fuliginosa	*Amblyomma* sp.	10		[31, 70]
		Deconychura longicauda	*A. longirostre*	4	1	[69, 71]
			Amblyomma sp.	39		
		Deconychura stictolaema	*A. longirostre*	1	1	[69]
			Amblyomma sp.	1		
		Dendrocolaptes hoffmannsi	*A. calcaratum*	1	2	[69]
			A. geayi	2	1	
			A. humerale	4	3	
			A. longirostre	39		
			Amblyomma sp.	59		
		Dendrocolaptes certhia	*A. longirostre*		1	[71]
		Glyphorynchus spirurus	*A. longirostre*		1	[71]
		Glyphorynchus spirurus	*A. geayi*	1	1	[69]
			A. longirostre	6	2	
			Amblyomma sp.	16		
		Hylexetastes brigidai	*A. geayi*	1	2	[69]
			A. longirostre	6		
			Amblyomma sp.	21		
		Sittasomus griseicapillus	*A. humerale*	2	1	[70]
			Amblyomma sp.			
		Xiphorhynchus guttatus	*A. longirostre*		2	[68]
			A. nodosum		1	
		Xiphorhynchus guttatoides	*A. humerale*	1	2	[70]
			Amblyomma sp.			
		Xiphorhynchus elegans	*A. longirostre*	3	2	[69]
			Amblyomma sp.	2		

Hosts			Ticks			References
			Species	**Stages**		
		Xiphorhynchus ocellatus	*A. calcaratum*		1	[71]
		Xiphorhynchus pardalotus	*Amblyomma* sp.	3		[31]
		Dendroplex picus	*A. longirostre* *A. nodosum*		3 3	[68, 70]
	Xenopidae	*Xenops minutus*	*A. nodosum*		2	[69]
	Pipridae	*Pipra filicauda*	*A. geayi* *Amblyomma* sp.		1	[68, 70]
		Pipra fasciicauda	*A. longirostre* *A. nodosum* *Amblyomma* sp.	1 8	7	[68]
		Machaeropterus pyrocephalus	*A. longirostre*		1	[68]
		Dixiphia pipra	*A. geayi* *A. longirostre*	1 1	1 1	[69]
		Lepidothrix nattereri	*Amblyomma* sp.	1		[71]
	Rhynchocyclidae	*Leptopogon amaurocephalus*	*A. geayi*	1		[68]
		Rhynchocyclus olivaceus	*A. geayi* *A. humerale* *A. longirostre* *A. nodosum* *Amblyomma* sp.	1 1 4	2 1 17	[68]
		Todirostrum maculatum	*A. nodosum*		1	[68]
		Poecilotriccus latirostris	*A. nodosum* *Amblyomma* sp.	6	3	[70]
		Hemitriccus flammulatus	*Amblyomma* sp.	1		[70]
	Tyrannidae	*Myiozetetes similis*	*A. nodosum*		1	[68]
		Empidonax alnorum	*A. nodosum* *A. longirostre*	4	1	[68]
		Platyrinchus saturatus	*Amblyomma* sp.	1		[69]
		Hemitriccus minor	*A. longirostre*	1		[69]
		Mionectes macconnelli	*A. longirostre* *Amblyomma* sp.	4 5		[69]
		Attila spadiceus	*A. geayi* *Amblyomma* sp.	1 3		[69]
		Myiobius barbatus	*A. longirostre*	3		[69]
		Onychorhynchus coronatus	*A. longirostre* *Amblyomma* sp.	3 6		[69]
		Ramphotrigon megacephalum	*A. nodosum* *Amblyomma* sp.	1	3	[70]
		Knipolegus poecilocercus	*Amblyomma* sp.	1		[71]

Hosts			Ticks			References
			Species	Stages		
	Troglodytidae	*Pheugopedius genibarbis*	*A. nodosum* *Amblyomma* sp.	363	1	[68]
		Microcerculus marginatus	*Amblyomma* sp.	3		[69]
	Tityridae	*Schiffornis turdina*	*A. longirostre* *Amblyomma* sp.	3 14		[69]
	Turdidae	*Turdus hauxwelli*	*A. longirostre*		2	[68]
		Turdus sanchezorum	*A. longirostre*		1	[68]
		Turdus ignobilis	*A. geayi* *A. nodosum* *A. longirostre* *Amblyomma* sp.	2	1 1 1	[68]
		Turdus amaurochalinus	*Amblyomma* sp.	1		[70]
		Turdus albicollis	*A. longirostre* *A. coelebs* *Amblyomma* sp.	2 3	3	[31, 69]
	Thraupidae	*Ramphocelus carbo*	*A. nodosum* *A. longirostre* *Amblyomma* sp.	44	15 4	[30, 68, 70, 71]
		Tachyphonus surinamus	*A. humerale* *A. longirostre* *Amblyomma* sp.	1 1 5		
	Cardinalidae	*Cyanoloxia cyanoides*	*Amblyomma* sp.	1		[71]
Columbiformes	Columbidae	*Columbina talpacoti*	*A. nodosum*		1	[68]
Cuculiformes	Cuculidae	*Crotophaga major*	*A. nodosum*		1	[68]
Coraciiformes	Momotidae	*Momotus momota*	*A. longirostre* *A. humerale*		1 1	[68, 70]
Piciformes	Capitonidae	*Capito auratus insperatus*	*A. nodosum* *A. longirostre*		2 1	[68]
	Ramphastidae	*Ramphastos vitellinus*	*A. geayi*		3	[31]
		Pteroglossus bitorquatus	*A. longirostre*		1	[30]
Galbuliformes	Bucconidae	*Monasa nigrifrons*	*Amblyomma* sp.	1		[70]
		Malacoptila rufa	*Amblyomma* sp.	1		[69]
		Galbula cyanicollis	*Amblyomma* sp.	2		[69]
Psittaciformes	Psittacidae	*Primolius maracana*	*A. dissimile*		1	[72]
Accipitriformes	Accipitridae	*Harpia harpyja*	*A. cajennense*		16	[41]
			H. juxtakochi		1	[41]
Cariamiformes	Cariamidae	*Cariama cristata*	*A. cajennense*		5	[39]
Galliformes	Cracidae	*Penelope superciliaris*	*A. cajennense*	1	1	[42]
Falconiformes	Falconidae	*Micrastur ruficollis*	*A. longirostre*		1	[71]
Total				**884**	**184**	

Table 2.
Ticks identified on wild birds in the Brazilian Amazon biome.

emphasis on Bucconidae with three species (**Figure 2** and **Table 2**). To date, approximately 1,068 specimens of ticks have been collected from birds in the Brazilian Amazon, in the stages of larvae (884/83%), nymphs (184/17%) and no adults (**Figure 3**). These are included in the genera *Amblyomma* and *Haemaphysalis*. The genus *Amblyomma* was the dominant with eight species (**Table 2**). The greatest diversity of ticks was reported for the Thamnophilidae family with seven species: *A. longirostre*, *A. nodosum*, *A. humerale*, *A. calcaratum*, *A. geayi*, *A. coelebs* and *Haemaphysalis juxtakochi*. The Dendrocolaptidae family was the second with five

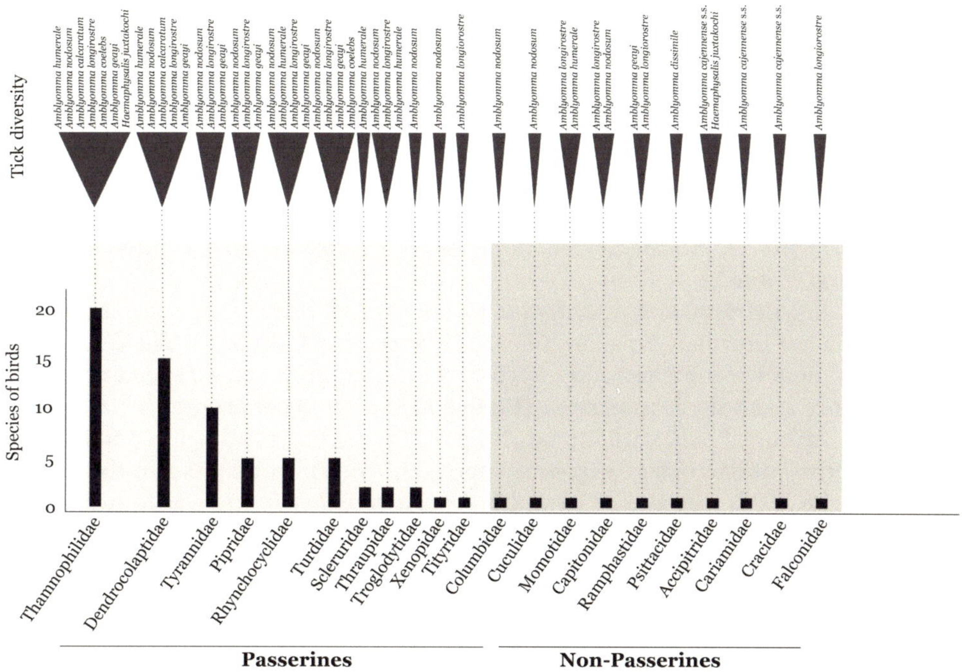

Figure 2.
Diversity of hard ticks parasitizing wild birds from the Brazilian Amazon.

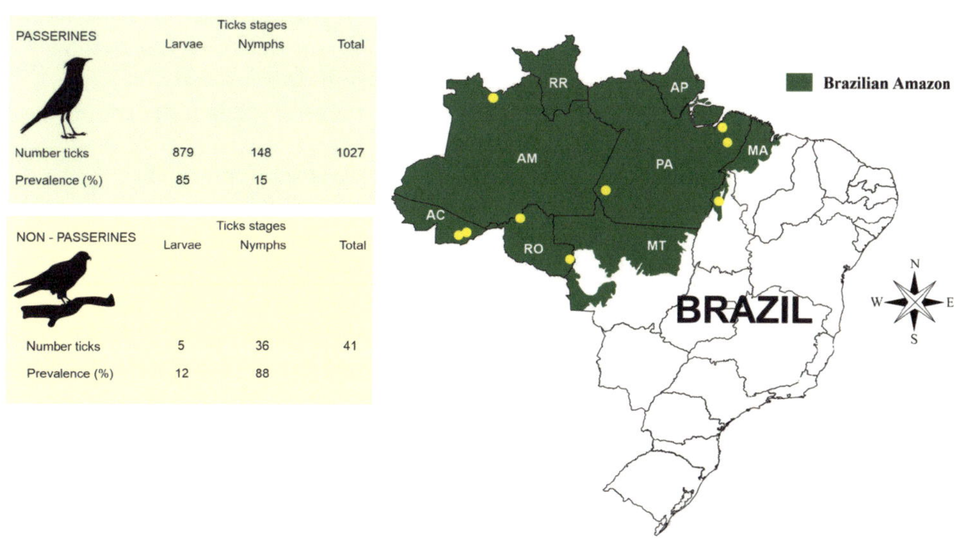

Figure 3.
Distribution of studies reporting the association of hard ticks on wild birds from the Brazilian Amazon.

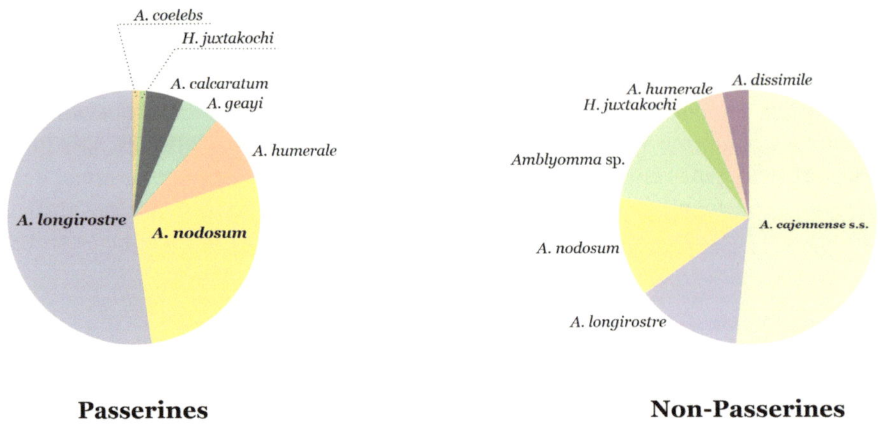

Figure 4.
Hard tick species parasitizing wild birds from the Brazilian Amazon.

species: *A. longirostre*, *A. nodosum*, *A. humerale*, *A. calcaratum* and *A. geayi* (**Figure 2** and **Table 2**).

Overall, *A. longirostre* and *A. nodosum* are the two most common species on wild birds in the Brazilian Amazon [68–71] (**Figure 4** and **Table 2**). The tick *A. longirostre* was the most common, recorded in 12 families and 40 species of birds (36 Passerines and 4 non-Passerines) (**Figure 2**). *Amblyomma longirostre* has been treated as an arboreal tick, with immatures parasitizing birds and adults parasitizing rodents Erethizontidae (e.g., *Sphiggurus* spp.) [73, 76]. This tick was also the most abundant with 110 larvae and 39 nymphs. *Amblyomma longirostre* was found in co-infestation with the following species: *A. calcaratum*, *A. nodosum*, *A. coelebs*, *A. humerale*, *A. geayi* and *H. juxtakochi* (**Table 2**). In addition, as it is frequent on birds, *A. longirostre* is popularly known as "bird tick" or "bird earring" [77]. *Amblyomma nodosum* was the second most common species collected on 12 families and 22 species of birds. This tick was recorded in co-infestation with *A. longirostre*, *A. coelebs*, *A. humerale* and *A. geayi* (**Table 2** and **Figures 2** and **4**). The birds most infested by *A. nodosum* were *Rhynchocyclus olivaceus* and *Ramphocelus carbo* with 17 and 15 nymphs, respectively. These birds inhabit the forest understory and visit the soil occasionally [67]. As Xenartha mammals (*Myrmecophaga tridactyla*, *Tamandua tetradactyla*) are the primary hosts of *A. nodosum* [17], it is believed that the low areas of the understory are a major source of infestation. Interestingly, *T. tetradactyla* may have arboreal habits [78], which helps to explain the presence of *A. nodosum* also on birds of different forest strata.

To date, of the total of ticks collected, 736 (70%) were larvae identified as *Amblyomma* sp. due to lack of reliable tools for larval identification, thus, indicating that the diversity of ticks on birds of the Amazon may be underestimated. Luz et al. [75], using molecular biology, identified more than 90% of the larvae collected from birds in the Atlantic Forest biome, reporting the greatest diversity of ticks on birds in a single study in Brazil and description of the new tick *Amblyomma romarioi* [14]. Therefore, the identification of all larvae by molecular biology in addition to morphological identification is extremely important to ascertain the diversity of ticks in the Amazon biome.

4. Hard ticks associated with wild and domestic mammals

The fauna of wild mammals in Brazil is quite diverse and more than half lives in the Amazon biome [79]. Like birds, amphibians and reptiles, mammals play an important role in preserved or anthropized ecosystems. Therefore, the knowledge of the local diversity of wild mammals, and their relationship with ticks is considered an important tool for public conservation policies and consequently for public health. The alteration of wild habitats can determine changes in the patterns of parasitic specificity, inducing tick species to seek new groups of hosts, increasing the risk of disease transmission [79, 80]. Some wild mammals (small, medium and large) are directly or indirectly involved in the transmission cycles of many tick-borne pathogens worldwide, including *Anaplasma* spp., *Babesia* spp., *Borrelia* spp., and *Rickettsia* spp. [80, 81]. In Brazil, the main zoonosis transmitted by ticks is Brazilian Spotted Fever caused by the bacterium *Rickettsia rickettsii*, which has the rodent *Hydrochoerus hydrochaeris* as its main amplifier. In addition, there is evidence that marsupials and small rodents can serve as amplifier hosts for *R. rickettsii* in nature [82].

In Brazil there are approximately 755 species of mammals distributed in all its six biomes, including the Amazon biome [78, 79]. Of the total mammal species, 41% ($\cong$ 311 species) occur in the Amazon biome (**Figure 1**) [2, 76]. These vertebrates are distributed in 11 orders, 51 families and 249 genera [78, 79]. Among the families, Cricetidae is the most diverse, with 144 species [78, 79]. The vast majority of tick species, including all life stages, in Brazil have records on wild mammals of different sizes [17]. In general, small mammals of the orders Rodentia and Didelphimorphia are those that have a greater number of studies in association with ticks, especially the families Cricetidae and Didelphidae [17, 82]. In general, medium and large mammals are parasitized by ticks in all stages (larva, nymph and adult), while in small mammals the stages of larva and nymph are more common. In this last group of hosts we can highlight the cricetids *Akodon* spp., *Calomys* spp., *Oligoryzomys* spp. and *Nectomys* spp. as the most parasitized by immature ticks in nature.

Nine orders and 24 families of wild mammals have representatives in association with ticks in the Brazilian Amazon (**Figure 5**). Rodentia was the most diverse with 16 species of mammals, followed by the orders Carnivora (13 species) and Didelphimorphia (12 species) (**Figure 5**). However, it was the family Didelphidae that presented a greater number of parasitized species, followed by Dasyproctidae (seven species) and Mustelidae (four species). These records corroborate with numerous studies of tick parasitism on wild animals from South America, with emphasis on the orders Rodentia and Didelphimorphia [17, 82, 83].

Hard ticks parasitizing wild mammals in the Brazilian Amazon are represented by five genera: *Amblyomma*, *Ixodes*, *Haemaphysalis*, *Dermacentor* and *Rhipicephalus*. Of these, the genus *Amblyomma* was more frequently recorded with 23 species (**Figure 5**). The orders Rodentia, Pilosa and Didelphimorphia are hosts for highest diversity to *Amblyomma* species 17, 15 and eight, respectively. All these species are also recorded on a variety of wild mammals in Brazil [83–87], except for *A. rotundatum* and *A. dissimile,* which are more specific ticks of cold-blooded animals (Amphibians and Reptiles), although there are occasional reports on mammals in South America [20, 83]. *Amblyomma humerale,* of which the adult stage is more specific to tortoises, immature stages have been found on a variety of small mammals, reptiles and birds [17, 68, 71, 83].

The second most common genus in the Amazon is *Ixodes*, with six reported species: *Ixodes amarali*, *Ixodes bocatorensis*, *Ixodes lasallei*, *Ixodes luciae*, *Ixodes schulzei* and *Ixodes spinosus* (**Table 3**). Although there are two exclusive species on birds in Brazil [65], all species recorded in the Brazilian Amazon parasitize mainly wild mammals [13, 16, 31, 32, 88, 99, 102]. *Ixodes* spp. were found on families Didelphidae, Dasyproctidae, Cricetidae, Myrmecophagidae, Bradypodidae and Cyclopedidae (**Figure 5** and **Table 3**). In general *Ixodes* ticks mainly parasitize

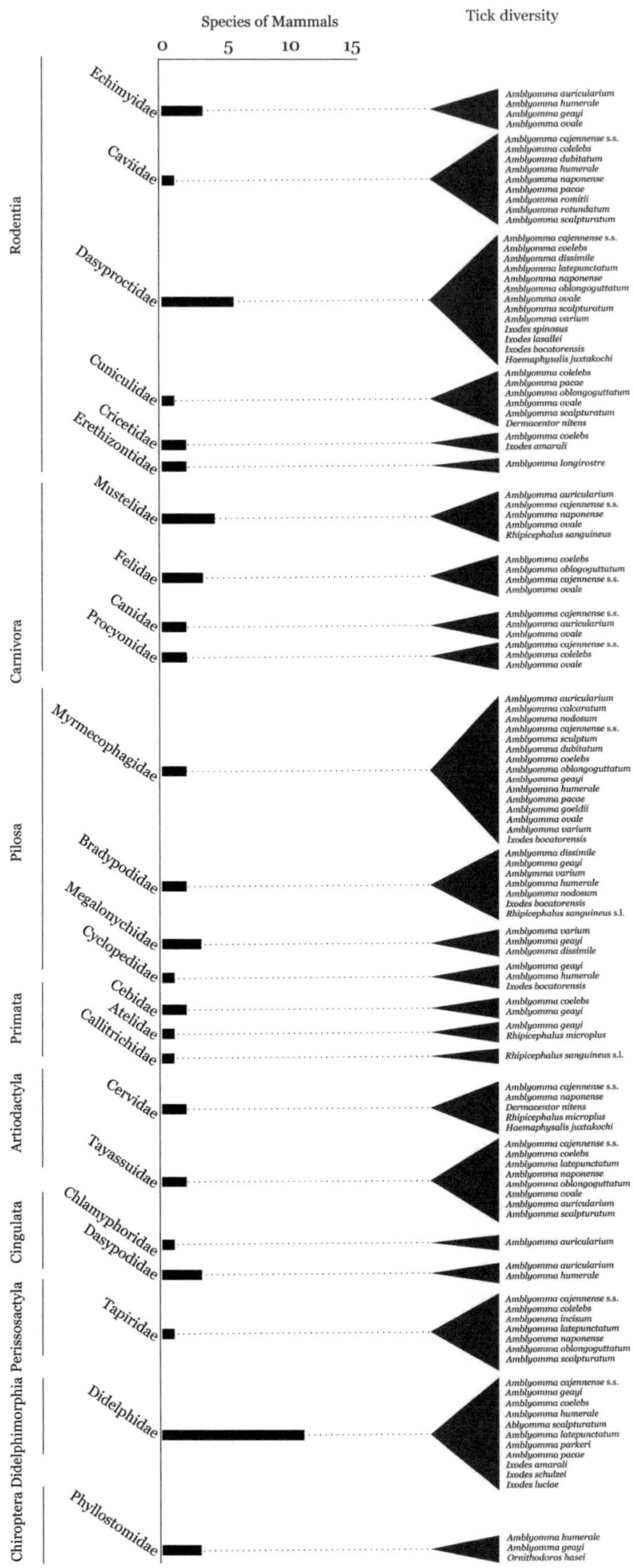

Figure 5.
Diversity of hard ticks parasitizing wild mammals from the Brazilian Amazon.

Tick species	Domestic mammals	Wild mammals	States	References
Amblyomma auricularium		*Cabassous unicinctus*[4], *Dasypus novemcinctus*[4,5,7], *Dasypus septemcinctus*[4,9], *Euphractus sexcinctu*[4,9], *Galactis cuja*[4,9], *Galictis vittata*[4], *Lycalopex vetulus*[9], *Myrmecophaga tridactyla*[9], *Pecari tajacu*[9], *Tamandua tetradactyla*[4], *Thrichomys inermis*[4]	MA[4], MT[5], RO[7], TO[9]	[30, 32, 36, 39]
Amblyomma cajennense sensu stricto	*Equus caballus*[4,5,6,7,8,9], *Sus scrofa*[4,9], *Canis familiaris*[4,9], *Bubalus bubalis*[7]	*Caluromys philander*[5], *Cerdocyon thous*[5], *Didelphis marsupialis*[5], [6] *Dasyprocta azarae*[5], *Hydrochoerus hydrochaeris*[5], *Mazama gouazoubira*[4], *Metachirus nudicaudatus*[4], *Myrmecophaga tridactyla*[5,9], *Nasua nasua*[5], *Panthera onca*[5], *Pecari tajacu*[5], *Pteronura brasiliensis*[3], *Tapirus terrestris*[5], *Tayassu pecari*[5]	AM[3], MA[4], MT[5], PA[6], RO[7], RR[8], TO[9]	[32, 36, 42, 88–91]
Amblyomma calcaratum		*Myrmecophaga tridactyla*[7], *Tamandua tetradactyla*[45,6]	MA[4], MT[5], PA[6]; RO[7]	[36, 38, 42]
Amblyomma coelebs	*Equus caballus*[7]	*Agouti paca*[7], *Cuniculus paca*[6], *Dasyprocta azarae*[5], *Dasyprocta* sp[7]; *Didelphis albiventris*[5], *Didelphis marsupialis*[3,5], *Hydrochoerus hydrochaeris*[7], *Metachirus myosuros*[5], *Nasua nasua*[5], *Neacomys spinosus*[5], *Panthera onca*[5], *Sapajus macrocephalus*[3], *Tayassu pecari*[6], *Pecari tajacu*[7], *Tapirus terrestris*[3,5,7]	AM[3], MT[5], PA[6], RO[7]	[30–32, 38, 42, 88]
Amblyomma dissimile		*Bradypus tridactylus*[3], *Choleopus didactylus*[3], *Coendou* sp[3], *Dasyprocta leporina*[3]	AM[3]	[31, 36, 40]
Amblyomma dubitatum		*Hydrochoerus hydrochaeris*[1,5,7], *Tamandua tetradactyla*[9]	AC[1], MT[5], RO[7], TO[9]	[30, 32, 39, 54]
Amblyomma geayi		*Alouatta nigerrima*[6], *Artibeus lituratus*[3], *Bradypus tridactylus*[3,6], *Bradypus variegatus*[1,3,4,6,9], *Caluromys lanatus*[3], *Choloepus hoffmanni*[7], *Choloepus didactylus*[3,7], *Cyclopes didactylus*[3], *Didelphis marsupialis*[6], *Philander opossum*[3], *Proechimys* sp[3], *Oecomys* sp[6], *Saguinus bicolor*[3], *Tamandua tetradactyla*[3]	AC[1], AM[3], MA[4], PA[6], RO[7], TO[9]	[30, 31, 33, 36, 39, 40, 42, 92]
Amblyomma goeldii		*Tamandua tetradactyla*,[3,6], *Bradypus tridactylus*[3]	AM[3], PA[6]	[31, 40, 42]
Amblyomma humerale		*Bradypus tridactylus*[3], *Chrotopterus auritus*[3], *Cyclopes didactylus*[7], *Dasypus novemcinctus*[5,6], *Didelphis marsupialis*[3,5,6], *Hydrochoerus*	AC[1], AM[3], MT[5], PA[6], RO[7]	[30–33, 38–40, 42, 54, 88]

Tick species	Domestic mammals	Wild mammals	States	References
		hydrochaeris[1], *Marmosa constantiae*[5], *Marmosa* sp[3], *Metachirus myosuros*[5], *Philander opossum*[3], *Proechimys roberti*[5], *Tamandua tetradactyla*[3,6]		
Amblyomma incisum		*Tapirus terrestris*[7]	RO[7]	[38, 93]
Amblyomma latepunctatun		*Tapirus terrestris*[3,6,7], *Tayassu pecari*[3], *Dasyprocta fuliginosa*[3], *Didelphis marsupialis*[3]	AM[3]; PA[6]; RO[7]	[31, 38, 93]
Amblyomma longirostre		*Coendou* sp[3], *Coendou prehensilis*[6,7], *Coendou nycthemera*[6]	AM[3], PA[6], RO[7]	[30, 31, 33, 40, 42]
Amblyomma naponense		*Dasyprocta fuliginosa*[3], *Galictis* sp[3], *Hydrochoerus hydrochaeris*[1], *Mazama gouazoubira*[4], *Myoprocta acouchy*[3], *Tapirus terrestris*[7], *Tayassu pecari*[3,4]	AC[1], AM[3], MA[4], RO[7]	[31, 36, 38, 54]
Amblyomma nodosum		*Tamandua tetradactyla*[4,5,6,7,9]; *Bradypus variegatus*[9], *Myrmecophaga tridactyla*[9]	MA[4], MT[5], PA[6], RO[7], TO[9]	[30, 32, 38–40, 42]
Amblyomma oblongoguttatum	*Canis familiaris*[4,6,9], *Sus scrofa*[7]	*Agouti paca*[7], *Cuniculus paca*[9], *Dasyprocta* sp[7], *Myrmecophaga tridactyla*[9], *Tayassu tacaju*[7], *Tayassu pecari*[3,5,6,7], *Dasyprocta fuliginosa*[3], *Mazama americana*[3], *Tapirus terrestris*[5,7], *Panthera onca*[7]	AM[3], MA[4], MT[5], PA[6], RO[7], TO[9]	[31, 32, 38, 39, 42, 91, 94, 95]
Amblyomma ovale	*Canis familiaris*[3,4,9], *Sus scrofa*[7]	*Panthera onca*,[3,5,7,9], *Tapirus terrestres*[3], *Leopardus pardalis*[3], *Proechimys quadruplicatus*[3], *Proechimys* sp[3], *Tapirus terrestres*[3], *Pauxi tuberosa*[5], *Puma concolor*[5], *Procyon cancrivorus*[7], *Tayassu tacaju*[7], *Tayassu pecari*[7], *Agouti paca*[7], *Dasyprocta* sp[7], *Eira barbara*[7], *Nasua nasua*[7], *Cerdocyon thous*[9]	AM[3], MA[4], MT[5], RO[7], TO[9],	[30–32, 38, 39, 42, 91, 94]
Amblyomma pacae	*Canis familiaris*[7]	*Hydrochoerus hydrochaeris*[1], *Cuniculus paca*[5], *Tamandua tetradactyla*[6], *Monodelphis glirina*[6], *Didelphis marsupialis*[7]	AC[1], MT[5], PA[6], RO[7]	[38, 42, 54, 92]
Amblyomma parkeri		*Didelphis marsupialis*[5]	MT[5]	[32]
Amblyomma romitii		*Hydrochoerus hydrochaeris*[5,6,7]	MT[5], PA[6], RO[7]	[32, 41, 96]
Amblyomma rotundatum		*Hydrochoerus hydrochaeris*[i], *Xenodon merremii*[9]	RO[7], TO[9]	[30–33, 36, 38, 39, 41, 42, 54]
Amblyomma scalpturatum	*Canis familiaris*[7], *Sus scrofa*[7]	*Agouti paca*[7], *Dasyprocta fuliginosa*[3], *Didelphis marsupialis*[5], *Pecari tajacu*[5,7],	AM[3], MT[5], RO[7]	[30–32, 38, 88]

Tick species	Domestic mammals	Wild mammals	States	References
		Tapirus terrestris[5,7], *Hydrochoerus hydrochaeris*[7]		
Amblyomma sculptum	*Equus caballus*[4,7,9]	*Myrmecophaga tridactyla*[9]	MA[4], RO[7], TO[9]	[88]
Amblyomma varium		*Bradypus tridactylus*[3], *Bradypus sp*[3], *Bradypus variegatus*[7] *Choleopus didactylus*[3,7], *Choleopus hoffmanni*[3,7], *Choloepus* sp[7] *Dasyprocta aguti*[3], *Tamandua tetradactyla*[3]	AM[3], PA[6], RO[7]	[30, 31, 33, 38, 40, 97]
Dermacentor nitens	*Bos taurus*[4], *Canis familiaris*[4,7], *Equus caballus*[4,7,8,9] *Capra hircus*[4], *Ovis aires*[4]	*Agouti paca*[7], *Mazama gouazoubira*[9]	MA[4], RO[7], RR[8], TO[9]	[36, 38, 39, 95, 98]
Ixodes amarali		*Hylaeamys megacephalus*[6], *Monodelphis glirina*[6], *Monodelphis touan*[6],	PA[6]	[99]
Ixodes bocatorensis		*Bradypus tridactylus*[3], *Cyclopes didactylus*[3], *Dasyprocta leporina*[3], *Tamandua tetradactyla*[3]	AM[3]	[16]
Ixodes spinosus		*Dasyprocta* sp[7], *Myoprocta pratti*[1]	RO[7,] AC[1]	[13]
Ixodes lasallei		*Dasyprocta variegata*[7]	RO[7]	[16]
Ixodes luciae		*Didelphis marsupialis*[1,3,4,7], *Marmosa murina*[6], *Marmosa demerarae*[3], *Monodelphis* sp[7], *Oecomys* sp.[7], *Orysomys* sp.[7], *Oligorysomys* sp.[7], *Philander opossum*[6]	AC[1], AM[3], MA[4], PA[6], RO[7]	[31, 36, 38, 100]
Ixodes schulzei		*Monodelphis touan* [6]	PA[6]	[92]
Haemaphysalis juxtakochi	*Canis familiaris*[4,7],	*Mazama americana*[3,9], *Mazama* p[7], *Dasyprocta azarae*[5]; *Dasyprocta fuliginosa*[3],	AM[3], MA[4], MT[5]; RO[7], TO[9]	[31, 36, 38, 39, 42]
Rhipicephalus microplus	*Bos taurus*[4,7,8,9], *Canis familiaris*[4,7,9], *Felis catus*[7], *Equus caballus,*[4,7,8,9], *Capra hircus*[4,7], *Ovis aires*[4]	*Alouatta puruensis*[7], *Mazama gouazoubira*[5,9], *Mazama sp.*[7], *Tapirus terrestres*[7,9], *Tayassu* sp[7]	MA[4], MT[5]; RO[7], RR[8]; TO[9]	[32, 33, 36, 38, 39, 95, 98]
Rhipicephalus sanguineus sensu lato	*Canis familiaris*[1,3,4,7,9]	*Bradypus tridactylus*[3], *Galictis cuja*[4], *Saquinus bicolor*[3]	AC[1], AM[3], MA[4], RO[7]	[30–33, 36, 38, 39, 95, 98, 101]

[1]Acre – AC.
[2]Amapá – AP.
[3]Amazonas – AM.
[4]Maranhão – MA.
[5]Mato Grosso – MT.
[6]Pará – PA.
[7]Rondônia – RO.
[8]Roraima – RR.
[9]Tocantins – TO.

Table 3.
Records of ticks in the Amazon biome, Brazil, according to tick species, hosts (domestic and wild animals), states, references.

rodents (e.g., Cricetidae) in the larvae and nymph stages, with adults feeding mostly on marsupials (e.g., *Didelphis* spp., *Monodelphis* spp.) [17, 83], and agoutis (e.g., *Dasyprocta*) [13, 16]. Most species were found parasitizing *Monodelphis glirina* (*I. amarali*, *I. schulzei*, *I. luciae*) and *Monodelphis touan* (*I. amarali*, *I. schulzei*), with some reports of *I. amarali* on the rodent *Hylaeamys megacephalus* [30, 38, 92, 99, 100]. Also on rodents the species *I. spinosus, I. lasallei* and *I. bocatorensis* have been recorded [30, 31, 38], although these previous records referred to them erroneously as *Ixodes fuscipes*, which according to more recent data, does not occur in the Amazon biome (**Figure 5** and **Table 3**) [13, 16].

In general, the greatest diversity of ticks was recorded on *T. tetradactyla* with 10 species, followed by the rodent *H. hydrochaeris* with nine species. These records indicates the importance of these hosts for maintaining local tick diversity, in addition to act as dispersers of these ectoparasites. Additional hosts have also been shown to be important in maintaining diversity of tick in the Amazon: *Tapirus terrestris*, *Tayassu pecari*, *Pecari tajacu*, *Agouti paca*, *Dasyprocta fuliginosa*, *Bradypus tridactylus*, *Bradypus variegatus*, *Panthera onca*, *Nasua nasua*, *Choleopus didactylus*, *Choloepus hoffmanni*, *Cyclopes didactylus*, *Dasyprocta azarae*, *Dasyprocta leporina*, *Philander opossum*, *M. glirina*, *M. touan*, *Mazama gouazoubira*, *Mazama americana* and *M. tridactyla* (**Table 3**), because they are parasitized by more than one species of ticks. There is also a record of *A. cajennense s.s.* on *Pteronura brasiliensis* a semi-aquatic animal [89].

Interestingly, three species of ticks commonly found on domestic animals have also been found parasitizing wild animals: *Rhipicephalus sanguineus* sensu lato, *R. microplus* and *D. nitens*. Although occasional, the encounter of these species of ticks on wild animals is possible, especially when wild animals coexist with domestic animals infested by ticks (e.g., pastures, corrals, houses). A total of 11 species of ticks have been recorded parasitizing domestic animals in Amazon (**Table 3**).

5. Hard ticks associate with humans

Spotted fevers caused by *R. rickettsii* and *Rickettsia parkeri* are currently the only confirmed tick-borne disease affecting humans in the country, and *A. sculptum*, *A. aureolatum*, *A. ovale* are the main vectors. The disease is severe and highly lethal when caused by the bacterium *R. rickettsii* (vectors *A. sculptum* and *A. aureolatum*) and moderate, non-lethal, when caused by *R. parkeri*, vectored mainly by *A. ovale* [10]. Only *A. sculptum* and *A. ovale* have been recorded in the Brazilian Amazon, the first in rare reports and the second frequently recorded and populations established in this biome. In fact, considering only the tick adult stage, *A. ovale* has been reported as the most common human-biting tick in the Amazon [38].

In the last century, papers about ticks on human beings have been published in the Brazilian Amazon, however these studies are fragmented and scarce. At least 14 species of hard ticks have already been found and documented parasitizing humans within the limits of the Brazilian Amazon. Of these, the vast majority (11 species) belong to the genus *Amblyomma*: *A. cajennense* s.s., *A. coelebs*, *A. dissimile*, *A. latepunctatum*, *A. naponense, A. oblongoguttatum*, *A. ovale*, *A. romitii*, *A. rotundatum*, *A. scalpturatum* and *A. sculptum*. The other three species belong to the genus *Rhipicephalus,* including *R. microplus* and *R. sanguineus* s.l., and the genus *Haemaphysalis* with only *H. juxtakochi*. The public health importance of human–tick associations in the Amazon biome is unclear and further research are needed to clarify the issue.

Nymphs of *H. juxtakochi* have been reported on humans in the states of Rondônia (municipality not precisely indicated) and Amazonas, in the municipality of Santa Isabel do Rio Negro [31, 38]. *H. juxtakochi* has been found infected with the bacterium *Rickettsia rhipicephali* in the Amazon region, more precisely in the municipalities of Monte Negro and Confresa, states of Rondônia and Mato Grosso, respectively [42, 93]. Human parasitism by *R. microplus* was expected, since this species can parasitize humans when it comes into direct contact with domestic cattle, as reported by [38]. Therefore, its importance is much more economical, causing severe losses to livestock in the country than for public health.

Despite sporadic records of *R. sanguineus* s.l. on humans in Brazil, this association in the Brazilian Amazon requires confirmation. Records of approximately 4,020 *R. sanguineus s.l.* (larva, nymph, and adult), supposedly collected on humans, in the state of Pará by Serra-Freire et al. [103] do not corroborate the common findings in country. The bioecology of this endophilic and introduced tick is well studied. In other countries, *R. sanguineus* s.l. is vector of some zoonotic agents for humans (*Rickettsia conorii*, *Rickettsia massiliae* and *R. rickettsii*) [83]. Adults of *A. cajennense* s.s. have been reported on humans in the municipality of Monte Alegre (Pará state) [90]. Reports of *Amblyomma cajennense* sensu lato parasitizing humans published by Martins et al. [90] in the municipalities of Sinop and Tucuruí in the states of Mato Grosso and Pará, is possibly *A. cajennense* s.s. taking into account the area of occurrence of this species. This tick is aggressive to humans. In the Amazon biome the *R. bellii* bacterium was detected in this species in Mato Grosso state and *R. amblyommatis* in Mato Grosso, Maranhão and Rondônia state [32, 42, 53, 83, 88, 90, 91, 104, 105].

Nymphs and adult of *A. coelebs* were found on humans in Rondônia state [38]. An adult of this tick was found on human in Roraima state, municipality of Caroebe [106]. The nymph and adult stages of *A. colelebs* were also found on humans in Amazonas state, in the municipalities of Coari, Fonte Boa, Presidente Figueiredo and Santa Isabel do Rio Negro [31, 107]. There are reports of infection by *R. amblyommatis* in *A. coelebs* in the Amazon region in the states of Rondônia and Mato Grosso, respectively [32, 53, 88]. Adult of *A. dissimile* was recorded on humans in the Maicurú River/Amazon biome, located in Pará state [106]. Adults of the species *A. latepunctatum* have been reported on humans in Coari and Santa Isabel do Rio Negro, both municipalities located in the state of Amazonas [31].

Adults of the tick *A. naponense* (published as *Amblyomma mantiquirense*) were found on humans in the state of Pará (municipality not reported) [108]. Human parasitism also by adults was later reported on the Maicurú River, located in this same state [106]. However, nymphs have been found on humans in the states of Rondônia (municipality not specified precisely) and Amazonas, in the municipalities of Coari and Santa Isabel do Rio Negro [31, 38]. The bacteria *R. bellii* and *Rickettsia* sp. strain PA were identified in this species in the Amazon region of the municipalities of Santarém and Rurópolis, in the state of Pará [42]. *Rickettsia* sp. strain Tapirapé was found in this tick in the municipalities of Confresa and Rio Branco, in the states of Mato Grosso and Acre, respectively [42, 54].

Adults of the species *A. oblongoguttatum* were recorded parasitizing humans in the state of Pará, in the Maicurú River and in the municipality of Uruará [106, 108, 109]. Human parasitism by nymphs and adults of this tick was recorded in the state of Rondônia (municipality not specified with precision) [38]. However, the nymphal stage of this species was later recorded on humans in the municipality of Monte Negro in the same state [106]. Human parasitism by adult of this tick was also recorded in the municipality of Caroebe, state of Roraima [106]. Nymphs and adults of this species have been recorded parasitizing humans in three municipalities (Coari, Manacapuru and Santa Isabel do Rio Negro), all located in the state of Amazonas [31, 106, 107]. On the other hand, the life cycle of this tick has already been studied under laboratory conditions [110]. The bacteria *R. bellii* and

R. amblyommatis were detected in this species in the Amazon in the municipalities of Governador Jorge Teixeira and Pimenta Bueno, in the state of Rondônia [53, 105].

Adults of the tick *A. ovale* have been reported to parasitize humans in several areas of the state of Rondônia [38]. As previously reported in this chapter, adults of *A. ovale* parasitize mainly domestic and wild carnivores, while immature (larva and nymph) parasitize rodents of the families Cricetidae and Echimyidae, with sporadic reports on wild birds [17, 75, 83]. This preference for such host groups was observed in the laboratory [111]. This tick is a vector of the protozoan *Hepatozoon canis* and the bacterium *R. parkeri*, agents of importance in veterinary and human medicine, respectively [83]. In the Amazon region, the bacterium *R. bellii* was identified in this species in the municipalities of Governador Jorge Teixeira and Monte Negro, in the state of Rondônia; and in the municipality of Cururupu, in the state of Maranhão [53, 91].

The adult stage of the species *A. romitii* (published as *Amblyomma tasquei*) was found parasitizing humans in the north of the state of Pará (municipality not reported) [108]. In this same state, the larval and adult stages of this tick were found on humans in the municipality of Rurópolis [96, 112]. The life cycle of this tick was studied under laboratory conditions [113].

Parasitism by adults of *A. rotundatum* was recorded in the municipality of Belém in the state of Pará on a human who worked in frequent contact with reptiles in the Zoobotanical Park of the Museu Paraense Emílio Goeldi [114].

Adults of the *A. scalpturatum* have been reported to parasitize humans in the state of Pará (municipality not reported) [108], and in the Maicurú River located in this state [93, 106]. Human parasitism by the nymphal and adult stages of this species has been reported in the states of Mato Grosso (municipality of Jauru) and Rondônia (municipality not precisely indicated), however in this latter state it was not specified whether these adult ticks were fixed or walking on humans [38, 93]. The nymphal stage was also collected from a human who was probably infested with this tick in the municipality of Porto Velho in the state of Rondônia [115]. The nymph and adult stages of this species were also found parasitizing humans in the state of Amazonas, in the municipalities of Coari, Fonte Boa, Jutaí, Presidente Figueiredo and Santa Isabel Rio Negro [31, 107]. In the Amazon biome, two species of rickettsiae (*R. bellii* and *R. amblyommatis*) were detected in *A. scalpturatum,* in the municipalities of Governador Jorge Teixeira and Sinop, states of Rondônia and Mato Grosso, respectively [53, 88].

Human parasitism by adults of *A. sculptum* was recorded in the municipality of Jauru in the state of Mato Grosso [90]. The Jauru territory comprises 90% of the Amazon biome and 10% of the Cerrado biome, so this human record in this region is in accordance with the transition area of the occurrence of this species in sympatry with *A. cajennense* s. s. [90]. The bioecology of this native tick has been extensively studied due to its aggressive parasitism towards domestic animals and humans in the Brazilian territory. In the geographical area in which this species occurs, it completes an annual generation, with larvae occurring mostly during autumn, nymphs during winter, and adults during spring and summer, with larval behavioral diapause being the main regulating factor of its life cycle [116, 117]. Besides being considered a pest for domestic animals (dogs, cattle and horses), the species can transmit to humans the Brazilian Spotted Fever caused by the bacterium *R. rickettsii*, being this the most important zoonotic disease transmitted by ticks within the limits of the National territory. Additionally, *A. sculptum* has also been identified with the bacterium *R. amblyommatis* in the municipality of Pium, state of Tocantins [118].

There is a trend of seasonal behavior for some species of *Amblyomma* in the Amazon biome, with immatures predominating in the dry period (June to October), and adults in the rainy period (October to March) [119]. Similar to the pattern observed for *A. sculptum* in areas outside the Amazon biome [116].

6. Soft ticks in the Brazilian Amazon

Taxonomy of Argasidae is currently questionable and relies on five schemes that divide the family in up to ten genera [120]. Achieving a consensus between soft tick taxonomists depends now chiefly on the molecular characterization of early collected type specimens for some genera. For instance, elucidating the status of pivotal taxa such as *Alectorobius* is mandatory if we are to understand the systematics of the Argasidae, particularly in the American Continent. Considering a practical approach, in this chapter we adopt but not necessarily endorse the classification of soft ticks into five genera, namely: *Antricola, Argas, Nothoaspis, Ornithodoros*, and *Otobius* [121].

The fauna of argasid ticks in Brazil is currently composed by 24 species [15]. Ticks of this family parasitize terrestrial vertebrates including amphibians in this country [25, 26]. With the exception of larvae from an undetermined *Ornithodoros* sp. collected on *Potus flavus* (Procionidae) [84], reports of soft ticks parasitizing mammals in the Brazilian Amazon are few and almost restricted to larvae collected on bats. Adults and nymphs have been collected either inside bat-inhabited caves or over massive rock formations where bats shelter (**Table 4**).

The first record of a soft tick in the Brazilian Amazon was published by North American entomologists Robert A. Cooley and Glen M. Kohls back in 1941. They received a tick collected inside a bat-inhabited three hole at Marajó Island (Pará state) and identified it as female of *Ornithodoros hasei* (mentioned as *Ornithodoros dunni*) [126]. Nowadays, we know that *O. hasei* is a wide spread species in Brazil and that three-roosting bats *Artibeus planirostris* and *Noctilio* spp. could act as main hosts [87, 130–132]. With the exception of recent collection of larvae on *A. planirostris* [87], knowledge on the distribution of *O. hasei* along the Brazilian Amazon is still poorly vague.

Tadarida laticaudata were the first bats reported to be parasitized by soft ticks in the Brazilian Amazon [133]. This report referred to *Ornithodoros setosus,* which was recently reclassified as *Nothoaspis setosus* [124]. In 1972, the bats *Noctilio labialis* and *T. laticaudata* were reported to be parasitized by *Ornithodoros stageri* in the Brazilian Amazon [129], which was recently confirmed [11]. A report in the Brazilian Amazon is a remarkable fact for *O. stageri*, since it also has distribution in Southern United States and Mexico [126, 134].

Bat inhabited caves constitute excellent niches to find argasid ticks. In particular, special, large colonies of insectivorous bats dwelling inside small chambers create high temperature conditions (28-40°C) where hundreds of *Antricola*, *Nothoaspis* and *Ornithodoros* ticks might occur [135]. Between 2004 to 2010 collections of ticks performed in hot caves from Porto Velho (Rondônia) fostered the description of two novel species (e.g., *Nothoaspis amazoniensis* and *Ornithodoros rondoniensis*) [122, 123]; the redescription of larvae and description of postlarval stages of *Ornithodoros marinkellei* [127], and the expansion of geographical distribution of *Antricola delacruzi* and *Antricola guglielmonei* into Brazilian Amazon [122]. At least for two species, *O. marinkellei* and *O. rondoniensis*, further collections performed in caves from Pará state underpined a larger distribution along the Amazon ecosystems [128].

Most amazing feature of soft ticks inhabiting hot caves, is that adaptation to this particular milieu seems to have modified their morphology and biology drastically. For instance, evidence showing that adults of *Antricola* ticks lack the capacity to

Species	State	Locality	Milieu of collection/Host	Reference
Antricola delacruzi	Rondônia	Porto Velho	Bat inhabited cave/*Pteronotus* spp.	[122]
Antricola guglielmonei	Rondônia	Porto Velho	Bat inhabited cave/*Pteronotus* spp	[122]
Nothoaspis amazoniensis	Rondônia	Porto Velho	Bat inhabited cave/*Pteronotus parnellii*	[123]
Nothoaspis setosus	Rondônia	Monte Negro	Rock formation/ *Molossus molossus*	[124]
Ornithodoros cavernicolous	Rondônia	Monte Negro	Bat inhabited caves/Unknown	[125]
Ornithodoros hasei	Amapá	unknown	Mistneted bat *Artibeus planisrostris*	[87]
	Pará	Marajó Island	Bat-inhabited three hole/ Unknown	[126]
Ornithodoros kohlsi	Rondônia	Monte Negro	Mistneted bat *Molossops mattogrossensis*	[30]
Ornithodoros marinekellei	Rondônia	Porto Velho	Bat inhabited cave/*Pteronotus* spp	[127]
	Pará	Carajás, Paraupebas, Canaã dos Carajás	Bat inhabited caves/Unknown	[128]
Ornithodoros peropteryx	Rondônia	Monte Negro	Bat inhabited cave/Unknown	[125]
Ornithodoros rondoniensis	Rondônia	Porto Velho	Bat inhabited cave/Unknown	[122]
	Pará	Carajás, Paraupebas, Canaã dos Carajás	Bat inhabited caves/Unknown	[128]
Ornithodoros stageri	Rondônia	Unknown	Unknown/*Noctilio albiventris*, *Nyctinomops laticaudatus*	[129] [11]

Table 4.
Soft ticks and their respective hosts occurring in the Brazilian Amazon.

digest blood has been gained after transcriptomic analyses of their saliva [136]. In fact, scoop-like short mouth parts suggest that adults of *Antricola* do not suck blood [137]. Moreover, adults of *Antricola*, *O. marinkellei* and *O. rondoniensis* possess huge spiracular plates [122, 127, 137], perhaps necessary to thrive in such hot and extremely humid environments.

Except for larvae of *Ornithodoros kohlsi* collected on the bat *Molossops mattogrossensis* [30], prospections performed by our group between 2016 and 2019 focused mainly in the search for soft ticks inside natural cavities over massive rock formations. During these expeditions, *Ornithodoros cavernicolous* and *Ornithodoros peropteryx* were collected for the first time in caves from Monte Negro (Rondônia), extending their distribution of both species to the Brazilian Amazon [125]. Furthermore, analyses performed on large larvae isolated from *Molossus molossus* bats clarified that *O. setosus* was incorrectly classified in its original description. Indeed, *O. setosus* matches morphologically and molecularly within the genus *Nothoaspis* therefore the statement of *N. setosus* n. comb. was proposed [124]. It is important to note that our last collections performed at Monte Negro included several morphotypes of *Ornithodoros* pending formal description. Consequently, the fauna of soft ticks occurring in the Brazilian Amazon is likely to increase soon.

7. Conclusion

Ticks parasitize a wide variety of vertebrates around the world such as amphibians, reptiles, mammals and birds, including humans. Although there are a variety of studies of parasitism by ticks on animals and humans in the different ecoregions of Brazil, in the Amazon biome they are scarce and fragmented. Because of this, it is possible to infer that the diversity of ticks in the Brazilian Amazon is underestimated. In the Amazon, amphibians and reptiles were important hosts for *A. rotundatum*, *A. dissimile* and *A. humerale*, hard ticks common in these hosts in other regions of the country. Birds and mammals, on the other hand, proved important for the maintenance and dispersion of over 30 species of hard ticks in the Amazon, but there are still many gaps between hosts and their ticks in this region. Interestingly, immature *A. humerale* appears to be frequent on wild birds in this biome. No soft ticks have been reported parasitizing amphibians, reptiles, birds and/or non-flying mammals (except for a single record of *Ornithodoros* sp. on *P. flavus*), possibly due to scarcity of more studies in this biome. Some studies show that bats seem to have an important role spreading soft tick populations along Amazonian caves and that the diversity described for this group is still very poor. Two vector species of spotted fevers were found in the Amazon biomes parasitizing humans (*A. sculptum* and *A. ovale*). However, to date, there are no reports of BSF in humans in the region. Finally, it is of paramount importance that researchers (acarologists and epidemiologists) direct their attention to the Amazon biome, in order to fill the numerous existing gaps in the diversity of ticks in Brazil and prevent possible outbreaks of diseases transmitted by these ectoparasites to animals and humans.

Conflict of interest

No conflict of interest declared.

Author details

Hermes R. Luz[1,2*], Thiago F. Martins[3,4], Sebastián Muñoz-Leal[5], Francisco B. Costa[6], Sérgio L. Gianizella[7], João Luiz H. Faccini[8] and Marcelo B. Labruna[3]

1 Faculdade Florence, São Luís, MA, Brazil

2 Programa de Pós-graduação em Biotecnologia (RENORBIO), Universidade Federal do Maranhão – UFMA, MA, Brazil

3 Departamento Medicina Veterinária Preventiva e Saúde Animal, Faculdade de Medicina Veterinária e Zootecnia, Universidade de São Paulo, SP, Brazil

4 Departamento de Laboratórios Especializados, Superintendência de Controle de Endemias, Secretaria de Estado da Saúde de São Paulo, São Paulo, SP, Brazil

5 Departamento de Patología y Medicina Preventiva, Facultad de Ciencias Veterinarias, Universidad de Concepción, Chillán, Ñuble, Chile

6 Departamento de Patologia, Curso de Medicina Veterinária, Universidade Estadual do Maranhão, São Luís, MA, Brazil

7 Laboratório de Zoologia, Departamento de Biologia, Instituto de Ciências Biológicas, Universidade Federal do Amazonas, Manaus, AM, Brazil

8 Departamento de Parasitologia Animal, Universidade Federal Rural do Rio de Janeiro, RJ, Brazil

*Address all correspondence to: hermesluz@globomail.com

References

[1] Silva JMC, Rylands AB, Fonseca GAB. The fate of the Amazonian areas of endemism. Conservation Biology. 2005; 19: 689-694

[2] MMA. Ministério do Meio Ambiente. 2018. Acessed: 03/10/2120

[3] Vieira ICG, Toledo PM, Silva JMC, Higuchi H. Deforestation and threats to the biodiversity of Amazonia. Brazilian Journal of Biology. 2008; Suppl 0.

[4] Evans KL, Newson SE, Storch D, Greenwood JJ, Gaston KJ. Spatial scale, abundance and the species-energy relationship in British birds. Journal of Animal Ecology. 2008; 77: 395-405

[5] Rylands AB. Amazonia. In: *Wilderness: Earth's last wild places*. Agrupación Serra Madre, S. C. Mexico: CEMEX, 2002; 56–107

[6] Parrotta JA, Wildburger C, Mansourian S. Understanding relationships between biodiversity, carbon, forests and people: the key to achieving REDD+ objectives. A global assessment report prepared by the Global Forest Expert Panel on Biodiversity, Forest Management and REDD. 2012; 31: 161, 2012

[7] Duscher GG, Fuehrer HP, Kübber-Heiss A. Fox on the run – molecular surveillance of fox blood and tissue for the occurrence of tick-borne pathogens in Austria. Parasites & Vectors. 2014; 7: 521

[8] de la Fuente J. Controlling ticks and tick-borne diseases looking forward. Ticks and Tick Borne Diseases. 2018; 9: 1354–1357

[9] Massard CL, Fonseca AH. Carrapatos e doenças transmitidas, comuns ao homem e aos animais. A Hora Veterinária. 2004. 135; 1: 15-23

[10] Szabó MP, Pinter A, Labruna MB. Ecology, biology and distribution of spotted-fever tick vectors in Brazil. Front Cell Infect Microbiol. 2013; 3: 27

[11] Dantas-Torres F, Martins TF, Muñoz-Leal S, Onofrio VC, Barros-Battesti DM. Ticks (Ixodida: Argasidae, Ixodidae) of Brazil: Update species checklist and taxonomic keys. Ticks and Tick Borne Diseases. 2019; 10: 101252

[12] Peñalver EA, Arillo X, Delclòs D, Peris DA, Grimaldi SR, Anderson PC. Nascimbene, and R. Pérez-de la Fuente. Ticks parasitised feathered dinosaurs as revealed by Cretaceous amber assemblages. Nature Communications.2017; 8: 1924

[13] LabrunaMB, OnofrioVC, Barros-Battesti, GianizellaSL, VenzalJM, GuglielmoneAA. Synonymy of *Ixodes aragaoi* with *Ixodes fuscipes*, and reinstatement of *Ixodes spinosus* (Acari: Ixodidae). Ticks and Tick-borne Diseases. 2020; 11: 101349

[14] Martins TF, Luz HR, Muñoz-Leal S, Ramirez DG, Milanelo L, Marques S, et al. A new species of *Amblyomma* (Acari: Ixodidae) associated with monkeys and passerines of the Atlantic rainforest Biome, southeastern Brazil. Ticks and Tick-borne Diseases, 2019; 10: 1-10

[15] Muñoz-Leal S, Martins MM, Nava S, Landulfo GA, Simons SM, Rodrigues VS, et al. *Ornithodoros cerradoensis* n. sp. (Acari: Argasidae), a member of the *Ornithodoros talaje* (Guérin-Méneville, 1849) group, parasite of rodents in the Brazilian Savannah, Ticks and Tick-borne Diseases. 2020; 11: Issue 5

[16] OnofrioVC, Guglielmone AA, BattestiDM, GianizellaSL, MarciliA, QuadrosRM, et al. Description of a new species of *Ixodes* (Acari: Ixodidae) and

first report of *Ixodes lasallei* and *Ixodes bocatorensis* in Brazil. Ticks and Tick-borne Diseases. 2020; 11: 101423

[17] Guglielmone AA, Robbins RG, Apanaskevich DA, PetneyTN, Estrada-Peña A, Horak IG. The hard ticks of the world: (Acari: Ixodida: Ixodidae). London: Springer; 2014; 738 p.

[18] Esser HJEA, Herre R, Kays YL, Jansen PA. "Local Host-Tick Coextinction in Neotropical Forest Fragments." International Journal for Parasitology 2019; 49: 225–233

[19] Ferreira GC, Sturaro MJ, Peloso PLV. Amphibians and reptiles from Floresta Nacional de Pau-Rosa, Amazonas, Brazil: an important protected area at the heart of Amazonia. Acta Amazonica. 2017; 47: 259 – 268

[20] Luz HR, Faccini JLH. Parasitim on anurans in Brazil (in portuguese). Vet Zootec. 2013; 20: 100-111

[21] Horta MC, Saraiva DG, Oliveira GMB, Martins TF, Labruna MB. *Rickettsia bellii* in *Amblyomma rotundatum* ticks parasitizing *Rhinella jimi* from northeastern Brazil. Microbes Infect. 2015;17: 856–858

[22] Alcântara EP, Silva CF, Ávila RW, Pacheco RC, Martins TF, Muñoz-Leal S, et al. Ticks (Acari: Argasidae and Ixodidae) infesting amphibians and reptiles in Northeastern Brazil. Systematic and Applied Acarology. 2018; 23: 1497–1508

[23] Oda FH, Kitagawa C, Noronha JC, Rodrigues DJ, Martins TF, Valadão MC, et al. *Amblyomma* species infesting amphibians and reptiles in the seasonally dry Amazon forest, with new host records for *Amblyomma rotundatum* (Acari: Ixodida: Ixodidae). Systematic and Applied Acarology. 2018; 23: 387-392

[24] Fonseca, MS, Campanharo TB, Borges AAS, Castilho VO, Doria TB, ParanáSBM, et al. Ticks and Associated Pathogens From Rescued Wild Animals in Rainforest Fragments of Northeastern Brazil. Frontiers in Veterinary Science. 2020; 7: 177

[25] Barros-Battesti DM, Landulfo GA, Luz HR, Marcili A, Onofrio VC, Famadas KM. *Ornithodoros faccinii* n. sp. (Acari: Ixodida: Argasidae) parasitizing the frog *Thoropa miliaris* (Amphibia: Anura: Cycloramphidae) in Brazil. Parasites & Vectors 2015; 8: 268

[26] Muñoz-Leal S, Toledo LF, Venzal JM, Marcili A, Martins TF, Acosta ICL, et al. Description of a new soft tick species (Acari: Argasidae: *Ornithodoros*) associated with stream-breeding frogs (Anura: Cycloramphidae: *Cycloramphus*) in Brazil. Ticks and Tick-Borne Diseases. 2017; 8: 682-692

[27] Labruna MB, Camargo LMA, Terrassini FA, Schumaker TTS, Camargo EP, Notes on Parasitism by *Amblyomma humerale* (Acari: Ixodidae) in the State of Rondônia, Western Amazon, Brazil. Journal of Medical Entomology. 2002; 39: 814–817

[28] Polo G, Luz HR, Martins TF, Regolin AL, Silva HR, Winker G, et al. Distribution modeling of *Amblyomma rotundatum* and *Amblyomma dissimile* in Brazil: estimates of environmental suitability. Parasitology Research. 2020; https://doi.org/10.1007/s00436-020-06924-9

[29] Luz HR, Silva-Santos E, Costa-Campos CE, Acosta I, Martins TF, Munhoz-Leal S, et al. Detection of *Rickettsia* spp. in ticks parasitizing toads (*Rhinella marina*) in the northern Brazilian Amazon. Experimental and Applied Acarology. 2018; 75: 309-318

[30] Martins TF, Venzal JM, Terrassini FA, Costa FB,; Marcili A, Camargo LMA, et al. New tick records from the state of Rondônia, western Amazon, Brazil. Experimental and Applied Acarology 2013; 62: 121-128

[31] Gianizella SL, Martins TF, Onofrio VC, Aguiar NO, Gravena W, Nascimento CAR, et al. Ticks (Acari: Ixodidae) of the state of Amazonas, Brazil. Experimental & Applied Acarology. 2018; 74: 177-183

[32] Witter R, Martins TF. Campos AK, Melo ALT, Correa SHR, Morgado TO, et al. Rickettsial infection in ticks (Acari: Ixodidae) of wild animals in midwestern Brazil. Ticks and Tick-borne Diseases. 2016; 7, 415–423

[33] Zimmermann NP, Aguirre AAR, Rodrigues VS, Garcia MV, Medeiros JF, Blecha IMZ, et al. Wildlife species, Ixodid fauna and new host records for ticks in an Amazon forest area, Rondônia, Brazil. Brazilian Journal Veteriany Parasitology. 2018; 27: 177-182

[34] Torres AC, Minervino AHH, Santos Júnior AP, Sarturi C, Martins TF, Vale WG, Morini AC, et al. *Amblyomma* ticks infesting amphibians and Squamata reptiles from the lower Amazon region, Brazil. Experimental & Applied Acarology. 2018; 75: 399–407

[35] Pedroso-Santos F, Santos ES, Sanches PR, Campos EC, Luz HR, Faccini JLH. First record of *Amblyomma dissimile* (Acari: Ixodidae) infesting the Granular Toad *Rhinella major* (Anura: Bufonidae) in the Eastern Amazon region. Herpetology Notes. 2020; 13: 385-387

[36] Costa FB, Martins TF, Muñoz-Leal S, Serpa MC, OgrzewalskaM, Luz HR, et al. Retrospective and new records of ticks (Acari: Argasidae, Ixodidae) from the state of Maranhão, an Amazon-Cerrado transition area of Brazil Veterinary Plarasitology: Regional Studies and Reports. 2020; 21: 100413

[37] Ogrzewalska M, Machado CR, Forneas T, Cunha DLE, Lemos, ERS. Microorganisms in the ticks *Amblyomma dissimile* Koch 1844 and *Amblyomma rotundatum* Koch 1844 collected from snakes in Brazil. Medical and Veterinary Entomology. 2019; 33: 154–161

[38] Labruna MB, Camargo LMA, Terrassini FA, Ferreira F, Schumaker TTS, Camargo EP. Ticks (Acari: Ixodidae) from the State of Rondônia, western Amazon, Brazil. Systematic & Applied Acarology, 2005; 10: 17–32

[39] Martins TF, Reis J L, Viana EB, Luz HR, Oda FH, Dantas SP, et al. Ticks (Acari: Ixodidae) on captive and free-ranging wild animals in Tocantins State, a Cerrado-Amazon transition region of northern Brazil. International Journal Acarology. 2020; 46: 254-257.

[40] Nascimento KKG, Veríssimo SMM, Raia VA, Guimarães RCS, Seade GCC, Azevedo ACP, et al. Tick fauna of wild animals received and attended at the Santarém Zoological Park, western Pará State, Brazil. Ciencia Rural. 2017; 47: 20170159

[41] Labruna MB, Barbieri FS, Martins TF, Brito LG, Ribeiro FDS. New tick records in Rondônia, Western Brazilian Amazon. Revista Brasileira Parasitologia Veterinária. 2010; 19: 192-194

[42] Soares HS, Barbieri ARM, Martins TF, Minervino AHH, Lima JTR, Marcili A, et al. Ticks and rickettsial infection in the wildlife of two regions of the Brazilian Amazon. Experimental & Applied Acarology. 2015; 65: 125–140

[43] Schumaker TTS, Barros DM. Note on the biology of *Amblyomma dissimile* Koch, 1844 (Acari: Ixodidae) on *Bufo marinus* (Linnaeus, 1758) from Brazil. Memórias do Instituto Oswaldo Cruz. 1994; 89: 29–31

[44] Keirans JE, Oliver Jr JH. First description of the male and redescription of the immature stages of *Amblyomma rotundatum* (Acari:

Ixodidae), a recently discovered tick in the U.S.A. Journal of Parasitology. 1993; 79: 860-865

[45] Labruna MB, Terrassini FA, Camargo LM. First report of the male of *Amblyomma rotundatum* (Acari: Ixodidae) from a field-collected host. Journal of Medical Entomology. 2005; 42: 945-947

[46] Silva TKS, Blanco CM, Lemos ERS, Ogrzewalska M. Notes on parasitism and screening for microorganism of ticks *Amblyomma* (Acari: Ixodidae), Amazon, Brazil. Virus Reviews & Research. 2016; 21:41–44

[47] Gianizella SL, Moraes JJr, Nascimento CAR, Martins TF. Primeiro registro de machos de *Amblyomma rotundatum* (Acari: Ixodidae) em jabuti-tinga (*Chelonoidis denticulatus*) no estado do Amazonas, Amazônia brasileira: relato de caso. Arquivo Brasileiro de Medicina Veterinária e Zootecnia. 2018; 70: 195-198

[48] Luz HR, Faccini JLH, Silva HR. Patterns of parasitism by *Amblyomma rotundatum* (Ixodidae) on *Rhinella schneideri* (Bufonidae) in islands of São Francisco River, Minas Gerais, Brazil. Pesquisa Veterinária Brasileira. 2015; 35: 579-582.

[49] Oliveira GMB, Araújo AC, Santos JR, Silva IWG, Labruna MB, Horta MC. Lack of seasonality of *Amblyomma rotundatum* (Acari: Ixodidade) on *Rhinella jimi* (Anura: Bufonidae) in a semi-arid region of northeastern Brazil. Ticks and Tick-borne Diseases. 2018; 9: 1350-1353

[50] Rodrigues DS, Maciel R, Cunha LM, Leite RC, Oliveira PR. *Amblyomma rotundatum* (Koch, 1844) (Acari: Ixodidae) two-host life-cycle on Viperidae snakes. Revista Brasileira de Parasitologia Veterinária. 2010; 19: 174–178

[51] Luz HR, Faccini JLH, Pires MS, Silva HR, Barros-Battesti DM. Life cycle and behavior of *Amblyomma rotundatum* (Acari: Ixodidae) under laboratory conditions and remarks on parasitism of toads in Brazil. Experimental & Applied Acarology. 2013; 58: 89-99

[52] Petit G, Landau I, Baccam D, Lainson R. Description et cycle biologique *d'Hemolivia stellata*, n.g., n. sp., hemogrégarine de *Crapauds brésiliens*. Annales de Parasitologie Humaine et Comparee. 1990; 65:3-15

[53] Labruna MB, Whitworth T, Bouyer DH, McBride J, Camargo LMA, Camargo P, et al. *Rickettsia bellii* and *Rickettsia amblyommii* in *Amblyomma* ticks from the state of Rondônia, Western Amazon, Brazil. Journal of Medical Entomology. 2004; 41: 1073-1081

[54] Gruhn KD, Ogrzewalska M, Rozental T, Farikoski IO, Blanco C, Freitas L S, et al. Evaluation of rickettsial infection in free-range capybaras (*Hydrochoerus hydrochaeris* Linneaus, 1766) (Rodentia: Caviidae) and ticks (Acari: Ixodidae) in Western Amazon, Brazil. Ticks and Ticks-Borne Diaseases. 2019; 10: 981-986

[55] Barros-Battesti DM, Jerozolimski A, Onofrio VC, Arzua M. 2002. *Amblyomma humerale* Koch, 1844 (Acari: Ixodidade) on tortoises *Geochelone denticulatus* (L.) and *G. carbonarius* (Spix) from the Kayapó Indian land, state of Pará, northern Brazil. XI International Congress of Acarology. Program and Abstract Book, Mérida, México. 160 p

[56] Souza AR, Gianizella SL, Roberto IJ. *Rhinoclemmys punctularia* (Eastern Spot-Legged Wood Turtle). Ectoparasites. Herpetological Review. 2018; 49: 713

[57] Morais DH, Carvalho VT, Bonora L, Kawashita-Ribeiro RA, Strussmann C.

Patterns of Parasitism by *Amblyomma humerale* (Acari: Ixodidae) on *Chelonoidis denticulatus* (Testudines: Testudinidae) in the Brazilian Midwest. Herpetological Review. 2017; 48: 316-319

[58] Martins TF, Teixeira RHF, Benatti HR, Minervino AHH, Soares HS, Soares JF, et al. Life cycle of the tick *Amblyomma humerale* (Parasitiformes: Ixodida) in the laboratory, International Journal of Acarology. 2020. https://doi.org/10.1080/01647954.2020.1802510

[59] Dantas-Torres F, Mascarenhas-Junior PB, Anjos HR, Santos EM, Correia JMS. Tick infestation on caimans: a casual tick-host association in the Atlantic rainforest biome?. Experimental & Applied Acarology. 2019; 79: 411–420

[60] Martins TF, Gianizella SL, Nunes PH, Faria DC, Nascimento CA, Abrahão CR, et al. New records of *Amblyomma goeldii* (Acari: Ixodidae) and description of the nymphal stage. Zootaxa. 2015; 3949: 439-444

[61] Luz HR, Faccini JLH. Ticks on brazilian birds: overview. In: Ruiz, L., Iglesias, F. (Eds.), Birds – Evolution, Behavior and Ecology 1. Nova Science Publishers, Inc, New York. 2013; 97–125.

[62] Hasle G. Transport of ixodid ticks and tick-borne pathogens by migratory birds. Frontiers in Cellular and Infection Microbiology. 2013; 3: 48

[63] Comstedt PS, Bergstrom B, Olsen U, Garpmo L, Marjavaara H. Mejlon, A. G. Barbour, and J. Bunikis. *Migratory passerine birds as reservoirs of Lyme borreliosis in Europe.* Emerging Infectious Diseases. 2006; 12: 1087-*1095*

[64] Ogrzewalska M, Pinter A. Ticks (Acari: ixodidae) as ectoparasites of Brazilian wild birds and their association with rickettsial diseases Brazilian Journal Veterinary Animmal Sciense. 2016; 53: 1-31

[65] Barros-Battesti DM, Arzua M, Pichorim M, Keirans JE. *Ixodes (Multidentatus) paranaensis* n. sp. (Acari: Ixodidae) a parasite of *Streptoprocne biscutata* (Sclater 1865) (Apodiformes: Apodidae) birds in Brazil. Memórias do Instituto Oswaldo Cruz. 2003; 98: 93–91

[66] CBRO – Comitê Brasileiro de Registros Ornitológicos. 2014. Listas das aves do Brasil. 10 ed. Disponível em<http://www.cbro.org.br>. Acessed in: 02.10.2020

[67] Sick H. Ornitologia Brasileira. Editora Nova Fronteira, Rio de Janeiro. 1997. 912 p

[68] Lima MA, Martins TF, Muñoz-Leal S, Guilherme E, Ogrzewalska M, Labruna MB. Ticks and tick-associated spotted fever group *Rickettsia* from birds in the Southwestern Brazilian Amazon. Revista Colombiana de Ciencias Pecuárias. 2018; 31: 26-35

[69] Ogrzewalska M, Uezu A, Labruna MB. Ticks (Acari Ixodidae) infesting wild birds in the eastern Amazon, northern Brazil, with notes on rickettsial infection in ticks. Parasitol. Res. 2010; 106: 809–816.

[70] Souza VL, Martins TF, Guilherme E, Santos FGA. New records of ticks (Acari: Ixodidae) infesting wild birds in a forest fragment in Acre, Brazilian Amazon. 2020. DOI: 10.5433/1679-0359.2020v41n5Supl1p2163

[71] Martins TF, Fecchio A, Labruna MB (2014) Ticks of the genus *Amblyomma* (Acari: Ixodidae) on wild birds in the Brazilian Amazon. Systematic and Applied Acarology. 2010; 19: 385–392

[72] Scofield A, Bahia M, Martins AL, Góes-Cavalcante G, Martins TF, Labruna MB. *Amblyomma dissimile* Koch

(Acari: Ixodidae) attacking *Primolius maracana* Vieillot (Psittaciformes: Psittacidae) in the Amazon Region, State of Pará, Brazil. Neotropical Entomology. 2011; 40: 509-511

[73] Labruna MB, Sanfilippo LF, Demetrio C, Menezes AC, Pinter A, Guglielmone AA. Ticks collected on birds in the state of São Paulo, Brazil. Experimental & Applied Acarology. 2007; 43: 147-160.

[74] Lugarini C, Martins TM, Ogrzewalska M, Vasconcelos NCT, Ellis VA, Oliveira JB, et al. Rickettsial agents in avian ixodid ticks in northeast Brazil. Ticks and Tick Borne Diseases. 2015; 6: 364–375

[75] Luz HR, Faccini JLH, McIntosh D. Molecular analyses reveal an abundant diversity of ticks and rickettsial agents associated with wild birds in two regions of primary Brazilian Atlantic Rainforest. Ticks and Tick Borne Diseases. 2017; 8: 657–665

[76] Luz HR, Martins TF, Teixeira RHF, Terassini FA, Soares HS, Soares JF, et al. Some biological and behavioral aspects of *Amblyomma longirostre* (Acari: Ixodidae) under laboratory and natural conditions. Systematic and Applied Acarology. 2018; 23:10. 1965

[77] Aragão HB. Ixodidas brasileiros e de algunos paízes limitrophes. Memórias do Instituto Oswaldo Cruz. 1936; 31: 759–843

[78] Reis NL, Perachi AL, Pedro WA, Lima IP. Mamíferos do Brasil. Universidade Estadual de Londrina, Brazil. 2006.

[79] Abreu-Jr EF, Casali DM, Garbino GST, Loretto D, Loss AC, Marmontel M, et al. Lista de Mamíferos do Brasil. Comitê de Taxonomia da Sociedade Brasileira de Mastozoologia (CT-SBMz). Disponível em: <https://www.sbmz.org/mamiferos-do-brasil/>. Acessed em: 03/10/2020.

[80] Bown KJ, Begon M, Bennett M, Woldehiwet Z, Ogden N. Seasonal dynamics of *Anaplasma phagocytophila* in a rodent-tick (Ixodes trianguliceps) system, United Kingdom. Emerging Infectious Diseases. 2003; 9: 63–70

[81] Zhan L, Cao WC, Chu CY, Jiang BG, Zhang F, Liu W, et al. Tick-borne agents in rodents, China, 2004–2006. Emerging Infectious Diseases. 2009; 15: 1904-1908

[82] Horta MC, Moraes-Filho J, Casagrande RA, Saito TB, Rosa SC, Ogrzewalska M, et al. Experimental Infection of opossums *Didelphis aurita* by *Rickettsia rickettsii* and evaluation of the transmission of the infection to ticks *Amblyomma cajennense*. Vector Borne Zoonotic Diseases. 2009; 9: 109-118

[83] Nava S, Venzal JM, González-Acuña D, Martins TF, Guglielmone AA. Ticks of the Southern Cone of America. Elsevier Academic Press, London, San Diego, Cambridge, Oxford. 2017. 348 p

[84] Labruna MB, Jorge RS, Sana DA, Jácomo AT, Kashivakura CK, Furtado MM, et al. Ticks (Acari: Ixodida) on wild carnivores in Brazil. Experimental & Applied Acarology. 2005; 36: 149-163

[85] Martins TF, Milanelo L, Krawczak FS, Furuya HR, Fitorra LS, Dores FT, et al. Diversity of ticks in the wildlife screening center of São Paulo city, Brazil. Ciência Rural. 2017; 47: 1-6

[86] Luz HR, Neto SFC, Weksler M, Gentile R, Faccini JLH. Ticks parasitizing wild mammals in Atlantic Forest areas in the state of Rio de Janeiro, Brazil. Brazilian Journal Veterinary Parasitology. 2018; 27:409–414

[87] Luz HR, Muñoz-Leal S, de Carvalho WD, Castro IJ, Xavier BS, Toledo JJ, et al. Detection of "Candidatus *rickettsia wissemanii*" in

ticks parasitizing bats (Mammalia: Chiroptera) in the northern Brazilian Amazon. Parasitology Research. 2019; 118: 3185–3189

[88] Colle AC, Mendonça RFB, Maia MO, Freitas LC, Martins TF, Muñoz leal S, et al. Rickettsial survey and ticks infesting small mammals from the Amazon forest in midwestern Brazil. Systematic and Applied Acarology. 2020; 25:78–91

[89] Rosas FCW, Sicsu da Cruz TM, Gianizella SL, dos Santos Ramalheira C, Martins TF. A first record of ticks in free-ranging giant otter (*Pteronura brasiliensis*) in the Brazilian Amazon. IUCN Otter Species Group Bulletim. 2016; 33: 3–7

[90] Martins TF, Barbieri ARM, Costa FB, Terassini FA, Camargo LMA, Peterka CRL, et al. Geographical distribution of *Amblyomma cajennense* (*sensu lato*) ticks (Parasitiformes: Ixodidae) in Brazil, with description of the nymph of *A. cajennense* (*sensu stricto*). Parasites & Vectors. 2016; 9: 14 p

[91] Costa FB, da Costa AP, Moraes-Filho J, Martins TF, Soares HS, Ramirez DG, et al. *Rickettsia amblyommatis* infecting ticks and exposure of domestic dogs to *Rickettsia* spp. in an Amazon-Cerrado transition region of northeastern Brazil. PLoS ONE. 2017; 12: e0179163

[92] Leal PF. Distribuição e relação trófica entre carrapatos e pequenos mamíferos da Ordem Rodentia e Didelphimorphia na floresta nacional de Carajás, estado do Pará, Dissertação. Belém, 2014

[93] Labruna MB, Keirans JE, Camargo LMA, Ribeiro AF, Soares RM, Camargo EP. *Amblyomma latepunctatum*, a valid tick species (Acari: Ixodidae) long misidentified with both *Amblyomma incisum* and *Amblyomma scalpturatum*. Journal of Parasitology. 2005; 91: 527–541

[94] Furtado MM, Metzger B, Jácomo ATA, Labruna MB, Martins TF, O'Dwyer LH, et al. Hepatozoon spp. Infect Free-Ranging Jaguars (*Panthera onca*) in Brazil. Journal Parasitology. 2017; 3: 243-250

[95] Martins TF, Spolidorio MG, Batista TCA, Oliveira IAS, Yoshinari NH, Labruna MB. 2009. Occurrence of ticks (Acari: Ixodidae) in the municipality of Goiatins, Tocantins. Revista Brasileira Parasitologia Veterinária. 2009; 18:50-52

[96] Sampaio JS, Martins TF, Labruna MB, Onofrio VC, Faccini JLH, Barros-Battesti DM. A second record of *Amblyomma romitii* Tonelli-Rondelli, 1939 (Acari: Ixodidae) from the State of Pará, Brazil. Systematic & Applied Acarology. 2010; 15: 184–186

[97] Marques S, Barros-Battesti DM, Faccini JLH, Onofrio VC. Brazilian distribution of *Amblyomma varium* Koch, 1844 (Acari: Ixodidae), a common parasite of sloths (Mammalia: Xenarthra). Mem Inst Oswaldo Cruz. 2002; 97: 1141-6

[98] Oliveira SVde, Pereira SVC, Pinna FV, Fonseca LX, Serra-Freire NM, Cardoso KM, et al. Vigilância de ambientes da febre maculosa: explorando as áreas silenciosas do Brasil. Revista Pan-Amazônica de Saúde. 2016; 7: 65-72

[99] Martins TF, Bernardi LFO, Munoz-Leal S, Marcili A, Labruna MB. New records of *Ixodes amarali* (Acari: Ixodidae) in the Amazon biome, with description of the male. Systematic and Applied Acarology. 2009; 24: 2552–2558

[100] Luz HR, Faccini JLH, Landulfo GA, Sampaio JS, Neto SFC, Famadas KM et al. New host records of *Ixodes luciae* (Acari: Ixodidae) in the

State of Pará, Brazil. Revista Brasileira Parasitologia Veterinária. 2013; 22: 152–154

[101] Fernandes MMP, Medeiros FER, Carvalho YK, Ribeiro VMF, Souza SF. Ectoparasitas de cães domiciliados e errantes do município de Rio Branco-Acre. Enciclopédia Biosfera. 2018; 15: 442-450

[102] Onofrio VC, Barros-Battesti DM, Labruna MB, Faccini JL. Diagnoses of and illustrated key to the species of Ixodes Latreille, 1795 (Acari: Ixodidae) from Brazil. Systematic and Applied Acarology. Parasitology. 2009; 72: 143–157

[103] Serra-Freire NM. Occurrence of ticks (Acari: Ixodidae) on human hosts, in three municipalities in the State of Pará, Brazil. Revista Brasileira de Parasitologia Veterinária. 2010; 19: 141–147

[104] Lopes LB, Guterres A, Rozental T, Carvalho de Oliveira R, Mares-Guia M, Fernandes J, et al. *Rickettsia bellii*, *Rickettsia amblyommii*, and Laguna Negra hantavirus in an Indian reserve in the Brazilian Amazon. Parasites & Vectors. 2014; 7: 191

[105] Aguirre A AR, Rodrigues VS, Costa IN, Garcia MV, Higa LOS, Medeiros JF, Andreotti R. Biological parameters of *Amblyomma coelebs* Neumann, 1906 (Acari: Ixodidae) under experimental conditions. Brazilian Journal Veterinary Parasitology. 2018; 27: 81-86

[106] Guglielmone AA, Beati L, Barros-Battesti DM, Labruna MB, Nava S, Venzal JM, et al. Ticks (Ixodidae) on humans in South America. Experimental & Applied Acarology. 2006; 40: 83–100

[107] Silva TKS, Blanco CM, Lemos ERS, Ogrzewalska M. Notes on parasitism and screening formicroorganism of ticks *Amblyomma* (Acari: Ixodidae), Amazon, Brazil. Virus Reviews & Research. 2016; 21: 41–44

[108] Aragão HB, Fonseca *F. nota* de ixodologia. VIII. Lista e chave para os representantes da fauna ixodológica brasileira. Memórias do Instituto Oswaldo Cruz. 1961; 59: 115–129

[109] Labruna MB, Homem VSF, Heinemann MB, Neto JSF. Ticks (Acari: Ixodidae) associated with rural dogs in Uruará, Eastern Amazon, Brazil. Journal of Medical Entomology. 2000; 37: 774–776

[110] Martins TF, Luz HR, Faccini JLH, Labruna MB. Life-cycle of *Amblyomma oblongoguttatum* (Acari: Ixodidae) under laboratory condition. Experimental & Applied Acarology. 2017; 71: 415–424

[111] Martins TF, Moura MM, Labruna MB. Life-cycle and host preference of *Amblyomma ovale* (Acari: Ixodidae) under laboratory conditions. Experimental & Applied Acarology. 2012; 56: 151–158

[112] Guglielmone AA, Robbins RG. Tick species found feeding on humans. In: Guglielmone AA, Robbins RG, editors. Hard ticks (Acari: Ixodida: Ixodidae) parasitizing humans: a global overview. 1st ed. London: Springer. 2018. 205 p

[113] Landulfo GA, Luz HR, Sampaio JS, Faccini JLH, Barros-Battesti DM. Life Cycle of *Amblyomma romitii* (Acari: Ixodidae) Under Laboratory Conditions. Journal of Medical Entomology. 2015; 53: 48-53

[114] Serra Freire NM, Peralta ASL, Teixeira RHF, Gazeta GS, Amorim M. *Amblyomma rotundatum* parasitando *Homo sapiens* no parque zoobotánico do MPEG e em Itaboraí. Arquivos da Sociedade de Zoológicos do Brasil. 1995; 16: 20

[115] Aguirre AAR, Rodrigues VS, Costa IN, Garcia MV, Csordas BG,

Andreotti R, Medeiros JF. *Amblyomma scalpturatum* Neumann, 1906 (Acari: Ixodidae): confirmation in Acre State, Brazil, and description of parasitism in a human. Brazilian Journal of Veterinary Parasitology. 2019; 28: 473-478

[116] Labruna MB, Kasai N, Ferreira F, Faccini JL, Gennari SM. Seasonal dynamics of ticks (Acari: Ixodidae) on horses in the state of São Paulo Brazil. Veterinary Parasitology. 2002; 19: 65–77

[117] Labruna MB, Amaku M, Metzner JA, Pinter A, Ferreira F. Larval behavioral diapause regulates life cycle of *Amblyomma cajennense* (Acari: Ixodidae) in Southeast Brazil. Journal of Medical Entomology. 2003; 40: 170-178

[118] Bitencourth K, Amorim M, De Oliveira SV, Caetano, RL, Voloch CM, Gazêta GS. *Amblyomma sculptum*: genetic diversity and rickettsias in the Brazilian Cerrado biome. Medical and Veterinary Entomology. 2017; 3: 1-11

[119] Labruna MB, Terassini FA, Camargo LM. Notes on population dynamics of *Amblyomma* ticks (Acari: Ixodidae) inBrazil. Journal Parasitology. 2009; 95: 1016-8

[120] Mans BJ, Featherston J, Kvas M, Pillay K, de Klerk DG, Pienaar R, et al. Argasid and ixodid systematics: Implications for soft tick evolution and systematics, with a new argasid species list. Ticks and Tick-borne Diseases. 2019; 10: 219-240

[121] Hoogstraal H. Mint: Argasid and nuttalliellid ticks as parasites and vectors. Advances in Parasitology. 1985; 24: 135-238

[122] Labruna MB, Terassini FA, Camargo LMA, Brandão PE, Ribeiro AF, Estrada-Peña A. New reports of *Antricola guglielmonei* and *Antricola delacruzi* in Brazil, and a description of a new argasid species (Acari). Journal of Parasitology. 2008; 94: 788-792

[123] Nava S, Venzal JM, Terassini FA, Mangold AJ, Camargo LMA, Labruna MB. Description of a new argasid Tick (Acari: Ixodida) from bat caves in Brazilian Amazon. Journal of Parasitology. 2010; 96: 1089-1101

[124] Muñoz-Leal S, Terassini FA, Marcili A, Oliveira GMB, Camargo LMA, Labruna MB. A third species of *Nothoaspis* Keirans & Clifford, 1975 (Acari: Argasidae): *Nothoaspis setosus* (Kohls, Clifford & amp; Jones, 1969) n. comb. Systematic Parasitology. 2019; 96 :595-602

[125] Muñoz-Leal S, Terassini FA, Luz HR, Fontana I, Camargo LMA, Labruna MB. First report of *Ornithodoros peropteryx* in Brazil, and the occurrence of *Ornithodoros cavernicolous* in the western Brazilian Amazon. Systematic & Applied Acarology. 2018; 23: 2113–2121

[126] Cooley RA, Kohls GM. The Argasidae of North America, Central America and Cuba, The American Midland Naturalist. Monograph N°1. 1944

[127] Labruna MB, Nava S, Terassini FA, Onofrio VC, Barros-Battesti D, et al. Description of adults and nymph, and resdescription of the larva, of *Ornithodoros marinekellei* (Acari: Argasidae), with data on its phylogenetic position. Journal of Parasitology. 2011; 97: 207-217

[128] Henrique-Simões M, Bernardi LFO, Ogrzewalska M, Labruna MB, Ferreira RL. New records of rare Ornithodoros (Acari: Argasidae) species in caves of the Brazilian Amazon. Persian Journal of Acarology. 2012; 1: 127-135

[129] Jones EK, Clifford CM, Keirans JE, Kohls GM. The ticks of Venezuela (Acarina: Ixodoidea) with a key to the species of *Amblyomma* in the Western Hemisphere. Brigham Young University

Science Bulletin - Biological Series. 1972; 17: 1-40

[130] Muñoz-Leal S, Eriksson A, Santos CF, Fischer E, Almeida JC, Luz HR, Labruna MB. Ticks infesting bats (Mammalia: Chiroptera) in the Brazilian Pantanal. Experimental & Applied Acarology. 2016; 69: 73-85

[131] Muñoz-Leal S, Barbier E, Soares FAM, Bernard E, Labruna MB, Dantas-Torres F. New records of ticks infesting bats in Brazil, with observations on the first nymphal stage of *Ornithodoros hasei*. Experimental & Applied Acarology. 2018; 76: 537-549

[132] Luz HR, Muñoz-Leal S, Almeida JC, Faccini LH, Labruna MB. Ticks parasitizing bats (Mammalia: Chiroptera) in the Caatinga Biome, Brazil. Brazilian Journal Veterinary Parasitology. 2018; 35: 484-491

[133] Kohls GM, Clifford CM, Jones EK. The systematics of the subfamily Ornithodorinae (Acarina: Argasidae). IV. Eight new species of *Ornithodoros* from the Western Hemisphere. Annals of the Entomological Society of America. 1969; 62: 1035–1043

[134] Guzmán-Cornejo C, Herrera-Mares A, Robbins RG, Rebollo-Hernández A. The soft ticks (Parasitiformes: Ixodida: Argasidae) of Mexico: species, hosts, and geographical distribution. Zootaxa. 2019; 4623: 485-525

[135] De La Cruz J. Mint: Bioecologia de las grutas de calor. Mundos Subterraneos, UMAE, Mexico, DF. 1992; 3: 7-21

[136] Ribeiro JMC, Labruna MB, Mans BJ, Marumaya SR, Francischetti IMB, Barizon GC, Santos IKFM. The sialotranscriptome of *Antricola delacruzi* female ticks is compatible with non-hematophagous behavior and an alternative source of food. Insect Biochemistry Molecular Biology. 2012; 42: 332-342

[137] Estrada-Peña A, Venzal JM, Barros-Battesti DM, Onofrio VC, Trajano E, Firmino JVL. Three new species of *Antricola* (Acari: Argasidae) from Brazil, with a key to the known species in the genus. Journal of Parasitology. 2004; 90: 490-498

Chapter 4

Ecology of the Seed Bank in the Amazon Rainforest

Natali Gomes Bordon, Niwton Leal Filho and Tony Vizcarra Bentos

Abstract

The seed bank is directly related to forest resilience because it contributes to the greatest number of regenerants after the occurrence of disturbances. Changes in seed density, floristic composition, and life forms completely alter the successional trajectory of forest environments. These changes are directly related to land use. For example, suppression of the seed bank can occur in pastures, that experience frequent fires with increase of density of seeds and predominance of herbs are typical of highly degraded areas, such as Poaceae, Rubiaceae, Asteraceae, and Cyperaceae. Melastomataceae seedlings are an important component of the seed bank in the Amazon rainforest. On the other hand, Urticaceae has greater representation in forests that exhibit low-impact land use. Any change in seed bank functionality is bound to compromise the diversity, regeneration potential and overall maintenance of tropical forests. Therefore, it is necessary to expand studies that investigate seed banks in the Amazon rainforest. It is as important to prioritize sampling methods and pursue standardization of data presentation, as well as improve the identification of species that occur in the seed bank.

Keywords: floristic composition, forest disturbance, anthropic changes, forest regeneration, land use

1. Introduction

The seed bank, or stock of viable seeds in the soil, can be defined as a set of latent, or dormant, seeds capable of originating adult plants [1, 2]. Studying the composition of the seed bank and understanding its role in regeneration are important to the conservation and management of tropical forests, as well as the control and eradication of invasive species in agrosilvopastoral systems [3–8]. The seed bank is influenced by the local plant community, history of land use, and forest matrix in general, it also has spatial and temporal variations [5, 9–12]. Spatial variations occur both horizontally and vertically; however, the greatest amounts of seed are observed in the upper layers [13–16]. Temporal variation occurs as a result of both loss and incorporation of seeds in the soil [13, 17]. The incorporation of seeds is the result of seed rain, which also presents seasonality owing to the different fruiting patterns of the species [9, 18, 19]. The rate of seed loss in the soil depends intrinsic loss of viability resulting from dispersal, environmental conditions, predation, and attack of pathogens [17, 20, 21].

Seeds of pioneer species are found in high density in the soils of tropical forests and constitute the main reserve of propagules for the regeneration of areas subject to disturbances [10, 22–24]. Most pioneer species have quiescent diaspores, owing to canopy light conditions or temperature variations [25–27], and compound the persistent seed bank [10, 22–24]. Species that compound the transitional seed bank have a lower density and are composed of late species of the forest succession [10, 14, 22], commonly forming a seedling bank [28, 29]. Consequently, floristic composition of the tropical rainforest seed bank does not reflect the composition of species in the arboreal, or regenerating, strata of old-growth forests [8, 14, 29]. Thus, seed bank serve to allow the establishment of a set of species that do not occur in vegetation or that present in low density in old-growth forests, but persist in the seed bank [18, 30–32].

After formation of a clearing by natural or anthropic disturbance, the quantity of seeds in the soil decreases as a consequence of recruitment rates or loss of seed viability [14, 29, 33]. After the establishment of pioneer species and subsequent fruiting, seed density in the soil increases in the initial stages [9, 21, 24, 29, 33, 34]. The, with the advance of forest succession, the number of seeds in the seed bank tends to decrease and return to pre-disturbance equilibrium [9, 21, 24, 29, 33, 34]. The seed bank plays a major role in the re-establishment of plant communities subjected to medium and high-intensity disturbances and can have a wide impact on the dynamics of plant communities during the process of ecological succession [35–38]. For example, in forest areas of the Amazon burned and converted to pasture, almost no vestiges of the seed bank, remain [39]. Nonetheless, pioneer species of *Vismia* were reported to dominate regeneration [40, 41]. In contrast, areas with some seed bank left intact were initially reported to already be occupied by pioneer species of *Cecropia,* allowing a larger set of plant species to regenerate under its canopy [36, 39, 42–44]. In terms of forest management, the role of the seed bank in the regeneration of forests increases in importance, when compared to the seed rain, both in clearings and trails generated by skidders [45].

The seed bank is known for its low contribution to the establishment of late species in the forest succession in which these groups derived from dispersion and stock seedlings [1, 14]. However, the seed bank can be considered highly diverse in life forms thus contributing to the restructuring of forest strata [33, 46]. The abundance of herbs and shrubs in the seed bank of forest environments can be a consequence of the surrounding matrix, as well as the history of land use [5, 14, 22]. It is a reflection of vegetation that has already undergone some type of anthropic or natural change [5, 29, 33, 47]. In general, herbaceous and shrub species are more commonly found in altered areas and secondary vegetation [14]. However, disturbances that occur around the forest also contribute to the entry of ruderal, or invasive, species in the seed bank [5]. Notwithstanding this phenomenon, forest areas surrounding pastures or agricultural areas change the density and floristic composition of the seed bank in these areas [5]. This gives rise to the entry of common trees and shrubs into the seed bank of forest areas [5]. The seed bank in tropical forests is, therefore, highly variable. At the same time, studies reporting on this natural component of the Brazilian rainforest are scarce. Therefore, this chapter aims to analyze variations in density, family abundance, and life forms of the seed bank in *terra firme* forest of the Amazon rainforest, as well as assess the impact of the main changes in land use in this region on seed bank characteristics.

2. Materials and methods

Data were obtained from published and unpublished scientific reports and monographs written by undergraduate students under the supervision of Dr. Niwton Leal Filho of the National Institute for Amazon Research (INPA). One dataset contains complete data on density and floristic composition. We used 17 datasets from a seed bank in *terra firme* forest of the Amazon rainforest, which dataset [48–50] was not included in the floristic composition. **Table 1** list all datasets

Study area location	Latitude and longitude	Type of disturbance	EA	N	A	D	DS	Reference
(1) Old-growth forest (with no evidence of anthropogenic changes in the last 60 years or more)								
Biological Dynamics of Forest Fragments Project (BDFFP), Amazonas, Brazil	2°25' S; 59°50' W	—	—	90	0.70	2	913 ± 1112	unpublished data
Experimental Station of Tropical Forestry (EEST), Amazonas, Brazil	2°37'38" S; 60°09'11" W	—	—	30	0.71	3	722	unpublished data
Biological Dynamics of Forest Fragments Project (BDFFP), Manaus, Amazonas, Brazil	2°25' S; 59°50' W	—	—	45	1.01	3	662 ± 741	[7]
Experimental Station of Tropical Forestry (EEST), Manaus, Amazonas, Brazil	2°36'50" S; 60°12'13" W	—	—	160	5.03	5	498 ± 437	[46]
Adolfo Ducke Forest Reserve, Amazonas, Brazil	02°53' S; 59°58' W	—	—	1440	11.31	2	460	[19]
River Capim Farm, Paragominas, Pará, Brazil	03°37'59.9" S; 48°32'46.8" W	—	—	60	3.75	5	423	[57]
Adolfo Ducke Forest Reserve (riparian or bottomlands forest, with periodic flooding), Amazonas, Brazil	02°53' S; 59°58' W	—	—	72	4.32	10	367	[58]
Adolfo Ducke Forest Reserve, Amazonas, Brazil	02°53' S; 59°58' W	—	—	30	0.24	5	299	unpublished data
Experimental Station of Tropical Forestry (EEST), Manaus, Amazonas, Brazil	2°37' S; 60°09' W	—	—	30	0.24	5	246	unpublished data
Experimental Station of Tropical Forestry (EEST), Manaus, Amazonas, Brazil	2°37' S; 60°09' W	—	—	40	0.31	3	194 ± 263	unpublished data
Ferreira Penna Scientific Station (ECFPn), Melgaço, Pará, Brazil	1°42'30" S; 51°31'45" W	—	—	100	6.25	5	94 ± 61	[50]
(2) Forest fragments								
Biological Dynamics of Forest Fragments Project (BDFFP), Amazonas, Brazil	2°25' S; 59°50' W	Remaining isolated forest fragments of 1 ha	25	90	0.70	2	4073 ± 3578	unpublished data
Biological Dynamics of Forest Fragments Project (BDFFP), Amazonas, Brazil	2°25' S; 59°50' W	Remaining isolated forest fragments of 10 ha	25	90	0.70	2	3829 ± 2565	unpublished data

Study area location	Latitude and longitude	Type of disturbance	EA	N	A	D	DS	Reference
Biological Dynamics of Forest Fragments Project (BDFFP), Amazonas, Brazil	2°25' S; 59°50' W	Remaining isolated forest fragments of 1 ha	30	45	1.01	3	1690 ± 2530	[7]
Biological Dynamics of Forest Fragments Project (BDFFP), Amazonas, Brazil	2°25' S; 59°50' W	Remaining isolated forest fragments of 10 ha	30	45	1.01	3	1309 ± 787	[7]
Science Grove of the National Institute for Amazon Research (INPA), Manaus, Amazonas, Brazil	03°08' S; 60°10' W	Urban fragment of 13 ha, with a history of selective logging before the creation of the preservation area	47	30	2.40	5	1264 ± 969	unpublished data
Science Grove of the National Institute for Amazon Research (INPA), Manaus, Amazonas, Brazil	3°05'50"S; 59°59'10" W	Urban fragment of 13 ha, with a history of selective logging before the creation of the preservation area	47	30	0.24	5	747	unpublished data
Mindú Park, Manaus, Amazonas, Brazil	03°07' S; 59° 05' W	Urban fragment of 31 ha, with a history of selective logging before the creation of the preservation area	30	30	0.24	5	633	unpublished data
Biological Dynamics of Forest Fragments Project (BDFFP), Amazonas, Brazil	2°25' S; 59°50' W	Remaining isolated forest fragments of 100 ha	30	45	1.01	3	576 ± 450	[7]
Petro Set, preservation area, Manaus, Amazonas, Brazil	03°04' S; 59°58'W	Urban fragment of 2 ha intensely altered, with a history of selective logging before the creation of the preservation area	40	30	0.24	5	410	unpublished data
Federal University of Amazonas (UFAM), Manaus, Amazonas, Brazil	03°4.34' S; 59°57.30' W	Urban fragment of 800 ha, with a history of selective logging before the creation of the preservation area	55	30	0.24	5	395 ± 68	unpublished data

Study area location	Latitude and longitude	Type of disturbance	EA	N	A	D	DS	Reference
(3) Forests with logging of wood species								
Experimental Station of Tropical Forestry (EEST), Manaus, Amazonas, Brazil	2°36′50" S; 60°12′13" W	The area explored in forest management, clearing of exploration	14	30	0.71	3	2219	unpublished data
Experimental Station of Tropical Forestry (EEST), Manaus, Amazonas, Brazil	2°36′50" S; 60°12′13" W	Explored area of forest management, tractor trail	14	30	0.71	3	1561	unpublished data
Experimental Genetic Resource Station "José Haroldo", Benevides, Pará, Brazil	01°10' S; 48°20' W	Old-growth forest with logging of wood species	17	25	6.25	8	1427 ± 729	[59]
Experimental Station of Tropical Forestry (EEST), Manaus, Amazonas, Brazil	2°36′50" S; 60°12′13" W	Explored area of forest management, tractor trail	21	30	0.71	3	1274	unpublished data
Experimental Genetic Resource Station "José Haroldo", Benevides, Pará, Brazil	01°10' S; 48°20' W	Old-growth forest with logging of wood species	30	25	6.25	8	756 ± 250	[59]
Experimental Station of Tropical Forestry (EEST), Manaus, Amazonas, Brazil	2°36′50" S; 60°12′13" W	The area explored in forest management, clearing of exploration	21	30	0.71	3	711	unpublished data
River Capim Farm, Paragominas, Pará, Brazil	03°37′59.9" S; 48°32′46.8" W	Logging and woody waste	1	120	7.50	5	317 ± 413	[57]
(4) Secondary forests (with evidence of natural and anthropogenic changes)								
Biological Dynamics of Forest Fragments Project (BDFFP), farm Esteio, Manaus, Amazonas, Brazil	2°24'48" S; 59° 52′21" W	With a history of abandoned pasture	7	32	1.28	3	8085	[39]
Experimental Genetic Resource Station "José Haroldo", Benevides, Pará, Brazil	01°10' S; 48°20' W	Abandoned pasture with burning history	6	25	6.25	8	2848 ± 537	[59]
Biological Dynamics of Forest Fragments Project (BDFFP), Manaus, Amazonas, Brazil (three topographic positions: plateaus, slopes, and bottomlands)	2°30' S; 60°10' W	Abandoned pasture with history of fires	20	21	2.65	5	2187 ± 1137	[60]

Study area location	Latitude and longitude	Type of disturbance	EA	N	A	D	DS	Reference
Experimental Station of Tropical Forestry (EEST), Manaus, Amazonas, Brazil	2°36′50" S; 60°12′13" W	Blowdown	6	160	5.03	5	704 ± 770	[46]
(5) Agriculture areas								
Manacapuru, Amazonas, Brazil	3°16′20" S; 60°33′07" W	Agroforestry systems	> 5	20	0.45	5	9540	[48]
Manacapuru, Amazonas, Brazil	3°16′20" S; 60°33′07" W	Agroforestry systems	> 5	20	0.45	5	8909	[48]
Manacapuru, Amazonas, Brazil	3°16′20" S; 60°33′07" W	Cassava cultivation	< 5	20	0.45	5	8329 ± 122	[49]
Manacapuru, Amazonas, Brazil	3°16′20" S; 60°33′07" W	Cassava cultivation	< 5	20	0.45	5	7471 ± 203	[49]
Manacapuru, Amazonas, Brazil	3°16′20" S; 60°33′07" W	Agroforestry systems	> 5	20	0.45	5	7173	[48]
Kilometer 2 of the road to Balbina Village (PAS3), near the clover on BR-174, Amazonas, Brazil	2°03′57" S; 60°01′20" W	Pasture	20	20	0.45	5	6153 ± 75	[61]
Manacapuru, Amazonas, Brazil	3°16′20" S; 60°33′07" W	Agroforestry systems	> 5	20	0.45	5	3320	[48]
km 50 of the BR–174 (PAS2), Manaus to Presidente Figueiredo, Amazonas, Brazil	2°03′57" S; 60°01′20" W	Pasture	14	20	0.45	5	3209 ± 48	[61]
Manacapuru, Amazonas, Brazil	3°16′20" S; 60°33′07" W	Cassava cultivation	< 5	20	0.45	5	2691 ± 116	[49]
Esteio Farm (PAS1), Manaus to Presidente Figueiredo, Amazonas, Brazil	2°03′57" S; 60°01′20" W	Pasture	9	20	0.45	5	2593 ± 59	[61]
Manacapuru, Amazonas, Brazil	3°16′20" S; 60°33′07" W	Cassava cultivation	< 5	20	0.45	5	1962 ± 27	[49]
Kilometer 2 of the road to Balbina Village (PAS4), near the clover on BR-174, Amazonas, Brazil	2°03′57" S; 60°01′20" W	Pasture	20	20	0.45	5	304 ± 5	[61]

Table 1.
Details of density, family abundance, and forms of seedlings that emerged from the seed bank in the terra firme forest of the Amazon rainforest. EA: Estimated age at the time of the study; N: Number of samples; A: Total area sampled (m^2); D: Depth of the sample collected (cm); DS: Density of seedlings (m^2) (mean ± standard deviation).

used in this stydy. The datasets involve different types of land use in the Brazilian Amazon rainforest, incluinding (1) old-growth forests, with no evidence of anthropogenic changes in the last 60 years or more; (2) forest fragments with different historical changes; (3) forests with logging of wood species; (4) secondary forests with evidence of natural and anthropogenic changes and (5) agriculture areas.

Seedling density emerging from soil samples is used in all datasets as an indirect estimate of seed density in the seed bank [51–56]. In addition to the highly variation found in the seed banks, even at small distances [20, 54, 55], we see variation in the methods of soil sampling and sampled area [56], sampling depth and spread of sample in the nursery [15], all of which could influence both density and floristic composition. It should be noted that the tropical region lacks seedling identification guides or floras, making this activity largely dependent on the expertise of parabotanics and researchers involved in the field. It is well known that the seedling stage is one of the most difficult stages to identify, as reflected in the floristic composition of the seed bank. Even the division of seedlings into a life form, is difficult to position and categorize. To compile a file form database, we followed the categories proposed by the authors, but with minor changes. We chose to group emerged seedlings into four major categories, i.e., tree, herb, shrub, and support-dependent plants, which included lianas, epiphytes, and hemiepiphytes. Some species like *Miconia serialis* DC. can be shrubby to small trees; however, the small tree life form is the most common, and this species was placed in the tree categories.

3. Results

In general, the lower density of seedlings that emerged from the seed bank samples was observed in old-growth forests, while the highest density in seedlings emerged in agricultural areas (**Figure 1**). Seedling density in old-growth forests had less variability, with numbers varying between 94 and 913 seedlings per m^2 (**Table 1**). In the other classes of land use, the density of seeds in the soil was found to be higher and had high variation (**Figure 1** and **Table 1**).

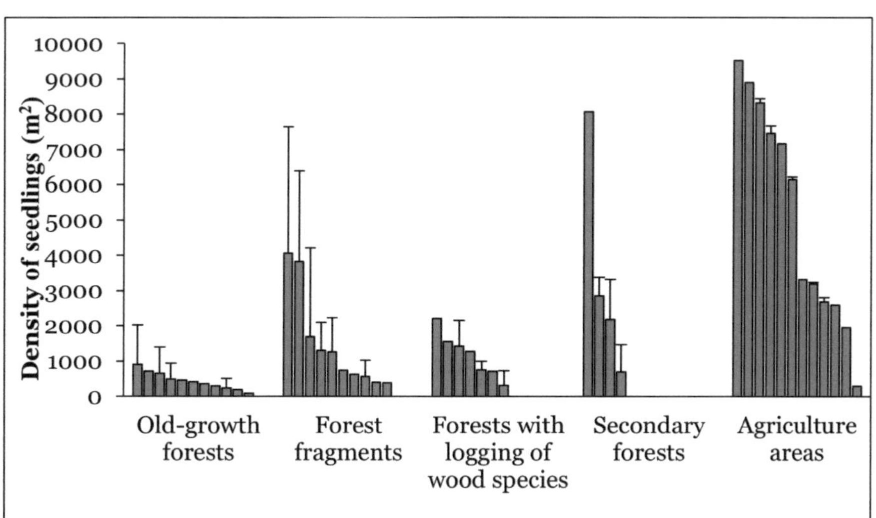

Figure 1.
Seedling density (m^2) emerged from the seed bank in different classes of land use. The vertical bar shows the standard deviation when cited. The datasets used are those described in table 1.

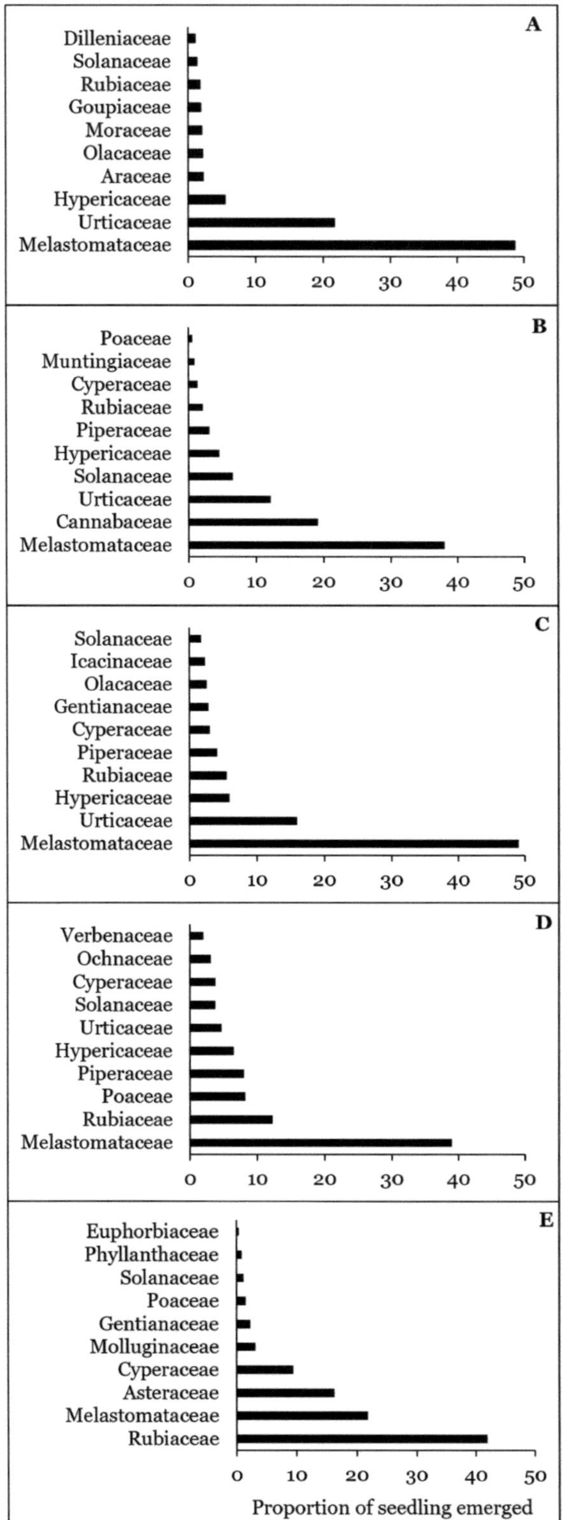

Figure 2.
The proportion of seedlings emerged from seed banks of the ten most abundant families according to different types of land use. A: Old-growth forests, with no evidence of anthropogenic changes in the last 60 years or more; B: Forest fragments with different historical changes; C: Forests with logging of wood species; D: Secondary forests with evidence of natural and anthropogenic changes; E: Agriculture areas.

The type of land use promotes changes in floristic composition (**Figure 2**). Melastomataceae seedlings predominated in all land uses, except for agricultural areas where Rubiaceae seedlings were the most abundant (**Figure 2**). Melastomataceae was represented by the following genera: *Aciotis*, *Adelobotrys*, *Bellucia*, *Clidemia*, *Henriettea*, *Leandra*, *Maieta*, *Miconia,* and *Tococa*. Urticaceae was the second most abundant family in the old-growth forests and the forests with logging of wood species. It was the third most abundant in forest fragments with different historical changes (**Figure 2**). Here, the following genera predominated: *Cecropia*, *Coussapoa,* and *Pourouma*, with only *Cecropia* occurring in agricultural areas and with low density. The families Dilleniaceae (*Davilla*, *Doliocarpus,* and *Tetracera*), Goupiaceae (*Goupia glabra* Aubl.), Moraceae (*Ficus*, *Bagassa*, *Helicostylis,* and *Maquira*), and Araceae (*Philodendron*) were present among the ten most abundant families, but only for old-growth forests (**Figure 2**). Hypericaceae seedlings, as represented by *Vismia* species, were among the ten most abundant families for all types of land use, except for agricultural areas, and, similar to Urticaceae, they occurred at low density (**Figure 2**). Cannabaceae seedlings represented an important component in forest fragments. It was represented by a single species, *Trema micranta* (L.) Blume, with wide distribution, and it serves as an indicator of degraded areas under anthropic use. The Piperaceae family was among the ten most abundant families in the category of intermediate change. It was absent from old-growth forests and agricultural areas. Cyperaceae and Poaceae were configured as a common component of altered areas. Poaceae, however, is not among the ten most abundant families for forests with logging of wood species. Solanaceae, as well as Rubiaceae, was present in all forest types; however, the latter had greater abundance in secondary forests and agricultural areas. Asteraceae (*Chromolaena*, *Rolandra*, and *Vernonia)* and Cyperaceae (*Cyperus*, *Rhynchospora*, and *Fimbristyllis*) had greater abundance in agricultural areas (**Figure 2**). Seedlings of Olacaceae (*Heisteria*) were among the ten most abundant families, but only for old-growth forests and forests with logging of wood species (**Figure 2**). Seedlings of Gentianaceae (*Coutoubea* and *Irlbachia*) were among the ten most abundant families, but only for forests with logging of wood species and agricultural areas. Muntingiaceae (*Muntingia*)

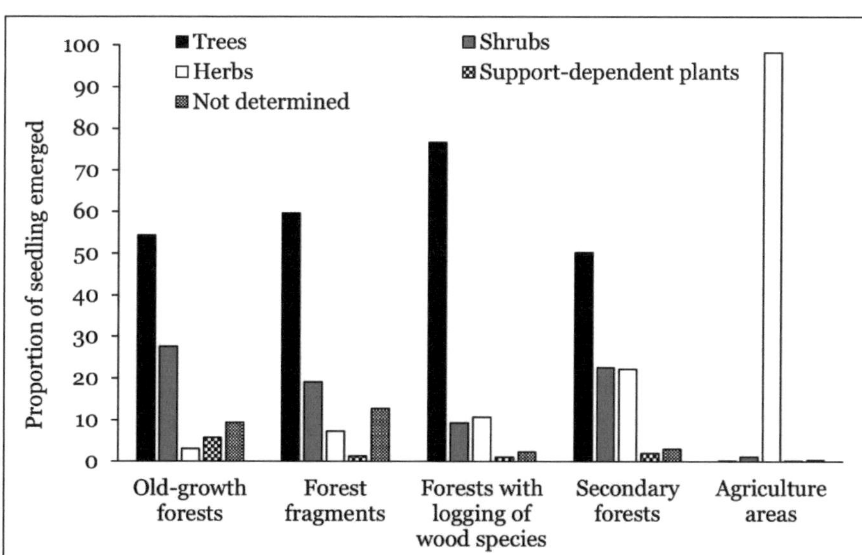

Figure 3.
The proportion of seedlings emerged from seed banks divided into life forms according to different types of land use.

was configured among the ten most abundant families only for forest fragments, Icacinaceae (*Dendrobangia*) only for forests with logging of wood species and Verbenaceae (*Stachytarpheta*) and Ochnaceae (*Lacunaria*, *Ouratea* and *Sauvagesia*) only for secondary forests. Euphorbiaceae (*Croton*), Phyllanthaceae (*Phyllanthus*) and Molluginaceae (*Mollugo*) were also among the ten most abundant families, but only for agricultural areas (**Figure 2**).

Tree seedlings predominated in all types of land use in the seed bank, except for agricultural areas (**Figure 3**). Herbs increased in frequency according to land use, with a high proportion in the seed bank in agricultural areas. Despite the low proportion of seedlings classified as support-dependent plants (lianas, epiphytes, and hemiepiphytes) they still showed a higher proportion in the old-growth forests. In the seed bank of agricultural areas, a suppression of other life forms was observed (**Figure 3**). In the seed bank of forests with logging of wood species, shrubs decreased, while the proportion of tree seedlings increased.

4. Discussion

The seed bank has been the subject of studies in different forest types. However, literature surveys carried out in the present study reveal that very few studies reporting on the Amazon rainforest have been published. This highlights the need to expand research on seed banks in natural and anthropized areas. The data used in this chapter account for regions close to the capitals of the states of Amazonas and Pará (**Table 1**), owing to easy access by highways, in addition to universities and research institutes with a long tradition in ecological studies.

Changing land use in the Brazilian Amazon threatens the extinction of a significant number of species and consequent loss of environmental functions and services of the largest tropical forest on the planet [35, 37, 62–64]. Despite repeated warnings and concerns of conservationists and the scientific community, deforestation continues at an accelerated rate [63, 65]. The replacement of the forest by pasture has been the main means of occupation and use of the land, as agriculture advances in the region [65–67]. The resilience of the forest and natural regeneration depends on several factors. Among them are type and intensity of the initial disorder, recurrence of disorders, topography, soil type, and the maintenance of accessible propagation sources [3, 36, 40, 68, 69]. The main mechanisms involved in the regeneration of these altered areas occur through the seed bank, dispersion of seeds from nearby areas, and vegetative regeneration, which includes surviving plants capable of sprouting from both the aerial part and the roots [3, 8, 69].

We generally do find a high density of seeds in the altered areas. Nevertheless, the type and intensity of disturbances and changes occurring around in forest areas contribute to corresponding changes in the floristic composition of the seed bank [5]. Moreover, invasive, or ruderal, species are common and cause the impoverishment of the seed bank [70–74]. Thus, understanding the effects of different types of land use on the seed bank is fundamentally essential to understand the evolution of the landscape, identify obstacles to the restoration of the forest, and, consequently, ensure the regeneration of forest environments and maintenance of environmental services [41, 64, 65, 69, 70, 74].

Our data support the results of other studies carried out in tropical regions where the density of seeds in the topsoil is highly variable [5, 10, 14, 21, 33, 34]. Seed density has increased from the old-growth forest to the altered areas (**Figure 1**). The observed variations in seed density in each class of land use (**Table 1**, **Figure 1**) reflect differences in forest typology, canopy opening, and sampling time among the areas [9, 14, 19, 21]. In addition, intrinsic variations are associated with the seed bank [20, 54], as well as methodological differences [15, 52, 56].

The seed bank is characterized by the occurrence and dominance of a limited number of botanical families. These families contribute markedly to common species and genera in secondary forests or the early stages of forest regeneration. Among the ten most abundant families in the seed bank, the presence of a high number of seedlings belonging to the Melastomatacea family stands out. This family has high diversity in the Neotropics, with approximately 3000 species, being composed of shrubs, lianas, herbs, epiphytes, and trees [75]. In the Amazon basin, the family is mainly composed of small tree species and shrubs, and it occurs in high abundance and diversity in the forest understory [76–78].

The Melastomataceae family is an important component of the seed bank of the Amazon rainforest [7, 19, 60], as well as other forest types in the Neotropical region [9, 23, 79, 80]. Its high abundance can likely be attributed to the number of small seeds produced per individual [18, 81], longevity [82], and photoblastic seeds, favoring the recruitment of seedlings in environments with greater luminosity [83–86]. The Melastomataceae family is composed of pioneer species that require high to low light, as well as species tolerant to shading [86–88]. Given the great importance of this family to the seed bank, more detailed studies need to be performed in order to better understand the spectrum and functionality of this group in the process of ecological succession.

Urticaceae seedlings consist of *Cecropia*, *Coussapoa*, and *Pourouma* configured as an important component in forest types with low land-use intensity, such as old-growth forests, forest fragments with different historical changes, and forests with logging of wood species (**Figure 2**). The pioneer species of *Cecropia* stand out for colonizing secondary areas that have suffered low impact disturbances, those are more important in the succession processes of these areas [39–43].

In the present study, seedlings of tree species predominated in the seed bank, except for agricultural areas where herbs predominated (**Figure 3**). A decrease in tree seedlings and an increase in herbs can already be observed in secondary forests. Herbs increased density with intensity of disturbance, with low density in old-growth forests (**Figure 3**). In general, the forest seed bank is dominated by trees (49% on average), while cultivated areas and secondary forests are dominated by herbs (75% on average) [14]. The high density of herbs in secondary forests and forest fragments results from the occurrence of anthropized areas around these areas [5, 14, 22, 29, 33]. The importance of shrubs and small trees is little studied in successional processes in tropical forests. Most studies focus on changes in the structure and floristic composition of the woody layer [89–94], but such studies exclude many groups that occur in high density in the seed bank, groups which can play a relevant role in the mechanisms of ecological succession. These groups also respond to different time scales in biological attributes, such as lifetime, reproductive age, and rate of evolution [95].

Secondary forests in the Amazon may result from the abandonment of areas previously used for different purposes, such as shifting agriculture, pastures, and mining [63, 65, 96, 97], which rarely originate from natural disorders [46]. Abandoned pastures occur after years of grazing and cleaning, usually by fire [65, 69]. These areas usually have a seed bank with high density and composition mainly consisting of locally produced herb seeds [41, 65, 69]. This seed bank is very similar to that with established vegetation cover [39–44], which is not seen in old-growth forests [14, 98, 99]. Among the ten most abundant families in the seed bank, common herbs from high-impact degraded areas, such as Poaceae, Rubiaceae, Asteraceae, and Cyperaceae predominate. However, seedlings of typical families from the seed bank of old-growth forests do occur (**Figure 2**). Floristic composition and seed density in agricultural areas suggest the need to use forest restoration techniques after abandonment, to facilitate and accelerate the return of the forest.

Forest fragments are stretches of forest inserted in a matrix of different types of land uses, typically of anthropic origin [38, 100–102]. The areas used in this study encompass a variety of forest fragments, requiring a more detailed analysis of the characteristics of each. Increase in seed density and changes in the floristic composition of the seed bank intensify in small fragments inserted in a matrix composed of pastures, as well as recurrence of disturbances in these forest fragments [5, 24, 34, 74, 100]. While large forest fragments over 100 ha have a density and floristic composition more similar to the seed bank of old-growth forests, the seed bank also contains species typical of anthropized areas [7, 19, 102, 103].

In the areas of forests with logging of wood species, we can find a mosaic of altered and unaltered areas [45, 57, 104–110] with marked differences between open canopy areas and those that suffered little or no impact [104, 107]. Thus, a greater number of seeds are found in the soil in the centers of exploration clearings and tractor trails [45, 57]. Later, with regeneration, seed density declines and approaches pre-exploratory conditions [45]. The density, as well as life forms, of these areas is closer to that of old-growth forests (**Figures 1** and **3**). For the three most abundant families in the seed bank, floristic composition is very similar to that of old-growth forests. On the other hand, in other families, such as Cyperaceae, Rubiaceae, and Piperaceae, we see higher density of seedlings characteristic of open areas (**Figure 2**).

The seed bank is directly related to forest resilience which contributes to a large number of regenerants, including species of ecological groups not present in the arboreal stratum of old-growth forests. This means that dramatic changes in the seed bank owing to the use and management of soil will, in turn, promote changes in floristic composition and density in a manner that favors the introduction of species not commonly found in the seed bank of old-growth forests. Ultimately, these conditions cause the impoverishment of the seed bank and consequent loss of its functionality. In extreme cases where total suppression of the seed bank has occurred, its absence completely alters the successional trajectory [39–44]. The seed bank is essential for resilience, forest regeneration, and forest diversity; therefore, any changes in its functionality compromise the diversity, regeneration, and maintenance of tropical forests.

Acknowledgements

This research is part of TVB's project of the Institutional Training Program of the National Institute for Amazon Research - INPA. TVB acknowledges the financial support of the National Council for Scientific and Technological Development – CNPq, Brazil [process #301481/2020-2]. To Thaiane Rodrigues de Sousa for providing the worksheets to prepare the review. For undergraduate students: Gisiane Rodrigues Lima, Tamires Ferreira Muniz, Gisele Rodrigues dos Santos and Pedro Cavalcante da Cruz under the supervision of Dr. Niwton Leal Filho, from the research group on forest seeds of the National Institute for Amazon Research - INPA, who proved unpublished data for the preparation of the chapter.

Conflict of interest

There are no conflicts of interest in the chapter entitled "Ecology of the seed bank in the Amazon rainforest" to be considered for publication in Open Access book at IntechOpen in the book Ecosystem and Biodiversity of Amazonia", ISBN 978–1–83,962-813-9. The data used in the chapter have no conflicting interests since there is no conflict of interest by the authors. All sources of funding were cited in the acknowledgments and all help received in the execution of the research was properly cited. The data used in this chapter refer to previously published articles and theses, as well as the unpublished data, refer to the result of scientific initiation under the supervision of Dr. Niwton Leal Filho.

Author details

Natali Gomes Bordon[1*], Niwton Leal Filho[2] and Tony Vizcarra Bentos[2]

1 University of Campinas (UNICAMP), Campinas, Brazil

2 National Institute for Amazon Research (INPA), Manaus, Brazil

*Address all correspondence to: natalibordon@yahoo.com.br

References

[1] Barker HG. 1989. Some aspects of the natural history of seed banks, p. 9-21. In: Leck MA, Parker T, Simpson RL, editors. Ecology of soil seed banks. San Diego: Academic Press; 1989. p. 9-21. DOI: 10.1016/B978-0-12-440405-2. X5001-5.

[2] Baskin CC, Baskin JM. 2001. Seeds: Ecology, Biogeography, and Evolution of Dormancy and Germination. San Diego: Academic Press; 2001. 659p.

[3] Castillo LS, Stevenson PR. Relative importance of seed-bank and post-disturbance seed dispersal on early gap regeneration in a Colombian Amazon Forest. Biotropica. 2010; 42: 488-492. DOI: 10.1111/j.1744-7429.2009.00605.x.

[4] López-Toledo L, Martínez-Ramos M. The soil seed bank in abandoned tropical pastures: source of regeneration or invasion? Revista Mexicana de Biodiversidad. 2011. 82: 663-678. DOI: 10.1111/j.1744-7429.2009.00605.x.

[5] Miranda IS, Mitja D, SILVA TS. Mutual influence of forests and pastures on the seedbanks in the Eastern Amazon. Weed Research. 2009; 49: 499-505.DOI: 10.1111/j.1365-3180.2009.00719.x.

[6] Thompson K. The functional ecology of soil seed banks. In: Fenner M. editor. Seeds: the ecology of regeneration in plant communities. Wallingford: CAB International; 1992. p. 231-258. DOI: 10.1079/SSR2003142

[7] Sousa TR, Costa FRC, Bentos TV, Leal Filho N, Mesquita RCG, Ribeiro IO. The effect of forest fragmentation on the soil seed bank of Central Amazonia. Forest Ecology and Management. 2017; 393: 105-112. DOI: 10.1016/j. foreco.2017.03.020

[8] Vieira ICG, Proctor J. Mechanisms of plant regeneration during succession after shifting cultivation in eastern Amazonia. Plant Ecology. 2007; 192: 303-315. DOI: 10.1007/sl 1258-007-9327-4

[9] Dalling JW, Denslow JS. Soil seed bank composition along a forest chronosequence in seasonally moist tropical forest, Panama. Journal of Vegetation Science. 1998; 9: 669-678. DOI: 10.2307/3237285.

[10] Dupuy JM, Chazdon RL. Long-term effects or forest regrowth and selective logging on the seedbank of tropical forests in northeastern Costa Rica. Biotropica. 1998; 30: 223-237.DOI: 10.1111/j.1744-7429.1998. tb00057.x.

[11] Menalled FD, Gross KL, Hammond M. Weed aboveground and seedbank community responses to agricultural management systems. Ecological Applications. 2001; 11:1586-1601. DOI: 10.2307/3061080.

[12] Wagner M, Mitschunas N. Fungal effects on seed bank persistence and potential applications in weed biocontrol: A review. Basic and Applied Ecology. 2008; 9: 191-203.DOI: 10.1016/j.baae.2007.02.003.

[13] Harper JL.1977. The Population Biology of Plants. London: Academic Press; 1977. 892 p. DOI: 10.1017/ S0376892900005774

[14] Garwood NC. Tropical soil seed banks: a review. In: Leck MA, Parker T, Simpson RL, editors. Ecology of soil seed banks. San Diego: Academic Press; 1989. p. 149-209. DOI: 10.1016/B978-0-12-440405-2.X5001-5

[15] Dalling JW, Swaine, MD, Garwood NC. Effect of soil depth on seedling emergence in tropical soil seed-bank investigations. Funct. Ecol. 1994; 9, 119-121. DOI: 10.2307/2390098

[16] Pearson TRH, Burslem DFRP, Mullins CE, Dalling JW. Germination ecology of neotropical pioneers: interacting effects of environmental conditions and seed size. Ecology. 2002; 83: 2798-2807. DOI: 10.1890/0012-9658

[17] Baskin CC, Baskin JM. Germination Ecology of seeds in the Persistent seed Bank. In: Seed Ecology, Biogeography, and Evolution of Dormancy and Germination. San Diego: Academic Press; 1998. P. 133-179. DOI: 10.1016/B978-0-12-080260-9.X5000-3

[18] Bentos TV, Mesquita RCG, Williamson GB. Reproductive Phenology of Central Amazon Pioneer Trees. Tropical Conservation Science. 2008; 1: 186-203. DOI: 10.1038/nature09273.

[19] Leal Filho N, Sena JS, Santos GR. Variações espaço-temporais no estoque de sementes do solo na floresta amazônica. Acta Amazonica. 2013; 43: 305-314. DOI: 10.1590/s0044-59672013000300006.

[20] Dalling JW, Swaine MD, Garwood NC. Dispersal patterns and seed bank dynamics of pioneer trees in moist tropical forest. Ecology. 1998; 79: 564-578. DOI: 10.1890/0012-9658(1998)079[0564,DPASBD]2.0.CO;2

[21] Fornara DA, Dalling JW. Seed bank dynamics in five Panamanian forests. Journal of Tropical Ecology. 2005; 21: 223-226. DOI: 10.1017/S0266467404002184.

[22] Medeiros-Sarmento PS, Ferreira LV, Gastauer M. Natural regeneration triggers compositional and functional shifts in soil seed banks. Science of the Total Environment. 2021; 753: 141934. DOI: 10.1016/j.scitotenv.2020.141934.

[23] Baider C, Tabarelli M, Mantovani W. The soil seed bank during Atlantic forest regeneration in southeast Brazil. Revista Brasileira Biologia. 2001; 61: 35-44. DOI: 10.1590/S0034-71082001000100006.

[24] Martins AM, Engel VL. Soil seed banks in tropical forest fragments with different disturbance histories in southeastern Brazil. Ecological Engineering. 2007; 31:165-174. DOI: 10.1016/j.ecoleng.2007.05.008

[25] Vázquez-Yanes C, Smith H. Phytochrome control of seed germination in the tropical rain forest pioneer trees *Cecropia obtusifolia* and *Piper auritum*, and its ecological significance. New Phytologist. 1982; 92: 477-485. DOI: 10.1111/j.1469-8137.1982.tb03405.x

[26] Vásquez-Yanes C, Orozco-Segovia A. Ecological significance of light controlled seed germination in two contrasting tropical habitas. Oecologia. 1990; 83: 171-175. DOI: 10.1007/BF00317748

[27] Pearson TRH, Burslem DFRP, Mullins CE, Dalling JW. Functional significance of photoblastic germination in neotropical pioneer trees: a seed's eye view. Functional Ecology. 2003; 17: 394:402. DOI: 10.1046/j.1365-2435.2003.00747.x

[28] Brokaw NVL. Treefalls, regrowth, and community structure in tropical forests. In: Pickett STA, White PS, editors. The ecology of natural disturbance and patch dynamics. San Diego: Academic Press, 1985. P. 53-69. DOI: 10.1016/C2009-0-02952-3

[29] Whitmore. T. C.1996. A review of some aspects of tropical rain forest seedling ecology with suggestions for further enquiry. In: Swaine, MD, editor. The ecology of tropical forest tree seedlings. Volume 17. New York: UNESCO; 1996. P. 3-39.

[30] Thompson K. Gap and Seedling Colonization. In: Fenner M, editor. Seeds The Ecology of Regeneration in Plant Communities. 2nd ed. London:

British Library; 2000. p. 375-395. DOI: 10.1079/9780851994321.0000.

[31] Uhl C. Factors Controlling Succession Following Slash-and-Burn Agriculture in Amazonia. The Journal of Ecology. 1987; 75: 377-407. DOI: 10.2307/2260425.

[32] Uhl C, Buschbacher R, Serrao EAS. Abandoned Pastures in Eastern Amazonia. I. Patterns of Plant Succession. The Journal of Ecology. 1988; 76: 663-681. DOI: 10.2307/2260566.

[33] Young KR, Ewel JJ, Brown BJ. Seed dynamics during forest succession in Costa Rica. Vegetatio. 1987; 71: 157-173. DOI: 10.1007/BF00039168.

[34] Quintana-Ascencio PFM, González-Espinosa; Mirez-Marcial, N.; Domínguez-Vasquez, G.; Martínez-Ico, M. 1996. Soil seed banks and regeneration of tropical rain forest from milpa fields at the Selva Lacandona, Chiapas, Mexico. Biotropica. 1996; 28:192-209. DOI: : 10.2307/2389074

[35] Jakovac CC, Peña-Claros M, Kuyper TW, Bongers F. 2015. Loss of secondary forest resilience by land-use intensification in the Amazon. J. Ecol. 2015; 103: 67-77. DOI: 10.1111/1365-2745.12298.

[36] Mesquita RDCG, Massoca PEDS, Jakovac CC, Bentos TV, Williamson GB. 2015. Amazon rain forest succession: stochasticity or land-use legacy? Bioscience. 2015; 65: 849-861. DOI: 10.1093/biosci/biv108.

[37] Chua SC, Ramage BS, Potts MD. Soil degradation and feedback processes affect long-term recovery of tropical secondary forests. J. Veg. Sci. 2016; 27: 800-811. DOI: 10.1111/jvs.12406.

[38] Arroyo-Rodríguez V, Melo FPL, Martínez-Ramos M, Bongers F, Chazdon RL, Meave JA, Norden N, Santos BA, Leal IR, Tabarelli M. Multiple successional pathways in human-modified tropical landscapes: new insights from forest succession, forest fragmentation and landscape ecology research. Biol. Rev. 2017; 92: 326-340. DOI: 10.1111/brv.12231.

[39] Monaco LM, Mesquita RCG, Williamson GB. Banco de sementes de uma floresta secundária amazônica dominada por *Vismia*. Acta Amazonica. 2003; 33: 41-52. DOI: 10.1590/1809-4392200331052.

[40] Longworth JB, Mesquita RC, Bentos TV, Moreira MP, Massoca PE, Williamson, GB. 2014. Shifts in dominance and species assemblages over two decades in alternative successions in Central Amazonia. Biotropica. 2014; 46: 529-537. DOI: 10.1111/btp.12143.

[41] Rocha GPE, Vieira DLM, Simon MF. Fast natural regeneration in abandoned pastures in southern Amazonia. Forest Ecology and Management. 2016; 370: 93-101. DOI: 10.1016/j.foreco.2016.03.057.

[42] Mesquita RCM, Icke SK, Ganade G, Williamson GB. Alternative successional pathways in the Amazon Basin. Journal of Ecology. 2001; 89:528-537. DOI: 10.1046/j.1365-2745.2001.00583.x.

[43] Williamson, GB, Bentos TV, Longworth JB, Mesquita RCG. Convergence and divergence in alternative successional pathways in Central Amazonia. Plant Ecol. Divers. 2014; 7: 341-348. DOI: 10.1080/17550874.2012.735714.

[44] Jakovac ACC, Bentos TV, Mesquita RCG, Williamson GB. Age and light effects on seedling growth in two alternative secondary successions in central Amazonia. Plant Ecology & Diversity. 2014; 7: 349-358. DOI: 10.1080/17550874.2012.716088.

[45] Leal Filho, N. Dinâmica inicial da regeneração natural de florestas exploradas na Amazônia brasileira [thesis]. São Paulo: Instituto de Biociências da Universidade de São Paulo. Departamento de Ecologia Geral; 2000.

[46] Bordon NG, Nogueira A, Leal Filho N, Higuchi N. Blowdown disturbance effect on the density, richness and species composition of the seed bank in Central Amazonia. Forest Ecology and Management, 2019; 453: 117633. DOI: 10.1016/j.foreco.2019.117633.

[47] Saulesi M, Swaine MD. Rain forest seed dynamics during succession at Gogol, Papua New Guinea. Journal of Ecology. 1998; 76: 1133-1152. DOI: 10.2307/2260639

[48] Costa JR, Mitja D. Bancos de sementes de plantas daninhas em sistemas agroflorestais na Amazônia Central. Revista Brasileira de Ciências Agrárias - Brazilian Journal of Agricultural Sciences. 2009; 4: 298-303. DOI: 10.5039/agraria.v4i3a12.

[49] Costa JR, Mitja D, Fontes JRA. Bancos de sementes de plantas daninhas em cultivos de mandioca na Amazônia Central. Planta Daninha. 2009; 27: 665-671. DOI: 10.1590/S0100-83582009000400004.

[50] Peçanha-Júnior FB. Avaliação do banco de sementes da floresta de Caxiuanã, município de Melgaço, Pará, Brasil [thesis]. Belém: Universidade de Federal Rural da Amazônia e Museu Paraense Emílio Goeldi; 2006.

[51] Gonzalez S, Ghermandi L. Comparison of methods to estimate soil seed banks: the role of seed size and mass. Community Ecology. 2012; 13: 238-242. DOI: 10.1556/ComEc.13.2012.2.14.

[52] Gross KLA. Comparison of Methods for Estimating Seed Numbers in the Soil. The Journal of Ecology. 1990; 78: 1079-1093. DOI: 10.2307/2260953.

[53] Price JN, Wright BR, Gross CL, Whalley WRDB. Comparison of seedling emergence and seed extraction techniques for estimating the composition of soil seed banks. Methods in Ecology and Evolution. 2010; 1: 151-157. DO: 10.1111/j.2041-210X.2010.00011.x.

[54] Plue J, Goyens G, Van Meirvenne M, Verheyen K, Hermy M. Small-scale seed-bank patterns in a forest soil. Seed Science Research. 2010; 20: 13-22. DOI: 10.1017/S0960258509990201.

[55] Dalling JW, Swaine MD, Garwood NC. Soil seed bank community dynamics in seasonally moist lowland tropical forest, Panama. Journal of Tropical Ecology. 1997; 13: 659-680, 10 Sep. 1997. DOI: 10.1017/S0266467400010853.

[56] Butler BJ, Chazdon RL. Species Richness, Spatial Variation, and Abundance of the Soil Seed Bank of a Secondary Tropical Rain Forest. Biotropica. 1998; 30: 214-222. DOI: 10.1111/j.1744-7429.1998.tb00056.x. A

[57] Quanz B, Carvalho JOP, Araujo, MM, Francez LMB, Silva USC, Pinheiro KAO. Exploração florestal de impacto reduzido não afeta a florística do banco de sementes do solo. Rev. Ciências Agrárias. 2012; 55: 204-211. DOI: 10.4322/rca.2012.055.

[58] França, AL. Similaridade florística e banco de sementes da zona ripária de igarapés de bacias hidrográficas distintas na Amazônia Central [thesis]. Manaus: Instituto Nacional de Pesquisas da Amazônia; 2018.

[59] Araujo MM, Oliveira FA, Vieira ICG, Barros PLC, Lima CAT. Densidade e composição florística do banco de sementes do solo de florestas sucessionais na região do Baixo Rio

Guamá, Amazônia oriental. Scientia Forestalis. 2001; 59: 115:130.

[60] Bentos TV, Nascimento HEM, Williamson GB., 2013. Tree seedling recruitment in Amazon secondary forest: importance of topography and gap micro-site conditions. For. Ecol. Manag. 2013; 287: 140-146. DOI: 10.1016/j.foreco.2012.09.016.

[61] Costa JR, Mitja D, Leal Filho, N. Bancos de sementes do solo em pastagens na Amazônia Central. Pesquisa Florestal Brasileira. 2013; 33: 117-125. DOI: 10.4336/2013.pfb.33.74.431.

[62] Zemp DC, Schleussner CF, Barbosa HMJ, Rammig A. Deforestation effects on Amazon forest resilience. Geophysical Research Letters. 2017; 44: 6182-6190. DOI: 10.1002/2017GL072955.

[63] Bullock EL, Woodcock CE, Souza C, Olofsson P. Satellite-based estimates reveal widespread forest degradation in the Amazon. Global Change Biology. 2020; 26: 2956-2969. DOI: 10.1111/gcb.15029.

[64] Luther DA, Cooper WJ, Wolfe JD, Bierregaard RO, Gonzalez A, Lovejoy TE. Tropical forest fragmentation and isolation: Is community decay a random process? Global Ecology and Conservation. 2020; 23: e01168. DOI: 10.1016/j.gecco.2020.e01168.

[65] Laurance WF, Sayer J, Cassman, KG. Agricultural expansion and its impacts on tropical nature. Trends in Ecology & Evolution. 2014; 29: 107-116. DOI: 10.1016/j.tree.2013.12.001.

[66] Fearnside PM. Amazonian deforestation and global warming: carbon stocks in vegetation replacing Brazil's Amazon forest. Forest Ecology and Management. 1996; 80: 21-34. DOI: 10.1016/0378-1127(95)03647-4.

[67] Ferraz SFB, Vettorazzi CA, Theobald DM, Ballester MVR. Landscape dynamics of Amazonian deforestation between 1984 and 2002 in central Rondônia, Brazil: assessment and future scenarios. Forest Ecology and Management. 2005; 204: 69-85. DOI 10.1016/j.foreco.2004.07.073.

[68] Seifan M, Seifan T, Jeltsch F, Tielbörger K. Combined disturbances and the role of their spatial and temporal properties in shaping community structure. Perspectives in Plant Ecology, Evolution and Systematics. 2012; 14: 217-229. DOI: 10.1016/j.ppees.2011.11.003.

[69] Robin LC. Tropical forest recovery: legacies of human impact and natural disturbances. Perspectives in Plant Ecology, Evolution and Systematics. 2003; 6: 51-71. DOI: 10.1078/1433-8319-00042.

[70] Wijdeven SMJ, Kuzee ME. Seed Availability as a Limiting Factor in Forest Recovery Processes in Costa Rica. Restoration Ecology. 2000; 8: 414-424. DOI: 10.1046/j.1526-100x.2000.80056.x.

[71] Silva ÚSR, Matos DMS. The invasion of *Pteridium aquilinum* and the impoverishment of the seed bank in fire prone areas of Brazilian Atlantic Forest. Biodiversity and Conservation. 2006; 9: 3035-3043. DOI: 10.1007/s10531-005-4877-z.

[72] Gioria M, Jarošík V, Pyšek P. Impact of invasions by alien plants on soil seed bank communities: Emerging patterns. Perspectives in Plant Ecology, Evolution and Systematics. 2014; 16: 132-142. DOI: 10.1016/j.ppees.2014.03.003.

[73] Gioria M, Pyšek P. The Legacy of Plant Invasions: Changes in the Soil Seed Bank of Invaded Plant Communities. BioScience. 2016; 66: 40-53. DOI: 10.1093/biosci/biv165.

[74] Alvarez-Aquino C, Williams-Linera G, Newton AC. Disturbance effects on the seed bank of Mexican cloud forest fragments. Biotropica. 2005; 37: 337-342. DOI: 10.1111/j.1744-7429.2005.00044.x.

[75] Clausing G, Renner SS. Molecular phylogenetics of Melastomataceae and Memecylaceae: implications for character evolution. American Journal of Botany. 2001; 88: 486-498. DOI: 10.2307/2657114.

[76] Tuomisto H, Ruokolainen K, Aguilar M, Sarmiento A. Floristic patterns along a 43-km long transect in an Amazonian rain forest. Journal of Ecology. 2003; 91: 743-756. DOI: 10.1046/j.1365-2745.2003.00802.x.

[77] Tuomisto H, Ruokolainen K. Yli-Halla M. Dispersal, Environment, and Floristic Variation of Western Amazonian Forests. Science. 2003; 299: 241-244. DOI: 10.1126/science.1078037.

[78] Duque AJ, Duivenvoorden JF, Cavelier J, Sánchez M, Polanía C, León A. Ferns and Melastomataceae as indicators of vascular plant composition in rain forests of Colombian Amazonia. Plant Ecology. 2005; 178: p. 1-13. DOI: 10.1007/s11258-004-1956-2.

[79] Pereira-Diniz SG, Ranal MA. Germinable soil seed bank of a gallery forest in Brazilian Cerrado. Plant Ecology. 2006; 183: 337-348. DOI: 10.1007/s11258-005-9044-9.

[80] Franco BKS, Martins SV, Faria PCL, Ribeiro GA. Densidade e composição florística do banco de sementes de um trecho de floresta estacional semidecidual no campus da Universidade Federal de Viçosa, Viçosa, MG. Revista Árvore. 2012; 36: 423-432. DOI: 10.1590/S0100-67622012000300004.

[81] Bentos TV, Mesquita RCG, Camargo JLC, Williamson GB. Seed and fruit tradeoffs – the economics of seed packaging in Amazon pioneers. Plant Ecology & Diversity. 2014; 7: 1-2. DOI 10.1080/17550874.2012.740081.

[82] Silveira FAO, Ribeiro RC, Soares S, Rocha D, Oliveira C. Physiological dormancy and seed germination inhibitors in *Miconia* (Melastomataceae). Plant Ecology and Evolution. 2013; 146: 290-294. DOI: 10.5091/plecevo.2013.817.

[83] Carreira RC, Zaidan LBP. Germinação de sementes de espécies de Melastomataceae de Cerrado sob condições controladas de luz e temperatura. Hoehnea. 2007; 34: 261-269. DOI: 10.1590/S2236-89062007000300001.

[84] Silveira FAO, Ribeiro RC, Oliveira DMT, Fernandes GW, Lemos-Filho JP. Evolution of physiological dormancy multiple times in Melastomataceae from Neotropical montane vegetation. Seed Science Research. 2012; 22: 37-44. DOI: 10.1017/S0960258511000286.

[85] Godoi S, Takaki M. Seed germination in *Miconia theaezans* (Bonpl.) Cogniaux (Melastomataceae). Brazilian Archives of Biology and Technology. 2007; 50: 571-578. DOI 10.1590/S1516-89132007000400002.

[86] Silveira FAO, Fernandes GW, Lemos-Filho JP. Seed and Seedling Ecophysiology of Neotropical Melastomataceae: Implications for Conservation and Restoration of Savannas and Rainforests. Annals of the Missouri Botanical Garden. 2013; 99: 82-99. DOI: 10.3417/2011054.

[87] Ellison AM, Denslow JS, Loiselle BA. Seed and Seedling Ecology of Neotropical Melastomataceae. Ecology. 1993; 74: 1733-1749. DOI: 10.2307/1939932.

[88] Putz FE. Treefall Pits and Mounds, Buried Seeds, and the Importance

of Soil Disturbance to Pioneer Trees on Barro Colorado Island, Panama. Ecology. 1983; 64: 1069-1074. DOI: 10.2307/1937815.

[89] Lu D, Mausel P, Brondízio E, Moran E. Classification of successional forest stages in the Brazilian Amazon basin. Forest Ecology and Management. 2003; 181: 301-312. DOI: 10.1016/S0378-1127(03)00003-3.

[90] Van Breugel M, Bongers F, Martínez-Ramos M. Species Dynamics During Early Secondary Forest Succession: Recruitment, Mortality and Species Turnover. Biotropica. 2007; 39: 610-619. DOI 10.1111/j.1744-7429.2007.00316.x.

[91] Pena-Claros M. Changes in Forest Structure and Species Composition during Secondary Forest Succession in the Bolivian Amazon. Biotropica. 2004; 35: 450-461. DOI 10.1111/j.1744-7429.2003.tb00602.x.

[92] Saldarriaga JG, West DC, Tharp ML, Uhl C. Long-Term Chronosequence of Forest Succession in the Upper Rio Negro of Colombia and Venezuela. The Journal of Ecology. 1998; 76: 938-958. DOI:10.2307/2260625.

[93] Finegan B. Pattern and process in neotropical secondary rain forests: the first 100 years of succession. Trends in Ecology & Evolution. 1996; 11: 119-124. DOI 10.1016/0169-5347(96)81090-1.

[94] Marra DM, Chambers JQ, Higuchi N, Trumbore SE, Ribeiro GHPM, Santos J, Negrón-Juárez RI, Reu B, Wirth C. Large-Scale Wind Disturbances Promote Tree Diversity in a Central Amazon Forest. PLoS ONE. 2014; 9: e103711. DOI 10.1371/journal.pone.0103711.

[95] Lanfear R, Ho SYW, Davies J, Moles AT, Aarssen L, Swenson NG, Warman L, Zanne, AE, Allen AP. Taller plants have lower rates of molecular evolution. Nature Communications. 2013; 4: 1879. DOI: 10.1038/ncomms2836.

[96] Lapola DM, Schaldach R, Alcamo J, Bondeau A, Msangi S, Priess JA, Silvestrini R, Soares-Filho BS. Impacts of Climate Change and the End of Deforestation on Land Use in the Brazilian Legal Amazon. Earth Interactions. 2011; 15: 1-29. DOI: 10.1175/2010EI333.1.

[97] Aldrich SP, Walker RT, Arima EY, Caldas MM, Browder JO, Perz S. Land-Cover and Land-Use Change in the Brazilian Amazon: Smallholders, Ranchers, and Frontier Stratification. Economic Geography. 2009; 82: 265-288. DOI: 10.1111/j.1944-8287.2006.tb00311.x.

[98] Cui L, Li W, Zhao X, Zhang M, Lei Y, Zhang Y, Gao C, Kang X, Sun B, Zhang Y. The relationship between standing vegetation and the soil seed bank along the shores of Lake Taihu, China. Ecological Engineering. 2016; 96: 45-54. DOI: 10.1016/j.ecoleng.2016.03.040.

[99] Hopfensperger KN. A review of similarity between seed bank and standing vegetation across ecosystems. Oikos. 2007; 116: 1438-1448. DOI: 10.1111/j.0030-1299.2007.15818.x.

[100] Hill JL, Curran PJ. Area, shape and isolation of tropical forest fragments: effects on tree species diversity and implications for conservation. Journal of Biogeography. 2003; 30: 1391-1403. DOI: 10.1046/j.1365-2699.2003.00930.x.

[101] Turner IM. Species Loss in Fragments of Tropical Rain Forest: A Review of the Evidence. The Journal of Applied Ecology. 1996; 33: 200-209. DOI: 10.2307/2404743.

[102] Laurance WF, Lovejoy TE, Vasconcelos HL, Bruna EM, Didham RK, Stouffer PC, Gascon C, Bierregaard RO, Laurance SG,

Sampaio E. Ecosystem Decay of Amazonian Forest Fragments: a 22-Year Investigation. Conservation Biology. 2001; 16: 605-618. DOI 10.1046/j.1523-1739.2002.01025.x.

[103] Lippok D, Walter F, Hensen I, Beck SG, Schleuning M. Effects of disturbance and altitude on soil seed banks of tropical montane forests. Journal of Tropical Ecology. 2013; 29: 523-529. DOI 10.1017/S0266467413000667.

[104] Greguš C. Principles of long-term sustainable forest development implemented as the background for ecological forest management. Folia Oecologica. 2013; 40: 146-152. DOI: 91836966.

[105] Lindenmayer DB, Franklin JF, Fischer J. General management principles and a checklist of strategies to guide forest biodiversity conservation. Biological Conservation. 2006; 131: 433-445. DOI 10.1016/j.biocon.2006.02.019.

[106] Macdicken KG, Sola P, Hall JE, Sabogal C, Tadoum M, Wasseige C. Global progress toward sustainable forest management. Forest Ecology and Management. 2015; 352: 47-56. DOI: 10.1016/j.foreco.2015.02.005.

[107] Hartshorn GS. Application of Gap Theory to Tropical Forest Management: Natural Regeneration on Strip Clear-cuts in the Peruvian Amazon. Ecology. 1989; 70: 567-576. DOI: 10.2307/1940208.

[108] Summers PM, Browder JO, Pedlowski MA. Tropical forest management and silvicultural practices by small farmers in the Brazilian Amazon: recent farm-level evidence from Rondônia. Forest Ecology and Management. 2004; 192: 161-177. DOI: 10.1016/j.foreco.2003.12.016.

[109] Bonilla-Bedoya S, Estrella-Bastidas A, Ordoñez M, Sánchez A, Herrera MA. Patterns of timber harvesting and its relationship with sustainable forest management in the western Amazon, Ecuador case. Journal of Sustainable Forestry. 2017; 36: 433-453. DOI 10.1080/10549811.2017.1308869.

[110] Piketty MG, Drigo I, Sablayrolles P, Aquino EA, Pena D, Sist P. Annual Cash Income from Community Forest Management in the Brazilian Amazon: Challenges for the Future. Forests. 2015; 6: 4228-4244. DOI 10.3390/f6114228.

Chapter 5

The Nymph Architect of the Cicada *Guyalna chlorogena*: Behaviours and Ecosystem

Claude François Béguin

Abstract

At the beginning of the last year of its larval life, the nymph of Guyalna chlorogena builds, from a vertical well, which is the result of a verticalization process from a deep horizontal gallery, a clay turret 20 to 40 cm high which appears as a regulating device of the physico-chemical conditions inside the burrow. The construction of the turret is remarkable for its finish. The nymph maintains, repairs and rebuilds it if necessary. It opens and closes it under certain circumstances. Before moulting, the nymph comes out at the top, opening it according to a set protocol and time schedule, using its chitins' forelegs. The burrow is associated in a commensal relationship with arborescent Fabaceae species (of the Tachigali genus) through its nutrition mode, the suction of the elaborated sap in fine roots, close to the meristems.

Keywords: behavior, building, burrow, cicada, clay, forelegs' hooked end, molt, nymph, regulating device, turret, urine, well, commensalism

1. Introduction

Over a large area of the Amazon rainforest, extending from the Brazilian state of *Para* to the south-east of *Peru* (*Madre de Dios* department), passing through *Surinam*, one can observe curious clay buildings, having the shape of turrets, or chimneys (**Figure 1a**), with a height of 20 to 40 cm and an internal diameter at their base of about 2 cm (**Figure 1b**). The turret surmounts a vertical well (**Figure 1c**) with a depth of up to about a meter, i.e. the thickness of the fertile soil layer. The surface inside the turret is perfectly smooth (**Figure 1d**).

Each turret is the visible part of the pupal burrow of the cicada *Guyalna chlorogena* (**Figure 2**), or *Fidicina chlorogena*, according to its old taxon [1]. Endoscopic exploration made it possible to observe the nymph in the well (**Figure 3**) and to verify that each burrow is occupied by a single nymph, male or female [2], which builds its turret (between December and February) a few months before moulting into a winged imago, and reproducing (between late July and early September).

Our research was conducted at the *Museu da Amazônia* (MUSA)[1], installed in the *Botanical Garden of Manaus*, on the edge of the *Adolfo Ducke Reserve* (**Figure 4**), in the northern area of Manaus, State of Amazonas, Brazil, in an area of about twenty hectares, around 59° 56′ 21.5" WO and 3° 0′ 19.7" S. A total of more than 250 burrows were observed, between November 2013 and September 2019. About 60 buildings

[1] Av. Margarita, 6305 - Cidade de Deus, Manaus - AM, 69088-265, Brésil.

Figure 1.
(a) View of a turret. (b) Turret removed. (c) Entrance to the well. (d) Inner surface of the summit.

were monitored daily. The observations made up to 2016 have been published [2]. The results obtained subsequently will be the subject of a second publication (Béguin, Gama and Ribamar Mesquita Ferreira, to be published).

2. Construction of the turret

The turret is constructed from the well, which has a double curvature near the ground surface (**Figure 5a**) and a single one in depth (**Figure 5b**), as revealed by cement casts.

From the moment it appears, the night-time growth of the turret is very rapid; 3–4 cm per night. The nymph uses a special technique that allows it to lengthen the top without ever opening it, so without exposing itself to predators. It softens the top with a mixture of its urine and clay drawn from the bottom of the well and loaded on its clypeus; then it pushes everything upwards [2].

When the nymph encounters obstacles, it continues its construction obliquely, but restores the verticality as soon as possible, manifesting an acute perception of gravity (**Figure 6**).

The monitoring of many buildings, after the construction of the turret and until the nymph abandons its burrow before moulting into a winged imago, has made it possible to identify various maintenance and rehabilitation behaviours, as well as to make hypothesis about the role of the turret.

Figure 2.
The cicada Guyalna chlorogena*; young imago after moulting (photography Vanessa Gama).*

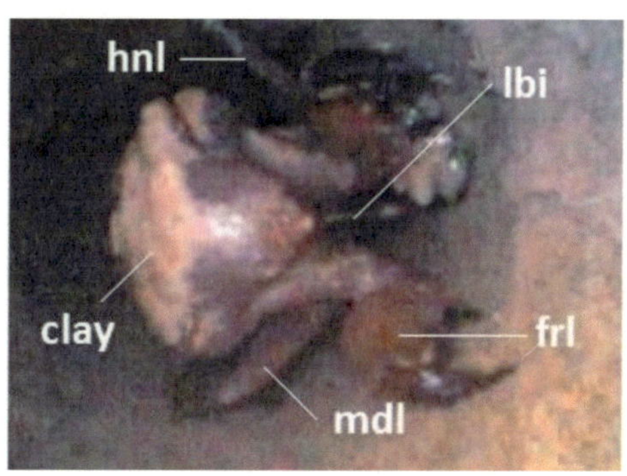

Figure 3.
Endoscopic images of the nymph in its well; lbi*: Labium,* frl*: Foreleg,*mdl*: Middle leg,* hnl*: Hind leg. Clay can be seen on the head of the nymph.*

3. Maintenance and rehabilitation

If a turret is damaged, the nymph fixes it without delay. It performs an occlusion with a mixture of clay and urine (**Figure 7**) if its turret was severed, before restoring the initial height (see below). Maintaining the sealing of the building appears to be a priority for the nymph; if one experimentally fractures a turret and then re-stack

Figure 4.
Manaus and the Adolfo Ducke Reserve. *Google Maps https://www.google.ch/maps/place/Reserva+Florestal+Adolpho+Ducke/@-2.8580657,-60.0242213,71,452 m/data =!3 m1!1e3!4 m5!3 m4!1s0x926c1ec4d40d48b7:0x897d42e519777eb2!8 m2!3d-2.9633439!4d-59.9228331.*

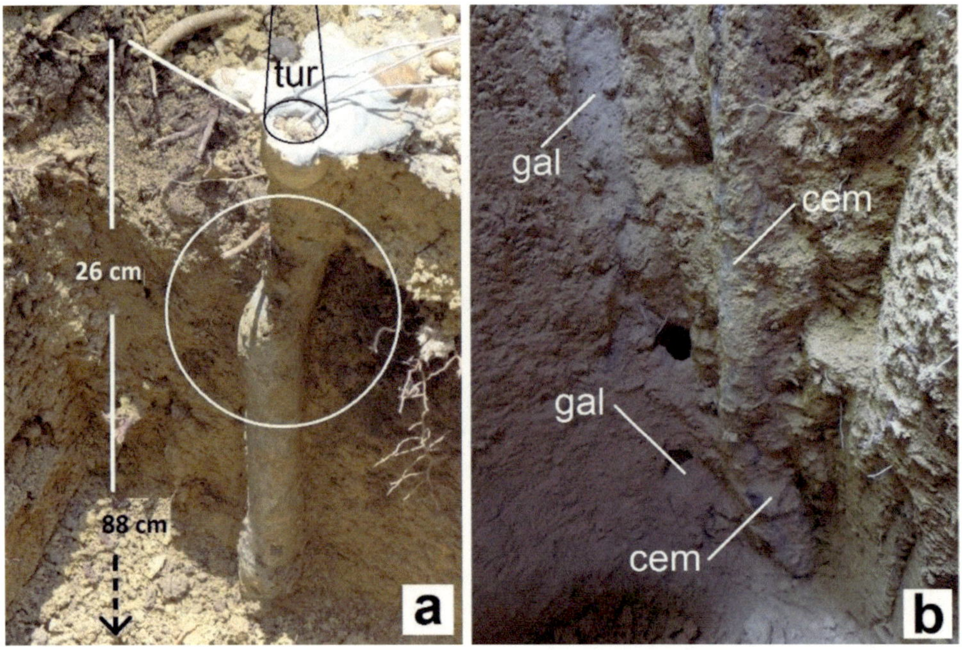

Figure 5.
(a) concrete moulding inside an 88 cm deep well, 26 cm of which have been excavated. The curvature (circle), just below the ground level, is clearly visible. The turret is traced (tur). (b) oblique galleries (gal) at the base of the well, oblique as well, up to a curvature at 15 cm from the bottom; cem: *Cement moulding.*

Figure 6.
A pile of stems and leaves became an obstacle during the construction of this turret; the nymph avoided it by raising the turret obliquely, but it reestablished the verticality as soon as the obstacle was passed.

Figure 7.
Occlusion in progress by the nymph with clay mixed with its urine (mcl) after its turret has been severed.

pieces over the base, the nymph plugs the interstices (**Figure 8**) by injecting soggy clay with its urine.

When a turret has been destroyed, or even when it tips over (**Figure 9**), the nymph rebuilds it completely.

Figure 8.
Sealing of interstices by injection of soaked clay (cl).

Figure 9.
Reconstruction of the turret after failover.

The experiment was carried out [2] to section a turret experimentally and to continuously monitor the repair; the nymph clogs the section and restores the original height within days using the same lengthening technique as when growing, and at the same rate of about 3 cm per night. The result of this experiment gives credence to the idea of the requirement of a minimum height necessary to maintain appropriate parameters (humidity, pressure, and O_2 and CO_2 levels) for the survival of the nymph in its burrow.

4. Spontaneous changes in height

A spontaneous increase in turret height to a value held constant thereafter has occasionally been observed, reflecting an increase in the minimum required height discussed above.

A spontaneous decrease in the height of the turret can also be observed, resulting from building a summit inside the turret, below the existing one, which, probably because it is no longer in contact with the moisture inside the burrow, dries up and crumbles (**Figure 10**), revealing the new top. After a few days, the parts above the new summit have completely disappeared and the result is a turret with reduced height.

Figure 10.
Spontaneous reduction in the height of the turret by construction of an internal top and drying out of the old one. The white line represents the old wall of the upper part of the turret.

The reasons leading to the performance of this operation by the nymph remain to be elucidated because, during experiments of artificial lengthening of a turret by transplanting fragments from another one, a nymph does not manifest a requirement of a maximum height; in the long run, it accommodates a higher turret than the one it has built itself; it does not practice the technique which has just been described, nor any other. On the other hand, it ensures the sealing of the modified turret by plugging with soaked clay the interstices between the base of its turret and the implants received, in the same way as described above (**Figure 8**).

5. Moving the turret

This action of the nymph is rarely observed, but its peculiarity makes it worthy of presentation. A turret emerges near another which later tips over and lies down on the ground. The communication with the well is either completely closed or in the process of being sealed with clay (**Figure 11a**). The hypothesis can be put forward (**Figure 11b**) that the new turret is built over a deviation made by the nymph from the exit of its well.

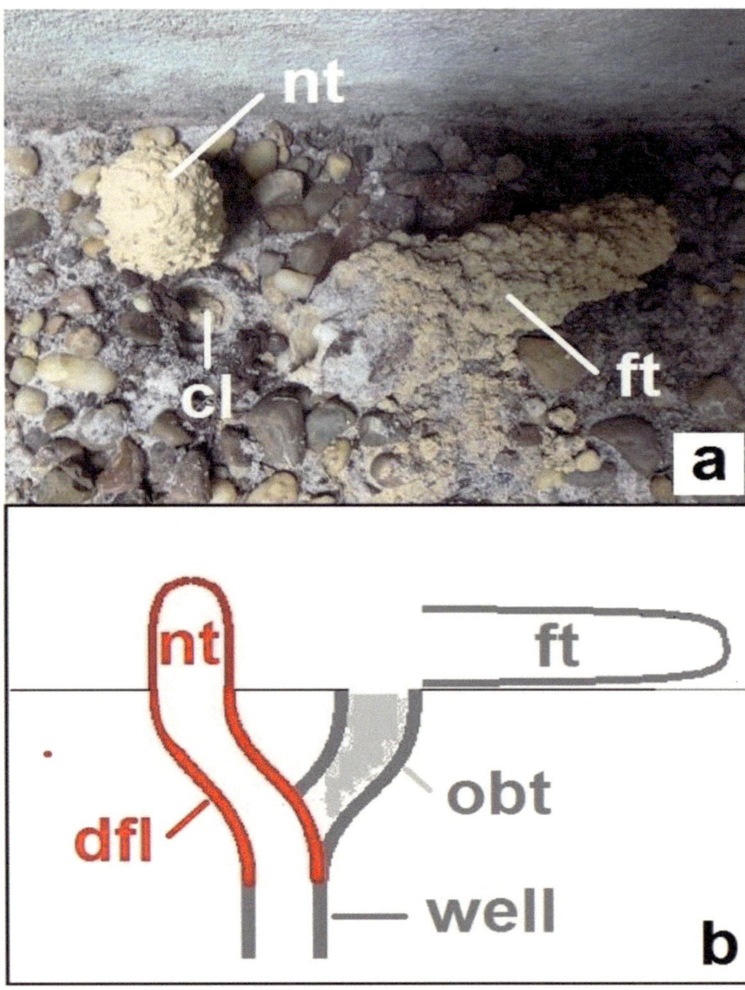

Figure 11.
(a) Spontaneous reconstruction of a new turret. (b) Diagram of deviation by the nymph of the outlet of the well: obt: *Filling with clay of the old outlet,* dfl: *Deviation made for a new outlet,* ft: *Former turret,* nt: *New turret.*

6. Role of the turret

A correlation has been established [2] and confirmed later (Béguin, Gama and Ribamar Mesquita Ferreira, to be published) between the appearance of intense rains and the simultaneous temporary opening of the turrets at their top. A succession of 3 episodes, between July 19 and 23, 2016, turned out to be particularly significant; many openings (**Figure 12**) appeared, which were then closed when the precipitation stopped (**Figure 13**). The openings vary in shape, from a small hole about 5 mm in diameter (**Figure 12a**) to a larger opening where the contour of the clypeus, eyes or anterior part of the pronotum (**Figure 12b**) can frequently

Figure 12.
(a) Circular orifice resulting from a slight pressure of the mass of soggy clay against the wall of the summit. (b) Orifice with imprint of clypeus (cly) and pronotum (pro) covered with soggy clay. (c–f) occlusion of the orifice shown in ***Figure 12a*** *after end of rainfall, by injection of soggy clay (captured images with the automatic camera* Brinno timelapseTLC200 Pro*).*

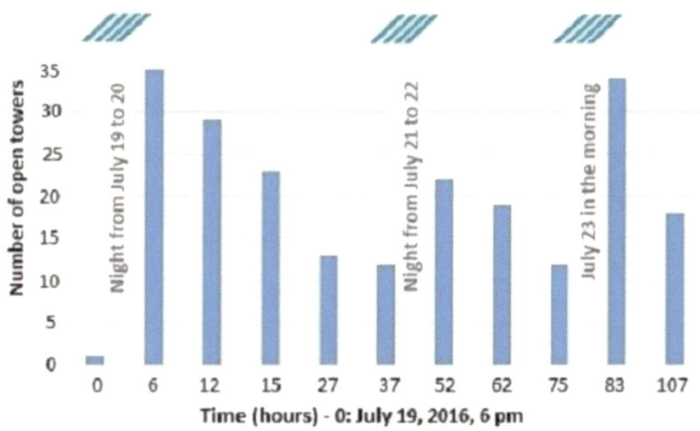

Figure 13.
Correlation between intense rainfalls (////) and temporary and simultaneous opening of 35 turrets, during 3 episodes, between July 19 and 23, 2016.

be observed, as a consequence of the liquefaction of the clay from the top wall by applying these parts covered with clay wet with urine. After the rain has stopped, the nymph closes these openings by injecting moist clay (**Figure 12c–f**).

When rainfalls are heavy, the soil becomes very wet and the moisture level in the burrow increases dramatically. In addition, rainwater in some places passes through a rapidly decomposing litter [3]. We can therefore imagine that the level of CO_2 also increases. The temporary opening of the turret would therefore lower these rates, as well as the concomitant increase in pressure. *The turret would thus appear as a regulating device of the physical–chemical conditions prevailing inside the burrow.*

7. Opening of the turret, exit and moulting

From the end of July begins the period of moults. The nymphs emerge from their turret after having opened it at the top, moult into a winged imago, on their own turret (**Figure 14a, b**) or on neighbouring vegetation (**Figure 14c**), then fly away for a brief adult life, during which males and females will mate.

We have identified three opening modes (Béguin, Gama and Ribamar Mesquita Ferreira, to be published), *each accomplished using the hook-shaped end of the forelegs.* It is therefore important to note that the opening behaviours of the top of the turret before moulting are totally distinct from those described above for the temporary opening of the top during intense rainfalls, and which are practiced by liquefaction of the top with soaked clay, without ever involving the front legs. *For the opening*

Figure 14.
(a) Exuvia attesting for the moulting of a nymph on its own building (photos Vanessa Gama). The white line represents the trip of the nymph between the opening at the top of the turret and the place where the moult took place. Note the rotation made by the nymph at the bottom of its turret. (b) imago from another nymph drying its wings after moulting on its own building. (c) Migration of a nymph from its turret to the branch where it has moulted. o: Opened turret.

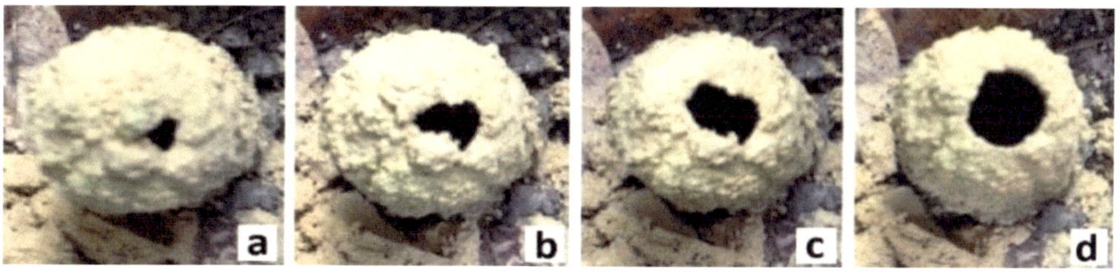

Figure 15.
Four stages of opening a turret. (a) Beginning of the draft work (6:21 a.m.). (b) draft in progress (9:58 a.m.). (c) Draft completed (10:43 a.m.). (d) Equalisation of the border completed (11:35 a.m.). Observation on 19.08.2019.

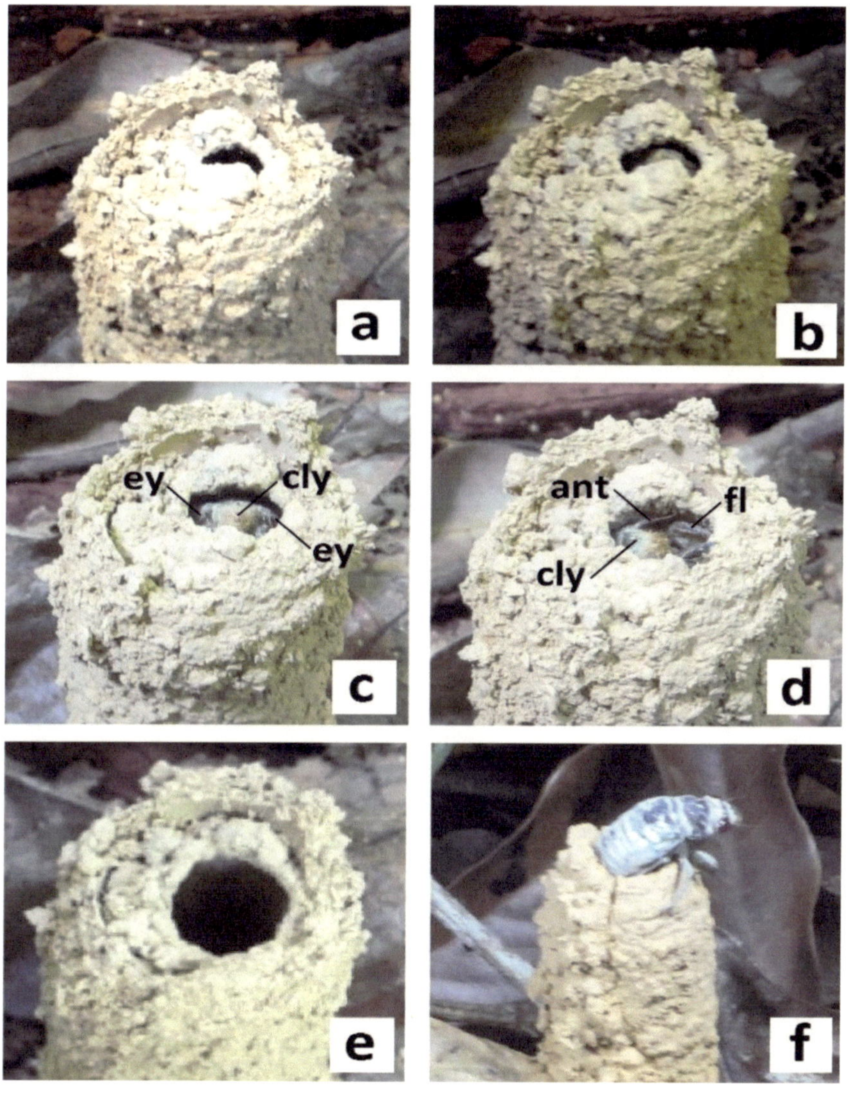

Figure 16.
(a) 6:59 a.m.: Opening resulting from the enlargement of a small perforation, such as that shown on ***Figure 15a****. (b) 7:03 a.m.: Continuation of the enlargement; the nymph is discernible. (c) 8:44 a.m.: The enlargement is finished; the nymph appears, on can distinguish its eyes (ey) and its clypeus (cly). (d) 9:38 a.m.: The nymph, gripped at the opening, is equalising the edge by scratching with the hook-shaped end of its forelegs (fl); one distinguishes also an antenna (ant) and the clypeus (cly). (e) 10:05 a.m.: The equalisation of the edge is complete, the opening is perfectly circular and the nymph withdrew. (f) 6:08 p.m.: The nymph leaves its building. Observation on 01.09.2019.*

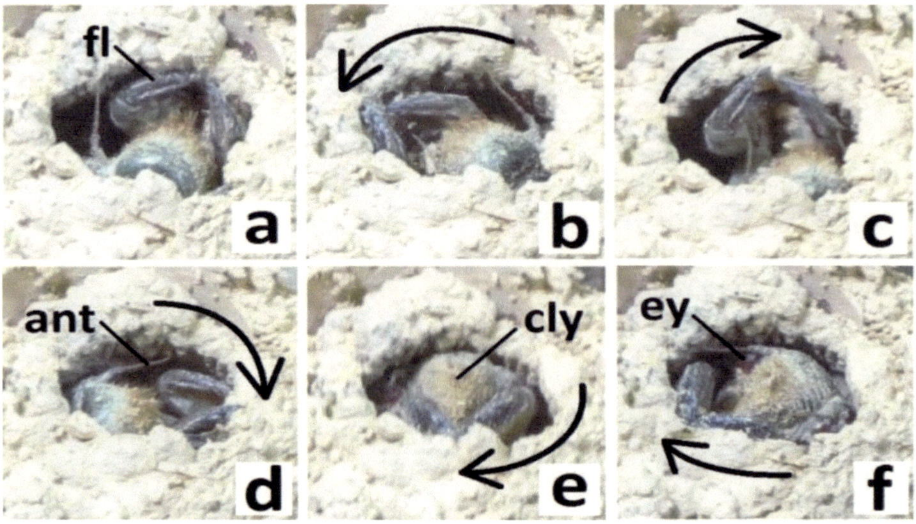

Figure 17.
Six snapshots illustrating the nymph's rotations when equalising the edge of the opening with its forelegs. (a) 9:15 a.m. (b) 9:35 a.m. (c) 9:37 a.m. (d) 9:38 a.m. (e) 9:41 a.m. (f) 9:45 a.m.. Arrows represent the rotations from the previous snapshot; ant: *Antenna,* fl: *Foreleg,* cly: *Clypeus. Observation on 01.09.2019.*

Moulting period	Late July - early September (maximum frequency during the first week of August)
First perforation for the opening of the top of the turret	Before 7:00 a.m.
End of opening (before equalising the edge) - program *A*	Between 8:30 and 11:00 a.m.
Equalisation of the edge completed - program *A*	Between 10:00 and 12:00 a.m.
Exit of the nymph through the opening at the top of its turret	Between 6:00 and 6:10 p.m.
Immobilisation before moulting	Between 6:45 and 7:00 p.m.
Downtime to moult	About 2 hours
Downtime after moulting (drying the wings)	About 5 hours
Taking flight	After 2 a.m.

Table 1.
Work schedule of the nymph during its moult to imago.

of its turret, the nymph therefore has a register of behaviours specific to the function performed.

The most common method of opening before moulting (named program A) is worth explaining here. It takes place in two phases: a draft of the opening (**Figure 15a–c**), then an equalisation until the opening is remarkably circular (**Figure 15d**).

Continuous monitoring makes it possible to describe in detail the behaviour of the nymph which creates the roughing by enlarging (**Figure 16a–c**) the initial perforation (**Figure 15a**) by a scraping carried out with the chitins' end of its forelegs, up to obtain an elongated opening with irregular edges (**Figure 16c**). Working from the inside, the nymph is almost invisible. When the roughing is complete, the nymph appears (**Figure 16c**) and undertakes the work of equalisation, which is accomplished by scraping the edge with the chitins' end of its forelegs (fl) also (**Figure 16d**), at the same time as it performs rotations. During this phase, the nymph's head is clearly visible, and one can easily recognise its *antennae* (ant), *front legs* (fl) and *clypeus* (cly). Once the equalisation is complete (**Figure 16e**), the nymph disappears until its exit (**Figure 16f**) which takes place with a remarkable precision between 6 p.m. and 6:10 p.m. (local time). The 6 pictures in **Figure 17** illustrate the rotations that the nymph performs at the same time as it levels the edge of the opening with its forelegs.

The monitoring of two individuals, from the exit of the nymph from its turret, until the flight of the imago, made possible to establish a detailed time schedule (**Table 1**).

8. Ecosystem

Each time an edifice of *G. chlorogena* was discovered on the site we studied, an arborescent *Fabaceae* of the genus *Tachigali* has been identified in the vicinity (Béguin, Gama and Ribamar Mesquita Ferreira, to be published). The building of the nymph is usually the much farther from the tree the larger this one is. This fact is not surprising when one takes into account: 1°) that a cicada nymph can plant its tiny rostrum [4] to suck up the elaborate sap on which it feeds, only in radicular extremities of comparable size, close to the meristems; 2°) that the root ends of trees, in the Amazon rainforest where the layer of fertile soil is very small (of the order of 1 meter), instead of plunging in depth, spread out [5], forming a "root disc" whose radius is all the greater as the size of the tree is large.

During the 2 years 2018 and 2019, we identified 132 buildings of *G. chlorogena* distributed among 14 trees of the genus *Tachigali*; 11 from the species *chrysophyllum* and 3 from the species *venusta*. We therefore measured, for each building, the distance separating it from the tree[2], as well as the diameter of the trunk of this latter[3].

The results were reported on two graphs relating the diameter of a tree with: a) the distances to all the buildings associated with it (**Figure 18a**); b) the average distance to them (**Figure 18b**). The trend lines calculated with *Excel* by the smallest squares indicate clearly, on each of the graphs, a proportionality between the diameter of the trunk of the tree and its distance to a building of *G. chlorogena*, thereby providing a logical-mathematical support to the association between the nymphs of *G. chlorogena* and these *Fabaceae*[4]. *G. chlorogena* apparently being the main, if not the only, beneficiary of this relationship, we admit that this is a case of *commensalism* [6].

The graph with plotting of all distances (**Figure 18a**) shows that part of the buildings are located near the trunk. We will be re-examining our statistics to determine if the location of the buildings is related to the *Tachigali* species (*chrysophyllum* or *venusta*). One thing is obvious; nymph buildings (i.e. nymphs feeding) close to the trunk involve root ramifications down to the level of the meristems from the proximal roots, particularly the buttresses. We recently discovered, at another place in the *Adolfo Ducke reserve*, a *Tachigali* station of a species (yet to be identified) which is neither *chrysophyllum* nor *venusta*, with nymph buildings (the species to which their hosts belong has not yet been confirmed as *G. chlorogena*) close to the trunk. In Santarém (*Pará*, Brazil), edifices of *Fidicina chlorogena*, a taxon since identified with *G. chlorogena* [1] have been described [7] near the base of the trunk; the

[2] More precisely of the estimated center of its implantation surface.

[3] If the trunk is connected to a buttress by its base, its diameter is measured above.

[4] It is due to this situation to meet, at the Botanical Garden of Manaus for example, buildings in a built area near the forest. This is the case with the one shown in **Figure 11**.

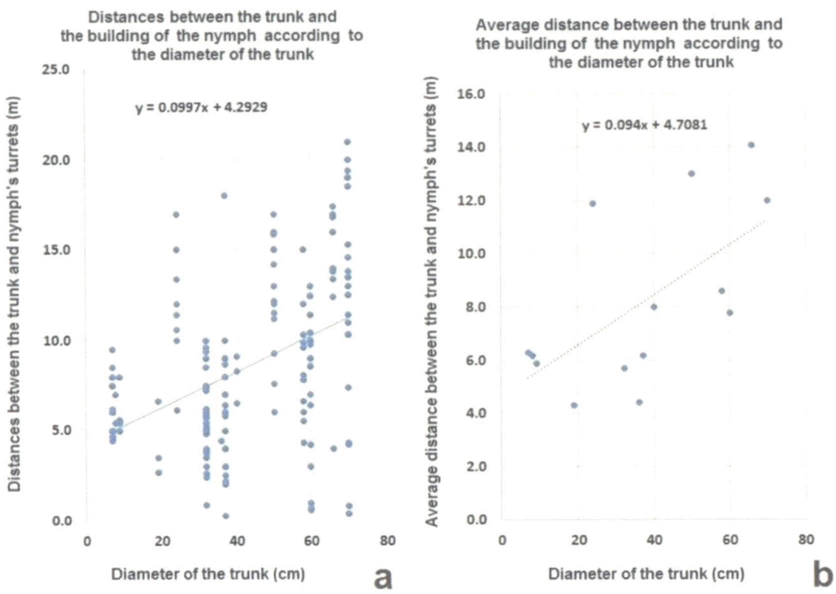

Figure 18.
Relationship between Tachigali's trunk diameters and distances to G. chlorogena's edifices. (a) Plotting distances to all the buildings associated with a given tree. (b) Plotting the average distance to the buildings associated with a given tree.

species name of the tree was not reported. The same disposition was observed [8] for *Orialella aerizulae* (Boulard), which could also feed on *Tachigali* sp. as well as possibly on *Tabebuia* sp. This aspect of the dependency relationship between cicadas and their foster trees has to be explored in more detail.

9. Reproductive cycle

The nymphs spend several years underground, moving to reach the fine roots (less than 2 mm) from which they suck, with their rostrum, the elaborate sap on which they feed (see above). They move forward by digging galleries with their forelegs, throwing the excavated material behind them.

On can consider [9] the beginning of the reproductive cycle at the time of mating (**Figure 19a**), after the nymph has emerged from its burrow and moulted into an imago (**Figure 19h, i**). The winged adults, stimulated by singing males, mate in trees. Then the females search for a stem in which they will plant their ovipositor several times [10] to deposit their eggs under the bark (**Figure 19b**). The larvae hatch on the twigs and then fall to the ground (**Figure 19c**) where they sink (**Figure 19d**). No one has yet observed their behaviour during this first phase of underground life. No one either knows the duration of the underground life until the exit to moult into a winged imago; it is estimated at several years.

From excavations undertaken after moulding with cement (see above), which revealed an inclination of the deepest zone of the well, as well as in its vicinity traces of oblique galleries of intermediate inclinations (**Figure 5b**), one can envisage the upwards digging of a vertical well from a deep horizontal gallery (**Figure 19e**) by digging successive oblique galleries more and more inclined (**Figure 19f**).

The digging of deep horizontal galleries giving access to fine roots that the nymphs can use to feed themselves (see above), is consistent with the presence of a

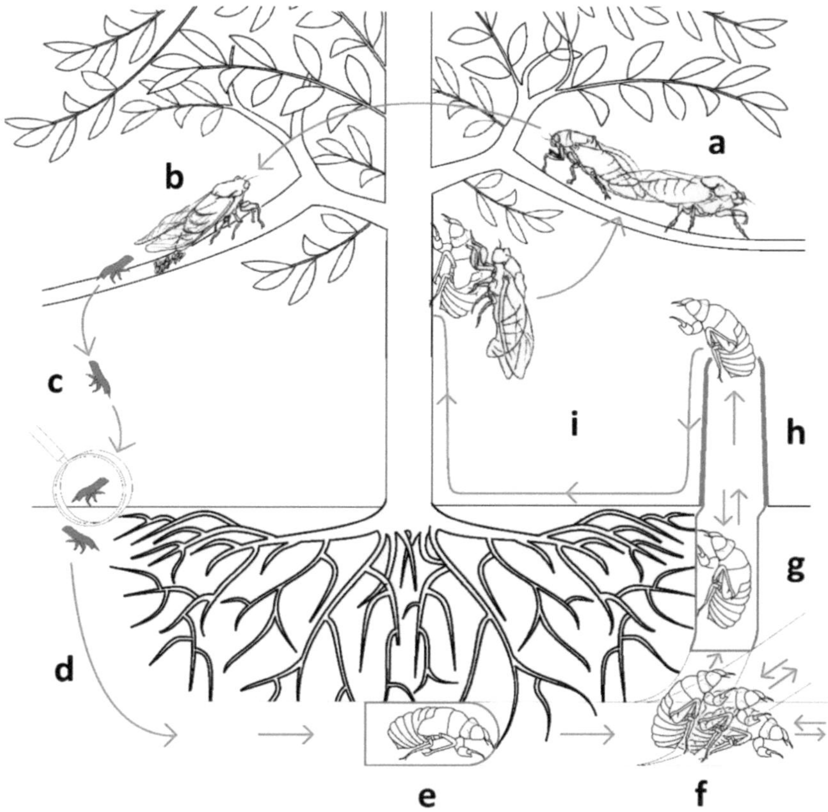

Figure 19.
Life cycle of G. chlorogena. (a) Mating. (b) Lay eggs. (c) After hatching, the tiny larvae (~ 2 mm) fall to the ground. (d) They bury themselves in the ground. (e) The nymph digs its gallery at the same time as it feeds on the sap of the roots. (f) As it moves forward and back, the nymph changes direction until it is vertically oriented. (g) The nymph digs the well which is a vertical gallery. (h) the turret is built from the well. (i) The nymph comes out of the turret to turn into an imago.

significant biomass of fine roots in the deep soil layer of the Amazon rainforest, as recent research has shown [11]. The digging of successive oblique galleries more and more inclined implies a capacity of the nymph to back up into an already dug burrow. Such a capacity is observed when one surprises, by the withdrawal of its turret, a nymph near the exit of its well. It can also be observed in experiments where the nymph is placed in a glass tube.

10. Role of urine

The key role mentioned above of urine in *G. chlorogena*, has been underlined in connection with the construction of the turret and the various maintenance activities described; in experiments where the turret was replaced by a glass tube [2], the observation was made of the rise of the urine soaked mass of clay that the nymph loads on its head. The role of urine, in European species of cicadas, was demonstrated [12] for digging underground burrows; abdominal gutters redirect urine to the anterior part of the body and allow the nymph to use it to soften the soil with its forelegs. In the case of *G. chlorogena*, urine is used to soften the clay [2]. Urine from cicadas also contains *mucin*, a *glycoprotein*, which strengthens the walls of burrows after drying [13].

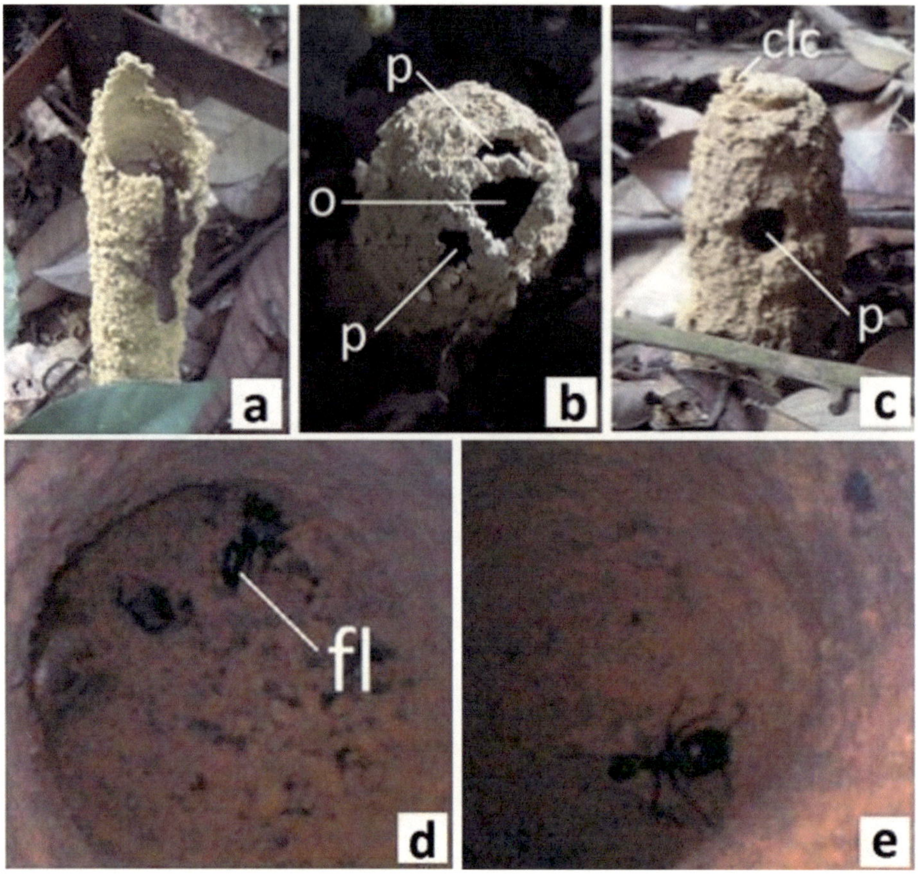

Figure 20.
(a) Lacertilian *on an open turret. (b) Perforations (p) following passage of* Legion ants *on an open turret after an episode of heavy rainfall;* o: *Opening in the shape of its clypeus practiced by the nymph in response to rainfall. (c) Perforations (p) following a passage of army ants on a turret that the nymph closed with a clay occlusion (clc) after the cessation of precipitation. (d) Bottom of a well after passage of ants; the remainder of the hook of the end of a front leg (fl) is visible. (e)* Ant *in the bottom of a well.*

11. Predators

Different reptiles, such as Lacertilians (**Figure 20a**), can be supposed as predators when a nymph opens or leaves its building. However ants, known by the vernacular name of *Legion ants*, or *Formigas de coreção* according to the local name, are the most dangerous (**Figure 20b–e**). They invade (**Figure 20e**) towers which have been opened (**Figure 20b**) following heavy rainfalls (see above: the role of the turret), and even manage to perforate the walls (**Figure 20c**). Entering a building, they devour the host, leaving only the chitins' organs, such as the hook-shaped ends of the forelegs (**Figure 20d**).

12. Conclusions

This article has presented the activities of the nymph of the cicada *G. chlorogena* during its last year of larval life, in relation to its burrow, the visible part of which is a clay turret built from a well whose depth can reach one meter. The clay with which the turret is built probably comes from the base of the well, from the volcanic bedrock on which the fertile soil rests. The nymph mixes this clay with its urine and transfers the mixture to its forelegs and then to its clypeus, through its abdominal and thoracic "gutters" [9]. It can then climb this mortar to the surface and build its turret. By replacing a turret with a glass tube, it was possible [2] to observe the rise of a nymph with a mass of soggy clay on its clypeus.

The existence of a one meter deep well, dug vertically from the bottom thanks to a process of verticalization of a horizontal gallery (**Figure 19f**), validates the representation of the digging by the nymph of paths below the root base of the tree with which it is associated, in order to find appropriate roots to introduce its rostrum and feed on elaborated sap [10], which is confirmed, as mentioned above, by the recent demonstration [11] of a significant biomass of fine roots in the deep soil layer of the Amazon rainforest. To dig, the nymph shovels with its chitins' forelegs, tears off earthy fragments which it impales on the bristles of its clypeus and deposits behind it [12]. It thus advances in a short gallery, which it opens in front of it and closes behind (**Figure 19e**). One question remains to be clarified; at what moment, and consecutively to which signals (external and/or endogenous), does the nymph begin its process of verticalization.

The well therefore has the status of a gallery, from which it is however distinguished by the fact that, after closing off the bottom, it is not closed as the nymph digs it, vertically and from bottom to top, until reaching the ground surface. The turret, for its part, is an additional device. Maintained sealed by the nymph (**Figure 8**), it appears to be devoted to maintaining appropriate conditions for the survival of the nymph in its burrow (see above; Role of the turret). As already mentioned, a minimum height of the turret is required by the nymph, which repairs or rebuilds it if necessary.

A behaviour, considered as motor coordination, involves [14, 15] stimuli (external and/or endogenous), as well as a recognition mechanism which is a neuronal structure. The observed richness of the nymph's behaviours is therefore concomitant with an important perceptual component in its nerve system. The realisation of similar tasks (opening the top of the turret, for example) by different behaviours according to the circumstance, can be considered as falling under an elementary cognitive system. The execution, without learning, of complex motor sequences, such as the opening of the top of the turret before moulting, according to a straight defined time schedule, is the result of highly perfected innate programming.

The dependence of *G. chlorogena* on arboreal *Fabaceae* of the genus *Tachigali* is a very important aspect of its reproductive cycle. In this regard, an important question arises: do the very young larvae which have just burrowed (**Figure 19b**), have the same mode of nutrition? If so, are they also related to *Tachigali* to find fine root ends near the soil surface?

Many questions still remain open about cicadas, insects as popular as their biology is poorly understood.

Acknowledgements

I wish to express my gratitude to *Prof. Ennio Candotti*, who entrusted me with this research and constantly supported me, to *Vanessa Gamma* and *José Ribamar Mesquita Ferreira* for technical assistance, to *Dr. Nállarett Dávila* for the determination of the *Tachigali*, and to *Olga Béguin* for critical reading. The work was supported by the *Conselho Nacional de Desenvolvimento Científico e Tecnológico* (CNpq), grants no 385612/2014-1 et no 310012/2015-5, by the *Fundação de Amparo à Pesquisa do Estado do Amazonas* (FAPEAM), grant no 462/2018, and by *Museu da Amazônia (MUSA)*, 69060-060 Manaus, AM, Brasil.

Author details

Claude François Béguin
Museu da Amazônia (MUSA), 69060-060, Manaus, AM, Brasil

*Address all correspondence to: claude.beguin@gmail.com;
claudefb@museudaamazonia.org.br

References

[1] Sanborn A F (2013) Catalogue of the Cicadoidea (Hemiptera: Auchenorrhyncha). Elsevier, London, p. 207

[2] Béguin C F (2017) An Architect Cicada in Brazilian Rainforest: Guyalna chlorogena (Walker). Neotropical Entomology 46 (2): 159-168 (DOI 10.1007/s13744-016-0464-0)

[3] Melo L (2014) Personal communication. Museu da Amazônia, 69060-060 Manaus, Brazil.

[4] Myers J G (1928) The morphology of the Cicadidae (Homoptera). Proceedings of the Zoological Society of London 98:365-472

[5] Coco Magnanville (2012) La forêt amazonienne : Une forêt tropicale humide. http://cocomagnanville.over-blog.com/ accessed 2020-08-25

[6] Breton S and Carrière A (2018) Notions sur la biodiversité. http://www.biodiversite-poitou-charentes.org/Relations-entre-les-especes.html / accessed 2020-08-25

[7] Ginzberger A (1934) Die bauten der larve der singzikade Fidicina chlorogena Wlk. Sitzungsberichte der Akad. d. Wissenschaften. Wien, math.-natunv. 143:91-94 A (1934). https://archive.org/details/sbaww_143_0091-0094/mode/2up / accessed 2020-08-31

[8] Wood Z (2020) Personal communication. University of California, Santa Barbara, CA 93106, USA

[9] Ibanez M (2003) Les Cigales, Cycles de vie, caractéristiques et symbolique. Invertébrés continentaux des Pays de la Loire. Gretia, Hérouville-Saint-Clair, France, pp 208-212

[10] Guyard A (2010) Les cigales du sud de la France. http://baladesnaturalistes.hautetfort.com/apps/search/?s=cigales / accessed 2020-08-25

[11] Magrini L (2020) Estoque de carbono subterrâneo da Floresta Amazônica é maior do que se pensava, UNICAMP. https://www.unicamp.br/unicamp/noticias/2020/08/11/estoque-de-carbono-subterraneo-da-floresta-amazonica-e-maior-do-que-se-pensava/ accessed 2020-08-25

[12] Boulard M (1988) Biologie et comportement des cigales de France. Insectes 69:7-13

[13] Boulard M (1991) L'urine des Homoptères, un matériau utilisé ou recyclé de façons étonnantes. Insectes 80:2-4

[14] Eibl-Eibesfeldt I (1972) Ethologie, biologie du comportement. Naturalia et Biologia Editions scientifiques, pp 15-44

[15] Tinbergen N (1980) L'étude de l'instinct. Petite bibliothèque Payot, Paris, pp 21-23

Section 2

Human Culture and History

Chapter 6

Shipibo Conibo and Chilean Diaguita Visual Art: Cognitive Technologies, Shamanism and Long-Distance Cultural Linkages

Paola González Carvajal

Abstract

Pre-Columbian art in all its varied forms offers rich terrain for furthering our insights into the cultural and symbolic lives of Amerindian peoples. This paper studies decorative patterns of Diaguita origin which present a visual logic characterized by the use of complex symmetries, illusory optical vibration, endless variability stemming from simple geometric forms, *horror vacui*, and gradual structural complication, among other techniques. The features described, the association of this visual art to an animal alter ego (jaguar), and the evidence of hallucinogen use together suggest a cultural link with specific ethnographic (Shipibo-Conibo) and archeological (Mojocoya) visual art. In this case, we are dealing not with a notation system but with visual "technologies of enchantment" (sensu Gell 1998) that are used to produce decorative patterns with social agency that captivate viewers with their visual artifice—the non-mimetic appearance of animation. In 2015, a large cemetery was excavated at the Diaguita site of El Olivar. The graves therein belong to an early Diaguita cultural period, during which the Diaguitas created a techno-stylistic material identity, expressed in visual languages rooted not only in the Andean world, but in ancient cultural traditions of the eastern lowlands of Bolivia and the Peruvian Amazon.

Keywords: Chilean Diaguita culture, Shipibo-Conibo culture, South American shamanic art, cultural symmetry, long-distance cultural linkages, indigenous art concept

1. Introduction

Structural analysis of Diaguita visual art [1] revealed certain decorative patterns (the predominance of abstract designs, the heavy use of symmetry, and the presence of spotted feline representations). The designs' minimal units and symmetrical structure have been shown to be sensitive to issues of group identity and cultural interaction processes [2], in which each culture selects and appropriates a limited number of symmetries that form a specific universe of socially recognized forms, and only these are approved for use.

Pre-Inca Diaguita art displays a series of traits that allow it to be considered representative of a specific South American shamanic art tradition. Notable among these traits are the association of the designs with an alter ego (jaguar or spotted feline), the practice of hallucinogenic consumption, and the development of non-figurative visual art featuring complex symmetry. The image of a spotted feline (jaguar) has been recorded from the Diaguita II phase on duck-shaped pitchers, zoomorphic serving bowls (**Figure 1**), and spatulas used for the consumption of hallucinogenic substances.

Evidence that Diaguita culture engaged in this practice includes tubes and spatulas, as well as the scallop shells that held the psychoactive substance itself (**Figure 2**).

The Diaguita culture developed in the semi-arid region of what is now Northern Chile, between the Elqui and Choapa rivers. The Diaguita were skilful potters who developed an abstract, symmetrical visual art style that was tremendously complex and beautiful. The variety, extraordinary complexity, and intricate composition of Diaguita ceramic designs comprise one of the most stunning visual universes of the pre-Colombian Americas (**Figure 3**). Two phases can be clearly identified in the development of this pre-Colombian people—a pre-Inca period (900–1470 A.D.) and a period of Diaguita-Inca contact (1470–1536 A.D.)—each with its own characteristic style of visual art.

Before Inca contact, the Diaguita were a peasant society with little social stratification, in which the family served as the basic social unit. The Diaguita of this phase practised small-scale farming and adapted to their environment without altering it significantly. They had a disperse settlement pattern, which suggests that they managed space through kinship-based residential units that occupied the transverse

Figure 1.
Diaguita zoomorphic serving bowl (spotted feline). Grave 62, El Olivar site.

Figure 2.
Spatula, tube, and shell used for consuming psychoactive substances. Grave 99, El Olivar site.

Figure 3.
Duck-shaped pitchers (Diaguita II phase). Museo del Limarí N°31.

Figure 4.
Male accompanied by spatula, tube, and shell for consuming psychoactive substances. Grave 99, El Olivar site.

(East-West) valleys of their territory. As the basic social units, individual families practised small-scale farming to provide for their members, forming communities that were both economically self-sufficient and politically autonomous.

Evidence from pre-Inca Diaguita burial contexts in the Choapa Valley [3] has shown that there was little social differentiation in these communities, at least in regard to burial practices, as the nature and distribution of ceramic grave goods was largely homogeneous across age and gender. However, the presence of offerings associated with the consumption of hallucinogenic substances (spatulas and scallop shells) indicates that certain males—who have been interpreted as shamans—seem to have occupied a slightly different social category (**Figure 4**).

In terms of social organization, researchers have affirmed [4] that the Diaguita had a dual social structure, with groups organized in complementary halves, similar to several pre-Hispanic and present-day Andean and Amazonian societies. Ethnohistoric evidence indicates that, at the time of the Spanish arrival, the Diaguita were organized into dual *señoríos* (chiefdoms).

2. Diaguita Pre-Inca visual art and South American shamanic art tradition

Aspects such as the optical illusion of movement and vibration (**Figure 5**); interminable variability based on simple geometric forms (**Figure 6a–c**); *horror vacui* (**Figure 7**); hypnotic attraction (**Figure 8**); positive/negative views (**Figure 9**), and

Figure 5.
Optical illusion of movement and vibration. Diaguita F2-2 wave pattern ([1], p. 118).

a

b

c

Figure 6.
Interminable variability based on simple geometric forms. (a) Diaguita C5 wave pattern ([1] p.111); (b) Diaguita C6 wave pattern ([1], p.111; (c) Diaguita C7 wave pattern ([1], p.112).

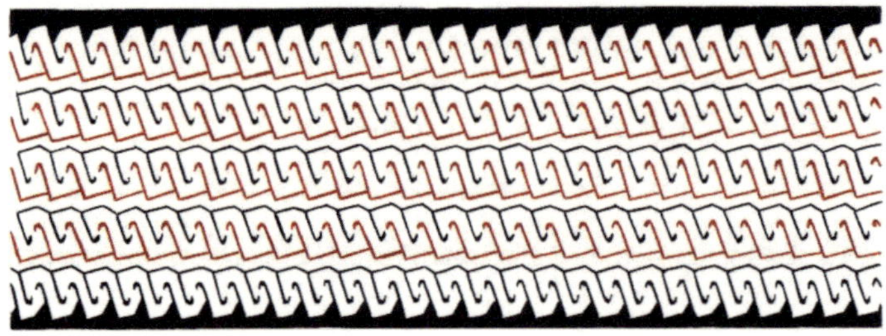

Figure 7.
Horror vacui. Diaguita J 1-1 wave pattern ([1], p. 123).

Figure 8.
Nearly hypnotic attraction. Diaguita A2 labyrinth pattern ([1], p. 152]).

Figure 9.
Positive/negative designs. Diaguita stepped motifs, reflection B ([1] p. 149).

designs organized according to the principle of gradual structural complexification, among other features.

Recent investigations have made headway in identifying Diaguita rock art in the Limarí and Choapa river basins [5, 6]. Certain rock art representations (petroglyphs) have been attributed to the Diaguita culture on the basis of their formal similarities with Diaguita ceramic designs, both in terms of motifs and minimal units (labyrinth, stepped designs and stepped fret borders, for example) as well as in regard to the complex symmetrical operations performed by the creators. On this rock support we find the same visual logic that is recorded on Diaguita ceramic decoration, which is marked by its abstract character and complex symmetry (**Figure 10**).

Our comparative examination of Diaguita art and other South American ethnographic visual art styles linked to shamanic religions, especially those of the Amazonian Tukano [7], Shipibo-Conibo [8], and Caduveo [9] cultures, reveals a set of shared characteristics [10]. Two common external associations of these visual art traditions are the reference to an animal *alter ego* (jaguar, anaconda, feline) and the practice of hallucinogenic consumption (**Figure 11**).

a

b

Figure 10.
Diaguita rock art. (a) Diaguita labyrinth pattern, Choapa Valley; (b) Diaguita double zigzag pattern, Choapa Valley.

Figure 11.
Polychrome vessel with feline representation. Diaguita-Inca phase. Grave 59, El Olivar site.

Diaguitas and Shipibos also share the practices of ceramic urn burials, intentional cranial deformation, and the use of "sacrificial" objects in burial contexts. But the most prominent similarities are found in the styles of these South American indigenous art traditions, which employ complex symmetry to articulate the motifs, using three or more inter-operational symmetrical principals, and the frequent use of mirror reflection [10] (**Figure 12**).

For instance, in relation to Shipibo art, De Boer [11] affirms

> *"...the basic cross is mastered first. This element is then successively submitted to a number of increasingly complex operations that entail concatenation, repetition, compounding- first in one, then in two dimensions- and finally various partitioning and orientational transformations. The strength of this sequential pattern suggest that a powerful maturational program of the kind famously diagnosed by Piaget [12] is at work."*

Another trait is the endless continuation or the self-generative power of the compositions, which endows them with an outstanding rhythmic quality. In contrast, the principle of *horror vacui* and the sectioning of the designs in delimited fields are also present. Yet another characteristic is the interplay between positive and negative perspectives. The use of (strictly abstract) geometric motifs,

Figure 12.
Mirror reflection. Shipibo textil, Panaillo community.

combined with complex symmetrical structures that makes them barely intelligible, is another characteristic feature of Shipibo visual art [10].

These are elaborate and systematic procedures of graphic creation, visual litanies that generate a hypnotic attraction (**Figure 13**) through designs executed in different media such as facial paint, textile decoration, ceramics, and home decor.

In relation to this specific class of South American visual art, present in ethnographic contexts and associated with shamanic practices, Reichel-Dolmatoff ([7], p. 292) has linked the ingestion of psychoactive substances to geometric designs. The author highlights how the hallucinatory experience offers a potentially rich array of symmetrical images, which are then reproduced by the Tukano people in their art. He comments:

> *one of the most important discoveries in the last decades, in the ethnology field consists in the confirmation that art and shaman religions are closely related to the use of hallucinogenic drugs"..."the ingestion of hallucinogenic drugs represents the principal mechanism that induced states of shaman visions, accompanied by the*

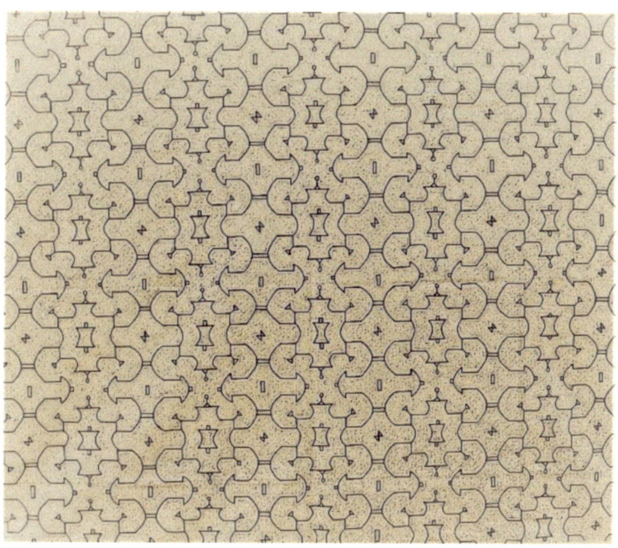

Figure 13.
Nearly hynotic attraction. Shipibo textil. Paola González collection.

> *illusion of flying through the air, animal transformations and the penetration into other cosmic dimensions, out of this world*

Similarly, among the Shipibo-Conibo people, *kené* designs left on a material support (ceramic, fabric, or wood) suggest designs perceived during shamanic visions resulting from the ingestion of Ayahuasca and believed to hold great healing power. Thus, these symmetrical, non-figurative designs have a therapeutic as well as an esthetic function. For the Shipibo people, the execution of these complex and extraordinarily beautiful designs constitutes a collaboration with the sacred, spiritual realms, erasing the dividing line between religious and esthetic experience. Thus, there is an evident link between shamanic practices, the consumption of hallucinogenic substances (Ayahuasca), and synesthetic healing therapies, which range from the artistic to the musical [8]. Gehbart-Sayer ([6], p. 161) points out:

> *the visions of the designs are described as non analytical impressions on "pages" or "sheets", organized in a pattern that flicker rapidly in front of the eyes of the shaman then disappear as soon as he tries to take a closer look.*

This synesthetic quality of South American shamanic art also includes the transition between visual art and music, with the use of symmetry being a common feature in both. Gehbart-Sayer ([8], p. 170) comments, *"symmetry prevails in the formal, melodic, and rhythmic characteristics of the songs. An example of a more complex consonance is the lateral symmetry of melodic inversion."*

In these communities, artistic activity goes hand in hand with the sacred; shamans and female artisans create magical-religious objects, in which esthetic beauty is not an end in itself. For example, the Shipibo-Conibo people believe that individuals are "marked" at birth with invisible decorative patterns. When an individual becomes ill, the healing designs are spiritually projected upon the patient by the shaman to bring them back to a healthy state. The attributing of agency to the symmetrical pattern is evident in the story of Netén Vita (a Shipibo shaman from Caimito), as related by Illius ([13], p. 197). Netén Vita affirms that, in order to heal, the shaman must contact the guardian spirits, who are "the bands of drawings from my heaven (*noco naina kené*)." Furthermore, recent studies of Amazonian peoples have revealed the existence of a series of pan-Amazon beliefs [14, 15], expressed in different stylistic traditions. Such commonalities include the representation of specific animals, particularly jaguars and snakes, as well as abstract art, shamanic practices, and the consumption of hallucinogenic substances. These findings point to a longstanding shared ideology, the common elements of which can be found in both present-day cultures and in the archeological record [14]. Also on the basis of archeological and ethnographic evidence, Roe proposes the existence of a common cosmology within the South American tropical forest, pointing out that, *"with its vast interconnecting web of waterways knitting together similar microenviroments, the Amazonian region is a perfect highway of ideas"* ([14], p. 26).

In terms of the temporal depth of this shamanic art tradition in South America, Taylor ([16], p. 16) affirms that *"it is possible that what we see in the contemporary Amazonian graphic tradition is only the preservation of a tendency present since pre-Colombian times, the fruit of an ancient cultural tradition, of an eminently intellectual, cerebral, elaborate and complex form of art"*.

In the paragraphs below, we will examine archeological information which suggests the existence of long-distance relations that played a determining role in Diaguita ethnogenesis. These cultural relations will also explain the cultural linkages that one observes between the Shipibo-Conibo and pre-Hispanic Diaguita peoples, despite their separation in both space and time.

3. The Mojocoya culture, a nexus for understanding the cultural linkages between the Chilean Diaguita and Shipibo-Conibo peoples

Affirming the linkages between the Chilean Diaguita and Peruvian Amazon Shipibo-Conibo peoples (which considers stylistic aspects, social practices, and material culture), and following the arguments of Lathrap [17], González [8] proposed the existence of a common antecedent between the two cultures, namely the Mojocoya culture of the Bolivian lowlands. As the cultural ancestor of the Shipibo people, Lathrap proposes the Cumancaya culture (Alto Ucayali), which also displays evident parallels with Chilean Diaguita culture.

González [8] confirmed the presence of patterns in Mojocoya iconography that were previously identified in Diaguita iconography, including the Wave A, stepped vertical reflection, and Chain C [1] patterns. The coincidences observed are not limited to the identification of minimal units (stepped borders, triangles combined with scrolling, simple stepped motifs, etc.), but also include symmetrical principles (or groups of them) that make up these units (reflection, displacement combined with translation, double specular reflection, horizontal reflection, rotation, and translation, among others). This structural similarity is a very good indicator of cultural interrelatedness [2].

According to Brockington et al. ([18], p. 4), the Mojocoya culture originated in the Amazon-Chaqueñas lowlands of Bolivia expanded into the areas around the highlands of Cochabamba and Chuquisaca. Based on their ceramics, two phases have been identified: an early, pre-Tiwanaku one (ca. 1-600 A.D.) and a later one (600 – 900 A.D.) that displays the influence of Tiwanaku. Mojocoya designs are executed in black and red on orange, with some use of white. The most frequent motifs include triangles; triangles with scrolling; stepped and triangular stepped motifs with scrolling; and zig-zagging lines in alternating colors, very often organized in bilateral or quadrilateral symmetry. We add to this list recorded examples of the labyrinth pattern A2 [1]. The main ceramic forms include simple convex bowls, tripod bowls, kero cups, and "effigy pieces" [18]. Monochrome ceramics include urns and other large vessels.

For the Mojocoya culture, archeologists have noted the practices of child burial in urns, cranial deformation, and the consumption of psychoactive powders. At the El Tambo site (1-600 AD), researchers have found copper bells with folded bases [18], which have also been recorded at Diaguita sites in Chile's semi-arid north. The authors emphasize that the Mojocoya sites are located near traffic routes connecting the Andean and Amazon regions, which would have facilitated the movement of people and goods among different ecozones. Along the same lines, Pereira and Brockington ([19], p. 2) hypothesize that the area southeast of Cochabamba would have operated as a focal point for social complexity, a place where a variety of social processes emerged over time, making it *"an area of historic interplay between environmental factors and Andean, Amazonian, and Chaqueño peoples, with a dynamic that had major repercussions and wielded significant influence even beyond its sphere of interaction as a zone."*

4. New research on the origins of Chilean Diaguita culture: El Olivar archeological site

In light of the above information about the Mojocoya culture and its ceramic tradition, it is interesting to examine analyses of recent discoveries at the Diaguita site of El Olivar [20, 21], which have identified more stylistic coincidences between the Diaguita and Mojocoya representational universes [8], while at the same time shedding light upon the genesis and development of the Diaguita culture itself in Chile's semi-arid north.

An extensive cemetery was discovered during construction of a highway between the Chilean cities of La Serena and Vallenar, and excavated between 2015 and 2017 [20, 21]. The funerary contexts recovered correspond to 212 individuals in primary graves, 39 of which contained one or two camelids, along with another 44 secondary graves. Among the whole, 86 burial contexts correspond to the period in which Diaguita culture was emerging, which we have called the Early Diaguita phase [21]. The evidence unearthed suggests that it was during this stage that the Diaguitas created their own techno-stylistic material identity. And although there was some degree of interaction with neighboring communities in the semi-arid North, Diaguita ceramics forged an identity all its own, developing visual languages with roots that were linked not only to the Andean world, but with the ancient cultural traditions of the Eastern Bolivian lowlands.

Included among the discoveries associated with those Early Diaguita funerary contexts were novel polychrome vessels, including an anthropomorphic bi-globular polychrome pitcher, which bears a rather crude version of the Wave A pattern on the back of its "head" (**Figure 14a,b**). Also outstanding among the grave goods found is a polych-rome anthropomorphic cup decorated with a labyrinth pattern on both sides of the face and body (**Figure 15a,b**). Yet another notable vessel is the polychrome globular pitcher with a flared neck, which displays the minimal units of the Chain B pattern (triangle combined with a rectangular hook), also present in Mojocoya iconography (**Figure 16**). Lastly, we wish to highlight the four-footed anthropomorphic-zoomorphic vessel (**Figure 17**) found at the site, for both its shape and the designs molded and incised on its "face" (nose, eyes, ears), both of which qualities bear evident similarities to the "effigy pieces" illustrated by Brockington et al. [18].

It is interesting to note that, as Diaguita culture consolidated over time, these early motifs gradually transformed and became more complex, until achieving the esthetic and geometric virtuosity characteristic of the Classic Diaguita style and the Diaguita-Inca phase. Clearly, the Wave pattern remained a point of reference throughout Diaguita development, acting as an "axis of formal coherence" [22], an essential formative aspect of Diaguita identity. In effect, when we look at how Diaguita ceramics has evolved over time, we observe the persistence of certain

a b

Figure 14.
Bi-globular anthropomorphic polychrome pitcher with wave A pattern A. Grave 182. El Olivar Site; (a) Front section; (b) Back section.

a b

Figure 15.
Polychrome anthropomorphic cup decorated with labyrinth pattern. Grave 182, El Olivar site. (a) Front section; (b) Lateral section.

Figure 16.
Globular polychrome pitcher with flared neck, displaying the minimal units of the chain B pattern. Grave 58, El Olivar site.

complexly structured designs, expressed in a simpler version in the Early Diaguita phase (the Wave A pattern, for example). These decorative patterns were taken up and enhanced time and time again by the Diaguita people in a kind of ongoing sequence that withstood the passage of time. This insistent reproduction of a set of specific designs is a clear expression of how memory constructs and establishes identity. In this sense, we propose that designs like the Wave A pattern, which can

Figure 17.
Four-legged anthropomorphic-zoomorphic vessel. Grave 186, El Olivar site.

Figure 18.
Evolution of Wave A pattern over time: (a) Mojocoya trichromatic design (Lathrap 1970, p.143); (b) Sherd of the Wave A pattern, Early Diaguita period, El Olivar site; (c) Wave A pattern in trichromatic cup, Diaguita pre- Inca period, Grave 149, El Olivar site; (d) Wave A1 pattern in felinomorphic bowl, Diaguita-Inca period, Grave 59, El Olivar site.

be traced back to the earliest times of the Mojocoya culture and is represented in Diaguita culture without interruption from its origin to the Inca stage, is an "axis of formal coherence" [21]. These particular forms and designs are used by a group of people as a means of their own social reproduction (**Figure 18a–d**).

Obviously, the Mojocoya and Chilean Diaguita cultures are separated by a major gap in both time and space, making it difficult to speak of a direct link between the two. Perhaps the latter can be thought of as an original source of the concepts and designs that spread, albeit with some variations, over a geographically extensive area that stretched to the eastern lowlands and likely into the Amazon uplands as well.

5. Pre-Inca Diaguita art: the agency of art and technologies of enchantment

A detailed examination of pre-Inca Diaguita visual art [1] from the perspective of its motifs, formal logic, and symmetrical structure suggests that this representational universe does not seek to communicate specific concepts of any kind. Notably, the hyper-concentrated visual litanies of that period were crafted by artisans highly skilled in intricate, complex symmetrical operations. For us, it appears that their aim was more likely the creation of what Gell [22] has termed "technologies of enchantment."

In turn, the same author considers decorative art a special form of technology. Gell comments [23],"the decoration of objects is a component of a social technology, which I have elsewhere called the technology of enchantment" [22], and continues, "this psychological technology encourages and maintains the motivations 6necessitated by social life. The world is filled with decorated objects because decoration is often essential to the psychological functionality of artifacts, which cannot be dissociated from the other types of functionality they possess, notably their practical or social functionality" ([23], p. 74).

As Gell explains, work of art captivate the viewer with the difficultly of their conception and execution. Their uniqueness and beauty transform them into active agents in the social sphere. In particular, abstract art exploits our perception of internal agency because

> *"Patterns, by their multiplicity and the difficulty we have in grasping their mathematical or geometrical basis by mere visual inspection, generate relationships over time between persons and things, because what they present to the mind is, cognitively speaking, always "unfinished business" ([23], p.23)*

The author goes on to argue that this "sets up a biographical relation – an unfinished exchange – between the decorated index and the recipient" ([23], p. 81), and thus attributes this "cognitive stickiness" of abstract patterns "to this blockage in the cognitive process of reconstructing the intentionality embodied in artefacts."

Thus, by employing these visual strategies, the abstract patterns acquire properties, such as the illusion of movement and vibration, that help to captivate the viewer and enhance the agency of the pattern. In the words of Gell ([23], p. 77), "nothing could be more animated than the tessellations (tiling patterns) devised by Islamic decorative artisans…it seems, to inspire ever more effective inducements to captivation by visual artifice, the non-mimetic appearance of animation".

In this regard, Taylor [16] notes that in the Amazonian representational universe, figurative representations of the entities of the cosmovision are virtually absent, while abstract and geometrical representations abound. The author concludes, "*any surface with geometric motifs can evoke subjectivity or personality. Even paintings themselves can be treated as persons. For example, the Jíbaro people have ritual songs that are sung directly to body paintings to ask them for a certain type of intervention*" [14 p. 6]. The author therefore proposes that the abstract body painting of Amazonian groups was intended to *"represent complete beings as "incarnations."* This suggests that this geometric, non-figurative Amazonian art was not "iconophobic," but rather played a major role "*in the perpetuation of this tradition of figuratively imagining supernatural beings*" ([16], p. 13).

6. Conclusions

In this article we have attempted to build our understanding of the visual logic deployed by the Chilean Diaguita people, as expressed in the designs of Diaguita polychrome ceramics. The use of symmetry in Diaguita culture is remarkable. In their pre-Inca designs the highly skilled Diaguita artisans managed to create complex abstract constructs that were highly symmetrical and capable of producing optical effects such as vibration and movement. Without a doubt, these reinforced the active social role of their visual art, in association with a shamanic belief system. Current reflections about Amazonian animism offer extensive opportunities for interpreting pre-Inca Diaguita art, which, as we have previously proposed, seems to be part of a larger cultural tradition that remains alive in different parts of South America, particularly among indigenous groups of the Amazon region.

Furthermore, the surprising similarities observed in both the visual art and certain social practices of the Chilean Diaguitas and the Shipibo-Conibo people, despite their spatial and temporal distance, can be better understood by observing the dynamic long-distance cultural contact that likely took place between peoples living in the Bolivian, Andean, and even Amazonian lowlands. If we are to better understand the pre-Hispanic past of the southern cone of the Americas as well as its ethnographic present, we must broaden our perspective and grasp that in the distant past, cultural traditions possessed of a clearly recognizable identity exercised their influence over extensive geographic areas and diverse ecosystems. 3 In the case presented, the analysis of symmetry was crucial in demonstrating that cultural contact did exist, and has proven to be an approach well suited to detecting processes of cultural inter-relatedness and exchange.

It is also interesting to examine the esthetic facet of symmetry, and how it helped define the elusive concept of *beauty.* In the pre-Inca Diaguita representational universe, its function in the creation of highly complex abstract images is evident. At the same time, the painstaking display of symmetry in a significant number of designs from across the entire Diaguita territory demonstrates the social importance that Diaguita artisans attributed to managing and comprehending symmetry. Indeed, the use of symmetry enabled communities in different valleys to affirm their individual identities, while marking Diaguita identity as a whole.

So, how did the Diaguita people conceive of such vessels loaded with symbolic content? We find it difficult to believe that they were simple, everyday pieces crafted to hold food. Instead, based on their complex iconography and rich esthetics, it would appear that they were invested with an active social role.

DOI: http://dx.doi.org/10.5772/intechopen.93984

Author details

Paola González Carvajal
Sociedad Chilena de Arqueología, Santiago, Chile

*Address all correspondence to: paoglez@gmail.com

References

[1] González P. Arte y cultura Diaguita Chilena: simetría, simbolismo e identidad. Santiago: Ucayali Editores; 2013, 349 p.

[2] Washburn D, Crowe D. Symmetries of Culture. Theory and practice of plain pattern análisis. 2nd ed. Seattle: University of Washington Press, 1992. 294 p.

[3] Becker C, Rodríguez J, González P, Troncoso A, Pavlovic. *Arqueología en el río Chalinga. Informe Final*. Proyecto Fondecyt N°1000039. 2003 MS.

[4] Hidalgo J. Diaguitas Chilenos Protohistóricos. In: Hidalgo J, Schiappacasse V, Niemeyer H, Aldunate C, Solimano I, editors. Prehistoria. Desde sus orígenes hasta los albores de la Conquista. Santiago: Editorial Andrés Bello, 1989. p. 289-293.

[5] Troncoso A. Arte rupestre, prácticas socio-espaciales y la construcción de comunidades en el Norte Semiárido de Chile (Valle del Limarí). In: Falabella F., Sanhueza L, Cornejo L, Correa I, editors. Distribución Espacial de Sociedades No Aldeanas. Serie Monográfica de la Sociedad Chilena de Arqueología N°4. Santiago: Ediciones On Demand; 2014. p. 89-115.

[6] González P. Arte rupestre de origen Diaguita-Inca en los valles de Choapa y Limarí: su importancia en las estrategias de interacción política Inca en territorio Diaguita. In: Actas XIX Congreso Nacional de Arqueología Chilena; 8-12 October 2012; Arica: Andros Impresores; 2015. p. 214-223

[7] Reichel-Dolmatoff G. Aspectos chamanísticos y neurofisiológicos del arte indígena In: Museo Chileno de Arte Precolombino, editors. Estudios en Arte Rupestre. Santiago: Editorial Museo Chileno de Arte Precolombino, 1985. p. 291-307

[8] Gehbart-Sayer A. The geometric designs of the Shipibo-Conibo in ritual context. Journal of Latin American Lore. 1985; 11 (2): 143-175.

[9] Levi-Strauss C. Tristes Trópicos. Barcelona: Ediciones Paidós, 1988 [1955]. 468 p.

[10] González P. La tradición de arte chamánico Shipibo-conibo (Amazonía peruana) y su relación con la cultura Diaguita chilena. Boletín del Museo Chileno de Arte Precolombino. 2016; 21 N°1: 27-47

[11] De Boer W. Interaction, imitation and communication as expressed in style: the Ucayali experience. In: Conkey M, Hastorf C, editors. The uses of style in Archaeology. Cambridge: Cambridge University Press; 1990. p. 82-104.

[12] Piaget J. Structuralism: Translated by Chaninah Maschler. New York: Basic Books. Harper & Roe; 1970. 16 p.

[13] Illius B. La Gran Boa: Arte y cosmología de los Shipibo-Conibo. Amazonía Peruana 1994; 24 (12): 185-212.

[14] Roe P. The cosmic zygote. Cosmology in the Amazon Basin. New Brunswick: Rutgers University Press; 1982. 400 p.

[15] Barreto C. Meios místicos de reproducao social: Arte e estilo na cerámica funeraria da Amazonia antiga [thesis]. Sao Paulo: Museo de Arqueologia e Etnologia, Universidade de Sao Paulo; 2008.

[16] Taylor A. Arte y mito en las culturas amazónicas. Universidad Pompeu Fabra. Centre Investigador en Art Primitiv i Primitivisme. Jornada Arte y Miti en las culturas primitivas. Conference on 10 March 2008. Available from:

http://www.upf.edu/ciap/_pdf/Conference_Taylor_Amazonas_peq.pd [Accessed 2015-08-15]

[17] Lathrap D. The Upper Amazon. Southampton: The Camelot Press Ltd; 1970. 256 p.

[18] Brockington D, Pereira D, Sanzetenea R. El sitio El Tambo y la tradición Mojocoya: Cambios, Tiempo y Espacio en el Sur-Este de Bolivia. In: Pereira D, y Brockington D, editors. Mojocoya y Grey Ware: Interacción espacial e intercambio entre la Amazonía, Chaco y Andes. Cuaderno de Investigación N°10. Serie Arqueología. Cochabamba: Universidad Mayor de San Simón. Dirección de Investigación Científica y Tecnológica, Instituto de Investigaciones Antropológicas y Museo Arqueológico. 2005. p. 3-58

[19] Pereira D, Brockington D. Mojocoya y Grey Ware: Interacción espacial e intercambio entre la Amazonía, Chaco y Andes. Cuaderno de Investigación N°10. Serie Arqueología. Cochabamba: Universidad Mayor de San Simón. Dirección de Investigación Científica y Tecnológica, Instituto de Investigaciones Antropológicas y Museo Arqueológico; 2005. 91 p.

[20] González P. *Sitio El Olivar: su importancia para la reconstrucción de las comunidades agroalfareras del norte semárido chileno.* Colecciones Digitales. Subdirección de Investigación. *Colecciones Digitales.* Subdirección de Investigación DIBAM. 2017. Available from: http://www.museoarqueologicolaserena.gob.cl/sitio/Contenido/Colecciones-digitales/83495:El-sitio-arqueologico-El-Olivar [Accesed: 2020/07/15]

[21] Cantarutti G, González P. Boletín de la Sociedad Chilena de Arqueología. Santiago: Nuevos antecedentes sobre la cultura Diaguita Chilena en el valle del Elqui a partir del sitio El Olivar. Journal article (forthcoming)

[22] Gell A. The Technology of Enchantment and the Enchantment of Technology. In: J. Coote J. Shelton A, editors. Anthropology, Art and Aesthetics. Oxford: Clarendon Press; 1992. p. 40-67.

[23] Gell A. Art and Agency: An Anthropological Theory. Oxford: Oxford University Press; 1998. 271 p.

Chapter 7

From the "*Terra Preta de Indio*" to the "Terra Preta do Gringo": A History of Knowledge of the Amazonian Dark Earths

Klaus Hilbert and Jens Soentgen

Abstract

The anthropogenic origin of the Amazonian dark earths (*Terra Preta de Índio*) was first verified more than 70 year ago. However, the last 30 years have seen a massive wave of scientific investigation, public interest and an ever-expanding intensification of commercial activity toward all things connected to "*Terra Preta*." Today, the dominant concept, which drives current research, is that of binding atmospheric carbon with artificially concocted dark earths. The large-scale production of *Terra Preta* is said to be an effective tool in efforts to mitigate global warming. This text attempts to provide a history of the knowledge on Amazonian dark earths. It not only focuses on scientific aspects but also considers traditional indigenous knowledge. The position is taken that without indigenous knowledge, modern *Terra Preta* research would not exist; a view, which has profound implications for the ethical evaluation of all further, applied *Terra Preta Nova* research and commercial endeavors.

Keywords: Amazonia dark earths, *Terra Preta de Índio*, traditional indigenous knowledge, biochar, history of knowledge, global warming, carbon storage

1. Introduction

Today, the term Amazonian dark earths (*Terra Preta de Índio*—Amazonian dark earths) refers to a dark soil most often found in limited zones in the lowland areas of Amazonia. These soils are, as a rule, concentrated near rivers and located on bluff zones above the floodplains (*várzea*) [1]. This unique soil has a significantly higher nutrient content, especially phosphorus and nitrogen, and a much higher pH value (around 6.7) when compared to other Amazonian soils [2, 3]. These soils are—as the name indicates—very visually dark and all of them contain some quantity of prehistoric material cultural remains. For this reason, all ancient *Terra Preta* sites are nowadays considered archeological in character and protected by Brazilian law.

According to Woods and Denevan [4], the *Terra Preta* sites cover an area of 0.1–0.3% (6000–18,000 km^2) of the wooded Amazonian lowlands. As these soils are characterized by a long-term fertility, many of them are used by the local population, known as *caboclos*, in their agricultural practices [5–7].

The baseline, "non-*Terra Preta*," primary forest soils (especially oxisols) are generally extremely nutrient-poor, unless modified significantly through the introduction of artificial fertilizer. These primary forest soils have a low pH value, which means they are extremely alkaline and suited for long-term agricultural use. The usual agricultural workaround, known as slash-and-burn, produces ash and coal, which provide for a temporary fertility. Immediately thereafter, the areas are generally abandoned as the soil fecundity declines and much waste in biomass and land use is the overall result in this highly inefficient approach [8].

The indigenous knowledge of *Terra Preta*, their location and qualities are still invaluable to the current discussion (for a more in-depth discussion on the subject see Manuela Carneiro da Cunha [9]). Simply taking into consideration solely peer-reviewed scientific journal publications is, in our opinion, insufficient [4]. Such a limited approach creates a distorted perspective. Such a selective corpus might inspire the conclusion that only Western-based scientists and 21st century business concerns discovered and developed the fundamental benefits of the *Terra Preta de Índio.* Such a clearly simplified historically misinformed view might even be used for legitimizing claims to the economic exploitation of this important indigenous knowledge [10].

2. Dark earth as anthropogenic soil: the legacy of Curt Nimuendajú

Foreign researchers and travelers first learned about the existence of this uniquely fertile soil by local informants. Herbert Smith [11], for example, had this description to give after his visit to Santarém, "...the rich terra preta, 'black land', the best on the Amazon (...) is a fine, black loam, a foot, and often two feet thick. Strewn over it everywhere we find fragments of indian pottery, so abundant in some places that they almost cover the ground." Friederich Katzer [12], a German geologist, gave the "*Schwarze Erde*" (black earths) a natural origin and interpreted them, as they were located near the rivers, as ancient lake deposits (*igapó).* He analyzed three soil samples from *Terra Preta* and was surprised by their exceptionally high content of organic matter. He noticed a great quantity of potsherds as well of indigenous origin and concluded from this observation that these "ancient lake deposits" were once cultivated by the ancient Amazonians, when the region was much more densely populated.

Curt Unkel (1883–1945) went beyond the mere confirmation of the existence of the Amazonian dark earths and the understanding that these were places in which archeological objects could be found. Beyond a doubt, Unkel (later receiving his Indian-sanctioned moniker "Nimuendajú") is one of the most important 20th century researchers of Amerindian cultures and has done invaluable work for the investigation and protection of Brazilian indigenous populations [13].

In his research on *Terra Preta*, Nimuendajú started from the widely-understood fact, that wherever a certain earth was found (locally called *Terra Preta de Índio*), ceramic fragments of earlier indigenous cultures were always present. Following this premise, near the city of Santarém, at the confluence of the Rio Tapajos and the Amazon, Nimuendajú registered the location of 63 previously unknown *Terra Preta* sites (**Figure 1**) [14].

In 1945, in a letter to the ethnologist Herbert Baldus, Nimuendajú outlined his ideas about the origin of the *Terra Preta*. This letter, summarizing the considerations of many earlier studies, was a watershed publication in *Terra Preta* knowledge and major hallmark of Nimuendajú's personal research legacy, as it contains nearly all the keys ideas of all subsequent *Terra Preta* research [15].

His systematic analysis of the locations of the *Terra Preta* led him to reject the previously prevailing opinion that the special soil was of a natural origin and the result of flooding, the remnants of lake sediments or even volcanic ash [16]. Nimuendajú concluded, by comparing the spatial distribution of the *Terra Preta* that the soils were totally anthropogenic, did not occur naturally and clearly

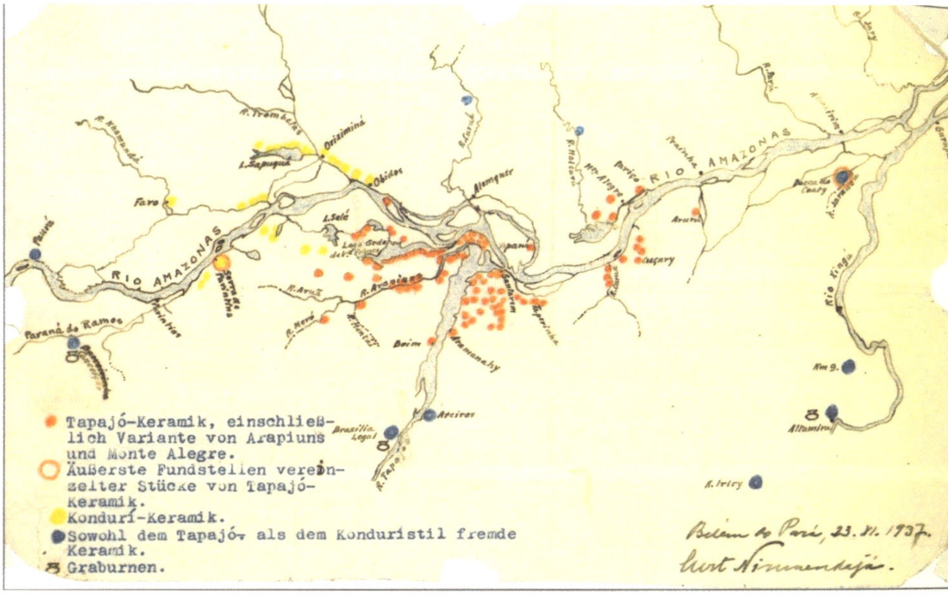

Figure 1.
Map of Terra Preta *sites located in the Rio Tapajós area, drawn by Curt Nimuendajú in 1937 (property: Klaus Hilbert).*

an outcome of indigenous production. Nimuendajú did not, however, mention whether he assumed this soil was created intentionally or unintentionally. What he did say with certainty and much authority was that all *Terra Preta* in Amazonia was of indigenous origin. Their formation, he explained, was due to the burning of wood in hearths and not the product of slash-and-burn techniques. He concluded that all dark earth sites were necessarily archeological sites, because of their clear association with Amazonian peoples. He was thereby squarely positioning himself in opposition to the prevailing interpretation that presumed a natural production process at work in *Terra Preta* genesis.

3. *Terra Preta* as an archeological site

Nimuendajú bequeathed, so to speak, *Terra Preta*, as a serious research topic to the archeological community. Nevertheless, the academic community was slow to ascertain the value of his contributions.

Betty Meggers and Clifford Evans [17], who eventually picked up where Nimuendajú research had left off, initially began their research inquiries following the work of Charles Hartt [18] and Orville Derby [19] by conducting their research on Marajó Island. They were more interested in the artificial mounds of the Marajoara culture, located on the eastern part of the island, than in *Terra Preta*.

Initially, the researcher to most fully grasp the gravity of Nimuendajú's conclusions regarding *Terra Preta* was Franciscan priest Protásio (Günther) Frikel. He, in turn, bestowed upon fellow researcher Peter Paul Hilbert [20] the location of more than 40 *Terra Preta* archeological sites found in the vicinity of the Nhamundá and Trombetas rivers, where Frikel happened to hold a parish seat. Both Hilbert and Frikel associated the potsherds found in and about these sites, again following Nimuendajú's groundwork, with the Konduri natives, mentioned in the earliest European chronicles. It is likely that this was the time and place when truly systematic archeological research on *Terra Preta* first began.

During the subsequent 10 years, Peter Paul Hilbert located, sample and excavated *Terra Preta* sites along the middle Amazon River [21]. Later, archeologists associated with the Museu Paranse Emilío Goeldi in Belém, such as Mario Simões, continued Hilbert's work and even intensified the systematic approach towards treating *Terra Preta* soils as bonafide archeological sites.

Under Betty Meggers' intellectual supervision and financial support, Mário Simões and his assistants, located and pit-tested hundreds of *Terra Preta* sites, mostly along the Amazonian tributary rivers, such as Rio Negro, Rio Madeira and Rio Tocantins [22]. Recognizing these *Terra Preta* sites mainly as deposits of discarded refuse (kitchen middens), these archeologists were largely interested in the material cultural remains, such as fragmented ceramic vessels and some rare polished stone implements "thrown away" by their original, ancient owners.

Quantifying and classifying the ceramic fragments by their diagnostic features, such as decoration, style, technique and form of vessel, this group of researchers created a chronology-based model of site categorization and orientation. According to the dominant typological pattern of the ceramic fragments, four cultural traditions were pronounced: Zone-Hachured, Incised Rim, Polychrome and Incised-Punctuated [23].

As most of these habitation sites had a multi-compositional sequence of archeological remains, it was concluded that the ancient Amazonians had a semi-nomadic strategy of survival. This was necessitated, it was argued, by the poor soil conditions [24] which could not support extended agricultural use and therefore, neither a complex society nor a high population rate. Despite the extraordinarily high biodiversity of the Amazon region, it was claimed, the habitants had to change their settled areas periodically, which on one hand, explained the great number of *Terra Preta* sites and, on the other, their frequent reoccupation by varying archeologically displayed cultures. Betty Meggers elaborates on this paradoxical situation contrasting the Amazonian environment and archeological record with the idea of a "counterfeit paradise" [25].

Donald Lathrop [26], José Brochado [27] and Anna Roosevelt [28] questioned this model of environmental restraint and soil poverty, and its inherent relationship to a low level of social complexity. They defended the opposite scenario of a rich and fertile Amazonian environment, which supported a large population and complex societies, based on powerful and interrelated chiefdoms.

Eduardo Neves, observing a gradually waning academic interest in these debates and marked shift away from topics involving ancient Amazonia as important research terrain, felt impelled to develop an interdisciplinary-driven and internationally-inclusive research strategy. Initially located in the lower Rio Negro and Rio Solimões region, he united a group of scientists from soil sciences, anthropology, biology, ecology, geography, geology and sociology and set out to reinvigorate the research topic of *Terra Preta*. This wave of researchers made pivotal contributions to the knowledge of Amazonian Archeology, focusing most recently on the question of the origins of early crop domestication [29–35].

Contemporaneously, Michael Heckenberger [36], Denise Schaan [37, 38] and Denise Gomes [39] reignited discussion on the social-power relations expressed by the construction of the monumental structures of the ancient Amazonians. Their primary focus being the massive vertical constructions and their reflection of a high social and cultural complexity. Some examples include the mound builders of Marajó Island, the road builders of the Tapajos, in *Alter do Chão*, near Santarém, the village builders of the upper Rio Xingu or the geoglyph builder in Rondônia and Acre.

Despite all of these advances in the knowledge of *Terra Preta*, the subject was generally, yet again, abandoned as a serious research focus and the idea of the "dark earth" as a socio-cultural phenomenon, a massive public works project of horizontal monumentality, or as an expression of high social organization and cultural complexity, was again, relegated to the academic shadows.

4. The scientific characterization of *Terra Preta de Índio*

While archeologists, and to some extent anthropologists, were becoming more deeply involved in the discussions about the relatively high or low fertility of the Amazon region, investigations into the soil's relation to social complexity, sedentary or semi-nomadic lifestyles or population density remained secondary at best. Other scientists took up the subject of *Terra Preta* and ran with it. Soil scientists more and more gained international public interest and headlines from prestigious and highly visible publications.

In the early 1980s, research on the "natural scientific character" of the *Terras Pretas* began in earnest, with the investigations of Eije Erich Pabst and Gerhard Bechtold. They demonstrated, for the first time, that the "black earth" differed dramatically in chemical composition from the typical, reddish oxisols of Amazonia. They re-confirmed the soil's higher pH value, abundance of organic substances, and much higher nitrogen and phosphorus levels [2, 3].

How the *Terra Preta de Índio* could have come about what the leading question of Eije Erich Pabst, who not only explored the soil properties, but also ethno-pedological aspects. Pabst posited that if one could determine how the *Terra Preta* was arose from the oxisol, then one could recreate such improved soil enhancements today. As part of his strategy, Pabst visited several ethnic groups in the Amazon region. Here he found quite diverging opinions expressed, including their natural provenance by the Assurini, divine origin, as expressed by an Arawete: "the forest God threw them down from heaven," and an anthropogenic or "man-made" origin. Finally, some Waiapi natives simply explained that they just did not know how the soil came to be [3].

Despite efforts, Pabst was not able to track down a definitive indigenous "recipe" for the manufacture of *Terra Preta* during his oral interview investigations. Pabst emphasizes, however, that this knowledge once existed, but was lost in the course of the population collapse the Amazonian people suffered during and since the European colonization. The indigenous populations were largely displaced from any favorable localities and generally forced into non-sedentary modes of survival. It is stressed that this historic and unmerited loss of cultural expertise should in no way be equated with native inventive capacity, agricultural competence or claims thereof.

Whether Amazonian black earths were intentionally produced or not, continues to fuel debate. On the one hand are the positions that exclude all intentionality and assume that the *Terra Preta* resulted to the indigenous lifestyle. This was more or less the opinion of the earliest archeologists and anthropologists conducting research in the twentieth century. On the other hand, positions exist that presume a very clever technique for the deliberate production of humus, even going so far as to suggest specific ceramics were created and then used in the process [40].

Ethnopedological research has shown that there still do indeed exist Amerindian groups whose lifestyle leads directly to the production of *Terra Preta* [41]. Even without this evidence, it would seem plausible to presume that at least some of the groups formerly living in the Amazon had consciously carried out soil improvement measures by adding a combination of plant charcoal, ceramics and organic matter [42]. Considering the many innovations of the Amazon Indians [43], it is highly unlikely that such a vital adaptive measure and its life-improving results would have totally escaped the conscious awareness of the indigenous population living in Amazonia prior to the arrival of the Europeans.

Moreover, historical studies have shown in other important cases that the ahistorical perspective, that the indigenous products are results of pure happenstance, stems from the faulty view of native peoples as "children of nature"—a perspective no longer seriously tenable [44, 45]. On the contrary, it is almost certain that at least some central components of the *Terra Preta* required deliberate actions with the purpose of soil enhancement at play. The introduction of charcoal on fields and plantings could be one such element which is current in Amazonia even today, where hearth ashes are strewn on house gardens.

The discovery of the black earths has far-reaching significance for the history of the American continent, as it forces a total revision in estimating number of inhabitants in Amazonia prior to 1492. Because of the high and enduring fertility of *Terra Preta*, it now seems plausible that the area, before European invasion, would have had a population of several times the currently accepted approximate figure of one million people [46]. Currently a pre-colonial population of 8–10 million is considered quite plausible, based on estimates of the currently known acreage of Amazonia *Terra Preta* [29].

This also corresponds to historical accounts of the conquistador Francisco de Orellana (1511–1546), whose chronicler Gaspar de Carvajal reported sighting "numerous huge settlements," in the first trip along the Amazon in 1542, especially above the confluence of the Rio Negro. Retrospectively Carvajal specifically praised the people of the Amazon for their extraordinary intelligence and inventiveness, demonstrated in their "wonderful works" [47].

5. The black earth as climate savior: Wim Sombroek's vision

In 1992 the Dutch soil scientist Wim Sombroek, who had previously devoted his doctoral thesis to the soils of the Amazon [48], outlined what would prove to be a watershed conceptual breakthrough in the on-going "*Terra Preta*" story.

In contrast to previous researchers who had enjoined *Terra Preta* research to ethnological, historical and archeological data, Sombroek infused climatic sciences and concerns into the *Terra Preta* dialog [49]. Sombroek sought to manufacture *Terra Preta Nova* or "new" black soil through the introduction of plant charcoal to nutrient-poor soils.

The addition of "Biochar" (plant charcoal) into the soil, sequesters carbon, diminishes the release of carbon dioxide and methane and provides for the soils' extended usefulness. The enhanced soils' new fertility, it was hoped, would provide a nutritional benefit to Amazonian farmers and could also be replicated elsewhere in the tropics.

Sombroek hoped this strategy could act as a counterweight to some of the pressures created by man-made global warming on extant agricultural practices. This combination of *Terra Preta* research with "climate change" concerns proved to be very ripe recipe indeed for both academic and public consumption and interest. It brought an otherwise obscure issue, a specific soil type, into the international spotlight.

Sombroek's new approach seemed to coincide with a new, global reckoning and awareness of the general climate change issue. In 1990 the United Nation's first IPCC (Intergovernmental Panel on Climate Change) report was issued, followed 2 years later by the U.N.'s Rio de Janeiro Earth Summit in June of 1992.

Existing elegantly at the intersection of environmental concern and hard science *Terra Preta* research took on a new life, birthing a robust research network and garnering unsuspected mainstream media attention. At the helm of this wave of interest was always the charismatic Wim Sombroek, who was duly recognized for bringing attention and funding to the fledgling world of *Terra Preta* research, he was even honored with a poem at his posthumous memorial service. He died in December 2003 [50].

Figure 2.
A Terra Preta *site used as papaya-plantation by the local population at the Rio Negro (Photo: J. Soentgen, 2015).*

At the core of Sombroek's research initiative was the goal of providing the inhabitants of Amazonia with a means to bring greater, more reliable harvests by reinstating native soil improvement techniques (**Figure 2**) [51].

6. Terra Preta Nova

Wim Sombroek would probably be a great deal befuddled, if not downright dismayed, by the surprising direction his groundbreaking research has since taken. The *Terra Preta Nova*, as both a concept and a physical substance, has now been seized up by commercial interests and its being offered on the home improvement/horticultural market right alongside the more widely known "enhanced" home gardening soils.

The rapid and profound proliferation of the *Terra Preta* model to large-scale commercial and agricultural uses has been supported by organizations like the International Biochar Initiative (BCI). Founded in July 2006 this body aims to support research on and commercialization of biochar suitable for the manufacture of black earth (www.biochar-international.com).

By promoting the spread of Biochar's use, BCI and its proponents, hope to limit the effects of climate warming through long-term carbon sequestration and increased soil fertility. Their strategy includes political lobbying in addition to their commercial investment. One principal aim is to insert biochar, as sequestered carbon, squarely into global political climate decisions, particularly in the Kyoto protocol and its successor agreements. Further discussion in Bruges [52].

Clearly, the *Terra Preta Nova* community and its adherents have expanded well beyond the ideas Sombroek originally outlined. Carbon in the form of plant charcoal became more and more the central focus, largely because of the relation to the conversations about the climate issue.

However, it is important to take note that the climate-saving *Terra Preta Nova* is in many respects much different from the *Terra Preta de Índio*. First and foremost, it is not simply a "special soil type" that has formed over hundreds of years, but rather a substance, a stuff, that can be deliberately produced through a very specific technical process. Both the incorporation of plant charcoal, but also ceramic components (potsherds), whose porous properties act as a water retainer, were essential to the original *Terra Preta* recipe. By highlighting only certain characteristics of the *Terra Preta* properties, (those strictly relatable to climate change) and reframing the soil as a universally applicable and easily "manufacturable" substance, *Terra Preta Nova* is less an archeologically localized soil and more a commodity.

Through this change of meaning, the redefined *Terra Preta* can be industrially produced in a manner similar to what Bavarian writer Walter von Molo once imagined in his Murnau diaries as "earth factories." These factories "would produce wonderful wheat soil or whatever was desired, that then would be poured out over swamps, deserts, barren or stony areas of all types. Rains of earth would be allowed to occur with great transport aircraft, earth cloudbursts" [53].

7. Mythologizing *Terra Preta*

Another major component and catalyst for the dispersal of *Terra Preta* recognition in the public arena was the concurrent "mythologizing" of *Terra Preta* itself. Some aspects of the history told in this paper have been previously been published in: '*Terra preta de índio*': Commodification and Mythification of the Amazonian dark earths [54]. Key phrases, with very little specific meaning, like "ancient knowledge," "from the rainforest" and the like were peppered throughout the commercialization and discussion process. Thus, a "black revolution from the rainforest" was announced, showing the "way out of the world-wide climate and hunger crisis [40]. Likewise, the website of the "international Biochar Initiative reads: "Sustainable biochar is a powerfully simple tool to fight global warming. This 2000-year-old practice converts agricultural waste into a soil enhancer that can hold carbon, boost food security, and discourage deforestation." (www.biochar.org, accessed August 25, 2018).

German producer PALATERRA emphasizes the aura of mystique and arcane wisdom in their advertising slogan: "The gold of the earth—after the example of the Indians." Another commercial text reads: "More than 1,000 years old—'*Terra Preta*', the most fertile soil in the world, was produced by the advanced Indian culture of the Amazon from organic material and charcoal. This ancient form of a sustainable recycling economy enables an effective foodstuff cultivation on the infertile rainforest soils (oxisol) for an enormously large population. The production method lapsed into obscurity with the disappearance of the ancient culture years ago. Only around the end of the 20th century archaeologists discovered this 'gold of the earth'" (**Figure 3**) [55].

Figure 3.
Terra Preta Nova, *produced by PALATERRA (BB-mbh): "The Gold from the Earth, created after the model of the Indians." Provides healthy plants, powerful growth, healthy soil, activates the building of humus, lasting effects, active climate protection, and peat-free soil (Photo: J. Soentgen, 2015).*

No doubt, one expects such an enigmatically rich dramatization to improve sales of any commercial product. Nevertheless, similarly "enchanted" language and descriptions are also found in scientific papers, usually in the abstract or introduction.

One of the most renowned *Terra Preta* researchers in Germany, Bruno Glaser, explains in a scientific article published in the Philosophical Transactions of the Royal Society, that *Terra Preta* could contribute simultaneously to three of the "Millennium Development Goals. These include mitigating desertification and global warming, as well as maintaining 'biodiversity hotspots" in tropical rainforests [56]. Large-scale use of the "black stuff" would reduce the pressure on primary forests, thereby limiting further degradation of the rainforest while also benefiting the climate.

North American soil researcher Johannes Lehmann likewise considers the introduction of biocharcoal into soils as a "win-win approach" [57] and his colleague David A. Laird goes a step further, describing it as a "win-win-win scenario" [58]. It is worth noting that biocarbon production itself is big business, especially when its manufacture and incorporation as CO_2-sequestration becomes part of the emissions certificates market.

Thus the "vision" of the soil pioneer Wim Sombroek has in some sense become a reality. Carbon storage via the production of *Terra Preta* is unique amongst strategies being implemented in the fight against climate. Whether it's pumping CO_2 underground or the deliberate dispersal of aerosols into the environment, the *Terra Preta* strategy holds a unique persuasive power by being a modern adaptive measure from ancient pool of knowledge. Not only is a practical strategy being promulgated, but a myth and narrative is constructed that functions alongside the research-directing paradigm [45].

For such narratives and meaningful tales, it is characteristic to frame the saving agent as a return or resurrection of something ancient, prehistoric. It harkens back, painting a picture of archaic conditions, people, and gods in order to orient and motivate the act and agency of present-day protagonists [59], (on the term "myth" see also Christine Schmitz [60]). In our case, it is the ancient Amazonian indigenous population who was in possession of the secret to make fertile soil out of infertile ground. However, the ability of the traditional societies was based not only on knowledge, but also on wisdom. Therefore, it is valid to reactivate this now in order to find a way out a potential ecological crisis.

The conundrum presented by suggesting *Terra Preta*, or any other technical means, as a quick-and-easy, one-size-fits-all way solution in the combat against the steady on-set of global warming is that is relegates the role of politics and political will to the sidelines. If the political will to implement these technical solutions is not extant, then they can do very little on their own merit. Nevertheless, at the same time, the mythologizing is functional and can have real practical results. For one, it might help in convincing funding agencies. However, more importantly can spur the research community and inspire younger generations of researchers to enter the academic fray (see Jürgen Frese [61] on the function of myths in groups).

A report on a *Terra Preta* symposium, published in Nature, one of the most prestigious scientific journals, commented on a meeting of soil scientists researching *Terra Preta* describing how: "more than one eye in the room had a distinctly evangelical gleam" [62].

The same journal, Nature, is gladly supportive of spinning the *Terra Preta* mythos. In 2015, for example, an extended article was published [63] in which *Terra Preta* was praised for not only being a remedy for climate change, but also a nutrient and water reservoir and even an absorbing medium for poisons. It worth noting that *Terra Preta Nova* is now being produced in the United States and Europe (a report by the International Biochar Initiative names of 326 firms that are active in production: http://www.biochar-international.org/State_of_Industry_2015, accessed on June 22, 2019).

As an area of scientific research, *Terra Preta* is firmly established. By the end of September 2018, 1.230 scientific papers with titles containing the keywords "*Terra Preta*" and "biochar" had been published (online research in the ISI Web of Science on 24 September 2018). The academic fervor is not groundless, for it is now well documented that the introduction of biochar does indeed have soil-improving traits and does increase agricultural yields. In a limited sense, it contributes to climate protection, especially in Europe, if biocharcoal replaces compost [64]. *Terras Pretas* have also inspired research into novel methods of waste-water treatment [65].

The claim that newly-produced *Terra Preta Nova* can sequester carbon in massive amounts has not yet been sufficiently confirmed by studies. Now, there are only a few concrete field studies proving that the carbon particles really do remain in the soil for a substantial period and that the carbon bonding is not overcompensated through other processes [66]. Health is another concern when discussing the implementation of biochar, due to increased smoke levels (see a new analysis of smoke from burned plant material: Iinuma et al. [67], and products from the pyrolysis processes can cause adverse health effects. In addition, there are other competing uses of biomass to consider [68, 69].

In addition to these aforementioned concerns there is the need to stress the enormous profit incentives many large companies inevitably take into consideration when championing the integration of *Terra Preta Nova* and, respectively, biocharcoal in the CO_2 markets [66].

However, one enormous critically concerning issues is that it is often not possible to verify that biochar production is sustainably produced and does not stem directly from sources of illegal deforestation [52]. Accordingly, the Biochar activists are criticized by other NGOs, most notably Biofuelwatch [70, 71].

From the perspective of a history of knowledge and responsible use thereof, in closing, a consideration of the ethics of knowledge is deemed appropriate.

8. Indigenous knowledge and ethics

The modern *Terra Preta* research would have never begun without indigenous knowledge. There is a clear and direct link between the *Terra Preta* archeological sites and their characteristics that inform the modern search developments. Even the specific plants most adaptable to *Terra Preta* practices is delivered to us from traditional indigenous knowledge. Although observational evidence supports the idea that the modern-day indigenous people of Amazonia produce and understand the *Terra Preta* process, the idea remains debated [41, 72]. In addition, it is highly unlikely that the technically proficient indigenous people of Amazonia would have left their soil quality and conditions, so important for their very survival, entirely to chance.

Given this state of affairs, it would justify and correct that the intellectual and financial benefit gained through the research and manufacture of *Terra Preta* goes to not only the researchers and manufacturers, but also to the local inhabitants. Although certainly individual players (in the *Terra Preta* research and commercial diaspora) have exhibited the moral responsibility and principled outlook of Wim Sombroek and have indeed contributed to the benefit and well-being of the present-day inhabitants of Amazonia, it unfortunately remains the exception in the industry.

As the locally sourced *Terra Preta de Índio* mutated into the globally distributed and produced *Terra Preta Nova*, the benefits to the local communities seems to have gone missing in the equation. Generally speaking, the present-day inhabitants of Amazonia are excluded from participating in the profit-sharing made possible by the industrialized production of *Terra Preta* and biochar. Once might argue that they do indeed also benefit from the fact that climate change is being combatted. This remains a very abstract and long-term benefit however and means much less to generations living today, regardless of the efficacy the native technology does hold for negating the onset of human-caused climate changes.

Participation in both promulgation of and benefits from *Terra Preta* practices by native groups is essential. The current exploitative and non-collaborative methods used both scientifically and commercially disregard the core principles that were established for the reasonable protection of indigenous knowledge (World Intellectual Property Organization: Wipo Publication 920E). Modern suppliers and producers of *Tarra Preta* products insist on emphasizing the fact that their product was produced with the insight of "traditional indigenous knowledge". In this instance, therefore, the principle of "equitable benefit sharing" [73] (Wipo Publication 920E, p. 23) must be applied, for there are still remaining native peoples along the Amazon who continue to know and apply this knowledge [74, 75].

9. Conclusions

At present, it seems that neither the commercial manufacturers nor the science lobbyists have developed any plan of "benefit sharing" with the local populations. The situation takes on an aura of neo-colonial tendencies under the guise of protecting the world climate. In this scenario, the *Terra Preta de Índio* is today a "*Terra Preta do gringo*"; a black earth of English-speaking and publishing Northern-hemispheric researchers and for-profit companies. Incredibly, the patent process is now well underway where several companies already applying with the European Patent Office to protect "their" *Terra Preta* products (see European Patent 2188230B1). In Germany, *Terra Preta*, and *Terra Preta Mulata* are already registered words or trademarks (registered with the Terra Preta GmbH, Berlin). Clearly, these are not the product of European ingenuity, but the intellectual achievements of the people of the Amazon.

In blatant, unblushing irony, the same companies racing to register with the patent office continue to advertise the mystical "rainforest" as their products' point of origin. Major adjustments must be in the current profit and knowledge-sharing exchanges surrounding *Terra Preta*. Although efforts against debilitating climate change is certainly necessary and noble, shorter term economic value-setting and the absolute tangible benefits of disseminating *Terra Preta* practices should not continue to only benefit a very small and well-educated citizens of North American and European industrialized countries.

Acknowledgements

We would like to thank Norman Madarasz and John G. O'Donnell for their valuable suggestions as well as the German Academic Exchange Service (DAAD). This study was financed in part by the Coordenação de Aperfeiçoamento de Pessoal de Nivel Superior—Brasil (CAPES)—Finance Code 001 and CNPq for funding research on *Terra Preta*.

Author details

Klaus Hilbert[1*] and Jens Soentgen[2]

1 Escola de Humanidades, Pontifícia Universidade Católica do Rio Grande do Sul (PUCRS), Porto Alegre, Brazil

2 Wissenschaftszentrum Umwelt (WZU), Universität Augsburg, Augsburg, Germany

*Address all correspondence to: hilbert@pucrs.br

References

[1] Denevan W. A bluff model of riverine settlement in prehistoric Amazonia. Annals of the Association of American Geographers. 1996;**86**(4):369-385

[2] Zech W, Pabst E, Bechtolt G. Analytische Kennzeichnung der Terra preta do índio. Mitteilungen der Deutschen Bodenkundlichen Gesellschaft. 1979;**29**:709-716

[3] Pabst E. Terra preta. Ein Beitrag zur Genese-Diskussion auf der Basis von Geländearbeiten von Tupí-Völkern Amazoniens [thesis]. Gesamthochschule University Kassel; 1993

[4] Woods W, Denevan W. Amazonian dark earths: The first century of reports. In: Woods W, Teixeira CG, Lehmann J, Steiner C, WinklerPrins A, editors. Amazonian Dark Earths. Wim Sombroek's Vision. Springer Science; 2009. pp. 1-14

[5] Fraser J, Clement CR. Dark earths and manioc cultivation in Central Amazonia: A window on pre-Columbian agricultural system? Boletim do Museu Paraense Emílio Goeldi Ciências Humanas, Belém. 2008;**3**(2):175-194

[6] Harris M. Life on the Amazon. The Anthropology of a Brazilian Peasant Village. Oxford: University Press; 2000

[7] Kawa NC. Amazonia in the Anthropocene. People, Soils, Plants, Forests. Austin: University of Texas Press; 2016

[8] Frikel P. Agricultura dos índios Mundurukú. Boletim do Museu Paraense Emílio Goeldi, nova série, Antropologia, Belém do Pará1959. p. 4

[9] Cunha MC. "Culture" and Culture. Traditional Knowledge and Intellectual Rights. Chicago: Prickly Paradigm Press; 2009

[10] Vale F, Schaefer C, Vieira da Costa JA. Etnopedologia e transferência de conhecimento: Diálogos entre os saberes indígena e técnico na Terra Indígena Malacacheta, Roraima. Revista Brasileira das Ciências do Solo. 2007;**31**(2):403-412

[11] Brazil SH. The Amazons and the Coast. New York: Charles Scribners; 1879

[12] Katzer F. Grundzüge der Geologie des unteren Amazonasgebietes (des Staates Pará in Brasilien). Leipzig: Verlag von Max Weg; 1903

[13] Baldus H. Curt Nimuendajú, 1883-1945. American Anthropologist, N.S. 1946;**48**:238-243

[14] Nimuendajú C. Os Tapajós. Boletim do Museu Paraense Emílio Goeldi, Belém do Pará. 1949;**10**:93-106

[15] Baldus H. Tonscherbenfunde in Nordparana. Archiv für Völkerkunde. 1951;**6**:1-19

[16] Falesi IC. Soils of the Brazilian Amazon. In: Wagley C, editor. Man in the Amazon. Gainsville: University Press of Florida; 1974. pp. 201-229

[17] Meggers BJ, Evans C. Archaeological Investigations at the Mouth of the Amazon. Washington D.C: Bureau of American Ethnologie; 1957. p. 167

[18] Hartt CF. Ancient Indian Pottery of Marajó, Brazil. American Naturalist, New York. 1871;**5**(5):259-271

[19] Derby O. The artificial mounds on the island of Marajó. Brazil. American Naturalist, New York. 1879;**13**(4):224-229

[20] Hilbert P P. A cerâmica arqueológica da região de Oriximiná. Belém, Pará, Brasil: Instituto de Antropologia e Etnologia do Pará. Museu Goeldi. 1955;9

[21] Hilbert PP. Archäologische Untersuchungen am mittleren Amazonas: Beiträge zur Vorgeschichte des südamerikanischen Tieflandes. (Marburger Studien zur Völkerkunde). Berlin: Reimer Verlag; 1968

[22] Simões MF. Pesquisa e cadastro de sítios arqueológicos na Amazônia Legal Brasileira 1978-1982. Belém: Publicações Avulsas. Museu Paraense Emilio Goeldi; 1983. p. 38

[23] Meggers BJ, Evans C. An experimental formulation of horizon styles in the tropical forest area of South America. In: Lothrop SK et al., editors. Essays in Pre-Columbian Art and Archaeology. Cambridge, Mass; 1961. pp. 372-388

[24] Sioli H, editor. The Amazon: Limnology and Landscape Ecology of a Mighty Tropical River and its Basin. Dordrecht: Kluwer Academic Publishers Group; 1984

[25] Meggers BJ. Amazonia. Man and Culture in a Counterfeit Paradise. Chicago-New York: Aldine Aterton; 1971

[26] Lathrap DW. The Upper Amazon. London: Thames and Hudson; 1970

[27] Brochado JP. An Ecological Model of the Spread of Pottery and Agriculture into Eastern South America [thesis]. Urbana-Champaign: University of Illinois at Urbana-Champaign; 1984

[28] Roosevelt AC. Moundbuilders of the Amazon. Geophysical Archaeology on Marajó Island, Brazil. Academic Press; 1991

[29] Clement CR, Denevan WM, Heckenberger MJ, Junqueira AB, Neves EG, Teixeira WG, et al. The domestication of Amazonia before European conquest. Proceedings of the Royal Society B: Biological Sciences. 2015;**282**(1812):32-40

[30] McEwan C, Barreto C, Neves EG. Unknown Amazon. Culture and Nature in Ancient Brazil. London: The British Museum Press; 2001

[31] Pereira E, Guapindaia V, editors. Arqueologia Amazônica. Belém: MPEG; IPHAN; SECULT; 2010

[32] Neves EG. Arqueologia da Amazônia. Rio de Janeiro: Jorge Zahar Editora; 2006

[33] Hilbert L, Neves EG, Publiese F, Whitney B, Shock M, Veasey E, et al. Evidence for mid-Holocene Rice domestication in the Americas. Nature: Ecology & Evolution. 2017;**1**:1693-1698

[34] Alves DT. Dark Earth Plant Management in the Lower Tapajós [thesis]. Exeter: University of Exeter; 2018

[35] Watling J et al. Direct archaeological evidence for southwestern Amazonia as an early plant domestication and food production Centre. PLoS One. 2018;**13**(7):e0199868

[36] Heckenberger MJ. The Ecology of Power. Culture, Place, and Personhood in the Southern Amazon, A.D. 1000-2000. New York and London: Routledge; 2005

[37] Schaan DP. The Camutins Chiefdom: Rise and Development of Complex Societies on Marajó Island, Brazilian Amazon [thesis]. University of Pittsburgh; 2004

[38] Schaan DP. Sacred Geographies of Ancient Amazonia: Historical Ecology of Social Complexity. Walnut Creek: Left Coast Press; 2011

[39] Gomes DC. Padrões de organização comunitária no Baixo Tapajós: o desenvolvimento do Formativo na área de Santarém, PA [thesis]. Universidade de São Paulo; 2005

[40] Scheub U, Pieplow H, Schmidt H-P. Terra preta. Die schwarze Revolution aus dem Regenwald. 4th ed. München: Oekom; 2013

[41] Schmidt M. Amazonian dark earths: Pathways to sustainable development in tropical rainforests? Boletim do Museu Paraense Emílio Goeldi, Ciências Humanas Belém. 2013;**8**(1):11-38

[42] Brochado JP. Alimentação na floresta tropical. A analogia Etnográfica na reconstrução da alimentação por meio de evidências indiretas, a mandioca na floresta tropical. In: Revista do Instituto de Filosofia e Ciências Humanas. Porto Alegre: UFRGS; 1977. p. 4

[43] Nordenskiöld E. The American Indian as an inventor. The Journal of the Royal Anthropological Institute of Great Britain and Ireland. 1929;**59**:273-309

[44] Soentgen J. Die Rolle indigenen Wissens in der Geschichte des Kautschuks. Technikgeschichte. 2013;**80**(4):295-324

[45] Soentgen J, Hilbert K. Terra Preta als politischer Mythos: Das Wunder aus dem Regenwald, Scheidewege. Zeitschrift für Skeptisches Denken. 2016;**45**:265-275

[46] Smith N. Anthrosols and human carrying capacity in Amazonia. Annals of the Association of American Geographers. 1980;**70**(4):553-566

[47] de Carvajal FG. In: Medina JT, editor. Descubrimiento del Río de las Amazonas. Según la relación hasta ahora inédita de Fr. Gaspar de Carvajal con otros documentos referentes á Francisco de Orellana y Sus Companeros. Sevilla: Imprenta de E. Rasco, Bustos Tavera; 1894

[48] Sombroek W. Amazon Soils. A Reconnaissance of the Soils of the Brazilian Amazon Region. Wageningen: Center for Agricultural Publications and Documentation; 1966

[49] Sombroek W. Biomass carbon storage in the Amazon ecosystems. Interciencia. 1992;**17**(5):269-272

[50] Woods W, Teixeira CG, Lehmann J, Steiner C, WinklerPrins A, Rebellato L, editors. Amazonian Dark Earths. Wim Sombroek's Vision. Springer Science; 2009

[51] Sombroek W, Kern DC, Rodrigues T, Silva Cravo M, Cunha JT, Woods W, et al. Terra preta and Terra Mulata, Pre-Colombian Kitchen Middens and Agricultural Fields, their Sustainability and Replication. In: Dudal R, editor. Symposium 18, Anthropogenic Factors of Soil Formation, 17th World Congress of Soil Science, Bangkok; August 2002. 2002. pp. 1-9

[52] Bruges J. The Biochar Debate. Charcoal's Potential to Reverse Climate Change and Build Soil Fertility. Foxhole Dartington, Totnes, Devon: Green Books; 2009

[53] von Molo W. Wo ich Frieden fand. Erlebnisse und Erinnerungen. München: Braun und Schneider; 1959

[54] Soentgen J, Hilbert K, Groote-Bidlingmaier C, Herzog-Schröder G, Pabst E, Timpf S. Terra Preta de Índio: Commodification and Mythification of the Amazonian Dark Earths. GAIA: Ecological Perspectives for Science and Society. 2017;**26**(2):136-143

[55] Palaterra Betriebs-und Beteilungsungsgesellschaft mbH 2014: Informationsprospekt. Royal Society: Geoengineering the Climate: Science, Governance and Uncertainty. London: The Royal Society; 2009

[56] Glaser B. Prehistorically modified soils of Central Amazonia: A model for sustainable agriculture in the twenty-first century. Philosophical Transactions of the Royal Society B. 2007;**362**(1478):187-196

[57] Lehmann J. Biological carbon sequestration must and can be a win-win approach. An editorial comment. Climatic Change. 2009;**97**:459-463

[58] Laird DA. The charcoal vision: A win-win-win scenario for simultaneously producing bioenergy, permanently sequestering carbon, while improving soil and water quality. Agronomy Journal. 2008;**100**(1):178-181

[59] Wülfing W. Mythen und Legenden. In: Küttler W, Rüsen J, Schulin E, editors. Geschichtsdiskurs: Die Epoche der Historisierung. Frankfurt am Main; 1997. pp. 159-172

[60] Schmitz C. Mythos. A. Begriffsbestimmung. In: Schöllgen G, Brakmann H, de Blaauw S, Fuhrer T, Leppin H, Löhr, W, Speyer W, editors. Reallexikon für Antike und Christentum. Stuttgart: Verlag Anton Hiersemann; 2013. pp. 471-474

[61] Frese J. Intellektuellen-Assoziationen. In: Faber R, HolsteC,editors.ZurSoziologiemoderner Intellektuellenassoziation, Kreise, Gruppen, Bünde. Würzburg: Königshausen und Neumann; 2000. pp. 441-462

[62] Marris E. Black is the new green. Nature. 2006;**442**(7103):624-626

[63] Cernansky R. State-of-the-art-soil. A charcoal rich product called biochar could boost agricultural fields and control pollution scientists is putting the trendy substance to the test. Nature. 2015;**517**:258-260

[64] Kammann C, Kühnel Y, von Bredow C, Gößling J. Abschlussbericht desProjekts:'C-Sequestrierungspotential und Eignung von Torfersatzstoffen, Hergestellt aus Produkten der Landschaftspflege und Biochar'. Justus-Liebig Universität Gießen, Institut für Pflanzenökologie (Manuscript); 2010

[65] Rakelmann U, Werner T, Li Z, Schonlau H, Giese T, Augustin K, et al. Die Abwasserentsorgung als Kohlenstoffsenke? Wissenschaftliche Zeitschrift für Technik und Ökonomik der Wasserwirtschaft. 2009;**6**:44-50

[66] Ernsting A. Biochar—a climate smart solution? Report. Aachen: Misereor. Available from: https://www.misereor.org/fileadmin//user_upload/misereor_org/Publications/englisch/report-2-biochar.pdf [Accessed: 31 August 2018]

[67] Iinuma Y, Brüggemann E, Gnauk T, Müller K, Andreae MO, Helas G, et al. Source characterization of biomass burning particles: The combustion of selected European conifers, African hardwood, savannah grass, and German and Indonesian peat. Journal of Geophysical Research. 2007;**112**:D08209. DOI: 10.1029/2006JD007120

[68] Scholz SM, Sembres T, Roberts K, Whitman T, Kelpie W, Lehmann J. Biochar Systems for Smallholders in Developing Countries. Leveraging Current Knowledge and Exploring Future Potential for Climate-Smart Agriculture. Washington DC: The World Bank; 2014

[69] Rattan L. Managing soils for food security and climate change. Journal of Crop Improvement. 2007;**19**:49-71

[70] Biofuelwatch. 2013. Factsheet. Available from: http://www.biofuelwatch.org.uk/wp-content/uploads/Biochar-3-pager7.pdf [Accessed: 28 August 2018]

[71] The BA, Solution B. Carbon Farming and Climate Change. Gabriola Island, Canada: New Society Publishers; 2010

[72] Hecht SB, Posey DA. Preliminary results on soil management techniques of the Kayapó Indians. In: Posey DA,

Balée W, editors. Resource Management in Amazonia: Indigenous and Folk Strategies (Advances in Economic Botany). Vol. 7. New York: New York Botanical Garden; 1989. pp. 174-188

[73] Wipo Publication 920E. World Intellectual Property Organization: Intellectual Property and Traditional Knowledge. Booklet No 2. Available from: http://www.wipo.int/edocs/pubdocs/en/tk/920/wipo_pub_920.pdf [Accessed: 23 June 2018]

[74] Posey DA. Biodiversity, genetic resources, and indigenous peoples in Amazonia:(Re) discovering the wealth of traditional resources of native Amazonians. In: Hall A, editor. Amazonia at the Crossroads. The Challenge of Sustainable Development. London: Institute for Latin American Studies; 2000. pp. 188-204

[75] Denevan W. Cultivated Landscapes of Native Amazonia and the Andes. Oxford: University Press; 2002

Chapter 8

Ecology as Cosmology: Animal Myths of Amazonia

Alexandre Guida Navarro

Abstract

Amazonian ethnography is relied on the creation of the world by Animals as the Anaconda, considered an ancestral creator. The history that myths tell is that the Amazon River is the metaphor of an Anaconda because it has meanders like this snake and it is a large reptile-like the river itself. As far as Archaeology, many of these myths were represented in archaeological ceramic as Cosmology metaphor. Some animals like birds were presented as a ceramic rattle which noise is a shamanic trait of Amazonian Indian Cosmology. This chapter shows how the Stilt Villagers of the Eastern Amazonian built their Cosmology based on the Animals of their environment's Ecology.

Keywords: cosmology, ecology, prehistoric art, stilt villages, Amazon

1. Introduction: myths and animals

Myths provide cultural explanations for understanding the world. It contains creation stories and explains all the elements necessary to understand the universe [1]. They are memory, a vehicle of cosmological message and identity [2]. In Amazon, the myths rely on the animal metaphor as the social component of the cosmology [3]. As far as Archaeology, the myth usually is represented in the ceramic support as Art.

Many archaeologists think that the art should be understood within a much broader semantic significance: (a) artistic manifestations, functioning as symbolism; (b) cosmological message vehicles serving to communicate social, political and religious values of a certain society; (c) or art to be on its own, as a social cohesion form or political control strategy, to demonstrate these same values to other peoples, as a form of ethnic identity [4].

Artistic manifestations filled the precolonial Amazon and invited the archaeologists to study these themes [5]. However, it is still difficult to define the diverse variations of the art types of these societies, partly because there is no conceptual standardisation [6], or partly as the variability is not fully known, yet, due to the complexity and enormous size of the precolonial Amazon [7].

However, many important paths have already been walked. Among the theoretical assumptions about art that found greater development in Amazon is the structuralist school and its variations, mainly through the works of Lévi-Strauss who regarded the art as an expression of communication and sociability [8]. Gell [9] saw the artefacts as social and non-static agents, which highlighted the activities they were involved, such as rituals. In those magic gives a more pronounced property to the objects, what the author called incantation technologies. To understand the indigenous art the Amerindian Perspectivism [10], also under structuralist influence, has been used successfully, especially making the shamanistic relationships between men and animals more noticeable.

2. The stilt houses of eastern Amazon

Having presented above the definition of art and studies in the Amazon, this text addresses the art in the Brazilian precolonial stilt villages of eastern Amazon societies, temporally between the 8th to the 10th A.D., thus, they no longer existed during the period of European conquest in the 16th century.

The stilt villages were ancient dwellings (palafittes) built on piles over a lake. Stumps or trunks of trees served as a support for the superior buildings of the

Figure 1.
Map of the stilt villages in eastern Amazon ad apparent stilts at the time of the drought. Photography: Alexandre Navarro.

villages [11]. Traces of these wooden stilts are located within rivers and lakes and are shown only at the time of drought. This normally corresponds annually to October to December [12]. Rest of the year, the palafittes are submerged. In some of these prehistoric sites, such as Coqueiro, there can be over three thousand stilts [13]. This type of archaeological sites has been described in isolated cases in South America, namely in the reports of Amerigo Vespucci in 1499 on communities that lived in Stilt Houses on the Venezuelan coast. The stilt villages also appear in the reports of Amazon river mouth and from the Upper Amazon, near Peru [14]. However, as known today, the prehistoric palafittes of Maranhão are the only ones preserved, in the entire American continent (**Figure 1**). The ancient stilt villages have been also common in the prehistory of Europe during the Neolithic age, for instance in northern Italy and Switzerland (Lake Constanza).

In the Amazonian eastern coast, the geographical area where the Stilt Houses are located is called Baixada Maranhense, in the State of Maranhão, Brazil. This region comprises approximately an area of 20,000 km^2 within the legal Amazon and has more than 500,000 inhabitants. People are very poor, with the lowest HDI indices not only in Maranhão state but also in Brazil as a whole. Baixada Maranhense population lives from the subsistence of traditional agriculture, fishing, keeping small animals and growing vegetables. Santa Helena, Penalva, Pinheiro, and Viana are the main cities in this area.

It is assumed that the easy availability of food in the form of a rich variety of fish created a favourable situation for the sedentary housing of the human groups that occupied the region. It is also likely that these dwellings had defensive purposes. It also notable that the oxygen-free aquatic environment preserved the artefacts in such a way that even allowed the ink on the artefacts to remain visible. The straws of the cottage walls and roofs and utensils made from the hardwood, like oars and indigenous weapons (bordunas), also resisted the decomposition process over time.

The most striking Stilt House artefacts are the small objects, like plates and bowls, which most likely were used in serving foods or liquids in rituals. These utensils were made of good quality clay burnt at a very high temperature, giving rise to high-quality vessels. These small ceramics had very complex painted decoration with geometric motifs made with precision, indicating also that these vessels were used in rituals.

3. The animal myths in ceramics

These pottery utensils formed an information channel of the social and ideological structure among the members of the society. As regards to their form and decoration, these vessels reflect mythical themes and/or were used in rituals [15]. Archaeologists agree that the ceramics reflect the culture of a society and that the main social changes affect the production and types of vessels [16]. So, the ceramics are vehicles of expression of ideological content. The most likely evidence of this is that they were painted, decorated, incised, modelled, with plastic decoration, aiming to reflect mythical or ideological themes.

Therefore, the art interpretations are cultural and in this sense adhere to rigid systems of social conviviality. In prehistory, its main material vehicles are small-scale art that is moveable (mobiliary art), such as decorated figurines and pottery and feather art; the graphic art, with parietal art and rock graphics and body painting, the latter very important, because, in general, it demonstrates the status of the individual, as in the case of chiefs and other leaders who had a high social position [17]. Indigenous art is therefore the result of the identity relationship [18] and social conviviality between the indigenous groups [19].

Animal forms are recurring in the art of the Stilt Houses, being the most persistently various birds, but especially owls. The monkey and the jaguar are mostly illustrated Mammals, but also Amphibians (especially frogs) and reptiles (mainly snakes) are common. These animals form the appliques and figurines and are associated with indigenous myths some of which have been described by the missionary chroniclers of the 16th and 17th centuries and by anthropologists when working among indigenous Amazonian communities from the 19th century onward.

The ethnographic analogy shows that myths associating the snake to the creation, such as the canoe-snake, mother of the fish, for example, are recurring in Amazonian cosmologies [20]. Whereas the king vulture is mentioned in Amazonian ethnography as a bird belonging to the realm of the dead.

Anaconda is a shamanic animal in the Amazon. Some peculiar ecological characteristics of these snakes could have attracted the attention of the indigenous people, as the specimens can reach up to 10 m in length and weigh more than 200 kg, is the heaviest animal in the Amazon. Also, the Anaconda has pale skin with black drawings served as an iconographic orientation, efficient camouflage, speed of attack in the water and lethargy on land, active at night, sensitive tongue for predate [21], powerful teeth and muscles and the females are larger and more aggressive than the males, predator of large mammals such as jaguars, deer and tapirs [22].

The indigenous peoples we mentioned in this text are the Tukano of the Tukano linguistic family who lives in the northwest of the Amazon; the Pano family Shipino who live between the Amazon and the Ucayali river in Peru; the Tupi peoples who have a wide dispersion in the lowlands of South America and countries like Bolivia, Uruguay and Argentina; the Waujá who live in the Xingu and belong to the Arawak family; the Panare and Timbira that are groups of the Jê family of Central Brazil and the already extinct groups of which we only have archaeological material such as the beautiful ceramic pots Tapajó and Konduri from the Lower Amazon in Brazil. Finally, there are the Warao who still live on stilts in Venezuela.

Regarding the cosmological aspects, the characteristics described above could contribute to the association of this powerful animal with the creation myths of the world and humanity. Among the Tukano, Hugh-Jones [23] the songs intoned by the shamans allude to the Amazon River as the terrestrial Anaconda and the Milky Way, the supernatural and creative Anaconda. Many Amazonian peoples compare the supernatural milk of the creative Anaconda with the milky-coloured sap of hallucinogenic plants of the genus *Banisteriopsis*, whose tree trunk is the metaphor of the body of the creative Great Anaconda [24].

In the polychrome ceramic material of the stilt villages, the principal iconographic element that stands out is the presence of curvy black or hook-shaped designs that fit together (**Figure 2**). The rim of these polychrome vessels is painted red. This iconography is recurrent in almost all the sites of this river. Although abstract, the iconography has a reading order in horizontal bands, filling the entire internal space of the vessel. The predominance of black colour could corroborate the suggestion that these images correspond to the black spots that the anaconda snake (*Eunectes murinus*) has on its back, as Roosevelt also interpreted in his study of anacondas and women-shamans in Marajo island.

This magical world is enhanced by music and dance. According to Barcelos [25], the snakes are part of a myth-musical repertoire among the Wauja, in which *Kamalu Hai* stands out, "the gigantic snake-canoe that carries on its back a long series of singing pots". These ceramic cookwares are of different sizes and have a varied polyphony according to their function. In this sense, the snake-canoe could imply the origin of the ceramic activity among the Wauja, a pristine myth, therefore.

In the Amazon, the anaconda is associated with both the male universe, among the Tukano, as well as the female. Anaconda's association with a shaman woman

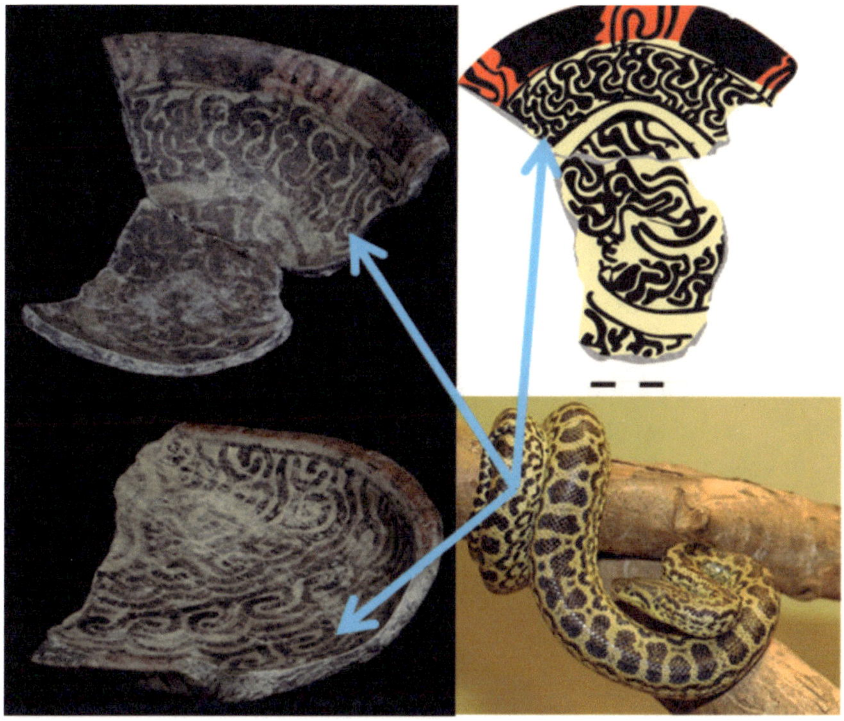

Figure 2.
Vessels with Anaconda skin iconographic pattern. Photograph: Fábio Matta.

and creative deity are common in oral traditions in the Amazon, as among the Shipibo, according to Roosevelt [22]. Indigenous peoples conceive of Anaconda as a dangerous ancestral spirit to the present day. She would be a master who governs the feminine part of the cosmos, which is the aquatic underworld. To Shipibo the anaconda is a woman shaman [26].

According to Roosevelt [22], many Amazonian peoples associate the Amazon River with Anaconda because these snakes dominate the aquatic landscape and because the meanders of the rivers imitate the movement of these reptiles. Thus, it is common for Anacondas drawings to be represented in community houses among the Tukano.

Roe [27] in his classic book *The Cosmic Zygote* reported among the Shipibo the mythological association of the creation of the world with the Anacondas, is also associated with the rituals of healing, divination, ceremonial dances and the creation of musical instruments. Among other groups, such as the Tupi-Gavião and Panare, Anaconda is a giant animal associated with the rainbow [28] or a celestial phenomenon among the Timbira in which the reptiles ends rest in the mouths of two anacondas [29]. The rainbow would be a symbol of disease [30] and for Weiss [31] it would represent "something demonic, repulsive and detestable, as well as the anaconda". Anaconda, therefore, inhabits a very large number of Amazonian myths, having as main characteristic the shamanic activities which include cosmological creation, under celestial aspects, and cultural properties associated with natural transformations of aquatic life and the water world.

The ceramic figurines indicate the presence of ritual [32]. They are characterised by the representation of animals, especially the owl, the monkey, the turtle and the frog. Some of them are anthropomorphic or zoomorphic design, the zoomorphic being the most recurring. Many of them have a sculptural standard: the legs are open in the shape of a half-moon and some of them possess the feminine genitalia on display. A figurine in particular, in the form of an owl, is a rattler and features a small handle that possibly had the function of being hung.

The shamanic nature of these archaeological materials is also evidenced by the production of figurines, generally female, where there are small clay balls inside, which could be a rattle, known in the lowlands of South America as *maracas*, as communication between the living and the dead. According to Zerries [33], the *maracá* has always been the most important shamanic instrument in non-Andean South American cultures, since "the noise of the little stone or *maracá* seeds inside is interpreted as the voice of the spirits". Thus, the *maracá* was considered an *idol* for the indigenous peoples of the lowlands of South America (**Figure 3**). Thus, in many cultures the owl is associated with death [34], and evil spirits [35].

These sound instruments are present in the ethnohistorical records of the colonial period such as Daniel [36], D'Abbeville [37] and D'Évreux [38] and also ethnography by anthropologists from the beginning of the 20th century. The *maracá*, therefore, is part of the shamanic paraphernalia since it can emit sound, thus, a form of communication between the different worlds in which the shaman acts. In this way, the rattle functions as a musical instrument whose sound together with the hallucinogens induce special sensations that alter their mental and psychological state. For the Warao, who still live in stilt houses in the Orinoco delta in Venezuela, the *maracás* have spiritual forces and their human forms refer to the ancestral shaman who visited heaven and was gifted with this instrument the Great Spirit of these water peoples [39].

The ceramic figurines were important products of the indigenous art in the Stilt Houses. Sometimes, they represented mainly hand-shaped geometrical figures. Often those represented also animals such as amphibians, fish, mammals and birds, which were similar to those among the Tapajó and Konduri peoples [40]. These animals were represented in a naturalistic style, thus preserving the identifiable traces of the species. A good example is the squirrel monkey (*Saimiri collinsi*) figurine, in which we can identify its furry ears, and in another piece, the torn mouth, typical of these primates [41] (**Figure 4**).

In turn, the frogs are associated with fertility, which is most likely due to the aquatic environment in which these societies lived. Themes associated with frogs are also common in the Mesoamerica [42] and the Caribbean [43] (**Figure 5**).

Figure 3.
*Rattler-figurine in the form of an owl, a maraca whose sound communicated the living with the dead. In the photo the head of the spectacled owl (*Pulsatrix perspicillata*). Photograph: Áurea costa.*

Figure 4.
*Figurine in the shape of squirrel monkey (*Saimiri collinsi*). Photograph: Fábio Matta.*

Regard to iconography, it happens through geometric lines or traces that delimit patterns within the stylistic composition of the vase: they are Greeks, zigzags or spirals taking up the interior of the pieces. We can see, that there are two opposite iconographic fields divided by one or two lines across the piece. Mostly the motifs differ in these two fields. For instance, when the square motifs were used in one artistic field, then the circular elements were selected for the opposite one. Red and black were the predominant colours, which were painted on cream engobe or white.

According to Prous [44], some of the motifs resemble the Tupi iconography and he associated them with the custom and practice of eating human flesh (anthropophagic) ritual, such as the representation of the intestine and the brain. However, the pottery shapes at the Stilt Houses, as well as their technologies, such as anti-plastic and quality of clay burning, are very distinct from those of the Tupi. Therefore, Stilt Houses' ceramics show better production control and technological quality.

Perspectivism deriving from structuralism serves the most fruitful theoretical and methodological discussions that apply to the study of archaeological pottery of the stilt villages. The study of the iconographic motifs, as well as their repetition and pattern, in addition to the ethnographic bibliographical revision of the Amazon, show that the geometric motifs of the ceramics are, in general, depictions of the skin or feathers of some animals, especially those of the top of the food chain, such as snakes and owls.

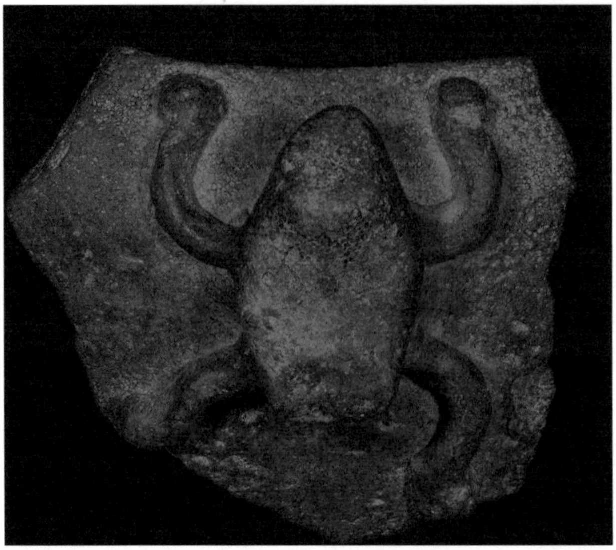

Figure 5.
Vessel frog appliqué of the stilt villages. Photograph: Fábio Matta.

4. Conclusion

This chapter has shown that the pre-colonial Amazon peoples were well adapted to the environment and produced a rich art of strong social cohesion. Two types of the artefact of mobiliary art and one of the graphic art stand out in the conception of the indigenous art of stilt villages. The mobiliary art corresponds to the figurines, generally representing animals (zoomorphic design) and sometimes human beings mixed with animals and the appliques showing different animal shapes, especially mammals, amphibians and birds. Therefore, the art reveals the Perspectivism associated with the cosmologies involving mythical concepts. From the graphic art stands out the black and red paintings on white and cream engobe of the ceramic vessels. They contain geometric shapes which take up two distinct geometric fields, forming motifs resembling the skin of predatory animals.

Guss [45] has pointed out when postulating the relationship between myth and artefacts, that objects act as "subtexts" that provide an understanding of the functioning of society bringing them closer to their origins. In this sense, the artefacts imitate primaeval objects, as they are copies of this primordial world. This concept refers to what Gell [9] would call the object's enchantment.

Very likely the representation of animals in appliques and painting of Stilt Houses ceramics have a close connection with the creation myths among the pre-colonial peoples in Amazon. Snakes are mainly dwelling in the river, and therefore associated with the origin of human life. The Amazonian rivers, on the other hand, have many meanders remembering the shape of the undulating movement of the snakes. Serpents are also associated with fertility and in an aquatic environment where stilts were found, these myths could be very important as social cohesion.

In the river environment of Amazon frogs are other important animals [46]. These amphibians were associated with the fertility and they produce hallucinogenic substances that allowed the shamans to make their spiritual journeys [47]. Many Indian ceramics have frog-shaped appliques and the small size of these vessels indicates their use for the consumption of liquid drinks. Some scholars have argued that the abstract drawings in ceramics paintings and appliques could originate from the view of the visual illusions (phosphenes) caused by hallucination.

On the other hand, the representation of mammals at the top of the food chain such as jaguars and monkeys may be associated with Amerindian perspectivism. Animals with aggressive characteristics in the ceramic appliques likely represented for the Indians a metaphor of power. It is also possible that these ceramics belonged to the chiefs. Often the animal itself was not represented, but the paintings of the vessels alluded only the skin of these animals.

The bright colours of the Anacondas, as well as their ecological characteristics such as constriction and their large size, played an important role in their choice to symbolise the canoe snake. Nothing better than a strong and large animal to be used as a transport vehicle to populate the villages. The fact of knowing how to swim, like a canoe, corroborates that Anaconda has adequate properties for its appropriation as a narrative. In this sense, the Anaconda had a higher status in the animal hierarchy, appearing only in painting and not in effigies.

Finally, it has to be considered that the contemporary Stilt Houses, most of them comprise the temporality of 800 to 1000 AD. So the long-range of their artistic ideologies in an extensive area, indicating firstly a cultural homogeneity of these societies, and, secondly, makes us think, even if hypothetically, that the existences of these chiefdoms of large regional scale between the 8th and 10th AD coincide with the pinnacle of the precolonial Amazonian societies.

Author details

Alexandre Guida Navarro
Professor of Archaeology at Federal University of Maranhão, São Luís, Brazil

*Address all correspondence to: altardesacrificios@yahoo.com.br

References

[1] Sahlins, M. D. (1968). *Tribesman*. Prentice-Hall, Englewood Cliffs.

[2] Geertz, C. (1989). *A interpretação das culturas*. Rio de Janeiro: Guanabara Koogan.

[3] Lévi-Strauss, C. (2008). *Antropologia estrutural*. São Paulo: Cosac & Naiff, [1958]

[4] Marconi, Marina de A. & Presotto, Z. M. (2009). *Antropologia: uma introdução*. São Paulo: Atlas.

[5] Schaan, D.P. (2004). *The camutins chiefdom. Rise and development of complex societies on Marajó island, Brazilian Amazon*. PhD thesis, Universidade de Pittsburgh.

[6] Neves, E. G. (2006). *Arqueologia da Amazônia*. Rio de janeiro: Zahar Editora.

[7] Gomes, Denise Maria Cavalcante. (2012). O perspectivismo ameríndio e a ideia de uma estética americana. *Bol. Museu. Paraense Emílio Goeldi*, vol. 7, n. 1, pp. 133-159. Belém.

[8] Lévi-Strauss, C. (2004). *O cru e o cozido*. Mitológicas 1. Tradução Beatriz Perrone-Moisés. São Paulo: Cosac & Naif, [1964].

[9] Gell, A. (1998). *Art and agency: an anthropological theory*. Oxford: Clarendon Press.

[10] Viveiros de Castro, E. (2002). *A inconstância da alma selvagem*. São Paulo: Cosac & Naif.

[11] Navarro, A. G. (2018). New evidence for late first-millennium AD stilt-house settlements in Eastern Amazonia. *Antiquity* 92 (366): 1586-603.

[12] Navarro, A. G., Gouveia Neto J.C., Marcondes L. da Costa, Abrahão S. N. F. da Silva, Rômulo S. Angélica, Suyanne S. Rodrigues. (2017). O muiraquitã da estearia da Boca do Rio, Santa Helena, Maranhão: estudo arqueológico, mineralógico e simbólico. *Boletim do Museu Paraense Emílio Goeldi* 12 (3): 869-894.

[13] Navarro, A.G. (2016). O complexo cerâmico das estearias, Maranhão. *Cerâmicas arqueológicas da Amazônia: rumo a uma nova síntese*, edited by C. Barreto, H. P. Lima and C. J. Bitencourt, 158-169. Belém: IPHAN/Museu Paraense Emílio Goeldi.

[14] Porro, A. (1992). *As crônicas do rio Amazonas. Notas etno-históricas sobre as antigas populações indígenas da Amazônia*. Petrópolis: Vozes.

[15] Arnold, Dean E. (1985). *Ceramic theory and cultural process*. Cambridge: Cambridge University Press.

[16] Grieder, T. (1975). The interpretation of ancient symbols. *American Anthropologist* 77:849-55.

[17] Ember, C.R., Ember, M. & Peregrine, P. (2004). *Antropología*. Madrid: Pearson Prentice Hall.

[18] Ribeiro, B. (1983). *O índio na história do Brasil.* São Paulo: Global Editora.

[19] Vidal, L. (1992). A pintura corporal e a arte gráfica entre os Kayapó-Xikrin do Cateté. Vidal, Lux (org.). *Grafismo indígena*. São Paulo: Studio Nobel/ Fapesp/Edusp, pp. 143-189.

[20] Reichel-Dolmatoff, G. (1971). *Amazonian cosmos*. Chicago: University of Chicago Press.

[21] Colthorpe, K. (2009). *Eunectes notaeus*. Yellow Anaconda. accessed July 11, 2020. https://animaldiversity.org/accounts/Eunectes_notaeus/

[22] Roosevelt, A.C. (2014). The great anaconda and woman shaman: A dangerous and powerful ancestral spirit from creation to today. In *Colocataires d'Amazonie: Hommes, animaux et plantes de part et d'autre de l'Atlantique*, edited by D. Barone-Visigalig, 1-20. Paris: Parution.

[23] Hugh-Jones, C. (1979). *From the Milk River: Spatial and Temporal Processes in Northwest Amazonia*. Cambridge: University of New York.

[24] Knoll, M. (1963). Effects of chemical stimulation of electrically-induced phosphenes on their bandwidth, shape, number and intensity. *Confinia Neurologia*, v. 23, pp. 201-26.

[25] Barcelos Neto, A. (2011). A serpente de corpo repleto de canções: um tema amazônico sobre a arte do trançado. *Revista de Antropologia* 54 (2): 981-1012.

[26] Gebhart-Sayer, A. (1984). *The Cosmos Encoiled: Indian Art of the Peruvian Amazon*. New York: Center for Inter-American Relations and Americas Society.

[27] Roe, P. (1982). *The Cosmic Zygote: Cosmology in the Amazon Basin*. New Brunswick: Rutgers U.

[28] Dumont, J-P. (1977). Musical Politics: On Some Symbolic Aspects of the Musical Instruments of the Panare Indians. In *Anthropology and the Climate of Opinion*, edited by S. Freed, 2016-214. New York: Annals of the New York Academy of Sciences.

[29] Nimuendajú, C. (1946). *The Eastern Timbira*. University of California Publications in American Archaeology and Ethnology 41. Berkeley: University of California Press.

[30] Basso, E. B. (1973). *The Kalapalo Indians of Central Brazil*. New York: Holt, Rinehart & Winston.

[31] Weiss, G. (1975). Campa Cosmology: The World of a Forest Tribe in South America. Anthropological Papers 52. New York: American Museum of Natural History.

[32] Roosevelt, A.C. (1988). Interpreting Certain Female Images in Prehistoric Art. Miller, Virginia E. (ed.). *The Role of Gender in Precolumbian Art and Architecture*. Lanham, MD.: U. Press of America, pp. 1-34.

[33] Zerries, O. (1981). Atributos e instrumentos do Xamã na América do Sul não-andina e seu significado. In *Contribuições a Antropologia em homenagem ao Prof. Egon Schaden*, edited by T. Hartmann and V. C. Penteado, 319-360. São Paulo: Coleção Museu Paulista, Série ensaios.

[34] Mikkola, H. (Ed.) (2020). Owls. 97 p. – InTech Open Access, London Print ISBN 978-1-78984-053-7; Online ISBN 978-1-78984-054-4; eBook ISBN 978-1-83880-461-9 http://dx.doi.org/10.5772/intechopen.80242

[35] Mikkola, Heimo. Owls of the World: A Photografic Guide. Firelfly Books, 2013.

[36] Daniel, J. (2004) [1757-1776]. *Tesouro descoberto no Máximo Rio Amazonas*: 1722-1776. 2 Vols. Rio de Janeiro: Contraponto.

[37] D'Évreux, Y. (2008) [1864]. *Continuação da História das coisas mais memoráveis acontecidas no Maranhão nos anos 1612 e 1614*. Brasília: Senado Federal.

[38] D'Abbeville, C. (1945) [1864]. *História da missão dos padres Capuchinhos na ilha do Maranhão e suas circunvizinhanças, em que se trata das singularidades admiráveis e dos costumes estranhos dos índios habitantes do país*. São Paulo: Livraria Martins Editora.

[39] Wilbert, J. (1963). Vestidos y ornamentos de los Indios Warao. *Antropologica* 12: 6-26.

[40] Guapindaia, Vera Lúcia Calandrini. (2008). *Além da margem do rio: a ocupação konduri e pocó na região de porto trombetas, PA.* PhD thesis. São Paulo: Universidade de São Paulo.

[41] Navarro, A.G. & Silva Júnior, J.S.E. (2019). Cosmologia e Adaptação Ecológica: o caso dos apliques-mamíferos das estearias maranhenses. *Anthropológicas*, Recife, vol. 30, n. 2, pp. 203-233.

[42] Coggins, C. C. (1992). *Artefacts from the Cenote of Sacrifice, Chichén Itzá, Yucatán.* Cambridge, Massachusetts: Peabody Museum of Archaeology and Ethnology, Harvard University.

[43] Boomert, A. (1987). Gifts of the Amazon: green stones pendants and beads as the item of ceremonial exchange in Amazonia and the Caribbean. *Antropologia,* núm. 67, Caracas, pp. 33-54.

[44] Prous, A. (2005). A pintura em cerâmica tupiguarani. *Ciência Hoje*, Rio de Janeiro, v. 36, n. 213, pp. 22-28.

[45] Guss, D. (1990). *To weave and sing. Art, symbol, and narrative in the South American rainforest*. Berkeley: University of California Press.

[46] Wassén, H. (1934). The frog-motive among the South American Indians. Antropos. Revue Internationale d'Ethnologie et de Linguistique, Freiburg, v. 29, n. 3-4, pp. 319-370.

[47] Reichel-Dolmatoff, G. (1976). O contexto cultural de um alucinógeno aborígine: *Banisteriopsis caapi*. *In*: Coelho, V.P. Os alucinógenos e o mundo simbólico. São Paulo: EPU/EDUSP, pp. 59-103.

Section 3

Development Opportunities and Problems

Chapter 9

Bioeconomic Potential of Sustainability Indicators in a Ceramic Production Center in the Western Amazon

Gelson Dias Florentino, Lucieta Guerreiro Martorano, Sandro Augusto Lima dos Santos, José Reinaldo da Silva Cabral de Moraes, Ires Paula de Andrade Miranda and Maria de Lourdes Pinheiro Ruivo

Abstract

The use of Amazonian biodiversity has great potential to produce bioproducts in diverse production chains and segments of industry. The combination of public policies with biotechnological development represents an important indicator for the implementation of sustainable production chains that adhere to the Sustainable Development Objectives (SDO). The ceramic industries in the Amazon region represent activities that promote local economic development through the use of biological resources that can be transformed into bioproducts that are considered a reference for sustainable production in world markets. The operations of these industries have great potential to incorporate technologies that can be used for fabrication of ceramic products on a biological base that is compatible with bioeconomic guidelines. The principle of a bioeconomy is centered on the possibility of transformation of natural resources into bioproducts that aggregate technologies and contribute to increase incomes and reduce environmental impacts. In this way, the integration of different fields of science should be stimulated to incorporate new technologies that favor business models that comply with the premises of sustainability.

Keywords: Amazon, biodiversity, production chain, bioeconomy, bioproducts

1. Introduction

The Amazon contains about a third of all tropical forests in the world and forms a mosaic of ecosystems in an area of about 6 million km^2 [1]. The extensive variety of species of flora and fauna spread out across widely differing landscape patterns, coupled with a hydrological system that is rich in biodiversity, makes this region the most biodiverse in the world.

The enormous biological diversity and territorial extension of the Amazon makes governance and sustainable management difficult considering the many different sources of anthropogenic pressure [2]. However, geological, hydrological, and climatic patterns of the Amazon, allied with its megadiversity favor sustainable use of its natural resources. Provision of inputs and raw materials for production of bioproducts and biofuels, energy generation, and phytotherapeutic products are examples of Amazonian bioeconomic potential.

Biodiversity is defined as the composition of all species of all forms of life and organizational levels present in ecosystems [3], a characteristic that is typical of the Amazon. The ecologic-economic interface at a large scale generally leads to persistent degradation of the natural environment and a loss of ecosystem services and biodiversity [4].

The value of ecosystem services is hotly debated due to the difficulty in establishing a monetary value for intangible goods [5, 6]. Studies have indicated that one of the largest gaps or deficits in knowledge of biodiversity is the real magnitude and quantity of microorganisms and species that have been formally described by science [7, 8]. However, the dynamics and multiple facets that characterize the trade-off between development and sustainability presupposes the necessity to discuss the bioeconomic potential of the Amazon, focused on balancing the magnitude of consumption of natural resources with the capacity to replace them in nature.

Even without knowing the exact quantity of species present in nature, biodiversity has great biological importance for all forms of life [9], such as:

- *Ecosystem functionality*: biodiversity facilitates the functioning of ecosystems, maintaining the planet habitable through carbon exchange, maintenance of surface- and groundwaters, protection and fertilization of soils, and regulation of temperature and climate.

- *A storehouse of values that are intangible and non-monetary*: biodiversity offers to humanity values that are universally recognized, such as those of esthetic, scientific and cultural character.

- *A storehouse of values that are tangible and monetary*: the beauty and uniqueness of many ecosystems provide value in the form of a diversity of recreational activities and ecotourism.

- *Provision of inputs and bioproducts that are destined for human consumption*: biodiversity is the base for agricultural crops and for the development and improvement of new varieties for human consumption.

- *Provision of raw material for production chains and industrial sectors*: biodiversity is the source of many products used by contemporary societies, such as fibers, pharmaceutical products, chemical, and is a source of information for the development of biotechnology.

Besides the economic value and biological importance to human survival, biodiversity provides ecosystem services in the form of provision, regulation, and support, and is also important for its contribution to cultures around the world [10–13]. This results in tangible and intangible benefits based on the diversity and functioning of ecosystems.

The challenge presented by the global market is to satisfy the assumptions of the Organization for Economic Cooperation and Development (OECD) in productive sectors in the Amazon, since the expansion of certain industrial activities necessary to meet national and international market demand generally results in loss of biodiversity, and consequently compromises ecosystem services.

Among the diverse initiatives that utilize biodiversity, there is great preoccupation with the sustainability of production chains, and if these adhere to the Sustainable

Development Objectives (SDO), as adopted by the United Nations Organization (UNO). The conceptual approach of sustainable development as elaborated in its most recent version, incorporates the necessity of adoption of sustainability parameters that take into account aspects that include efficiency (economically sustained), inclusion (socially desirable), and equilibrium (ecologically prudent).

It is understood that the concept of sustainable development is complex, and it is often interpreted imprecisely, ambiguously, or in a contradictory manner based on a complex and dynamic perspective rooted in the relationship between societies and nature [14]. This concept has been gaining strength and political expression, especially during the last quarter century of the twentieth century, due to a worldwide environmental crisis [15].

The trajectory of sustainability demands ethical compromises, ecological conscience, and social respect, and the demand for sustainable products has been intensifying [16]. Consumers now question how a product was made, if it is in compliance with the legislation of its country of origin, and what were the benefits generated for the agroextractivists and communities that obtained these products.

These requirements impose the necessity of introducing technologies and business models that are compatible with the tools of the bioeconomy.

The bioeconomy is organized around the concept of sustainable and innovative use of natural resources and biological knowledge in order to produce food, industrial products, bioenergy, and ecological services [17]. This concept arose as a new paradigm through the extension of new ideas and knowledge in the fields of biology, economics, and the environment in general, with the goal of providing solutions to complex problems. The concept of the bioeconomy integrates different fields of science, sectors of society, and market demands as a function of providing products that are sustainable, competitive and that adhere to policies of sustainability and applied biotechnology.

Ceramic industries in the Amazon are examples of businesses that use biodiversity resources and promote local economic development. However, the manufacturing operations of these companies have been engaged in a growing effort to achieve a pattern of development that incorporates economic, social, and environmental aspects [18].

Preoccupation with depletion of natural resources has increased interest in biologically based products. Productive activities in the Amazon, especially those derived from biological resources, as is the case of ceramic industries, have created a positive perspective with respect to the implementation of developmentalist models that incorporate innovation, technology, and sustainability.

The advance of the "Fourth Industrial Revolution", or Industry 4.0, has integrated technology and sustainability and created new practices and standards for products and services which are based on concepts of a sustainable economy [19].

Market strategies that search for alternatives that can differentiate a product based on its quality and on attributes that respect the environment normally allow for access to new markets that have higher aggregated value and purchasing power. Cluster analysis shows the potential to track and identify the origin of products, thus aggregating value and confidence in products from the Amazon region.

In this context, this research has as objective the analysis of the potential of the principal bioeconomic variables inherent to the production and marketing of ceramic products in the ceramic production center of Iranduba, state of Amazonas.

2. Characterization of ceramic production in the western Amazon

In the Amazon, the development of ceramic production techniques was a pioneering effort by indigenous people that was engaged in before agricultural activities and plant domestication [20]. The production of utensils and other artifacts was not done to sell them in a market, but rather the focus of this production was to make vases, jars, and plates for domestic use. Beginning with the industrial revolution and the introduction of machines and equipment to production processes, the production of ceramics ceased to be merely an artisanal activity and took on an industrial scale. Population growth and the expansion of large urban centers in the region created a new commercial niche, the civil construction market. The demand created by the civil construction production chain altered production models and required the fabrication of new products and consolidated the ceramic industries involved in production of specialized ceramic products.

In the state of Amazonas there are records that show that the first ceramic production activities were artisanal, especially with respect to production of plates, cups, and other utensils for domestic use by the local indigenous population [21]. Up to the decade of the 1970s, the ceramic sector of the state of Amazonas was concentrated in the region of the city of Manaus. Starting in the 1980s these companies migrated from the capital of Manaus to the district of Cacau-Pirêra, city of Iranduba [22–25].

The factors that contributed to the installation and strengthening of the ceramic sector in this region of the Amazon were: a) the existence of extensive deposits of clay raw material; b) proximity to the principal consumer center, connected by the Manoel Urbano – AM-070 highway; and c) large supply of unskilled labor that receives low salaries [18, 26].

2.1 Geographic and geologic aspects of the study area

The study area is located in the city of Iranduba, metropolitan region of Manaus, consisting of an aggregation of 18 industries situated on the right bank of the Negro River, in front of Manaus, with access and shipping of production done using the Negro River bridge (**Figure 1**).

Geologically, the study area is on a dissected plain, where sedimentary rocks of the Alter do Chão formation predominate. This sedimentation occurred in continental, fluvial, and lacustrine environments, constituted by intercalation of sandstones, mudstones, and conglomerates, which are principally composed of clay [27], which is used in the production of red ceramic products. The sandstones have a fine to medium texture and are red in color; the mudstones, massive or laminar,

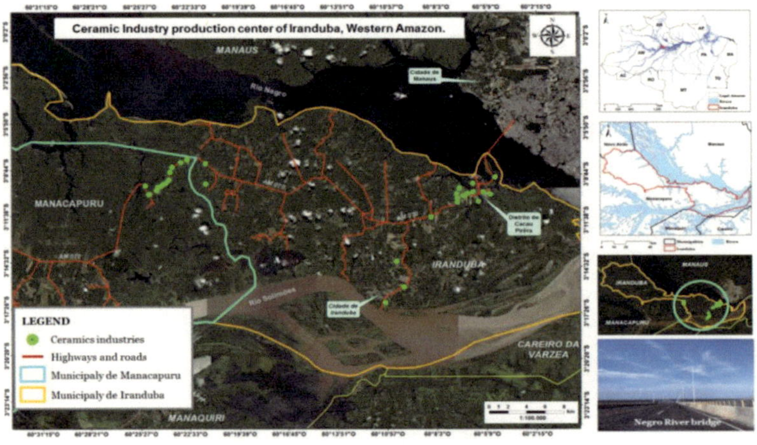

Figure 1.
Location of the ceramic production center of Iranduba, Amazonas, Brazil. Source: Geocartographic data from the Brazilian Institute of Geography and Statistics (IBGE) and ArGis (2020).

Figure 2.
Landscape heterogeneity of Iranduba: Lateritic profile (A); Sandy deposits (B); fluvial plain (C); floodplain (D). Source: The authors (2020).

are poorly consolidated, and have clasts of sand that are irregularly distributed; the conglomerates are quartz pebbles and silicified sandstone [28].

The ceramic industries are inserted in the Holocene Várzeas pedological sector, which is composed of várzeas (floodplains along riverbanks) that developed over Holocene sediments with a variable mineralogical composition. Geoenvironmental studies have indicated that the city of Iranduba is characterized by a heterogeneous landscape, with different soil types that are classified as a lateritic profile, sandy deposits, fluvial and flood plains [29] (**Figure 2**). The clay deposits, which are white to rose colored, together with the yellow Oxisols that overlay the Alter do Chão formation, are used as raw material for the production of red ceramics in the ceramic production industries of Iranduba. The principal products of this red ceramics industry are blocks, bricks, roof tiles, hollow ornamental bricks, slabs, floor tiles, encaustic tiles, tubes, ornamental objects, and domestic utensils, all using common clay as raw material.

Recent research conducted in the ceramic production center of Iranduba indicates that the area has a high level of geological diversity and potential for clay extraction at depths of 15 cm, 30 cm and 60 cm, and that this extraction has been occurring without interruption during the last four decades [18]. Even with the obstacles and the technological limitations confronted by this sector the ceramic industry in Amazonas has actively contributed to social and economic development in the region. It has been estimated that at the height of the real estate boom in Brazil between 2008 and 2013 the annual production of ceramics in Amazonas was approximately 35 tons of bricks per month, with the cities of Iranduba and Manacapuru being the principal producers of red ceramics. Just the city of Iranduba was responsible for about 75% of all brick production in the state, producing between 1700 and 2000 direct and indirect jobs [18, 30].

3. Bioeconomic and sustainability potential of the ceramic production center of Iranduba

The market for the ceramics industry in Brazil is highly heterogenous and is formed mostly by small companies that are located near the consumer markets in their regions. These companies basically use common clay as their principal source of raw material.

In the Iranduba ceramic production center, production is concentrated in red structural ceramics, with 8-hole bricks being the principal product that is made. With competition and an increasing demand for environmentally sustainable products, it has become of fundamental importance that the ceramics sector promotes biotechnological innovations in its production chain through integration of companies, technology, and sustainability (**Figure 3**).

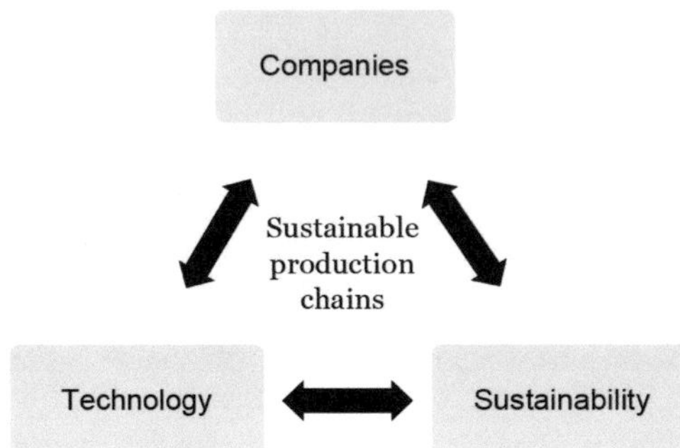

Figure 3.
Structure of sustainable production chains. Source: The authors (2020).

Sustainable production chains seek to improve business models by adding economic and social value to the goods that are produced. For this reason, the bioeconomy is seen as an opportunity to enhance production standards and strengthen relationships between agroextractivists, transformation industries, public authorities and research and development institutions. Companies and governments around the world have been acting to create public policies that strengthen the commitment to use biological resources in a way that favors a more sustainable economic model [31, 32].

An important principle underlying sustainable production chains is that they contemplate changes and or a reorientation of productive systems through necessity or opportunity (entrepreneurship), resulting in lower environmental impact (social sustainability), generation of work and income (economic sustainability), and maintenance of urban and rural characteristics (spatial sustainability) [33]. The combination of these elements requires a set of actions that incorporate scientific knowledge, innovation, and organizational experience (tacit knowledge) to produce, market, and distribute products that are competitive and that are sustainable.

As a result of globalization of markets, a growing number of companies has been striving to create technological tools that are able to track commodities and products, with the objective of supplying a consumer market that is becoming more and more demanding with respect to sustainability of productive processes [34]. The creation of tracking mechanisms that display the routes and management schemes of the entire production chain, from the extraction of raw material (inputs) to the final consumer is highly challenging. However, these initiatives incorporate practices and attitudes that represent a technological and sustainable advance for the industries of the ceramic production center at Iranduba.

3.1 Structural steps that integrate the production chain of the ceramic production center at Iranduba

The ceramic production chain at Iranduba encompasses three different steps of macroprocesses: extraction of raw material, production of ceramic artifacts, and marketing of the finished product (**Figure 4**).

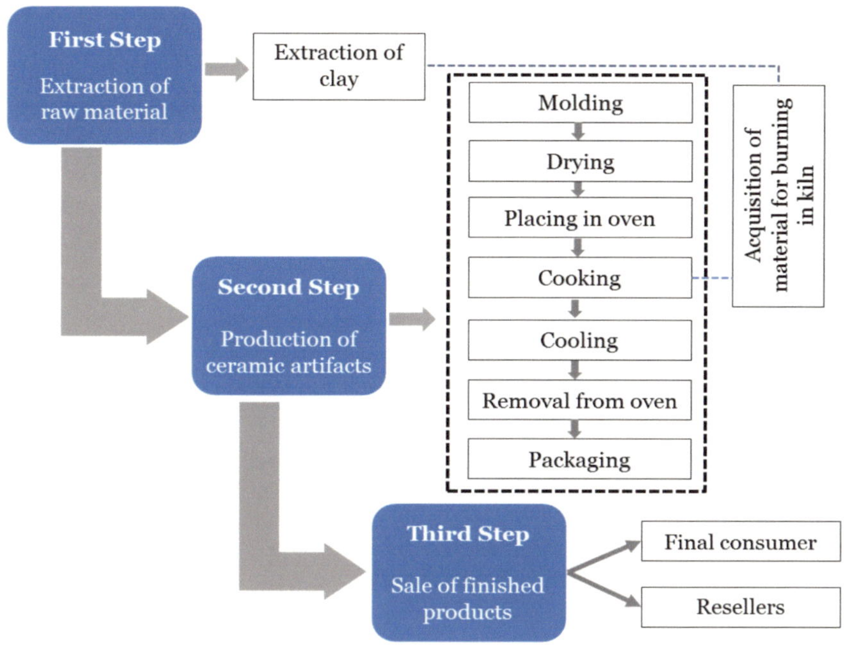

Figure 4.
Structure of the ceramic production chain at Iranduba. Source: The authors (2020).

The first step in the production chain consists of acquisition of raw material, which in this case is clay. The extraction of clay is one of the most important links in the production chain, and is done using machines and motorized equipment, principally a hydraulic backhoe, excavator shovel, and dump trucks (**Figure 5A** and **B**). It is important to highlight that before beginning the clay excavation process the excavation sites must be properly licensed to have permission for these activities. In general, these industries do not conduct any specific treatment on the extracted clay, and the clay is simply deposited under the roof of a large hangar or outside in large piles with no covering where it will remain for up to three months (**Figure 5C** and **D**). It is thought that this up to three-month period serves to eliminate impurities and microorganisms that can compromise the quality of the final product.

Clay extraction is permitted by law in article 3° of Law n° 12.651/2012 (Forest Code) and the process of concession/extraction is authorized by different federal agencies, principally the National Mining Agency – ANM, and state and municipal environmental agencies. In the state of Amazonas, licensing of exploitation of mineral substances is done by the Institute of Environmental Protection of the State of Amazonas (IPAAM), by means of three types of licenses that are issued either isolated or in succession, based on the National Environmental Council's (CONAMA) Resolution n° 237/1997 [35], which outlines the rules that organize this activity:

- Preliminary License (LP): issued in the preliminary phase of planning, analysis, and viability of the project in its basic form.

- Installation License (LI): issued in the execution phase of the project and the installation of the infrastructure, as well as during the phase of implantation of an Environmental Control Plan (PCA) which outlines methods of recuperation.

- Operation License (LO): issued during the extraction phase of the project and during the phase of execution of the proposed measures in the Plan for Recuperation of Degraded Areas (PRAD).

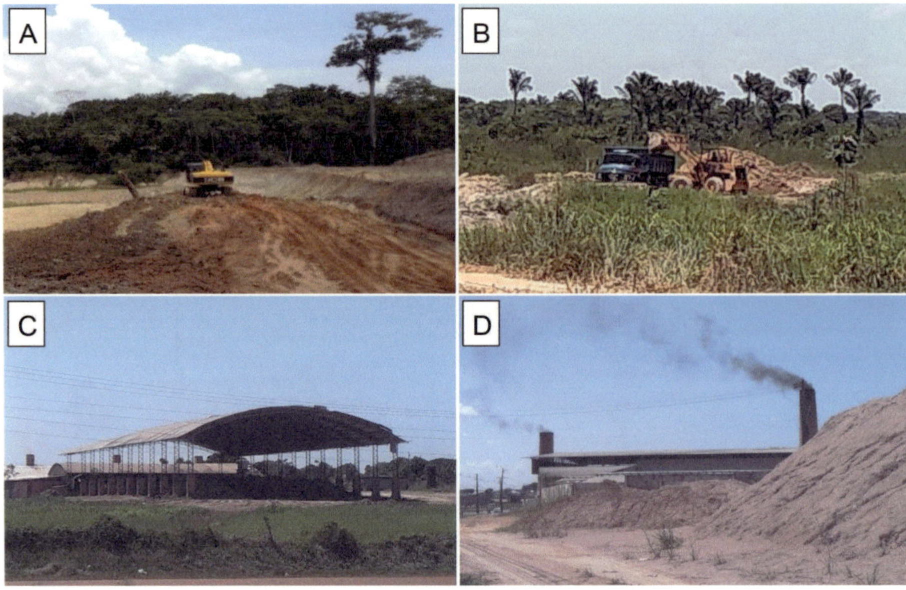

Figure 5.
Processes of extraction (A), transport (B), and storage of clay raw material (C, D) in the study area. Source: The authors (2020).

From the point of view of sustainability, the extraction of clay in open-air conditions without any subsequent processing represents an activity that promotes environmental degradation with significant impacts on biodiversity, which require mitigating actions and/or the reestablishment original ecosystem conditions.

Although the extraction of raw material for the fabrication of tiles, bricks, and other artifacts made of clay cooked in kilns can be considered an activity that causes environmental degradation of medium to strong intensity, the geographic location of the clay extraction sites used by the ceramic industries of Iranduba have high potential for recuperation using strategic mechanisms from bioeconomic principles. The clay extraction sites licensed by IPAAM are essentially situated in areas prone to flooding or are perpetually under water, which therefore have potential to be used for bioeconomic activities such as pisciculture (**Figure 6A** and **B**).

Most of the clay deposits are located on the land of the ceramic factory or near them (**Figure 6A** and **B**). Clay extraction is done between the months of August to November, which is the dry season where there is a large reduction in rainfall (considered the Amazonian summer). The principal clays that are extracted are kaolinite, with traces of gibbsite, and rarely illite, feldspar, and quartz.

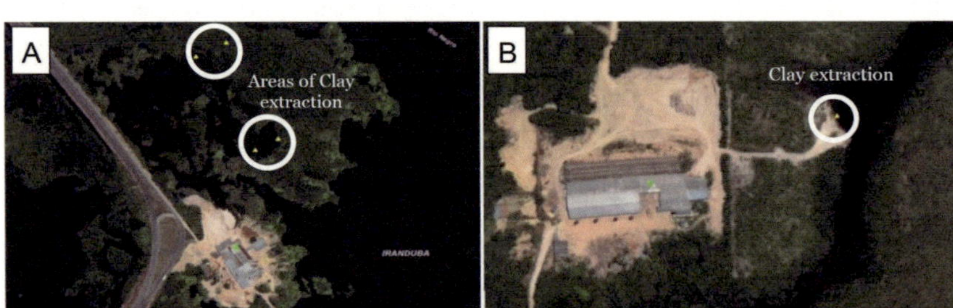

Figure 6.
Clay extraction sites used by the ceramic industries of Iranduba, Amazonas, Brazil: RN Ceramic (A), RO Ceramic (B). Source: Geocartographic data from the Brazilian Institute of Geography and Statistics (IBGE) and ArGis (2020).

Besides the compensatory environmental recuperation measures, these areas have excellent potential for pisciculture, especially for the species called tambaqui (*Colossoma macropomum*). Raising tambaqui in captivity is perfectly compatible with the culture, infrastructure, and technology available in the local market. In this context, during the rainy season when river levels are high (Amazonian winter), principally between January to June, the areas used for clay extraction could be reused following a new economic and social perspective. This option would foster additional insertion into local markets incorporating practical aspects of the bioeconomy and would mitigate the environmental impacts of the degraded area (**Table 1**).

It is important to emphasize that the fishing sector represents one of the most important parts of the local economy, since freshwater fish are one of the natural resources that is most abundant and consumed in the Amazon region [36, 37]. Between the years 2018 to 2019, pisciculture production increased by 34% in the region, especially for native species such as tambaqui (*Colossoma macropomum*), matrinxã (*Brycon amazonicus)* and pirarucu (*Arapaima gigas*). This points to a scenario of an increase in productive capacity during the rainy season when water levels are high in the region (**Figures** 7 and **8**).

Studies indicate that fish is the principal protein source for the populations that live along the river margins in Amazonas, with the upper, middle, and lower regions of the Solimões River having one of the highest levels of fish consumption, a reflection of the strong link that Amazonians have with regional fish populations [37, 38]. A person from Amazonas consumes, on average, 60 $kgyear^{-1}$ of fish, which represents more than ten times the national average per $capitayear^{-1}$, estimated at 5.8 kg [36, 37]. The high rate of fish consumption in the region, besides providing the basic necessities for survival, there are many natural benefits of fish to the health of consumers since fish is rich in protein, mineral, fatty acids, and omega 3 [37]. Furthermore, the introduction or intensification of pisciculture in areas of clay extraction at the ceramic production center at Iranduba will decrease the pressure on natural stocks of fish and will help to meet the guidelines set out in the sustainable development objectives of the UNO.

In this context, the regime high and low water fluctuation over the course of a year in Iranduba represents a bioeconomic alternative that has great potential reconciling several needs in the same natural environment by providing raw material for ceramics production as well as production of fish for subsistence and sale of the excess (**Figure 9**). These actions will stimulate an increase in income, access to new technologies and a reduction in impacts caused by the extraction of biodiversity resources [25].

The second step in the ceramic production chain at Iranduba occurs internally at the site of the company where the ceramic products are made. The principal changes in the physical–chemical characteristics of the raw material to obtain the finished product occur during this step of the production process. The operational activities are conducted using industrial equipment such as a laminator, extruder, conveyer belt, and cutter, which are operated by the factory workers. Socioeconomic studies done involving managers and workers from ceramic companies in Iranduba have indicated that these activities began in a completely artisanal manner and were then modified over the course of years of operation, to the point today where this process can be considered semiautomatic [21].

The ceramic companies in Iranduba are responsible for about 50% of all red ceramics produced in the state of Amazonas, generating approximately 1200 direct jobs and 6000 indirect ones, with an estimated production of 4.5 million pieces/month. The portfolio of products consists of bricks (80%), structural ceramic blocks (12%), roofing tiles (5%) and floor tiles and accessories (3%), as shown in **Figure 10**.

Bioeconomic potential	Months												Extraction of natural resources
	January	February	March	April	May	June	July	August	September	October	November	December	
Operational cycle of fish production												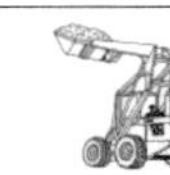	Operational cycle of clay extraction

Source: The authors (2020).

Table 1.
Bioeconomic proposal for areas of clay extraction used by the ceramic industries of Iranduba, Amazonas, Brazil.

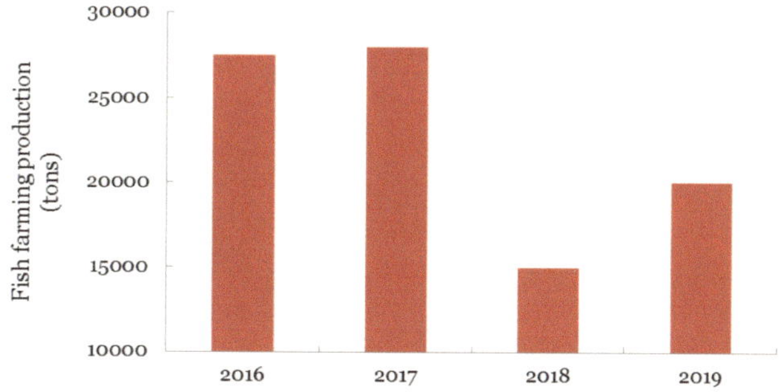

Figure 7.
Pisciculture production in the state of Amazonas (2016 to 2018). Source: Brazilian yearbook of fish farming (2020).

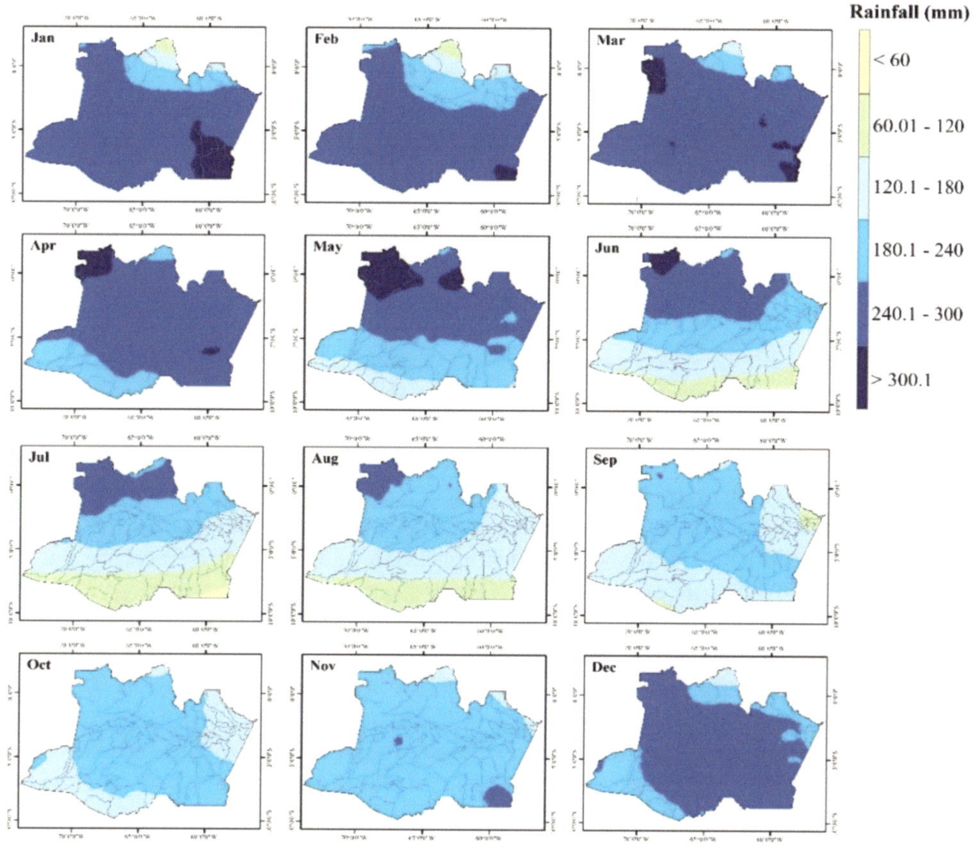

Figure 8.
Characterization of monthly pluviometric in the state of Amazonas (1989 to 2017). Source: European Centre for Medium-Range Weather Forecasts (2020).

As is the case for raw material extraction (clay), the manufacture of ceramic products requires environmental licensing and proof of origin of the input materials used in the ceramic kilns. The generation of thermal energy for heating and baking the ceramics demands creative strategies that adhere to bioeconomic principals. Wood from residuals generated from native forest timber processing is currently the principal fuel used by companies in the ceramic processing sector in the region [18, 39, 40]. However, bioeconomic principals require inclusion of new resources used as fuel for kilns to be based on renewable biomass [17]. In this sense, biomass is generally understood to be

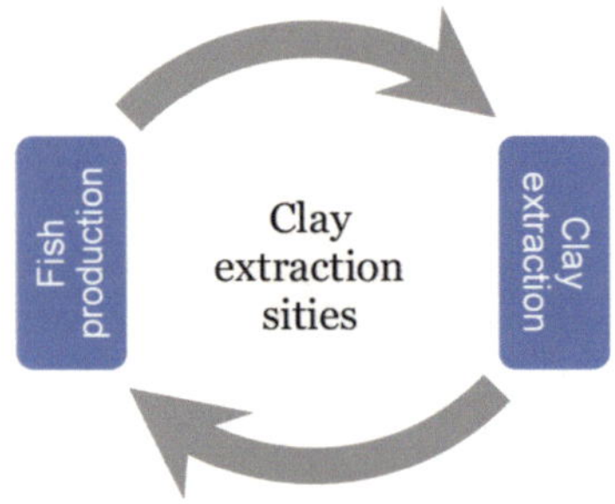

Figure 9.
Operational and bioeconomic cycle of clay extraction at the ceramic production center at Iranduba, Amazonas, Brazil. Source: The authors (2020).

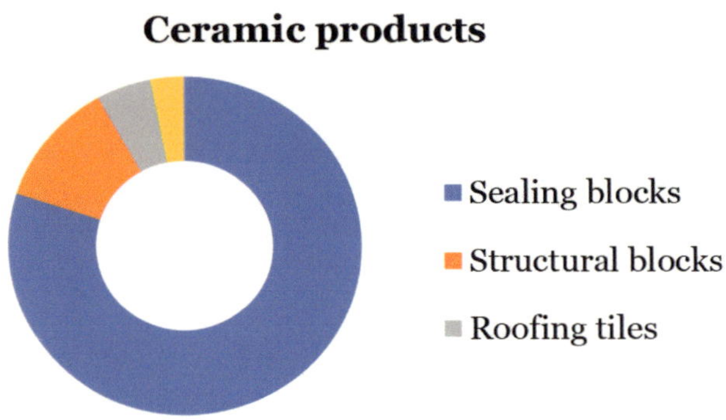

Figure 10.
Variety of products made at the ceramic production center at Iranduba, Amazonas, Brazil. Source: The authors (2020).

any non-fossil organic matter from plants, animals, or microbes that has economic potential and that can be used to generate energy [41, 42]. For example, residuals generated from silvicultural management in agroforests are normally left on the soil surface and will eventually contribute to CO2 release to the atmosphere. Alternatively, this biomass could be reutilized and technologically adapted into an energy source by industries, including those that produce ceramics.

The Amazon possesses an extensive variety of sources of residual biomass, whose reutilization could result in tangible and intangible benefits for transformational industries. In the geographic area of the ceramic production center at Iranduba there are agroindustries that produce and process plant fibers such as malva (*Urena lobata L.*) and juta (*Corchorus capsularis L.*), whose organic residual wastes are underutilized and discarded into the environment. A mix of these plant fibers and other sources of biomass such as seeds of açaí-do-Amazonas (*Euterpe precatória* Mart.) could be used as an innovative alternative for burning in the ceramic kilns of the ceramics industry in the region. Despite the rudimentary production system that uses only a low level of technology, the residuals of these plant fibers represent a promising biotechnological solution for thermal energy generation. During the fiber extraction process about 95% of the plant is discarded at the collection site, and this represents an opportunity to create an alternative destination for this material. A similar situation occurs during the processing of açaí, wherein approximately 83% of the fruits are discarded due to a lack of an alternative use in the market. These residuals, which are generated in abundance in the Amazon and subsequently disregarded, could become subproducts, such as briquettes, for transformational industries. The manufacture of briquettes through mixing different types of residual biomass such as pellets used for products made from açaí, malva and juta, is an innovative biotechnological solution that can provide an alternate destination for residual solids as well as generate thermal energy in the ceramic kilns.

Another regional residual biomass that is commonly and abundantly discarded into the environment is the outer shell that contains the nuts of the Brazil nut tree (*Bertholletia excelsa Bonpl.*), whose chemical composition is primarily lignin and cellulose. This residual biomass rapidly enters combustion and has a high calorific value, thus being easily converted into thermal energy [43–46]. The use of this source of biomass in the ceramics production process as an alternative source of thermal energy will reduce the demand for wood from residuals of native forest and aid in decreasing deforestation in the western Amazon [18].

From the perspective of innovation, the adoption of a production system based on new types of inputs for energy generation should occur in an integrated manner, with active participation of research institutions, agroextractivist communities, and ceramic industries.

A production chain based on bioeconomic principles requires knowledge and technology to incorporate biomass resources used as an energy source and to manufacture ecological products (bioproducts), and also needs a sustainable business model. **Figure 11** summarizes the sequence of activities of a ceramics production chain that will result in bioproducts and a sustainable business model.

The third step in the production chain consists of the process of marketing and distribution of the products to resellers or final consumers.

In terms of market competition, the structural characteristics of a sector are generally affected by an efficient combination of the processes of production, distribution, and sale. In the case of the ceramic industries of Iranduba, there is a certain level of homogeneity in the characteristics of internal processes, suppliers, and clients. These ceramic companies have operational structures like those categorized as Microcompanies (Microempresas (ME)) or small businesses (EPP) by the federal government. The principal characteristics of companies classified as ME or EPP are levels of revenue and tax bracket. According to Brazilian legislation, a company classified as ME must have gross annual revenue less than or equal to R$ 360,000,00 and pay taxes through one of the following options: (National Simple Method, Actual Profit Method, or Deemed Taxable Income). The legislation permits that the ME be registered in one of its four subcategories: Ordinary Partnership, Individual Limited Liability Company, Enterprise Corporation, and Sole Proprietorship. Companies classified as EPP gross annual revenue less than or equal to R$ 4,800,000,00, and can opt for either the National Simple, Actual Profit, or Deemed Taxable Income tax payment methods.

The consumers of the products made by the ceramic industry in Iranduba are principally the resellers, with physical store locations in the city of Manaus, and on a smaller scale in the upper Negro River region (**Figure 12**). These clients are responsible for about 65 to 75% of the purchases of ceramic production made in Iranduba. The remainder of the production is sold and distributed to endpoint consumers, principally civil construction companies that are responsible for public and private construction projects that used financial resources from the federal government. Among the endpoint consumers is a small proportion of purchases made by individual people who buy products directly from a ceramic company for use in their own homes.

Figure 11.
Aspects of the bioeconomy that should be incorporate into a production chain. Source: The authors (2020).

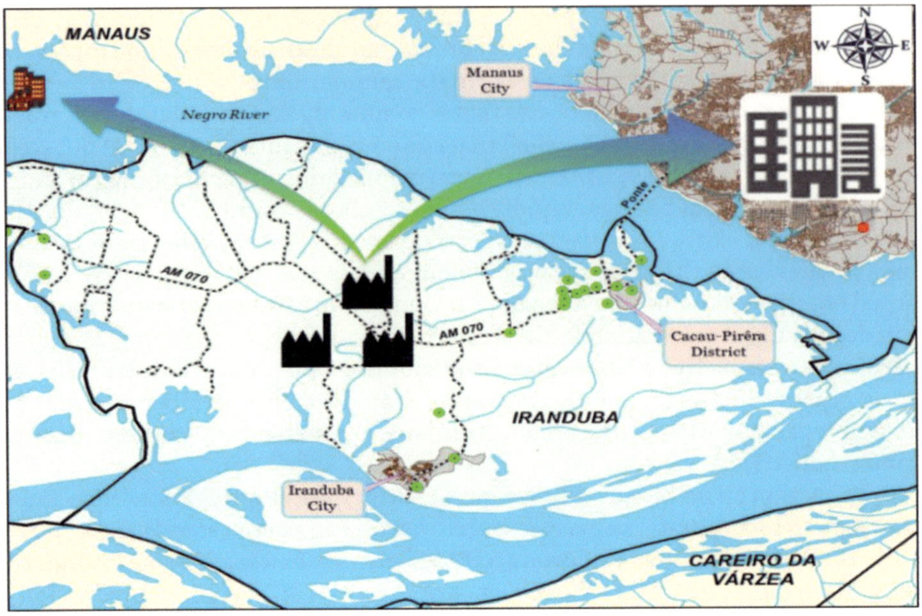

Figure 12.
Routes used to distribute products made in the ceramic production center at Iranduba. Source: ArGis (2020).

Normally, the sectorial or competitive environment in which a company operates is characterized by competing forces that drive companies to reinvent themselves through organizational strategies with the goal of increasing profits. A strategy can be defined as a set of operations that can create a competitive advantage by generating greater economic value compared to competitors. A competitive advantage is perceived through greater profitability, liquidity, leverage, and activity [47].

In the bioeconomy field, a competitive advantage can be related to the perception of the consumer that a certain company is more ecologically correct than its competitors, or that it is completely dedicated to producing and marketing products in harmony with principles of sustainability (**Figure 13**). This characteristic of ecological products imposes upon production systems the necessity of incorporating technologies that can aggregate value and attenuate impacts on biodiversity.

The search for sustainable competitive advantages has become an objective of all managers who are focused on the global market. The challenge consists in efficiently maintaining the skills that have been acquired over time while searching for novel strategies that will facilitate the manufacture and offer of ecologically

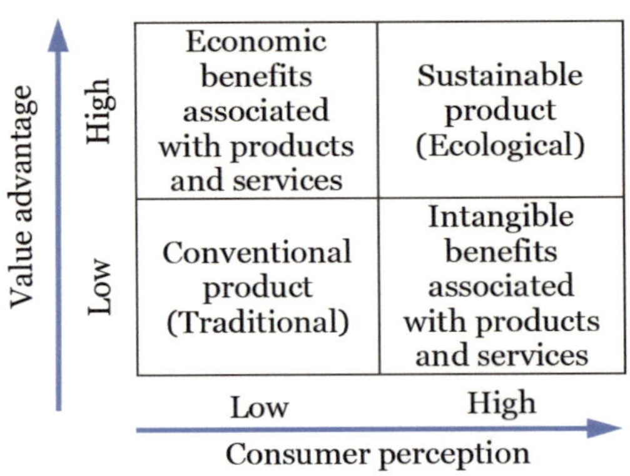

Figure 13.
Competitive advantages associated with biologically-based products. Source: The authors (2020).

sustainable products. For the ceramic industries of Iranduba, the commercialization and distribution of ceramic products is conducted through a simple, unsophisticated process, without innovative characteristics that would differentiate a product from its competition. In general, the ceramic products that are sold are stored on pallets and wrapped in thin, transparent plastic film (**Figure 14**).

To maintain a sustainable competitive advantage, companies need to innovate and acquire more highly developed marketing skills than their competitors. Products that have a seal of origin or any characteristic that conveys ecological correctness will normally aggregate intangible values that are translated into a competitive advantage.

In strategic terms, the ceramic industries of Iranduba represent an activity that efficiently transforms natural resources into bioproducts that are a reference for the entire Amazon region. On the other hand, structural changes in the production chain through addition of biotechnology and manufacture of sustainable products, requires a detailed analysis and identification of possibilities and vulnerabilities that involve the ceramics sector in the region.

Figure 14.
Finished product, wrapped and ready for sale. Source: The authors (2020).

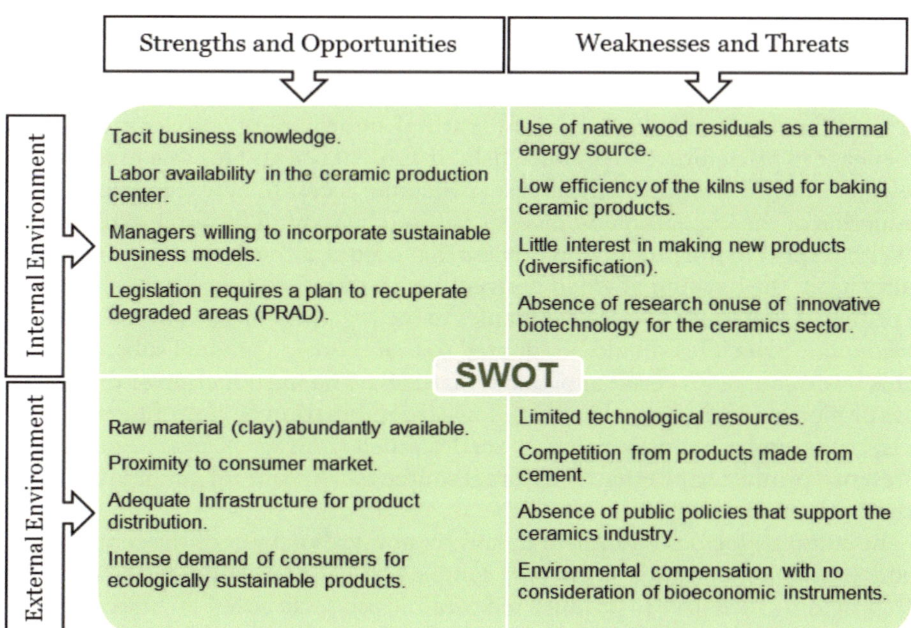

Figure 15.
Possibilities and vulnerabilities identified using the SWOT matrix. Source: The authors (2020).

In a geopolitical context, the ceramic production center at Iranduba has great economic and social importance for regional development. A redefinition of strategies and objectives together with the creation of participative public policies focused on bioeconomic principles will result in positive results in markets and will generate business models that will have a global impact. An important incentive to foster the implementation of new market strategies is the elaboration of strategic planning that identifies the possibilities and vulnerabilities that involve the ceramics sector. A tool that is commonly used for the elaboration of strategic planning is the SWOT (Strength, Weaknesses, Opportunities and Threats) matrix. The SWOT analysis allows for evaluation of how a company or group of companies behave in relation to competitors, identification of the factors that are responsible for this difference, fragilities, and the market opportunities that can be explored.

The SWOT matrix below synthesizes the principal environmental, economic, and social aspects necessary for the implementation of production of bioeconomic products in the ceramic production center at Iranduba (**Figure 15**).

Each segment of industry acts in a dynamic environment where companies must concentrate their efforts to establish a set of activities that can explore new markets, balancing development with conservation of biological resources.

The simplified strategic diagnostic in **Figure 13** will help ceramic industries identify points of rupture in their production chains and the bioeconomic alternatives necessary for organizational and technological restructuring.

4. Conclusions

In the red structural ceramics production chain, there are biotic and abiotic elements that interfere in the conditions of biodiversity. The industries in the ceramic production center in Iranduba have high potential to improve their production standards, strengthen relationships between stakeholders, and offer products that are competitive and ecologically correct.

The geological diversity of clays and the strategic location of the companies favors the manufacture and distribution of sustainable products, principally bricks and tiles, that are based on bioeconomic resources and principles.

The areas where clay is extracted to produce red ceramic products present excellent conditions to be able to, in the same natural environment, extract raw material, and engage in pisciculture to produce fish for subsistence and for sale in markets. These actions will stimulate an increase in income, access to new technologies, and a reduction of the impacts on biodiversity caused by natural resource extraction.

With respect to the production process that occurs at the ceramics manufacture facility itself, the burning of wood derived from native forest timber is currently the principal fuel used by these companies in the region. Strategies based on bioeconomic principles should be adopted so that there is a gradual substitution of this wood source for residual biomass materials. The introduction of different types of biomass including a biotechnological combination of plant fibers with seeds, fruits, and other materials that are discarded from agroforestry operations represent a promising alternative as a fuel source for the kilns of ceramic industries in the Amazon.

The cultural, socioenvironmental, and economic dynamics of the ceramic production center at Iranduba are very complex and challenging. The logic involved with providing incentive to produce red ceramic products based on bioeconomic strategies is not only rooted in the need to increase productive efficiency, but also in the necessity to respect the environment and develop a series of products made in the Amazon that are considered highly sustainable.

Business strategies that seek excellence and product differentiation normally succeed in obtaining a competitive advantage and greater profits. In this context, well-structured, robust planning should be conducted that strives to make the operational activities of the ceramic industries of Iranduba conform to sustainability criteria and thus make their products a reference for the global market.

Acknowledgements

The authors thank the Syndicate of the Ceramic Kiln Industry of the State of Amazonas (SINDICER/AM) for the excellent information given on the current situation of the production chain of the ceramic industry center in Iranduba. The authors are also very grateful of the Biodiversity and Biotechnology Network of the Legal Amazon (Bionorte) for financial resources for publication.

Author details

Gelson Dias Florentino[1*], Lucieta Guerreiro Martorano[2],
Sandro Augusto Lima dos Santos[3], José Reinaldo da Silva Cabral de Moraes[4],
Ires Paula de Andrade Miranda[5] and Maria de Lourdes Pinheiro Ruivo[6]

1 Doctoral Student, Biodiversity and Biotechnology Network of the Legal Amazon (Bionorte) Belém, College La Salle, Manaus, Brazil

2 Brazilian Agricultural Research Corporation (EMBRAPA) Eastern Amazon (CPATU/NAPT), Santarém, Pará, Brazil

3 Cerâmica Montemar Indústria e Serviço Ltda., Iranduba, Amazonas, Brazil

4 Doctoral Student, Plant Production, University of the State of São Paulo, São Paulo, Brazil

5 National Institute for Amazon Research (INPA), Manaus, Amazonas, Brazil

6 Emilio Goeldi Museum, Belém, Pará, Brazil

*Address all correspondence to: dias.gelson@gmail.com

References

[1] Reis P, Ruschel R, Coelho A, Luz A, Martins da Silva V. Avaliação do potencial madeireiro na floresta nacional do tapajós, após 28 anos da exploração florestal. Pesquisa Florestal Brasileira, v. 30, n. 64, p. 265-281, 2010. DOI: 10.4336/2010.pfb.30.64.265

[2] Blanco-Gutié I, Manners R, Varela-Ortega C, Tarquis M, Martorano L, Toledo M. Examining the sustainability and development challenge in agricultural-forest frontiers of the Amazon Basin through the eyes of locals. Natural Hazards and Earth System Sciences. 2020;**20**:797-813. DOI: 10.5194/nhess-20-797-2020

[3] Gaston K, Spicer J. Biodiversity. 2. Ed. Oxford. Blackwell Publishing; 2004

[4] De Groot R, Fisher B, Christie M. Integrating the ecological and economic dimensions in biodiversity and ecosystem service valuation (chapter 1). In: Kumar P, editor. TEEB Foundations, the Economics of Ecosystems and Biodiversity: Ecological and Economic Foundations. London: Earthscan; 2010

[5] Costanza R. Social goals and the valuation of natural capital. Environmental Monitoring and Assessment, v. 2003;**86**:19-28. DOI: 10.1023/A:1024045221992

[6] Gomes A, Dantas Neto J, Silva V. Serviços ecossistêmicos: conceitos e classificação. Revista Ibero Americana de Ciências Ambientais. 2018;**9**(4):12-23. DOI: 10.6008/CBPC2179-6858.2018.004.0002

[7] Hopkins M. Modelling the known and unknown plant biodiversity of the Amazon Basin. Journal of Biogeography. 2007;**34**(8):1400-1411. DOI: 10.1111/j.1365-2699.2007.01737.x

[8] Boubli J, Hrbek T. Introdução à Biodiversidade Amazônica. In: Marcon J [et al.] (Org.). Biodiversidade Amazônica: caracterização, ecologia e conservação. Manaus: Edua. 2012

[9] Veiga J, Ehlers E. Diversidade biológica e dinamismo econômico no meio rural. In: Peter MAY. (Org.). Economia do Meio Ambiente. 3.ed. Campus: Rio de Janeiro; 2018. pp. 289-308

[10] Reyers B, Polasky S, Tallis H, Mooney H, Larigauderie A. Finding common ground for biodiversity and ecosystem services. Bioscience. May 2012;**62**(5):503-507. DOI: 10.1525/bio.2012.62.5.12

[11] Mace G. Approaches to defining a planetary boundary for biodiversity. Global Environmental Change. Sep 2014;**28**:289-297. DOI: 10.1016/j.gloenvcha.2014.07.009

[12] Balvanera P [et al.]. Ecosystem Services (Chapter 3). In: Walters M, Scholes R (Eds.), The GEO Handbook on Biodiversity Observation Networks, SpringerOpen, 2017. ISBN 978-3-319-27288-7 (eBook).

[13] Parron L [et al.]. Pesquisa sobre serviços ecossistêmicos no Brasil: uma revisão sistemática. Rev. Ambient. Água [online]. 2019, vol.14, n.3, e2263. Epub May 20, 2019. DOI: 10.4136/ambi-agua.2263.

[14] Raynaut C. Meio ambiente e desenvolvimento: construindo um novo campo do saber a partir da perspectiva interdisciplinar. Desenvolvimento e Meio Ambiente, n. 2004;**10**:21-32, jul./dez. Editora UFPR

[15] Michelsen G, Adomßent M, Martens P, Hauff Michael. Sustainable development – Background and context (chapter 2). In Heinrichs H et al. (eds.), Sustainability Science. London: Springer, 2016.

[16] Nascimento E. Trajetória da sustentabilidade: do social ao ambiental, do ambiental ao econômico. Estudos Avançados (USP. Impresso), v. 2012;**26**: 51-64

[17] Lewandowski I, Gaudet N, Lask J, Maier J, Tchouga B, Vargas-Carpintero R. Context (chapter 2). In: Lewandowski I, editor. Bioeconomy - Shaping the Transition to a Sustainable, Biobased Economy. University of Hohenheim, Stuttgart, Germany: Springer; 2018

[18] Florentino G [et al.]. Dynamics of Space and Time of the Production Chain of the Ceramic Industry Production Center of Iranduba, Amazonas, Brazil. Sustainability, v. 11, p. 5576, 2019. DOI: 10.3390/su11205576

[19] Becker B, Stenner C. Um futuro para a Amazônia. Série inventando o futuro. São Paulo: Oficina de Textos; 2008

[20] Neves E. Duas interpretações para explicar a ocupação Pré-Histórica na Amazônia. In: Tenório M (org.). Pré-História da Terra Brasilis. Rio de Janeiro: Editora UFRJ; 1999

[21] Pinheiro H. Oleiros da vida: trabalho, ambiente e o futuro dos trabalhadores do barro em Iranduba (AM) [Tese] Universidade Federal do Amazonas: Manaus-AM, Brasil, 2015.

[22] AMAZONAS. Plano de desenvolvimento preliminar: APL de base mineral cerâmico-oleiro. Núcleo Estadual de Arranjos Produtivos Locais – NEAPL/ Secretaria de Estado de Planejamento e Desenvolvimento Econômico – SEPLAN/ AM. In: Governo do Estado. Manaus: NEAPL/SEPLAN; 2009

[23] Chaves S. Estudo de Mercado da Produção de tijolos no Município de Iranduba. Monografia, Faculdade de Estudos Sociais da Universidade Federal do Amazonas, 2001.

[24] Maciel C, Valle M, Moura J. Globalização, reestruturação produtiva e controle do trabalho no Polo oleiro-cerâmico de Iranduba - AM. Novos Cadernos NAEA. 2012;**15**:311-331

[25] Florentino G, Martorano L, Miranda I, Silva-Rocha S. Socioeconomic impacts caused by a center of ceramic manufacturing in the Amazon: Sustainable agrotechnological strategies based on the IrrigaPote technology. In: 7th Brazilian Biotechnology Congress and 2nd Biotechnology Ibero-American Congress. Brasília-DF, Brazil, November 18-21. 2018

[26] D'Antona R, Reis N, Maia M, Nava D. Projeto materiais de construção na área Manacapuru – Iranduba – Manaus – Careiro (domínio baixo Solimões). CPRM – Serviço Geológico do Brasil: Manaus; 2007. Available from: http://rigeo.cprm.gov.br/jspui/handle/doc/1737 Accessed: 12 Jun 2020

[27] BRASIL. Ministério de Minas e Energia. Projeto Argila: Regularização e levantamento ambiental do setor oleiro nos municípios de Iranduba e Manacapuru. Relatório Final. DNPM/MME, Manaus, 2000.

[28] Campelo N [et al.]. Ensaios tecnológicos em cerâmica vermelha do pólo oleiro dos municípios de Iranduba e Manacapuru, AM. In: Anais do 48° Congresso Brasileiro de Cerâmica. 28 de junho a 1º de julho de 2004. Curitiba-PR, 2004.

[29] Abreu N, Silva D, Carvalho D, Sampaio F. Uma abordagem geográfica dos aspectos físicos e dos potenciais econômicos do solo de Iranduba – AM. Eixo Temático: Geomorfologia e cotidiano. Revista Geonorte, Edição Especial, v.2, n.4. 2012:659-667

[30] Maciel C, Valle M, Maciel J. Caminhos cruzados: relação entre empresa, trabalhadores e sindicato em Polo oleiro-cerâmico da Região Metropolitana de Manaus-AM.

Novos Cadernos NAEA. Suplemento. 2013;**16**:127-162

[31] Morrison B, Golden J. An empirical analysis of the industrial bioeconomy: Implications for renewable resources and the environment. BioResources, v. 10, n. 3, p. 4.411-4.440. 2015

[32] Silva M, Pereira F, Bomtempo J. A Bioeconomia brasileira em números. BNDES Setorial. 2018;**47**:277-332

[33] Leripio A, Leripio D. Cadeias Produtivas Sustentáveis. Mix Sustentável. 2015;**1**:142-159

[34] O'Rourke D. The science of sustainable supply chains. Science. Jun 2014;**344**(6188):1124-1127. DOI: 10.1126/science.1248526

[35] BRASIL. Ministério do Meio Ambiente. Resolução CONAMA n° 237, de 19 de dezembro de 1997. Dispõe sobre a revisão e complementação dos procedimentos e critérios utilizados para o licenciamento ambiental.

[36] SUFRAMA. Superintendência da Zona Franca de Manaus: Projeto Potencialidades Regionais - Estudo de Viabilidade Econômica. Vol. 8. SUFRAMA: Piscicultura. Manaus-AM; 2003. Available from: http://www.suframa.gov.br/publicacoes/proj_pot_regionais/piscicultura.pdf Accessed: 15 Jun 2020

[37] Florentino G. Caracterização do extrativismo e comércio de pescado no Médio Rio Juruá. [Dissertação] Mestrado em Desenvolvimento Regional. In: Universidade Federal do Amazonas - Manaus: UFAM. 2012

[38] Petrere (Org.). O Setor Pesqueiro na Amazônia: situação atual e tendências. Manaus: IBAMA. ProVárzea; 2004

[39] Azevedo [et al.]. Produção de lenha na região de Iranduba e Manacapuru - Amazonas: *Acacia mangium* e *Acacia auriculiformis*. Embrapa Amazônia Ocidental, 2002. Available from: <https://www.embrapa.br/busca-de-publicacoes/-/publicacao/674334/producao-de-lenha-na-regiao-de-iranduba-e-manacapuru---amazonas-acacia-mangium-e-acacia-auriculiformis> [Accessed: 21 Jun 2020].

[40] Barros S [et al.]. Avaliação do potencial energético das espécies florestais *Acacia auriculiformis* e Ormosia paraensis cultivadas no município de Iranduba/Amazonas, Brasil. Madera y Bosques, v. 15, n. 2, p. 59-69, 2009.

[41] Dias J [et al.]. Produção de briquetes e péletes a partir de resíduos agrícolas, agroindustriais e florestais. Brasília, DF: Embrapa Agroenergia, 2012. 130 p.

[42] Perea-Moreno M, Samerón-Manzano E, Perea-Moreno A. Biomass as renewable energy: Worldwide research trends. Sustainability. 2019;**11**:863. DOI: 10.3390/su11030863

[43] Oliveira J, Lobo P. Avaliação do Potencial Energético de Resíduos de Biomassa Amazônica. In: 4° AGRENER, 2002, Campinas-SP. CD-Rom, v. CD-Rom. p. 0-6, 4° Encontro de Energia no Meio Rural – AGRENER, 2002. Available from: <http://www.proceedings.scielo.br/scielo.php?pid=MSC0000000022002000100026&script=sci_arttext> [Accessed: 20 Fev 2020]

[44] Freitas J, Naves M. Composição química de nozes e sementes comestíveis e sua relação com a nutrição e saúde. Revista de Nutrição (Impresso). 2010;**23**:269. DOI: 10.1590/S1415-52732010000200010

[45] Carmona I, et al. Potencial energético da biomassa e carvão vegetal de resíduos de castanha-do-pará (*Bertholletia excelsa Bonpl.*). III Congresso Brasileiro de Ciência e Tecnologia da Madeira, Florianópolis, 2017. Available from: https://proceedings.science/cbctem/papers/

potencial-energetico-da-biomassa-e-carvao-vegetal-de-residuos-de-castanha-do-para-%28bertholletia-excelsa-bonpl.%29 [Accessed: 10 Ago 2020]

[46] Alves A, et al. Thermal Diagnosis of Heat Conduction and Combustion Time of Fruits of the Brazil Nut Tree (*Bertholletia excelsa Bonpl.*). Advances in Bioscience and Biotechnology (Impresso), v. 11, p. 60-71, 2020. DOI: 10.4236/abb.2020.112005

[47] Barney J, Hesterly W. Administração estratégica e vantagem competitiva. 5. São Paulo: Pearson Prentice Hall; 2017. pp. 7-13

Chapter 10

Neo-Developmentalism and Regional Integration: IIRSA Impact in the Environmental Agenda in the Amazon

Pedro Pablo Cardoso Castro and Nirvia Ravena

Abstract

The regional integration policy in Latin America was based on the premises of the Initiative for the Integration of the Regional Infrastructure of South America (IIRSA), seen as a development strategy adopted by the regional political elites and as an instrument of Brazilian foreign policy. Focusing on the projects undertaken in the Amazon Hub, the article analyzes the integration strategy and its impact on environmental regulation. The paper highlighted the role played by the developmentalism goals in the region regarding the environmental regulation, adopting a model of integration encompassing a reduction of natural resources stock, which is a strategic feature for sustainable development policies. This chapter results from descriptive research, based on secondary data and official documents available from the involved institutions. They analyze the Brazilian government's neo-developmentalism goals to understand how the environmental agenda in Pan Amazon was affected.

Keywords: regional integration, neo-developmentalism, Amazon, environmental agenda

1. Introduction

The foreign policy of the Initiative for the Integration of the Regional Infrastructure of South America (IIRSA) is one of the key instruments for repositioning South American countries in an environment of global change [1, 2]. In this context, IIRSA emerged as an initiative of the Inter-American Development Bank (IDB) in August 2000, in partnership with the Andean Development Corporation (CAF) and the River Plate Basin Financial Development Fund (Fonplata), based on a proposal of the then President of Brazil, Fernando Henrique Cardoso (FHC), at the First Summit of South American Presidents (2000).

Originally launched by presidents of neoliberal affiliation (e.g., Andres Pastrana – Colombia; Fernando De la Rua – Argentina; Fernando Cardoso – Brazil); IIRSA was subsequently supported by the governments of progressive orientation (e.g., Ecuador, Bolivia, Peru, Chile). In December 2004, the twelve participant Presidents confirmed their commitment to IIRSA initiatives during the Cuzco meeting, approving the "Consensual Implementation Agenda for 2005-2010" that selected the priority projects to be concluded in 2010.

IntechOpen

In that sense, when IIRSA was created, there was a scenario determined by the Washington Consensus that outlined a neoliberal agenda. It recommended (with intense pressure from multilateral lending agencies and developed countries) the privatization of state companies, deregulation of the economy, and unilateral liberalization of the foreign trade in Latin American countries [1, 3, 4]. The declared goals of IIRSA were:

"To promote the development of the regional infrastructure within a framework of increasing competitiveness and sustainability, in order to generate the necessary conditions to achieve a stable, efficient and equitable development pattern in the region; identifying the necessary physical, regulatory and institutional requirements and seeking implementation mechanisms that promote physical integration at the continental level" [5].

IIRSA was conceived to execute the regional integration based on ten "integration hubs" based on territorial planning. Four of them pass through the Amazon Region - comprehending all territories located in the Amazon River basin's watershed area. From a geopolitical perspective, this definition includes the following Amazonian countries: Bolivia, Brazil, Colombia, Ecuador, Guyana, Peru, Suriname, and Venezuela. Thus, in Latin America, Brazil acting as a global player [6, 7], together with the Amazon countries, performed the regional dynamics coordination where IIRSA project integration occurred.

This chapter discusses the regional integration in the initial phase implemented by IIRSA in the period between the years 2013 and 2014, as well as the positioning of the integration projects concerning the environmental regulation created by the Amazonian countries highlighting the role played by the sustainability concept regarding the existing environmental regulation. The IIRSA adopted a model of integration encompassing a reduction of natural resources stock, a strategic feature for long term sustainable development policies. The central question that this article wants to answer is: How the implementation of the IIRSA integration projects affected the governance of the sustainability and environmental agenda in the Brazilian Amazon?

The study associates the analysis of the IIRSA Project hubs related to Brazilian Amazon areas and the neodevelopmentalism to understand this economic policy's impacts in the Brazilian environmental scenario [7, 8–14]. The empiric reference of this study, the IIRSA Project, is considered a material expression of the development strategies executed by the Latin American political elites from 2011 to 2014 when an accelerated implementation of the IIRSA took place.

This chapter is organized into three sections: The first one describes the neo-developmentalism approach and the recent reflections on the post-neoliberal period in the Brazilian Amazon Region. The second section presents the issues related to environmental regulation and sustainability as an arena of interest in the Amazon region. The last section describes and analyzes Amazon's IIRSA projects, the place of the environmental agenda, and the Brazilian foreign policy role in this context.

2. Neo-developmentalism and the pan-Amazonian context

The neo-developmentalism supports its interpretation in the governmental strategies assumed by the Latin American governments post-Consensus of Washington, being the broader goals of this strategy summarized by the generic term "macroeconomic" [15, 16]. The tactics are focused on the combination of stability, which includes inflation control, exchange rate and balance of payments, fiscal sustainability, low-interest rates, and reduction of uncertainties related to future demand, which should provide a more stable environment for investment decisions of these governments [12, 13].

The achievement of these goals would require complementary actions of monetary, fiscal, exchange, and wage policies [8–11], aimed at restoring state power to control the currency, facilitate policy implementation, promote competition, and support improvements in income distribution [8, 11, 15]. Also, the adoption of a development strategy that allows domestic firms to seize global economies of scale, and infrastructure and technological updating processes, supported by innovation policy and an activist trade policy targeted at strong intellectual property regimes and investment opportunities for domestic firms. These elements entail a commitment to mobilizing all available labor resources, increasing productivity in each industry, and the steady transfer of finance to high wage and high value-added sectors.

However, neo-developmentalism government strategies insisted on the implementation of neoliberal policies. The politicians conceived the Brazilian economy from 2003 to 2016 as potentially underutilized due to unrealized productivity gains that could be captured through infrastructure development and economic growth. This approach turned areas as Amazon in economies of scale; consequently, opening the region via state support to higher private-sector investment (shifting workers of lower productivity), and expanding the participation in foreign markets as a government goal [7, 17, 18].

2.1 IIRSA: Regional integration and developmentalism tool

The beginning of the XXI century represented a set of global changes, including socialist-oriented governments in many South American countries, which gained influence in the regional political landscape. This was evidenced by the general elections' electoral results in Venezuela, Chile, Brazil, Argentina, Uruguay, Bolivia, Peru, and Ecuador in the first decade of 2000. Programmatically, the elected candidates in these countries had left-wing programs in which the environmental agenda was associated with other electoral strategies based on developmental actions whose legitimacy was based on popular acclaim. The Latin American leaderships, as elected, combined developmentalism policies not considering environmental problems accrued from development strategies and relegated sustainability to a secondary policy level [19, 20].

The IIRSA implementation plan did not have a strong administrative and operational bureaucratic body at the multi-local levels. Consequently, after the delay of the integration agendas, some organizational changes occurred at the bureaucratic level in each of the involved territories, and administrative instances emerged to correct systemic inefficiencies in the execution of several projects. In response, the Brazilian government - as the IIRSA paymaster in Latin America - created the South American Council for Infrastructure and Planning (COSIPLAN) to coordinate and accelerate the IIRSA and the developmental premises of leftist governments in the Pan-Amazon countries. The implementation of IIRSA was developing by hubs, and this was how developmentalism politicians could associate diplomatic issues with the necessities of implementation.

Four of the ten IIRSA hubs of action were located in the Amazon territory. Among these, the Peru-Brazil-Bolivia hub concentrates much of the internationalization of the Madeira Complex. From the total IIRSA projects for this area, twenty-one are related to ports and waterways works; twelve to roads; three to works in seaports; five to air transportation and border crossings, and the other two deal with electrical interconnection for various hydroelectric power plants built along the Madeira River. The hubs materialize this complexity and demonstrate the magnitude of the integration and the coordination problems arising from this supranational proposal, as seen in **Figure 1**.

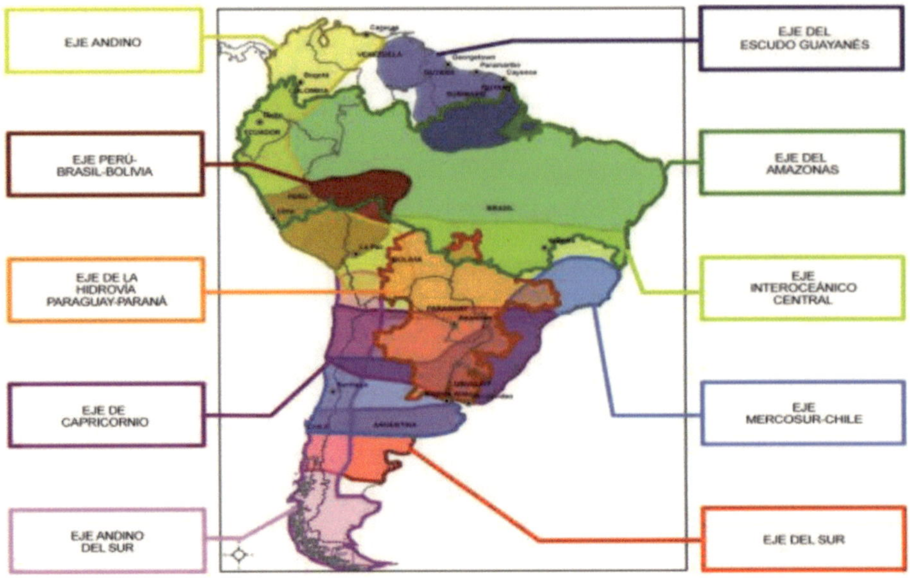

Figure 1.
Map of the IIRSA hubs. Source: GEOSUR, 2018.

2.2 Integration projects in the Amazon and environmental regulation

The lack of environmental regulation coordination among the Amazonian countries is notorious in dealing with transboundary water resources management issues. In this regard, an integrated analysis of environmental regulations for the Amazon detects absences and impossibilities of consolidated regulatory arrangements for effective cooperation in the region [21]. This fact certainly has had an impact on the possibility of developing infrastructure on a sustainable basis, as in Amazonian areas, the command and control instruments used by environmental regulation norms have systemic efficiency limits.

In addition to the increasing environmental deregulation in the Brazilian political scenario, IIRSA became an effective instrument for government interests in expanding jobs and increasing economic growth rates and market agents seeking to ensure their investment return in electoral campaigns. An example of this deregulation is the Bill N° 1876/1999, which "provides for Areas of Permanent Preservation, Legal Reserve, forest exploitation and other measures" approved in 2013. According to their environmental licensing situation, **Table 1** shows the infrastructure projects in the four IIRSA integration hubs located in the Amazon region.

Of the total 191 projects, 54.5% do not comply with environmental licensing norms. Noteworthy are many unlicensed waterway and railroad projects in the Amazon Hub: 18 of 27 projects, i.e., 66.6% of those in the same condition. Another situation worth mentioning is the hydroelectric plants and fiber optic transmission lines, mostly found without environmental licenses.

The data presented in the 2013 and 2014 reports - organized in **Table 2** - show the projects' typology. Systematized data of each type of integration show the infrastructure projects' situation according to the execution stage and the environmental license condition. The environmental licenses were obtained only for 45.5% of the projects, and several of them are still being executed or are at the pre-execution stage without an environmental license.

The data presented in **Tables 1** and **2**, and **Figure 2**, indicate the fragility of the regulatory actions undertaken at the beginning of the implementation of

Projects	IIRSA hubs in Amazonian territories									
	Peru-Brazil-Bolivia		Guyanas		Amazon		Andean		Total	
	Licensed	Not licensed	Licensed	Not licensed	Licensed	Not licensed	Licensed	Not licensed	Licensed	Not licensed
Ports and bridges	02	02	04	04	08	13	03	03	17	22
Waterways and railways	04	00	00	01	09	18	0	01	13	20
Roads and border crossings	03	06	03	06	11	12	25	19	42	43
Hydroelectric plants and fiber optic transmission lines	03	02	00	02	01	00	03	08	07	12
Airport and logistics centers	03	00	00	00	03	07	02	00	08	07
Total	**15**	**10**	**07**	**13**	**32**	**50**	**33**	**31**	**87**	**104**
%	**60.0**	**40.0**	**35.0**	**65.0**	**39.0**	**61.0**	**51.5**	**50.8**	**45.5**	**54.5**

Created by the authors. Source: IIRSA – Data Bank of COSIPLAN Project Portfolio (http://iirsa.org/proyectos/Principal.aspx).

Table 1.
Projects of the IIRSA hubs in the Amazon, according to their environmental licensing (EL) condition - 2013 and 2014.

IIRSA projects. Especially projects related to energy generation (63% without EL), waterways, and railroads (60.6% without EL) – according to Graphic 1 – express the dysfunctional character between IIRSA and the environmental agenda.

Projects	Project stage	IIRSA hubs in Amazonian territories				Total
		Peru-Brazil-Bolivia	Guyanas	Amazon	Andean	
Ports and bridges	Pre-execution	2	1	8	1	**12**
	AP. to begin	1	4	4	—	**9**
	Execution	—	—	6	3	**9**
	Completed	1	3	3	2	**9**
	Total	**4**	**8**	**21**	**6**	**39**
	With EL	**2**	**4**	**8**	**3**	**17**
Waterways and railways	Pre-execution	3		8		**11**
	AP. to begin	1	1	6	1	**9**
	Execution	—	—	5	—	**5**
	Completed	—	—	8	—	**8**
	Total	**4**	**1**	**27**	**1**	**33**
	With EL	**4**	**0**	**9**	**0**	**13**
Roads and border crossings	Pre-execution	2	2	5	8	**17**
	AP. to begin	2	1	3	11	**17**
	Execution	3	4	11	15	**33**
	Completed	2	2	4	10	**18**
	Total	**9**	**9**	**23**	**44**	**85**
	With EL	**3**	**3**	**11**	**25**	**42**
Hydroelectric plants and fiber optic transmission lines	Pre-execution	1	—	—	—	**1**
	AP. to begin	2	1	—	4	**7**
	Execution	1	—	—	3	**4**
	Completed	1	1	1	4	**7**
	Total	**5**	**2**	**1**	**11**	**19**
	With EL	**3**	**0**	**1**	**3**	**7**
Airport and logistics centers	Pre-execution	—	—	1	—	**1**
	AP. to begin	—	—	6	—	**6**
	Execution	3	—	2	1	**6**
	Completed	—	—	1	1	**2**
	Total	**3**	**—**	**10**	**2**	**15**
	With EL	**3**	**—**	**3**	**2**	**8**
Grand total		**25**	**20**	**82**	**64**	**191**
Total projects with EL		**15**	**07**	**32**	**33**	**87**
% Projects with EL		**60.0**	**35.0**	**39.0**	**51.5**	**45.5**

Source: IIRSA – Data Bank of COSIPLAN Project Portfolio (http://iirsa.org/proyectos/Principal.aspx).

Table 2.
IIRSA hubs projects in the Amazon, according to the project stage and environmental licensing (EL) condition - 2013 and 2014.

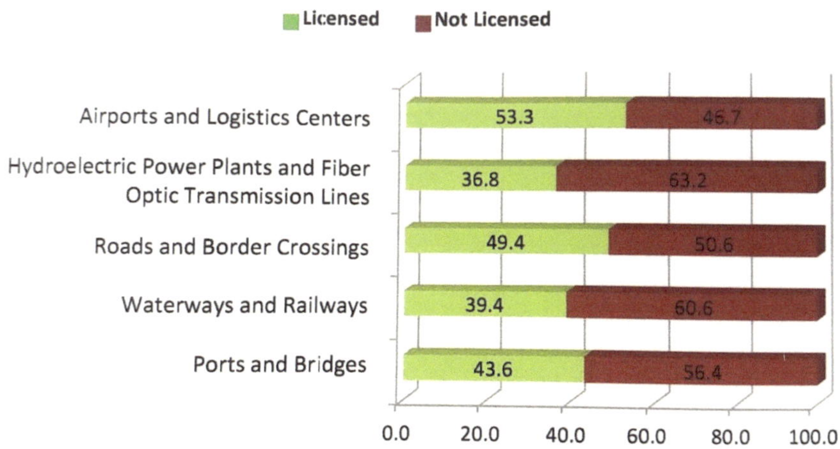

Figure 2.
IIRSA projects in the Amazon hubs, according to the environmental license condition. Source: IIRSA – Data Bank of COSIPLAN project portfolio (<http://iirsa.org/proyectos/Principal.aspx>).

The graphic shows that between the years 2013 and 2014, none of the activities that characterize the integration of IIRSA respected the environmental dimension and sustainability goals, as advertised by the governments. It is important to notice that in Bolivia and Brazil, many voters believed in the proposed administration's developmental agenda, and these popular governments took compliance with environmental laws to second place.

Specifically, in the Amazon countries, environmental regulatory instruments exist without a coordinating scope of environmental policies and have a low level of institutionalization [21, 22]. Brazil's actions in the coordination of regional integration projects show this country as a hegemonic power who imposed, at that moment, models of environmental regulation. In the case of IIRSA in the Amazon, these projects would operate throughout the territory where regulatory compliance projects were already underway. For instance, in the Andean Hub runs the Project: Armonización regulatoria: eléctrica, gasífera y petrolera; operating in Bolivia, Perú, Colombia, Ecuador, and Venezuela; and the Amazon Hub, where the largest stocks of natural resources on the planet are located, and the lowest degrees of institutionalization of environmental regulations were imposed (see **Figure 3**).

The connection of the Amazon region with the execution of projects with preliminary studies of social and environmental impact to subsidize licensing processes and the respective socio-environmental conditions for the projects' execution puts this territory in a degree of vulnerability never experienced before. The conception of development follows an ideology that relegates the environmental agenda to a secondary place. Companies that used to capture the agenda of construction works for IIRSA hubs during the FHC administration were the same at the time of the Latin American popular governments and influenced the decisions of left and center-left governments elected. These companies changed their performance operating in close connection to state structures, introducing their demands and exercising their influence through a plural strategy in supranational arenas as a movement of political capture. They have more flexibility to act due to the redefinition of Latin America's role in the global crisis of 2008 and the conversion of Brazil into a global player in the world scenario in that period [6, 23]. The dualism environmental regulation and developmentalism started to be obscured, and sustainability stayed as a corporation's discourse.

The World Investment Report indicated that after the 2008 crisis, the most significant increase in foreign investment occurred in South America. Investments were in the order of US$ 86 billion, with Brazil accounting for 56% of this amount.

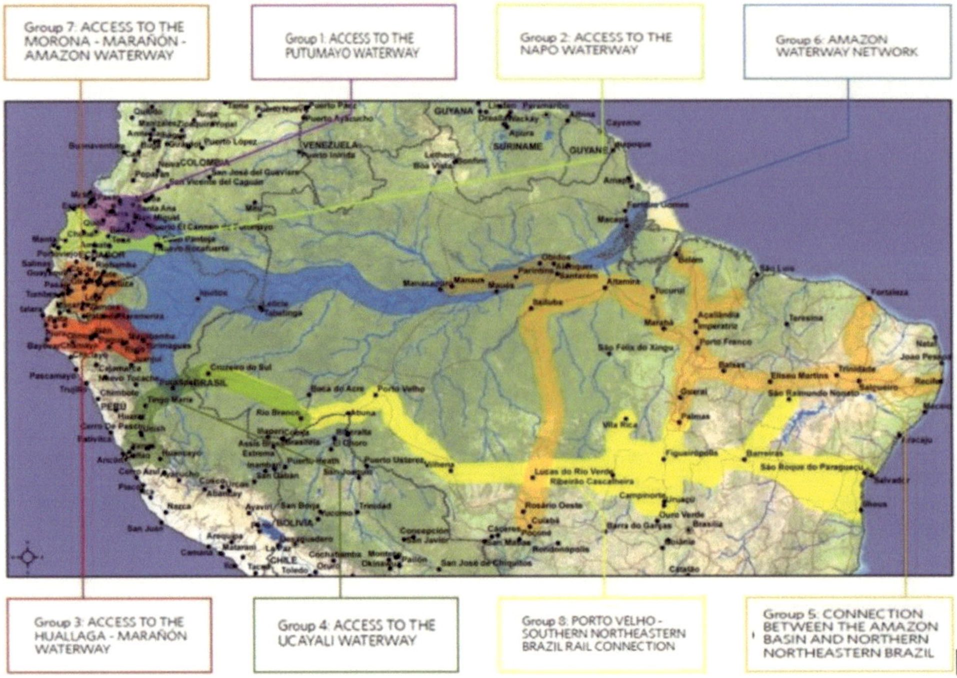

Figure 3.
Map of the IIRSA area of influence in the Amazon. Source: IIRSA, 2014.

Companies such as Vale, Gerdau, Camargo Correa, Votorantim, Petrobras, and Braskem made acquisitions in the iron ore, steel, food, cement, chemicals, and oil refining sectors, as well as other industries in Latin American countries [24]. From a domestic perspective in Brazil, Brazilian companies have met their demands through the association between the Growth Acceleration Program (PAC), a developmental program, and IIRSA actions. In this sense, hydroelectricity gains momentum, given its centrality as a structuring element of regional integration, simultaneously with the violation of environmental sustainability.

2.3 Foreign policy, South-South cooperation, and IIRSA

Latin American cooperation started with the Amazon Cooperation Treaty (ACT). The ACT started after the meeting of Brazil and Peru heads of state, who committed to elaborating the initial outline for the cooperation. Since its elaboration in 1978, the ACT was enforced under the premise of preservation and the development of the Amazon region, within the context of territoriality - not the environmental sense. In the same period, the priority sectors were defined for establishing cooperation: Territorial occupation, the development of technology, and scientific knowledge [25].

In 2002, the Amazon Cooperation Treaty Organization (ACTO) creation reinforced the integration purposes initiated with the ACT, emphasizing the environmental dimension. In this context, the ACTO Executive Secretariat - based in Brasilia and disconnected with the Amazonian reality - is now another body that has intensified regional inequalities, emphasizing the competitive dimension between signatories of projects funded by multilateral agencies for the Amazon [22].

This observation is relevant as the IIRSA operated within this context. The neo-developmental legacy that remained in some Brazilian sectoral bureaucracies influenced foreign policy towards the Latin American neighbors [26], making the Brazilian Amazon a region where sectoral policies were defined with a high degree of concentration when considering the federative pact. In the other signatory countries of the ACTO, the distance dynamics grows more and more from the centers where the decision-making processes took place [27, 28].

Within this context, the IIRSA actions and the course of the South-South cooperation in the Amazon were materialized by the left government's administrations until 2016 through ACTO's coordination. This organization's role was redefined, leaving it as a standard of regional sustainability associated with a change of political agenda that included development from a different perspective from the one that originated its creation in the ACT.

The development assumptions in the agenda of IIRSA's priority projects in execution and the lack of related licenses also attest that the Brazilian foreign policy indicates to its Amazonian neighbors the extent to which the environmental arena must be included in the South-South cooperation. Against this trend were the social movements emerging from Latin American political institutions. These movements rely on the defense of the environment and on the Amazon's environmental asset [29], where new information and communication technologies gave voice to the groups affected by the actions of IIRSA. The case of the Movimento Xingu Vivo Para Sempre (Xingu Lives Forever movement) illustrates the activism of social movements in the environmental arena. For instance, the Movimento Xingu Vivo Para Sempre – opposing the Belo Monte Hydroelectric Complex installation - qualifies itself as "for the awareness of the Brazilian society in defense of the Xingu River and against the hydroelectric of Belo Monte." Through its activism, the Xingu Vivo Movement put pressure on the Public Ministry of the Union, and 13 sentences to stop the works were issued. However, the judicial sentence succumbed to the power that companies that funded the candidates of the turn-left in Latin America. Many decisions were taken by different judges over ten years to stop licenses due to demonstrate illegalities. However, they were all suspended by the Federal Regional Court of Brasilia in the years of ruling left-wing Latin American candidates from 2005 [30].

3. Conclusion

The imbalance and the profound differences between Brazilian government developmentalism goals and the actions to promote sustainability in amazon areas can be understood by the place given to the environmental agenda during the IIRSA implementation. The proposal for regional integration resulted in the process of deregulation and systematic deconstruction of policies for sustainability in the areas of influence of IIRSA. In this perspective, the implementation of IIRSA made evident severe institutional weaknesses, such as independent institutions' existence, as COSIPLAN, enforcing the authorization to implement unsustainable projects. Out of the global environmental regulation, these organizations conducted the rupture of Brazil's environmental legislation - and the IIRSA region in general. The flexibilization and environmental deregulation resulting from the influence of IIRSA contractors and government agents interested in its implementation made the project an example of an environmentally unsustainable regional integration strategy in the Pan Amazon Region.

The dimension of the environmental impacts of the projects carried out are of an intensity never before experienced in the region. The consequences resemble the scenario described by [31] regarding the impact of the Industrial Revolution on the English fields and society. The intensity and frequency of the changes imposed by the submission of peasant social logics to the market have profound similarities with the impacts promoted by IIRSA in the Amazon region: Traditional communities, responsible for protecting and maintaining forest integrity, were submitted to lose their autonomy and being pushed to urban areas, with the negative social and economic consequences that such kind of migration brings to both, the forest and the cities; displacement and desolation.

The use of regional institutional arrangements biased towards interest groups promoted the environmental agenda's deconstruction. In this context, regional integration operates as a neo-developmentalist strategy in which the state acts as a coordinator of the market actions and simultaneously works in the regulatory arena by deregulating sectors in which the current and forthcoming projects are not in compliance with the environmental law. Instead of creating synergies between sustainability and regional integration, the Brazilian government began to induce actions at the regional level that bypassed regulatory incompatibilities between pan-amazon countries. The general consequence for the region was the generalization of a shared approach to environmental legislation to facilitate the integration in the supranational perspective, but blind towards the many international commitments made towards the amazon and its global environmental importance. IIRSA lost its fundamental principles and was converted into a framework to carry out environmentally unsustainable projects, with negative consequences for the governability of the amazon countries where it was implemented.

Acknowledgements

We want to express our gratitude to the Erasmus+ program which facilitated the cooperation between the researchers from Brazil and UK, as well as all the institutions who facilitated access to the documentation of the IIRSA development initiative.

Conflict of interest

The authors declare no conflict of interest.

Author details

Pedro Pablo Cardoso Castro[1*] and Nirvia Ravena[2]

1 Leeds Beckett University, Leeds, UK

2 Federal University of Para, Belem, Brazil

*Address all correspondence to: p.p.cardoso-castro@leedsbeckett.ac.uk

References

[1] Couto LO. Horizonte Regional do Brasil e a Construção da América Do Sul. Revista Brasileira Política. 2007. 50: 159-176. DOI: 10.1590/S0034-73292007000100009

[2] Caballero S. Brasil y La Región: Una Potencia Emergente y La Integración Regional Sudamericana. Revista Brasileira de Política Internacional. 2011. 54:158-172. DOI: 10.1590/S0034-73292011000200008

[3] Quintanar S, Lopez RO. Plano de Ação para a Integração da Infraestrutura Regional Sul americana (IIRSA): Oportunidades e Riscos. Seu significado para o Brasil e a Argentina. Revista Brasileira de Política Internacional, 2003. 46: 213-221. DOI: /10.1590/S0034-73292003000100011

[4] Pecequillo C. A Política Externa do Brasil no Século XXI: Os Eixos Combinados de Cooperação Horizontal e Vertical. Revista Brasileira de Política Internacional. 2008. 51: 136-153. DOI: 10.1590/S0034-73292008000200009

[5] IIRSA. COSIPLAN: From old to new developmentalism in Latin America. In: Ocampo, J. A. and Ros, J. Editors. The Oxford Handbook of Latin American Economics, Oxford University Press, London and New York. 2015. pp. 108-129. Available at: http://www.iirsa.org/Event/Detail?Id=145 (Accessed: 20th August, 2020).

[6] Malamud A. A leader with our followers? The growing divergence between the regional and global performance of Brazilian foreign policy". Latin American Politics and Society. 2011. 53: 1-24.

[7] Santos FL. Neo-development of underdevelopment: Brazil and the political economy of South American integration under the Workers' Party". Globalizations. 2018. DOI: 10.1080/14747731.2018.1479016

[8] Bresser-Pereira L. O Novo Desenvolvimentismo.", Folha de São Paulo. 2004. available at: http://www.bresserpereira.org.br/articles/2004/412.NovoDesenvolvimentismo.pdf .

[9] Bresser-Pereira L. Developing Brazil: overcoming the failure of the Washington Consensus. 1nd ed. Lynne Rienner Publishers, Boulder, CO. 2009.

[10] Bresser-Pereira L. From old to new developmentalism in Latin America. In: José Antonio Ocampo and Jaime Ros. Editors., The Oxford Handbook of Latin American Economics. 1ed.Oxford: Oxford University Press, London. 2011. 108-129. DOI: 10.1093/oxfordhb/9780199571048.013.0005

[11] Bresser-Pereira L. Reflexões sobre o Novo Desenvolvimentismo e o Desenvolvimentismo Clássico. Brazilian Journal of Political Economy. 2016. 36: 237-265. DOI: 10.1590/0101-31572015v36n02a01

[12] Sicsu J, Paula L, Michel R. Introdução, Novo-desenvolvimentismo: Um projeto nacional de crescimento com equidade social. Jornal dos Economistas. 2005. 186: 3-5.

[13] Sicsu J, Paula L, Michel R. Por que um novo-desenvolvimentismo?" Revista de Economia Política. 2007. 27: 507-524. DOI: 10.1590/S0101-31572007000400001

[14] Schutte G. Neodesenvolvimentismo e a busca de uma nova inserção internacional. Revista Brasileira de Estratégia e Relações Internacionais. 2012. 1: 61-97.

[15] Wade R. The Developmental State: Dead or Alive? Development and Change. 2018. 49: 518-546. DOI: 10.1111/dech.12381

[16] Paiva-Silva J. Pragmatism as a pillar of the New Developmentalism. Brazil. J.

Polit. Econ. [online]. 2020. 40: 376-397. DOI: 10.1590/0101-31572020-3099

[17] Morais L, Saad-Filho A. Da economia política à política econômica: o novo-desenvolvimentismo e o governo Lula. Revista de Economia Política. 2011. 31: 507-527.

[18] Carrillo I. The New Developmentalism and the Challenges to Long-Term Stability in Brazil. Latin American Perspectives. 2014. 41: 59-74. DOI: 10.1177/0094582X14543791

[19] Ravena N, Texeira E. Usina de Belo Monte: quando o desenvolvimento viola direitos. Ponto de Vista. 2010. 10: 1-16.

[20] Vergara-Camus L, Kay C. The agrarian political economy of left-wing governments in Latin America: Agribusiness, peasants, and the limits of neo-developmentalism. Journal of Agrarian Change. 2017. 17: 415-437. DOI: 10.1111/joac.12216

[21] Sant'Anna F. Inequality and water in the Amazon Basin: the difficulties of cooperation and sustainable management. Kieler Geographische Schriften 2012. 123: 237-261.

[22] Ravena N. Canete, R. Reflexões sobre a Integração Pan-Amazônica: O Papel da Organização do Tratado de Cooperação Amazônica (OTCA) na Regulação da ÁGUA. Revista Brasileira de Estudos Urbanos e Regionais. 2007. 9: 131-144. DOI: 10.22296/2317-1529.2007v9n1p131

[23] Malamud A, Rodriguez J. Com um Pé na Região e outro no Mundo: O Dualismo Crescente da Política Externa Brasileira". Estudos Internacionais. 2013. 1: 167-183.

[24] UNCTAD (2011), Word Investment Report, Geneva, SWI, Available at: https://unctad.org/en/PublicationsLibrary/wir2011_en.pdf (Accessed: 10th August, 2020).

[25] Roman, M. The Implementation of International Regimes. The case of the Amazonian Cooperation Treaty. Uppsala, Sweden, SE. 1998.

[26] Soares de Lima M. A política externa brasileira e os desafios da cooperação Sul-Sul. Revista Brasileira de Política Internacional. 2005. 48 (1):24-59.

[27] Silva L, Ravena N. A Padronização das Políticas de Desenvolvimento na Amazônia. Revista Desenvolvimento Regional. 2012. 17: 168-191.

[28] Neves E. Institutions and environmental governance in Brazil: The local government's perspective. Revista de Economia Contemporânea. 2016. 20: 492-516.

[29] Comegna M, Luchino M. Novas territorialidades e conflitos na Amazônia: a IIRSA e o Eixo Peru-Brasil-Bolívia. Paper presented at the Actas del XI Coloquio Internacional de Geocrítica, Universidad de Buenos Aires, Buenos Aires, AR. 2010. Available at: http://www.filo.uba.ar/contenidos/investigacion/institutos/geo/geocritica2010/672.htm (Accessed: 10th February, 2020).

[30] Costa A, Ravena, N Oliveira, I. Voz Coletiva para Mobilização pela Amazônia: @ XINGUVIVO E BELO MONTE. Contemporanea. 2015. Vol. 13, pp. 542-565. Available at: https://portalseer.ufba.br/index.php/contemporaneaposcom/article/view/14614/10886.UFBA (Accessed: 15th June 2020).

[31] Polanyi K. A grande transformação: as origens de nossa época. Fanny Wrabel. 1980.

Chapter 11

Social Changes in the Peruvian Amazon Due to Foreign Influence

Zoran Stiperski and Tomica Hruška

Abstract

The prehistoric Amazon had low numbers of hunter-gatherers due to poor soil and harsh landscape conditions, due to which it was not able to support advanced cultures. The arrival of Christian missionaries, oil companies, and farmers changed the lifestyle of a specific portion of the population, although some indigenous groups still avoid contact with the outside world. Missionaries stimulated changes in the indigenous medical-religious-political systems. In the Peruvian Amazon, the local government is too weak to carry out the usual functions of the state, and therefore oil companies have replaced the state in terms of various functions such as employment, building wells for the drinking water, healthcare, donation of electric generators, and aircraft transport of local indigenous authorities to meetings in Iquitos or Lima. The policies of the national government are turning the Peruvian Amazon into a productive area and are exploiting its natural raw materials. In modernising the Amazon region, however, the world is permanently and irreparably losing valuable knowledge regarding the nature of tropical areas.

Keywords: Peruvian Amazon, foreign influence, indigenous people, missionairies, oil industry, plantation agriculture, deforestation, loss of knowledge regarding the tropics

1. Introduction

Migrations of people, especially of the intense variety, bring change to societies and landscapes. The Amazon has historically been sparsely settled, mostly by hunter-gatherers and primitive farmers. Its settlements are small and spread far and wide throughout the vast forests. The largest South American cultures and empires arose in the Andes, along the Pacific Ocean coast, and in the lowlands in the southern part of the continent, while the Amazon was bypassed. Numerous researchers have searched the jungle for signs of highly-developed Amazonian civilisations, but none succeeded in finding anything—their attempts have resulted rather in great stories of adventure and valuable ethnic documentation. Franciscan monks founded a church and monastery in Santa Rosa de Ocopa in the Peruvian Andes 300 years ago to serve as a missionary school and destination for missionary work in the Amazon. During the 20th century, companies and states "discovered" the Amazon as a new area of development. This primarily meant exploitation of natural resources such as oil, natural gas, hydro energy, wood, and agriculture. Today, road networks are spreading along with pipelines, and harbours and airports are being built. Along with the development of business, workers from the Andes and other parts of the world are flocking to the Amazon. Concurrently, indigenous peoples are losing their territory and strong deforestation processes are underway. The world also regards the Amazon in ecological terms, referring to it as the "lungs of the world". Numerous politicians of Amazonian states look at the jungle in terms of what it could mean for economic development of their respective states. Many local politicians think that what is happening to the Amazon now is what European settlers did to both Europe and North America over the last several centuries: cut down the forests; developed agricultural land; opened new business opportunities; and spread their civilisation.

In this chapter, our goal is to confirm the reasons for the arrival of settlers in the Amazon, and the significance of individual activities and how they transform nature and society. Special emphasis is placed on logging and the widespread process of deforestation, whereby we are interested in the consequences of cutting down the forest from various aspects: the effectiveness of agricultural activity; maintenance of biological diversity; and the influence of climate change on indigenous communities. Special attention is given to the adoption of the principle of market exchange, i.e. the exchange of money for goods/services, in indigenous communities, as well as key related sociological changes within indigenous societies themselves. Important landscape changes in the Amazon often happen due to deforestation, but in this chapter we explore other types of landscape change that are rarely mentioned in scientific and profession literature: the emergence and spread of networks of modern settlements that are planned in a completely different way than traditional indigenous settlements. In our analysis of modern plans of the Peruvian government, as well as similar projects of Peruvian and foreign companies, we touch on both the strategic and geopolitical dimensions of exploitation of the Amazon. Finally, we also cover the vanishing Amazon forests and indigenous way of life, and with them knowledge regarding the tropics and the Amazon itself—a great loss for the entire world.

2. Study area

The territory of Peru can be divided into three parts: the Pacific coast (11%); the Andes (29%), and the Amazon (60%). The area of the Pacific coast (slope) is home to 65% of the population, while only 5% reside in Amazonian Peru (roughly 1.6 million). The Peruvian Amazon can be divided into three main landscape wholes: (1) humid, weak drainage areas; (2) drainage areas; and (3) high altitude areas.

1. Humid, weak drainage areas stretch from east of the Ucayali River and the eastern slopes of the Andes, upriver of Atalaya and into the northern Peruvian Amazon. Here the tropical rainforest (**Figure 1**) is dense and impenetrable, and the area is home to traditional agriculture-practicing tribes such as the Shipibo.

2. Drainage areas stretch from the eastern slopes of the Andes, at an altitude of 200–400 m a.s.l. to west of the upper and middle courses of the Ucayali River. The soil here is easier for hunters to walk on, and better for agriculture and general health. This area is inhabited by agricultural-practicing tribes such as the Ashaninka.

3. High altitude areas are Amazonian drainage areas of low hills, slopes of the Andes below 2000 m a.s.l. and plateaus such as El Gran Pajonal, which extends at 1000 m a.s.l. westwards from the point where the Tambo River flows into the Ucayali River (**Figure 2**).

Figure 1.
Tropical rainforest in a drainage area in the Peruvian Amazon along the Tambo River. Picture by the author Zoran Stiperski.

Figure 2.
The Ucayali River near Atalaya. Picture by the author Zoran Stiperski.

3. Settlement

Numerous archaeological papers describe how the prehistoric Amazon was sparsely settled by hunter-gatherers because of its poor soil, difficult landscape characteristics, and generally difficult living conditions, which limited population growth and the development of more advanced civilisations [1–5]. Past research of the settlement structure indicated how the permanent population was settled along the main rivers [6, 7] due the abundance of fish, which served as a source of animal protein [8]. An additional reason for the presence of population along the main rivers was fertile soil [9]. Archaeological data show that the contemporary population of the Amazon is smaller than that of the pre-Colombian era [10–15]. Certain researchers also claim that the population of the Amazon was ten times greater prior to the arrival of Europeans than it is today [16]. The pre-Colombian Amazon was the home of peoples who developed complex forms of social organisations and familial structures [17–40]. Currently, the notion among most archaeologists is that natural limitations preventing the development of agriculture are responsible for the lack of highly-developed civilisations [41], along with insufficient settlement.

The population of the Peruvian Amazon is a complex mix of different peoples and ethnicities: 64 indigenous nations speaking languages from 16 different language families [42], numbering roughly 333,000 according to data from 2007, i.e. roughly 21% of the population of the Peruvian Amazon. The most numerous indigenous tribes in the Peruvian Amazon have up to 90,000 members, while the smallest nations in remote parts of the forest have as few as 100. Even such small groups have the characteristics of a nation, as they speak their own language. Most such smaller tribes live deep in the forest, and are often nomadic. The two most important autochthonous Amazonian ethnic groups are the Ashaninka (88,700 members) (**Figures 3** and **4**), and the Aguaruna (55,400 members), followed by the Shipibo and the Chayahuita (both nations have more than 20,000 members), and the Quechuas and the Cocama (each nation has more than 10,000 members). The Peruvian Amazon is a massive area of new colonisation and numerous newcomers from the Peruvian Andes that have come to make a living. These newcomers are not indigenous Amazonian peoples, rather people from the Andes, Pacific coast, or other parts of the world. An important part of the colonisation is led by Peruvian and foreign companies and their respective enterprises. Highly-educated workers from these companies are newcomers to the area and are, as a rule, temporary residents of the Amazon. A smaller share of Peruvians have been sent by the state to work as teachers, doctors, and the like, and come from the Andes or Pacific coast. Missionaries also make up a share of the newcomers.

The main push factors on the Pacific coast and in the Andes are the lack of agricultural land, high poverty rates, and poor prospects for employment, while

Figure 3.
The indigenous, traditional village Buenos Aires along the Tambo River. Picture by the author Zoran Stiperski.

Figure 4.
A stilt house in an indigenous settlement along the Tambo River. Picture by the author Zoran Stiperski.

the main pull factors in the Amazon are cheap and available land, ability to gain ownership of land, perception of economic "climbing", lack of workforce, support from credit programs and tax easements, spread of road infrastructure, and general security [43].

In this vast area, there is a strong process of urbanisation underway (**Figure 5**). In the Loreto Region and its capital Iquitos (380,000 residents), which is also the largest city in the Peruvian Amazon, a higher share of urban population in relation to rural was first documented in the early 1970s; this trend has continued and in 2017 there were 600,000 urban residents and 280,000 rural residents in the region. Iquitos is the largest city in the world that has no road connections to the outside,

Figure 5.
The City of Atalaya. Picture by the author Zoran Stiperski.

transit takes place rather by boat and airplane. Likewise, in the Ucayala Region there are 400,000 urban and 100,000 rural resudents (2017). A similar ratio of urban-rural population is also found in the Madre de Dios and San Martin regions [43]. Urbanisation is linked to the development of certain professions and markets, as well as with new settlers arriving in the Amazon.

4. The arrival of Europeans

4.1 Missionaries

The Spanish put the Peruvian Amazon on the back burner while they were spreading their empire. The Pacific coastal desert area and the Andes were home to advanced indigenous civilisations, while the Amazon showed few such signs. There were researchers who believed that remains of highly-developed indigenous civilisations were hidden in the forest, but no archaeological remains have been found to prove their theories. Near Atalaya, at the headwaters of the Ucayali River, their might be evidence of the remnants of the Inca civilisation in the Amazonian jungle (according to some Peruvian experts) in popular hunting area for local indigenous people. The place in question is called Canuja, and it is a natural stone formation where people carved the likeness of some large tiger-like cat at some point in the distant past, long removed from common memory (**Figure 6**).

The first Europeans that came to the Peruvian Amazon were missionaries. Franciscan monks built a church and monastery at the beginning of the 18th century in Santa Rosa de Ocopa, near the city Huancayo. Santa Rosa de Ocopa became the base for missionaries who were active in what would come to be called the Peruvian Amazon (**Figure 7**), but also in parts of Brazil, Bolivia, and Ecuador. Iquitos, the largest city in the Peruvian Amazon, was founded by Jesuit missionaries from Quito in the mid-18th century. Evangelical missionaries arrived in the territory of the Piaroa people during the 1940s [16]. The missionaries set into motion

Figure 6.
Canuja stone monoliths are, according to some Peruvian experts, the remains of the Inca civilization. Picture by the author Zoran Stiperski.

Figure 7.
The interior of a Roman Catholic church in an indigenous village in the Peruvian Amazon. Picture by the author Zoran Stiperski.

radical changes in the indigenous medicinal-religious-political systems [44]. The contemporary influence of missionaries can be seen in the founding of universities and radio stations for broadcasting agricultural advice and farm reports [45]. There are still, however, indigenous tribes that avoid all contact with the outside world [45]. The missionary centre Puerto Ocopa on the Perene River has a church, a youth home, and a school. Until a few decades ago, local shamans would, from time to time, declare that a devil had entered into some poor child, who must be banished from the village into the jungle. Parents were obligated to accept the word of the tribal shaman, but often couldn't bring themselves to leave their child to the mercy of the jungle. Instead, they left them with the Franciscans of the missionary centre where they got a roof over their heads and education. The first encounter between missionaries and the Huaorani people in eastern Ecuador was during the 1950s [46], and the arrival of missionaries and oil companies started a series of changes in the lifestyle of a significant share of the Huaorani population. Some of them now live in newly-founded cities, while some live along rivers, far from roads [45].

4.2 Oil companies

In 1939, oil extraction in the Peruvian Amazon began [47]. An intense era of oil hunting took place in the 1970s [48]. Demand and high prices of oil spurred a new boom in oil extraction in the Peruvian Amazon. Today, oil and natural gas extraction is allowed throughout the Peruvian Amazon, except for within the boundaries of national parks and reserves that make up roughly 10% of the total area of Amazonian Peru. In neighbouring Brazil, the population grew by 10.9% from 2000 to 2010 (from 170 million to 191 million), while energy demands over the same period grew by 40.7% [50]. Hydro energy is the dominant energy source in Brazil, but recently-discovered oil and natural gas are also important sources of energy [50].

Lack of sufficient involvement on the part of state agencies and other institutions has led to local communities becoming dependent on oil companies [49]. The Achuar community, for example, is heavily dependent on the Pluspetrol oil company. Oil companies have replaced the state in various fields such as employment, construction of drinking water wells, health care centres, donations of electric generators, autostop, air transport for local indigenous authorities to meetings in Iquitos or Lima, and emergency medicine services [50, 51].

Via investment, the oil industry has sped up the process of migration towards areas with western services and goods, which are often located on the edge of traditional indigenous territory [16]. Illegal gold mining, such as the mine in the Madre de Dios region with its 30,000 workers [52], also attracts newcomers to the Peruvian Amazon. Connections to global networks are made with new roads, rail lines, harbours, and communications technology [53], much of which is built for the needs of the oil industry and agricultural activities (ranches, plantations). Activities such as mining, oil and natural gas extraction, and plantation agriculture in Amazonian states—especially Brazil and Peru—have become areas of interest for investment from Chinese state companies. In order to exploit natural wealth, large infrastructural projects like the construction of a highway that would link Brazil to the Pacific Ocean via Peru have been announced [54].

4.3 Plantation farmers and ranchers

Bosquesino (forest man) is a name for someone who lives off the forest in rural parts of the Amazon. It is interesting that indigenous Amazonians have named all rivers, creeks, and water sources, but rarely have names for mountains or large hills—waterways are clearly more important to them than mountains. A Bosquesino hunts, fishes, gathers fruit, plants, minerals, nuts, honey, insects, various water creatures, crabs, lizards, and amphibians. The difference between a farmer and a Bosquesino is that a farmer practices cultivation and uses less land, while a Bosquesino does not only grow food, they also hunt and gather over a much wider area [55]. Alluvial plains, which make up 7% of the area of the Amazon [55, 56], are desirable for cultivation agriculture (horticulture, monoculture), while flood plains and slopes are not. There are exceptions to this such as peanut, rice, corn, and cassava, which are sown in the sand in flood plains along rivers in the dry season.

Agricultural activities are the strongest factor of landscape and social change in the Amazon. Many workers settled the area to extract latex from caoutchouc trees for the rubber industry, creating a vast amount of wealth. This turned parts of the Amazon into a wasteland, useable to indigenous communities, and led to a general cultural and social transformation. The main centres of the rubber boom in the Amazon were Manaus and Belem in Brazil, and Iquitos in Peru. Numerous settlers arrived in Iquitos between 1880 and 1914 to work in the rubber industry.

Traditional Amazonian communities are small and they depend on hunting, fishing, gathering, and primitive agriculture. In various Amazonian communities, the bulk of households practice agriculture (84–99%), fishing (61–91%), hunting (4–25%), aqua culture (14–41%), and gathering (11–67%) [57]. Cassava is the main crop for many Amazonian communities, regardless of whether they are indigenous or descendants of European colonists [57]. Indigenous people along the Tambo River, downriver from Atalaya, practice hunting and fishing, and cultivate only bananas—they don't practice any sort of gardening. The last 50 years have seen major changes in food production: rice and jute cultures became common at the beginning of the 1960s, spreading throughout the 1970s and 1980s until the beginning of the 1990s. Production fell drastically thereafter as a consequence of the closure of the Banco Agrario (agricultural bank) and the lack of available credit for rice and jute farmers while Alberto Fujimori was president of Peru [55]. Accordingly, indigenous people from the area of Ampiyacu stopped sowing rice and jute. The fall of the Banco Agrario resulted in a drastic reduction in agricultural production in the Amazon, but reasonable prices for rice were maintained in Iquitos' markets, so production in the area around the city continued. Rice production grew significantly in 2000 when Caritas Internationalis offered short-term loans to small businesses [55]. The example of rice, in this case, is a good example of the vulnerability of local producers vis-à-vis availability of financing.

Traditional shifting agricultural cultivation systems practiced by indigenous peoples are sustainable long-term [58] in contrast to typical farm agriculture [59]. Shifting agriculture and cattle ranches are responsible for 80–85% of deforestation in the Amazon [60] (**Figure 8**). During the process of making new areas for agriculture at the expense of tropical rainforest, we must bear in mind the high value of the Amazon in terms of bio-diversity, which far exceeds regions with temperate climates. In just 100 ha of the study area in the Peruvian Amazon there are more species than in most individual states of the U.S.A. [61]. This means that roughly half of the planet's species are found in tropical forests. Areas along arterial thoroughfares are more exposed to logging than remote areas, and agricultural activity along trans-Amazon highways is shifting toward herding and multi-year crops [38]. The average cattle rancher from a dynamic village along a trans-Amazon highway covers 20 times more area than a family with a smaller operation [62]. Both Indigenous people and settlers cleared sections of the Amazonian forests during the 20th century [63]. Some research has indicated a negative link between exploitation of the forest and wealth, such that richer families depend less on extracting raw material from the forest [64, 65]. There are key differences in the forest clearing techniques of Bosquesinos and plantation farmers. Typically, plantation farmers clear larger areas than Bosquesinos, who only clear enough to allow sunlight to cover the area they want to plant (chakra). Plantation farmers cut trees down and dig up their roots, which makes the soil less fertile, while Bosquesinos leave the roots. Chakras are used for three years and left for nature to reclaim thereafter (**Figure 9**). Chakras are typically used for one banana harvest and two jute harvests, while other fruit can be harvested longer. Accordingly, this is a form of horticulture with quick and easy recovery, while the typical plantation practices horticulture without land renewal because the degradation of the

Figure 8.
An area of forest that has been cut down for the purposes of cattle herding along the Perene River. Picture by the author Zoran Stiperski.

Figure 9.
Tropical rainforest renewal in the area of a fallow chakra. Picture by the author Zoran Stiperski.

forest around the plantation continues. One of the consequences of deforestation is a reduction in precipitation [52]. Climate change has led to increasing average temperatures, larger and more frequent fires during the dry season, more frequent flooding along rivers, and more landslides. In terms of agricultural production amid current climate change trends, there has been a significant reduction in corn and coffee harvests.

4.3.1 Logging/forest clearing

Traditional communities like the Ashaninka depend on the land to survive. The greatest threat to traditional ways of life in the Peruvian Amazon comes from farmers who clear the forest to make arable land [58]. Cleared land is quickly exhausted, and cannot be used for longer than two to three years [66]. Ranching, commercial logging, and agricultural activities are the main reasons behind forest clearing in the tropical Amazon, but an important share of deforestation is linked to infrastructure and roadbuilding investment (**Figure 10**). Deforestation also contributes to uncertain ownership rights, because it encourages the transformation of forests into agricultural and pastoral areas [67]. Poor forest management also encourages logging and export of wood [68, 69].

In total, between 8.9 and 10.5 million ha of Peruvian forests, i.e. 11.3–13.4%, have been cut down [42]. Deforestation in the Peruvian Amazon was below average in relation to other Amazonian states until 2012, but thereafter the rate of deforestation doubled [42]. In the period of 1988 to 2007, an average of 18,000 km^2 of forest were cut down yearly in the neighbouring Brazilian Amazon [67]. Only 5.6% of the original forest areas in the Bolivian Amazon had been cut down for agricultural use up to 1990, but the rate of deforestation grew afterwards [70]. Lower deforestation rates in the Bolivian Amazon are the result of the fact that settlers gravitate toward cities and areas of coca production, rather than toward the Amazon [70]. A positive and significant link has been established (from 0.35 to 0.47) between household income and forest clearing among poor communities in rural areas of the Amazon. The forests in Bolivia's Tsimane area are also faced with growing pressure

Figure 10.
The road between Puerto Ocopa and Mazamari on the slopes of the Andes in the Peruvian Amazon. Picture by the author Zoran Stiperski.

from indigenous peoples, not just from settlers [63]. Research has shown that the commercialisation of wood and forest products creates good opportunities for local stakeholders and contributes to forest preservation [71].

4.4 Shotguns

Game is an important dietary element for traditional communities [72]. Widespread hunting throughout the Amazon threatens the survival of large primates (mammals) and other vertebrates. Shotguns have caused a much greater reduction in the numbers of hunted animals than traditional hunting tools (bow and arrow) of the indigenous Machiguenga, for example. This provides evidence of why pre-Colombian peoples didn't wipe out all the large animals that they hunted—bows and arrows are much less effective than shotguns. Furthermore, spider monkeys are quickly being driven toward extinction by shotguns, even around smaller settlements [73]. Hunting with shotguns strips areas of wildlife and consequentially hurts the local population because they end up stripping an ever-larger area of its normal fauna [73]. The usefulness of shotguns in hunting is short-term [74]. The Machiguenga have overhunted their lands near the Urubamba River, where the shotgun is the main means of hunting. This provoked a response from the Peruvian government, which has implemented aquaculture on a small scale to preserve local settlements and reduce the need for hunting [73].

4.5 Tourism

The economic effect of strong tourism development is rather small for locals because a large share of the earnings goes to external organisers, who pay indigenous folk meagrely [75]. There are only three areas in the Peruvian Amazon that are seriously involved in tourism: two resorts in the jungle and the City of Iquitos. Manu National Park is touristically attractive and is found on the UNESCO list of Natural Heritage. The Park, however, is faced with rapid population growth as a consequence of settlers from the isolated Machiguenga nation. The consequence of this overcrowding is the erosion of bio-diversity and emigration of Europeanised Machiguengas [73, 76]. From this example, we can see that it isn't always clear whether people living in national parks are allies or threats to bio-diversity. 'The forest is full of people and empty of fauna' insist conservationists of bio-diversity of tropical areas who argue that overpopulation or high-volume tourism is a threat to national parks. [77, 78].

In many areas, tourism is not yet practiced. After I returned from the Peruvian Amazon, tourist agents from Lima asked me what it was like—for them it was a totally unknown area. Outside of established paths and large cities, indigenous people are wary of foreigners—particularly white people, who they often associate with criminals, terrorists, or dubious characters. It is not rare for indigenous people to shoot at unknown foreigners, as they have had bad experiences with Shining Path revolutionaries and people involved in the cocaine trade. As there is no presence of state police to protect them, indigenous peoples have to protect themselves with their own weapons and warriors. After indigenous people get to know a foreigner (even a white one), however, they become very warm, friendly, and open.

4.6 The Market

A new socio-cultural phenomenon that has been observed over the last 50 years is the specialisation of cooperatives in relation to plurality among people. The activities of the polyvalent bosquesino have begun to commercialise. The use of money has entered into indigenous communities. Huaorani from the upper Amazon trade more with money than barter around larger markets [45]. Existential self-sourced economy and sharing food is common among hunter-gatherer communities [75], but market economy is spreading into traditional areas. In some areas both market and sharing systems operate simultaneously: first comes sharing among the community and the surplus is sold on the market. The advantage of the sharing system is food and material security for those who are unable to provide for themselves. Therefore, the poor favour the sharing system in order to minimise the risk that they might go without [79], while the more affluent easily accept the market system of economy because it maximises income [80].

5. Modern vs. traditional settlements

Planimetrics of settlements are a reflection of people's ways of life and the things that they create. Planimetrics of settlements of the Peruvian Amazon show mutual differences between two basic groups of settlements: traditional settlements and modern settlements. Modern settlements have emerged in areas with numerous and varied flows of foreign influence and communication. Traditional settlements are ubiquitous throughout the Amazon, with a noticeable distance from areas of modernity (**Figures 11** and **12**). The persistence of modern settlements is a sign of the spread of new lifestyles brought by settlers, and represents a step away from the traditional lifestyle of indigenous Amazonians, which is characterised by fishing, hunting, gathering, and basic agriculture. Newcomers from outside of the Amazon make up nearly 90% of the total population of the Peruvian Amazon [81]. Settlement density of Amazonian nations is traditionally very low, due to the prevailing hunter-gatherer lifestyle.

Planimetrics of modern settlements is the same as in the rest of the West, i.e. settlements with streets, stores, parks, and squares. Streets are usually laid out in a grid, which is also typical for modern city planning. The planimetric practice

Figure 11.
A traditional indigenous settlement with stilt houses in the lowland area of the Peruvian Amazon along the Ucayali River. Picture by the author Tomica Hruška.

Figure 12.
Gathering in the "village council chamber" in Buenos Aires along the Tambo River. Picture by the author Zoran Stiperski.

of "right angle" planning was brought to Peru by the Spanish. Planimetrics of traditional settlements, however, is reminiscent of an enlarged garden with irregularly-spaced houses. The houses do not have clearly-defined yards or gardens with fences. Modern Amazonian urban settlements name and arrange their central squares according to the standard Peruvian "Plaza de Armas" pattern, which indicates the spread of Peruvian identity in the Amazon (**Figure 13**). This pattern emerged after the arrival of the Spanish in the 16th century. The first thing built in newly-founded modern Amazonian settlements was often the Plaza de Armas (**Figure 14**). Traditional settlements have no squares and have not taken up this part of Peruvian identity [82].

Figure 13.
Business buildings on the main square (Plaza de Armas) in Atalaya. Picture by the author Zoran Stiperski.

Figure 14.
The main square (Plaza de Armas) in the design of the emerging settlement of Maldonadillo, being built for the needs of local plantation farmers. Picture by the author Zoran Stiperski.

Larger settlements, in terms of population, are modern, while smaller settlements can be either modern or traditional. Hunters, fishermen, gatherers, and traditional farmers live in traditional settlements. This sort of settlement only works in areas with very low population density, as this kind of lifestyle demands large areas of land, forest, and water resources per resident. Traditional settlements are typically populated by indigenous folk, with very few, if any, outsiders. The few outsiders that do live in traditional settlements are usually married to an indigenous resident or are school teachers. Newcomers to the Amazon mostly live in modern settlements, but there are also numerous indigenous people who have accepted the lifestyle [82]. Modern settlements are found along transit corridors, in this case rivers and roads. Traditional settlements are found in all areas, from rivers to very remote areas. Traditional settlements that lack any functions are often semi-permanent. These are mostly newer settlements that become abandoned over time. Hunting, fishing, and gathering require mobility, and the agricultural lifestyle is sedentary. The mobility of the settlement largely depends on the shifting of large river beds, in addition to other logical factors. Settlements that are threatened by river erosion move further from the banks. Another reason for changing the location of settlements can be death within the family, old houses, and the need to create new chakras. The arrival of the first functions, foremost schools, rendered many traditional settlements permanent. With time, such traditional settlements can become modern settlements [82].

6. Modern plans of the Peruvian government

Numerous development agencies and national governments around the Amazon encourage market economy for production growth, especially of wood, while simultaneously attempting to sustain the "environment" [83–85]. The policies of national governments are often visible in incentivising the transformation of the Peruvian Amazon into productive land for national or foreign investors, to be used for the purpose of extraction of raw materials and energy, forest plantations, and the production of biofuel [49]. The vast majority of deforestation is a direct consequence of state and corporate policies, aimed to colonise and develop agriculture in the Peruvian Amazon. These policies include road building. The majority of deforestation happens within 20 km of an arterial road [42]. State planners rarely consider forest communities when creating plans that serve for the extraction of wealth for corporate use; meaning that such plans often come at the direct (or indirect) expense of forest communities. This sometimes leads to conflict in indigenous areas, and indigenous political organisations are emerging and gaining strength. The majority of deforestation (75%) happens outside of protected indigenous lands. This indicates that the protected status of indigenous lands is also an effective measure against deforestation [42].

Yearly production of oil in the Peruvian Amazon is less than 4 hours of the world's total oil consumption [49] and, even so, the damage to the Peruvian Amazon is immense: indigenous human rights; climate change; loss of biodiversity; and loss of indigenous knowledge of the Amazon itself [49]. From 1970 to 2009, 84% of the Peruvian Amazon was spoken for via contract or under negotiation, and permits were issued for oil and natural gas extraction on 55% of indigenous land in remote, pristine areas of the Peruvian Amazon [49]. Oil extraction is a threat to the lifestyle of indigenous folk. Apart from the oil industry, the construction of hydro-electric power plants and powerlines is envisioned. According to expert opinion, the projects for building new hydro-electric power plants will have a negative impact on the local population and biodiversity along the Tambo River. There are also plans to mine, improve and widen the rail and road networks, widen and maintain river channels for water transport, intensify agriculture, and increase exploitation of the forest—in this case export of wood [86]. The most damage is caused by ignoring existing laws, which leads to the negative consequences that exceed what the forest can sustain. Growth in social conflicts over human rights violations, especially of unorganised indigenous folk, is predicted [86].

The Peruvian public on the Pacific coast and in the Andes is generally not interested in and poorly informed about government project in the Amazon. Planners to not attempt to reach harmonious, just, useful, and sustainable development for all stakeholders (including indigenous people). The public sector and regional government bodies develop projects in isolation from one another, without mutual agreement. The majority of projects are consequences of inertia in public administration, which recycles old ideas from days past. Projects are mainly taken advantage of by political opportunists, construction firms, bankers, foreigners, and certain interest groups and companies.

The numerous ambitiously-planned, anticipated projects for development of the Peruvian Amazon for the period of 2009 to 2021 include 52 hydro-electric power plants, 53 oil fields, numerous oil and natural gas pipelines, 24,818 permits for new mines (extracting ore), 4486 re-built roads including 880 km of new roads and 2089 km of asphalt roads, as well as around 2000 km of railway, 4213 km of river channel works, and new plantations for the production of bio-fuel [86]. There have never been so many projects underway at once in Peru's history, but it is likely that some will not come to fruition.

The consequences will be severe and will quickly lead to drastic changes in the Peruvian Amazon. By 2041, it is likely that deforestation and degradation will come to cover at least 56% of the forest—the more pessimistic projections predict up to 91%. Only 10% of the Peruvian Amazon is protected. CO_2 emissions are expected to rise in proportion to deforestation, putting Peru in a tough position in relation to promises made to the international community. The degradation of natural ecosystems in the process of estuarine water circulation, however, is much more dangerous than rising CO_2 levels, as it will cause increasingly expressed droughts and floods, as well as increases in violent erosive phenomena (landslides, torrents, gullies) [86]. The drastic negative influence will be experience foremost by indigenous peoples living in voluntary isolation, who will likely disappear entirely. Social conflicts over violation of human rights and ownership rights are predicted to rise in the future. Five main generators of conflict are often mentioned: (1) conflicts between indigenous leaders and the national government over protection of indigenous territory; (2) conflicts between illegal miners and the greater population; (3) conflicts between illegal coca producers and narco-traffickers; (4) conflicts between indigenous people and oil companies on indigenous territory; and (5) conflicts between those who would be effected by the construction of reservoirs and supporters of hydro-electric power plant projects.

There are also geopolitical motivations found within the Peruvian government's projects, as some of them are intended to forge better connections between the Pacific and Atlantic oceans, which is especially favourable for Brazil due to easing Brazilian foreign trade, producing electricity on the slopes of the Andes, widening the markets of Brazilian firms, and the possibility of investment from the Brazilian National Bank. This all comes, however, at the expense of nature itself, Peruvians who will pay for projects that give them no direct benefit, residents of the forest, and indigenous people. The resulting climate change, reductions in biodiversity, and loss of knowledge of the Amazon itself threaten the entire world.

7. Conclusion

The main conclusion of this chapter is that foreign influences have deeply changed the landscape and local societies in the Amazon over the last 60 years. The Amazon is traditionally a sparsely-settled area, in which more-advanced urban civilisations were not able to develop due to harsh landscape conditions. Indigenous people traditionally practice hunting, fishing, gathering, and basic agriculture, as well as a communal "sharing" system in place of a modern system of monetary exchange.

Missionaries were the first Europeans to spread Christianity and alter traditional indigenous Amazonian society (from the beginning of the 18th century). Oil companies have also left their mark on the Peruvian Amazon by extracting oil and natural gas, building pipelines, employing numerous settlers from the Andes and Pacific coast, and luring indigenous folk to the foreign lifestyle. The strongest causal factor of deforestation and societal change in Peruvian Amazonian societies is modern agriculture: plantation farmers and ranchers. By 2012, a total of between 11.3% and 13.4% of the total forest area had been cut down, and this process has only strengthened since then. There is also a risk of overpopulation and over-tourism in some of the area's few tourist resorts and highly-visited areas.

Assuming the arrival of numerous immigrants, the entry of the market, and strong change in the activities of the population, we expect the emergence of new modern settlements and the concentration of the population therein. Reduction in size and disappearance of traditional settlements, and their relocation into remote protected areas away from rivers and arterial roads is also expected. Contemporary settlements in the Amazon planimetrically resemble other settlements in the West with streets and stores, in contrast to traditional indigenous settlements which resemble extended gardens with multiple dwellings. The area of foreign influence, and intense landscape and societal transformation, follows flows of communication: primarily roads and rivers used for transport.

Plans of the Peruvian government are very ambitiously laid out and envision the construction of numerous hydro-electric power plants, oil and natural gas fields, mines, roads, railways, plantations, and river channel works. There are predictions of widening deforestation and forest degradation (estimates range from 56% to 91%) of the Peruvian Amazon by 2041. In this context, it is important to stress that only 10% of the Peruvian Amazon currently has protected status.

The loss of traditional knowledge is linked to the inclusion of the indigenous population in modern market economics, along with the simultaneous exodus from existential hunter-gatherer subsistence [53]. The spread of modernity and the Europeanisation of the population of the Amazon will end up costing the world generations of indigenous knowledge regarding tropical areas.

Author details

Zoran Stiperski[1*] and Tomica Hruška[2]

1 Department of Geography, Faculty of Science, University of Zagreb, Zagreb, Croatia

2 The Franciscan Monastery of the Holy Spirit, Požega, Croatia

*Address all correspondence to: zstiper@geog.pmf.hr

References

[1] Gross D. Protein capture and cultural development in the Amazon Basin. American Anthropologist. 1975;**77**(3):526-549. DOI: https://doi.org/10.1525/aa.1975.77.3.02a00040

[2] Meggers BJ. Environmental limitation on the development of culture. American Anthropologist. 1954;**56**(5):801-824

[3] Meggers BJ. Amazonia: real or counterfeit paradise? The Review of Archaeology. 1992;**13**(2):25-40

[4] Meggers BJ. Biogeographical approaches to reconstructing the pre-history of Amazonia. Biogeographica. 1994;**70**(3):97-110

[5] Meggers BJ. Prehistoric cultural development in Amazonia: an archaeological perspective. Research and Exploration. 1994;**10**(4):398-421

[6] Denevan WM. A bluff model of riverine settlement in prehistoric Amazonia. Annals of the Association of American Geographers. 1996; 86(4):654-681. DOI: https://doi.org/10.1111/j.1467-8306.1996.tb01771.x

[7] Lombardo U, Prümers H. Pre-Columbian human occupation patterns in the eastern plains of the Llanos de Moxos. Bolivian Amazonia. Journal of Archaeological Science. 2010;**37**:1875-1885. DOI: 10.1016/j.jas.2010.02.011

[8] Carneiro R. The history of ecological interpretations of Amazonia: does Roosevelt have it right? In: Sponsel L, editor. Indigenous Peoples and the Future of Amazonia: an Ecological Anthropology of an Endangered World. Tucson: University of Arizona Press; 1995. pp. 45-70

[9] Lathrap D. The Upper Amazon – Ancient Peoples and Places. Southampton: Thames & Hudson; 1970 256 p

[10] Denevan W. The pristine myth: the landscape of the Americas in 1492. Annals of the Association of American Geographers. 1992;**82**(3):369-385. DOI: https://doi.org/10.1111/j.1467-8306.1992.tb01965.x

[11] Denevan W, Treacy J, Alcorn J, Padoch C, Denslow J, Flores S. Indigenous agroforestry in the Peruvian Amazon: Bora Indian management of swidden fallows. Interciencia. 1984;**9**(6):346-357

[12] Heckenberger MJ. The enigma of the great cities: body and State in Amazonia. Tipiti: Journal of Society for the Anthropology of Lowland South America. 2003;**1**:17-48

[13] McEwan C, Barreto C, Neves E, editors. Unknown Amazon. London: The British Museum; 2001

[14] Uhl C, Nepstad D, Buschbacher R, Clark K, Kauffman B, Subler S. Studies of ecosystem response to natural and anthropogenic disturbances provide guidelines for designing sustainable land-use systems in Amazonia. In: Anderson AB, editor. Alternatives to deforestation: steps toward sustainable use of the Amazon rain forest. New York: Columbia University Press; 1990. pp. 24-42

[15] Zent S. Historical and ethnographic ecology of the Upper Cuao River Wothiha: clues for an interpretation of native Guianese social organization [thesis]. New York: Columbia University; 1992

[16] Freire GN. Indigenous Shifting Cultivation and the New Amazonia: A Piaroa Example of Economic Articulation. Human Ecology. 2007;**35**:681-696. DOI: 10.1007/s10745-007-9120-y

[17] Dougherty B, Calandra HA. Excavaciones arqueológicas en la Loma

Alta de Casarabe, Llanos de Moxos, Departamento del Beni, Bolivia. Relaciones de la Sociedad Argentina de Antropología. 1981-1982;**14**(2):9-48

[18] Dougherty B, Calandra HA. Prehispanic human settlement in the Llanos de Mojos, Bolivia. In: Rabassa J, editor. Quaternary of South America and Antarctic Peninsula. Vol. 2. Rotterdam: Balkema; 1984. pp. 163-200

[19] Erickson C. The domesticated landscapes of the Bolivian Amazon. In: Balée W, Erickson C, editors. Time and Complexity in Historical Ecology: Studies in the Neotropical Lowlands. New York: Columbia University Press; 2006. pp. 235-278

[20] Erickson C. Amazonia: the historical ecology of a domesticated landscape. In: Silverman H, Isbell W, editors. Handbook of South American Archaeology. New York: Springer; 2008. pp. 157-183

[21] Heckenberger MJ. Manioc agriculture and sedentism in Amazonia: the Upper Xingu example. Antiquity. 1998;**72**(277):633-648. DOI: https://doi.org/10.1017/S0003598X00087056

[22] Heckenberger MJ. History, ecology, and alterity: visualizing polity in Amazonia. In: Balée W, Erickson C, editors. Time and Complexity in Historical Ecology: Studies in the Neotropical Lowlands. New York: Columbia University Press; 2006. pp. 311-340

[23] Heckenberger MJ. Amazonian mosaics: identity, interaction, and integration. In: Silverman H, Isbell W, editors. Handbook of South American Archaeology. New York: Springer; 2008. pp. 941-961

[24] Heckenberger MJ. Lost garden cities: pre-Columbian life in the Amazon. Scientific American. 2009:64-71

[25] Heckenberger M, Kuikuro A, Kuikuro UT, Russell C, Fausto C, Franchetto B. Amazonia 1492: pristine forest or cultural parkland? Science. 2003;**301**(5640):1710-1714. DOI: 10.1126/science.1086112

[26] HeckenbergerMJ, RussellJC, FaustoC, Toney JR, Schmidt MJ, Pereira E, et al. Pre-Columbianurbanism, anthropogenic landscapes, and the future of the Amazon. Science. 2008;**321**(5893):1214-1217. DOI: 10.1126/science.1159769

[27] Mann CC. Ancient Earthmovers of the Amazon. Science. 2008;**321**(5893): 1148-1148. DOI: 10.1126/science.321.5893.1148

[28] Neves EG. hanging perspectives in Amazonian archaeology. In: Politis GG, Alberti B, editors. Archaeology in Latin America. London and New York: Routledge; 1999. p. 219-249.

[29] Barreto C, McEwan C, Neves EG. Unknown Amazon – Culture in Nature in Ancient Brazil. London: British Museum; 2001. 304 p.

[30] Pärssinen M, Schaan D, Ranzi A. Pre-Columbian geometric earthworks in the upper Purús: a complex society in western Amazonia. Antiquity. 2009;**83**:1084-1095. DOI: https://doi.org/10.1017/S0003598X00099373

[31] Porras PI. Investigations at the Sangay mound complex. National Geographic Research. 1989;**5**(3):374-381

[32] Roosevelt AC. The rise and fall of the Amazon chiefdoms. L'Homme. 1993;**33**(126-128):255-283

[33] Roosevelt AC. The development of prehistoric complex societies: Amazonia, a tropical forest. Archaeological Papers of the American Anthropological Association. 1999;9(1):13-33. DOI: https://doi.org/10.1525/ap3a.1999.9.1.13

[34] Roosevelt AC. Late Amazonian. In: Peregrine PN, Ember M, editors. Encyclopedia of Prehistory. New York: Kluwer Academic/Plenum Publishers and H.R.A.F; 2002. pp. 289-293. DOI: 10.1007/978-1-4615-0525-9

[35] Salazar E. De vuelta al Sangay – investigaciones arqueológicas en el Alto Upano. Amazonía Ecuatoriana. Bulletin de l'Institut Français d'Études Andines. 1998;**27**(2):213-240

[36] Schaan DP. The nonagricultural chiefdoms of Marajó Island. In: Silverman H, Isbell W, editors. Handbook of South American Archaeology. New York: Springer; 2008. pp. 339-358. DOI: 10.1007/978-0-387-79407-5

[37] Walker J. The Llanos de Mojos. In: Silverman H, Isbell W, editors. Handbook of South American Archaeology. New York: Springer; 2008. pp. 927-939. DOI: 10.1007/978-0-387-79407-5

[38] Walker R, Perz S, Caldas M, Silva LGT. Land use and land cover change in forest frontiers: the role of household life cycles. International Regional Science Review. 2002;**25**:169-199. DOI: https://doi.org/10.1177/016001760202500202

[39] Wüst I. Continuities and discontinuities: archaeology and ethnoarchaeology in the heart of the Eastern Bororo terrritory, Mato Grosso, Brazil. Antiquity. 1998;**72**:663-675. DOI: https://doi.org/10.1017/S0003598X0008707X

[40] Wüst I, Barreto C. The ring villages of central Brazil: a challenge for Amazonian archaeology. Latin American Antiquity. 1999;**10**(1):3-23. DOI: 10.2307/972208

[41] Grann D. Izgubljeni grad Z – Priča o pogubnoj opsesiji Amazonijom. Profil: Zagreb; 2011 281 p

[42] Valqui M, Feather C, Espinosa Llanos R. Revealing the hidden – Indigenous perspectives on deforestation in the Peruvian Amazon – The causes and the solutions. Lima: AIDESEP (Inter Ethnic Association for the Development of the Peruvian Amazon) and Moreton-in-Marsh: FPP (Forest Peoples Programme); 2015. 127 p.

[43] Menton M, Cronkleton P. Migration and forests in the Peruvian Amazon. Bogor City: Center for International Forestry Research (CIFOR); 2019. p. 25. DOI: 10.17528/cifor/007305

[44] Zent S. Donde no Hay Médico: las consecuencias culturales y demográficas de la distribución desigual de los servicios édicos modernos entre los Piaroa. Antropológica. 1993;**79**:43-84

[45] Sierra R, Rodriguez F, Losos E. Forest resource use change during early market integration in tropical rain forests: the Huaorani of upper Amazonia. Ecological Economics. 1999;**30**:107-119. DOI: https://doi.org/10.1016/S0921-8009(98)00101-3

[46] Cabodevilla M. Misión Capuchina y Huaorani. In: Jass J, editor. Naúfragos del Mar Verde. La Resistencia de los Huaorani a una Integración Impuesta. Quito: Editorial Abya-Yala. 1992. pp. 74-76

[47] MEM. Anuario Estadístico de Hidrocarburos 1976-1985. In: Lima: Ministerio de Energía y Minas de Perú. 1986

[48] Finer M, Orta-Martínez M. A second hydrocarbon boom threatens the Peruvian Amazon: trends, projections, and policy implications. Environmental Research Letters. 2010;**5**(014012):1-10. DOI: 10.1088/1748-9326/5/1/014012

[49] Orta-Martínez M, Finer M. Oil frontiers and indigenous resistance in the Peruvian Amazon. Ecological Economics. 2010;**20**:207-218. DOI: 10.1016/j.ecolecon.2010.04.022

[50] Pottmaier D, Melo CR, Sartor MN, Kuester S, Amadio TM, Fernandes CAH, Marinha D, Alarcon OE. The Brazilian energy matrix: From a materials science and engineering perspective. Renewable and Sustainable Energy Reviews. 2013;19:678-691. DOI: http://dx.doi.org/10.1016/j.rser.2012.11.063

[51] Pluspetrol Norte SA. Pampa-Hermosa. Convenio suscrito entre la empr esa PLUSPETROL NORTE S.A. y la Comunidad Nativa PAMPA HERMOSA. CN Pampa Hermosa. 2003:4

[52] Hoffman M, Grigera AI. Climate Change, Migration, and Conflict in the Amazon and the Andes – Rising Tensions and Policy Options in South America. Washington, DC: Center for American Progress; 2013 110 p

[53] Kramer DB, Urquhart G, Schmitt K. Globalization and the connection of remote communities: A review of household effects and their biodiversity implications. Ecological Economics. 2009;**68**:2897-2909. DOI: 10.1016/j.ecolecon.2009.06.026

[54] Wegner RC, Fernandes MP. The Amazon and the Internationalisation of Chinese Companies. Contexto Internacional. 2018;40(2):361-385. DOI: http://dx.doi.org/10.1590/S0102-8529.2018400200006

[55] Suess JG, Mendoza NV. Sociedad Bosquesina – Tomo I: Ensayo de antropologia rural amazónica, acompañado de una crítica y propuesta alternative de proyectos de desarrollo. Iquitos: Jong DW (collaborator) Instituto de Investigaciones de la Amazonía Peruana (IIAP), Lima: Consorcio de Investigaciones Económicas y Sociales (CIES), Kyoto: Center for Integrated Area Studies, Kyoto University (CIAS); 2011. 292 p.

[56] Coomes OT, Barham BL, Takasaki Y. Targeting conservation-development initiatives n tropical forests: insights from analyses of rain forest use and economic reliance among Amazonian peasants. Ecological Economics. 2004;**51**:47-64. DOI: https://doi.org/10.1016/j.ecolecon.2004.04.004

[57] Pinton F. Traditional knowledge and areas of biodiversity in Brazilian Amazonia. International Social Science Journal. 2003;**55**(178):607-618. DOI: https://doi.org/10.1111/j.0020-8701.2003.05504010.x

[58] Brady NC. Alternatives to slash-and-burn: a global imperative. Agriculture, Ecosystems & Environment. 1996;**58**:3-11. DOI: https://doi.org/10.1016/0167-8809(96)00650-0

[59] Serrao EAS, Nepstad D, Walker R. Upland agricultural and forestry development in the Amazon: sustainability, criticality and resilience. Ecological Economics. 1996;**18**(1):3-13. DOI: https://doi.org/10.1016/0921-8009(95)00092-5

[60] Salisbury DS, Schmink M. Cows versus rubber: Changing livelihoods among Amazonian extractivists. Geoforum. 2007;**38**:1233-1249. DOI: 10.1016/j.geoforum.2007.03.005

[61] Wilson EO. The current state of biological diversity. In: Wilson EO, editor. Biodiversity. Washington, DC: National Academy Press; 1988. 538 p. DOI: https://doi.org/10.17226/989

[62] Godar J. The environmental and human dimensions of frontier expansion at the Transamazon highway colonization area [thesis]. León: University of León; 2009

[63] Godoy R, Reyes-Garcia V, Vadez V, Leonard WR, Tanner S, Huanca T, et al. TAPS Bolivia Study Team. The relation between forest clearance and household income among native Amazonians: Results from the Tsimane' Amazonian panel study, Bolivia. Ecological Economics. 2009;**68**:1854-1871. DOI: 10.1016/j.ecolecon.2009.01.001

[64] Godoy R, Brokaw N, Wilkie D. The effect of income on the extraction of non-timber tropical forest products: model, hypotheses, and preliminary findings from the Sumu Indians of Nicaragua. Human Ecology. 1995;23(1):29-52. DOI: https://doi.org/10.1007/BF01190097

[65] Gavin MC, Anderson GJ. Socioeconomic predictors of forest use values in the Peruvian Amazon: A potential tool for biodiversity conservation. Ecological Economics. 2007;**60**:752-762. DOI: 10.1016/j.ecolecon.2006.01.010

[66] Sanchez PA. Properties and management of soils in the tropics. New York: Wiley and Sons; 1976. 686 p. DOI: https://doi.org/10.2307/1297555

[67] Araujo C, Araujo Bonjean C, Combes J-L, Combes Motel P, Reis EJ. Property rights and deforestation in the Brazilian Amazon. Ecological Economics. 2009;**68**:2461-2468. DOI: 10.1016/j.ecolecon.2008.12.015

[68] Pimental D, Floyd B, Teel W, Bourns J. Deforestation, biomass depletion, and land degradation: Linkages to policy reform in sub-Saharan Africa. Ithaca: Cornell University, Department of Natural. Resources. 1991

[69] Tchoungui R, Gartlan S, Mope Simo JA, Sikod F, Youmbi A, Ddjatsana M, et al. Structural adjustment and sustainable development in Cameroon. London: Overseas Development Institute; 1995 132 p

[70] Kaimowitz D, Thiele G, Pacheco P. The Effects of Structural Adjustment on Deforestation and Forest Degradation in Lowland Bolivia. World Development. 1999;**27**(3):505-520. DOI: 10.1016/S0305-750X(98)00146-6

[71] Pokorny B, Johnson J, Medina G, Hoch L. Market-based conservation of the Amazonian forests: Revisiting win-win expectations. Geoforum. 2012;**43**:387-401. DOI: 10.1016/j.geoforum.2010.08.002

[72] Peres CA, Lake IR. Extent of nontimber resource extraction in tropical forests: accessibility to game vertebrates by hunters in the Amazon basin. Conservation Biology. 2003;**17**:521-535. DOI: https://doi.org/10.1046/j.1523-1739.2003.01413.x

[73] Levi T, Shepard GH Jr, Ohl-Schacherer J, Peres CA, Yu DW. Modelling the long-term sustainability of indigenous hunting in Manu National Park, Peru: landscape-scale management implications for Amazonia. Journal of Applied Ecology. 2009;**46**:804-814. DOI: 10.1111/j.1365-2664.2009.01661.x

[74] Peres C. Effect of hunting on Western Amazonian primate communities. Biological Conservation. 1990;**54**:47-59. DOI: https://doi.org/10.1016/0006-3207(90)90041-M

[75] Franzen M, Eaves J. Effect of market access on sharing practices within two Huaorani communities. Ecological Economics. 2007;**63**:776-785. DOI: 10.1016/j.ecolecon.2007.02.001

[76] Terborgh J. Requiem for Nature. Washington, DC: Island Press; 1999. 246 p. DOI: 10.2458/v6i1.21439

[77] Redford KH. The empty forest. BioScience. 1992;**42**:412-422. DOI: 10.2307/1311860

[78] Terborgh J, van Schaik C. Minimizing species loss: the imperative of protection. In: Kramer R, van Schaik C, Johnson J, editors. Last stand: protected areas and the defense of tropical biodiversity. New York: Oxford University Press; 1997. p. 15-35.

[79] Fafchamps M. Cash crop production, food price volatility, and rural market integration in the Third World. American Journal of

Agricultural Economics. 1992;**74**:90-99. DOI: https://doi.org/10.2307/1242993

[80] Angelsen A, Kaimowitz D. Rethinking the causes of deforestation: lessons from economic models. The World Bank Research Observer. 1999;14:73-98. DOI: https://doi.org/10.1093/wbro/14.1.73

[81] Mikkelsen C, editor. The Indigenous World 2012. Copenhagen: International Work Group for Indigenous Affairs; 2012 591 p

[82] Stiperski Z, Hruška T, Heršak E, Lorber L. Planimetry of settlements as a confirmation of transformation in the Atalaya region in Peruvian Amazonia. Anthropological Notebooks. 2018;**24**(2):75-96

[83] Homma AKO. Amazônia: como aproveitar os benefícios da destruição. Estudos Avançados. 2005;19(54): 115-135. DOI: https://doi.org/10.1590/S0103-40142005000200007

[84] Almeida E, Sabogal C, Brienza S. Recuperação de áreas alteradas na Amazônia Brasileira: Experiências locais, lições aprendidas e implicações para políticas públicas. Belém: CIFOR; 2006. 204 p. DOI: https://doi.org/10.17528/cifor/002119

[85] Hoch L, Pokorny B, de Jong W. How successful is tree growing for smallholders in the Amazon? International Forestry Review. 2009;**11**(3):299-310. DOI: https://doi.org/10.1505/ifor.11.3.299

[86] Dourojeanni M, Barandiaran A, Dourojeanni D. Amazonia Peruana En 2021. Explotación de recursos naturales e infraestructuras: Qué está pasando? Qué es lo que significan para el future? Lima: ProNaturaleza – Fundación Peruana para la Conservación de la Naturaleza; 2009. 162 p.

Chapter 12

Quilombo Communities and Opportunities for Market-Driven Mechanisms for the Protection of the Amazon Forest

Anna Kanele and Pedro Pablo Cardoso Castro

Abstract

Quilombos are communities created since colonial times by emancipated African-descendants who located their free communities in the deep forest north of the Amazon in Brazil. These vulnerable communities are still oppressed and neglected. Paradoxically, their actual economic activities based on Amazon products' extraction are becoming key for protecting the rainforest in the north of Brazil (state of Para). In this chapter, the analysis of Honey production illustrates their socio-economical context, organizational capabilities, and the potential of their economic activities to preserve the rainforest's integrity. It also illustrates the challenges and limitations that they face to access markets where the value of their products' attributes is appreciated based on the profile of blue businesses (e.g., fair trade, organic, rainforest friendly).

Keywords: Quilombo, Amazon, Para, Rainforest, Honey

1. Introduction

The state of Para in the north of Brazil is part of the official Amazon region [1, 2]. Its territory is mostly composed of savannas, wetlands, and hundreds of islands in the Amazon River's delta, including the world's largest fluvial island, attributes that, together with its seasonal weather, made it iconic for the study of the Amazon region [3, 4]. In terms of population, out of the 211.38 million habitants, 0.4% are indigenous, and 2.5% Quilombola [5–7]. However, – and yet as another form of neglection and discrimination - despite being the Quilombola six times higher than the Indigenous people, this community is underrepresented in the academic literature. This fact was evidenced using Google Scholar under the search term "Indigenous Communities," obtaining 2,990,000 results. In contrast, the search term "Quilombola Communities" only showed 17,400 results in a search made in July 2020 [8, 9].

Historically, the state was occupied by Indigenous tribes. Later, with the arrival and interbreed of Portuguese colonists, different communities developed, varying on the kind of labor and stile of living they developed (e.g., riveirinhos, occupying the riversides of the Amazon delta, working on fishing and seasonal crops). Adding to the racial mixture, the import of African slaves was made early during the colonial times to substitute the decreasing population of Indigenous people.

Figure 1.
Location of the district of Santarem – where many Quilombolas are located, and ubication of the archetypical Quilombo of "Aldeia-Das-Abelhas" (modified from [1, 14]).

In this colonization process, African slaves revealed and emancipated, escaping their captors and moving from the plantations to forest areas of difficult access, funding free autonomous colonies/communities (Quilombos, inhabited by Quilombolas), in a process that lasted until the abolition of slavery in 1888 [7, 10–12]. Consequently, many villages are located along the Upper Trombetas, Erepecuru, and Cumina Rivers [13], at the north of Santarem. To consolidate the description of the numerous Quilombos in the region, in this document, we will present an archetypic - fictional - Quilombo named "Aldeia-Das-Abelhas" located in the middle of the Trombetas River (**Figure 1**).

Due to their location and the nature of their social organization, these communities developed productive activities that did not demand large areas of terrain, with positive side effects for preserving the forest, as its integrity was key for the protection of the community. Also, their use and commercialization of local products is favored by good levels of (bio)diversity, which provides a diverse range of products. In this context, ironically, the discrimination towards these communities – and consequently, their limited access to technology – impose constrains to the level of transformation of their products. Consequently, their offer to the market is composed of natural products that can fit within the categories or organic or rainforest friendly, being Honey one of those with high potential for their socio-economic inclusion and forest production.

2. The Quilombola context

From their origins, the Quilombola founded various smaller villages as it made it more difficult for the government troops to find them, making the Quilombola more flexible and mobile as they could quickly relocate individual villages if they were in danger [10]. The National Institute for Colonization and Agrarian Reform has figures from Quilombola communes regarding the number of large families and hectares used. In these figures, it is noticeable that almost the same number of families in Belem and Santarem are listed, but Belem's area is less than 10% of Santarem's area. Within this context, the "Aldeia-Das-Abelhas" composed of 70 families, would have as much land as the municipalities in Santarem, occupying 4177 ha [15]. A large part of this Quilombo can hardly afford to eat and therefore live in deplorable conditions [16]. Some houses of the Quilombo have been built provisionally from mud and bamboo [17]. Also, the village partly shows problematic sanitary conditions, with limited access to sanitation [18].

In terms of land, the Quilombola community "Barro Vermelho" illustrates these communities' general situation when only after a long struggle, they received the official status as Quilombola land [19]. The recognition of the Quilombola land is

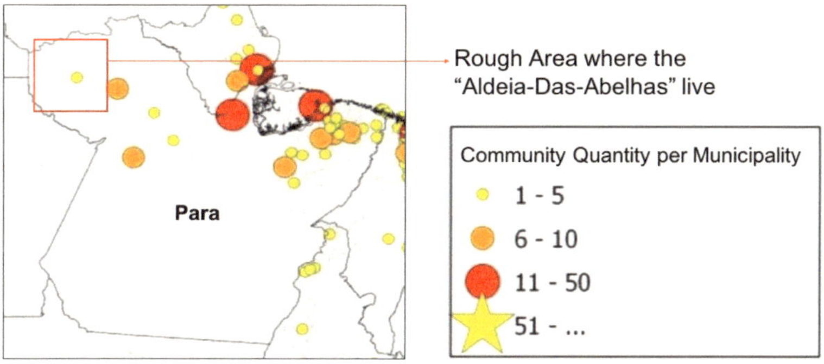

Figure 2.
Quilombola communities certified by municipalities (modified from [23]).

sometimes a very long process. Only 9.4% of the applications were recognized, and 44% of them still-open applications, which have been pending for more than ten years [6]. However, this title does not automatically imply the recognition of official land titling, as the example of the "Comunidade Quilombola do Saco Barreiro" shows [20]. The relevance of the land title comes from the fact that through the land title, the area where the Quilombola lives would be officially protected so that no clearing or mining would be allowed. According to title conservation, deforestation has decreased by 66% in areas where legal rights on the land have been granted [21]. However, Para is the only state in Brazil with more titles of Quilombola land than the land-titling processes of Quilombola lands; about a third (66 out of 181) of all pronounced Quilombola land titles fall on Para [22]. However, in the registers of CAISAN, the "Aldeia-Das-Abelhas" area has hardly been recognized as an official Quilombola community - see **Figure 2** [23]. Although the recognition of land in Para is generally relatively high, based on the data of CAISAN, for this document, we assumed that the "Aldeia-Das-Abelhas" have neither the official recognition as Quilombola or a land title.

It should be noted that even villages that have been granted the land continue to struggle for it, and their livelihood is still threatened. No one, except the government, can dispute the land after receiving the land title [17]. In this sense, Bolsonaros' administration plans to dispute these lands to introduce mining and agriculture in protected areas [21], as well as the legalization of deforestation in the Amazon. However, Bolsonaros' administration is not the first to adopt and to implement this approach. Previously, President Michel Temer abolished the Renca Reserve, which lies on the border between Amapa and Para [24]. The consequences of such systemic clearance of land by deforestation were defined by Mauricio Voivodic (executive director of WWF-Brazil) as irreversibly damaging local cultures, and it would also "... lead to an intensification of land conflicts and threats to indigenous peoples and traditional populations" [24]. Within this context of conflict between the communities protecting their land and the political and criminal interests who want to make the most lucrative use of the land and its resources in the short-term, gun violence would not deter economic interests. On the contrary, as reported by the nut collector Antonio Marcos de Lima: tractors, chainsaws, and gunfire now dominate a once quiet and peaceful jungle [25]. There has been a steady increase in direct violence against the villages and their inhabitants in the recent past [26].

3. Current situation

Bolsonaros' administration policy is to use the rainforest area to expand industrial agriculture developments, incentivizing the deforestation – legal and illegal – of the

Amazon [27]. In this context, and adding the historical process of displacement from their lands and neglection from state protection and recognition, the traditionally living people who claim their land [22] are a thorn in the side. This policy was first illustrated in Bolsonaro's election campaign in which he spoke out for the possession of weapons and against the lands claiming of the Indigenous and Quilombola [25, 28]. Bolsonaro said, "You can be sure that there will not be money for NGOs if I get there (Presidency of the Republic). If it is up to me, every citizen will have a firearm. There will not be a centimeter marked for an Indigenous reservation or a Quilombola." [29].

After winning the presidential election, Bolsonaro's policy focused on short-term profit maximization. To this end, the policy uses the high yield of rainforest land without regard to the long-term consequences for the environment and people [30]. Concerning as it is, Bolsonaros' administration continued to pursue this goal weakening the support of Indigenous and Quilombola. Accordingly, the Brazilian office FUNAI, which protects primarily the Indigenous and supports the Quilombola, has had its funds cut. Its management changed several times and was moved under the Ministry of Agriculture administration, which in line with governmental direction, has no interest in protecting the territories of traditional peoples living in the rainforest [31, 32]. Bolsonaro also tried with the appointment of Ricardo Lopes Dias (an evangelical missionary who has set himself the goal of evangelizing everyone, including the Indigenous and Quilombola) as head of FUNAI in February 2020, to weaken the social organization of Indigenous and Quilombola communities. However, in May 2020, the court ruled that the appointment of Lopes Dias was illegal because the judges have seen an apparent conflict of interest between Lopes Dias's believes and the mission of FUNAI. Consequently, Lopes Dias was dismissed with immediate effect [11]. Furthermore, as registered by Sarah Shenker (from Survival International): "It was an essential part of Bolsonaro's explicit policy to destroy the country's Indigenous peoples - to dissolve the teams protecting their territories and sell their land to loggers, miners, and ranchers." [11].

Russau describes Brazilian Indigenous/Quilombola politics as submissive assistentialist welfare state thinking and explicit paternalism, further saying that the current situation of the Brazilian Quilombola can be summarized as poverty, dependence, and exclusion [33]. These affirmations ignore - particularly for the Quilombolas – their long history as emancipated communities and their development of autonomous governance and social structures that made them self-sufficient communities. Since colonial times, they have been prosecuted, neglected, and systematically excluded from socio-economic development, and yet, able to coordinate their activities, survive and preserve their identity, culture, and traditions.

4. Quilombola production and logistics

Many Quilombola living in rural areas work on collectively-owned land with natural resources [7]. Within this form of productive organization, some Quilombos produce traditional jewelry and sell the surplus of their agricultural production to third parties [18, 34]. Their offer is mostly composed of Cassava (*Manihot esculenta*) – and domestically processed from it, farinha; raw fruits like Acai (*Euterpe oleracea*), Cupuassu (*Theobroma grandiflorum*), Nuts (*Bertholletia excelsa*); Honey, and other seasonal Amazon fruits. The product in many Quilombolas that is less dependent on the tenure of the land and seasonal changes and more suitable to the communities' sporadic mobility is Honey. Despite the popularity of products as Acai in the local (and now in the international) market, and Brazil Nuts that demand intense labor to collect, peel and prepare to present to the end consumer.

Honey production requires little labor and maintenance, low technology for the production and processing - adding value - of the final product, and can be delivered almost directly to the end consumer without processing, making it less susceptible to intermediaries' occurrence. Opposite to Acai, that needs treatment/processing to add value to the product, and refrigeration once processed. Also, due to the Acai's fragility and the lack of the communities' logistic resources, the product makes them more susceptible to depend on intermediaries to deliver to the market. It also demands a long-term settlement to develop the crop – hence, stability and clear rights on the land, if investment in adding value technologies is required. The case of Brazil Nuts demands a considerable extension of territory to provide commercial volumes of raw product, being a disadvantage, the intensive labor required to peel the Nut, and prepare it for the final consumer.

Concerning Honey production, mostly the two native bee species Melipona flavoneata and Melipona fasciculate, are used in Para for Honey production, as well as Apis mellifera [35], which is one of the most frequently used Honeybees worldwide. Of these, Melipona fasciculate has been identified as the best adapted and most productive in the Amazon context, standing out, among the native bees, for Honey production [36]. The native stingless bees are also highly adapted to "... the different soils, geographical location, good climatic conditions and the large diversity of plants and, in particular, flowering plants" [35], p. 1136. With these references, we can assume that the "Aldeia-Das-Abelhas" use Melipona fasciculate for Honey production.

Within this context, the Quilombola are partly dependent on selling their products in the nearest town or even directly in Belem, traveling by boat on a journey that can last several hours. For instance, for the "Aldeia-Das-Abelhas," Porto Trombetas is the nearest town, covering a distance of about 300 km on the Trombetas River, on a trip that can last up to 8.5 hours. Considering this logistic challenge and the characteristics summarized for the "Aldeia-Das-Abelhas," the community cannot afford a small fleet of boats to mobilize large amounts of products and keep fishing or other subsistence activities. Thus, they sell a significant part of their production to intermediaries; and trade directly in the city products that demand less labor, have more added or perceived value per unit of volume, and do not require refrigeration or any special treatment (e.g., Honey). These facts lead to the assumption that the "Aldeia-Das-Abelhas" have experience in selling to the middleman, limited experience in direct sales in the city. Therefore, their trading know-how is limited to local trade without the administrative or logistical experience for regional, national, or international trade.

As a counter-example, the "Itunuvico" historical development is presented. The "Itunuvico" started producing Honey in 2006 with nine people, and in thirteen years, they have grown to 180 small producers in Oaxaca-Mexico. The "Itunuvico" exported about 600 tons of organic Honey to Germany in 2018. The Mexican community further reports that they had their Honey certified organic through "Naturland," which has significantly increased yield. However, such certification is very costly [37]. This case also suggests developing skills in international trade and logistics - via partnerships with international ONGs, and the existence of logistics that facilitate the consolidation of production and its preparation for international distribution.

Without a reliable census in the region, due to the convenient size of the "Aldeia-Das-Abelhas," it is safe to assume that the community has around seven small producers. Each producer consists of the union of two families. About 20% of the population of the "Aldeia-Das-Abelhas" are trained in beekeeping. A total of 3 tons of Honey can be sold per year. This figure assumes that "Aldeia-Das-Abelhas" uses part of the Honey for own consumption, and they are not yet as efficient as the "Itunuvico"; as the "Itunuvico" can fall back on many years of experience, and have a different logistic context for the commercialization of their

Figure 3.
Wildfire in Para, August 2019. Each red dot represents a fire (modified from [43]).

production. Considering the production context, we can assume that the Honey from the "Aldeia-Das-Abelhas" would satisfy the certification criteria for fair trade, rainforest friendly, and organic, but such certifications do not exist.

However, there is a danger that Honey production is unsteady, and even the bees are endangered. From December 2018 to February 2019, high bee mortality (about 500 million) became known, especially in the south and adjacent regions of the southeast and central west of Brazil [38, 39]. Fast-tracked approval of almost 300 pesticides, together with the acceleration of the green frontier's expansion, might cause high bee mortality [39]. Particularly when known, these pesticides were already banned in other countries for their toxic effect on bees - among other effects on other species. In contrast, the counter-case of "Itunuvico" demonstrates that they do not have any problems with bee mortality due to strict organic certification regulations. This normative control restricts the boundaries and growth of the expansion of commercial crops, based on the trade-off that Honey production brings to the planning of the local rural economic development [37]. Adding to the threats to Honey production in the region, dos Santos and de Oliveira have found that temperatures in Para have risen in recent years and see the change in the natural landscape as the cause for the increased temperatures [40]. Clearing of the rainforests will increase the dry seasons [41], which harms the flora and, therefore, impacts the Honey production negatively [19].

Looking at the vast wildfires at the end of August 2019, although Para was severely affected, the "Aldeia-Das-Abelhas" area was not under risk (**Figure 3**). An early assessment of the impact of these recent changes in the region determined that weather conditions have become more extreme, which leads to the assumption that this trend has further increased in recent years [42]. Consequently, the wildfires do not directly threaten the "Aldeia-Das-Abelhas," but that they are already slightly affected by the extent of the rising temperatures and the resulting prolongation of the drought periods. In turn, they are being at risk of not being able to increase or maintain their Honey production.

5. The Quilombola case for honey as economic activity

Between 29 and 39% of the carbon dioxide (CO_2) released by deforestation is caused by international trade [44]. Forests are cut to make room for pastures and agricultural land. Based on the current demands of international investors [45] and in line with the Quilombola culture [25], clearing the rainforest for economic use and thus as a source of income is not an option. Additionally, Manoel Santiago (a Quilombola resident of "Pedras Negras") reports that some villages have lost their livelihoods due to the pandemic because they lived on tourism, selling homemade jewelry or traditional food [18]. Therefore, in the sense of the Quilombola, a sustainable basis of income should be created, where they can live independently from tourism.

Beekeeping offers an excellent alternative source of income compared to the sale of woodland to commercial enterprises or the commercialization of non-timber forest products. In connection with climate change, forest destruction can have devastating effects such as droughts or the deterioration of water supply [46, 47]. Accordingly, beekeeping is not only protecting the rainforest, but it is also an important income source, especially for small communities [48, 49]. The environmental richness emerging in Brazil, including pollen and nectar, is favorable to Honey production and positively impacts the socio-economy development of a state living in impoverished circumstances - considering the local low human development index of 639 [36]. In this context, the Honey production chain has a favorable trade balance based on the low capital input and the activity's fast financial profitability. This activity generates income for small and medium-sized farms [50, 51]. Successful beekeeping, as demonstrated by the communities of "Barro Vermelho" in northeast Brazil [19] and "Itunuvico" Mexico [37], both exporting their Honey to the EU; can provide effective economic integration of marginal communities to - even international - markets.

Since previous publications do not indicate which areas are particularly suitable for Quilombola beekeeping in the rainforest, a suitable area is determined based on the Quilombo's location and the conditions there. Unlike the Indigenous people, who are widely distributed in Brazil [52], the Quilombolas are relatively centralized. The Quilombola, with official titles of their land living primarily in Maranhao and Para [7]. Silva et al. see Para as "... one of Brazil's most promising regions for the production of distinct types of Honey..." [35], p. 1135. Another argument for Para is that the population density in Para (about 8.67 million inhabitants on 1,248,000 km^2) is lower than in Maranhao (about 7.10 million inhabitants on 331,982 km^2) [53] the area per inhabitant is more significant in Para and therefore offers more space for beekeeping. Para's significantly larger area is crucial concerning other farmers who may use pesticides, which may cause a significantly increased bee mortality rate [38, 39].

Furthermore, the latest figures from INPE show that Para requires especially strong support, as it accounted for around 40% (3862 km^2) of all clearings in Brazil in 2019 (**Figure 4**) [54]. The impact of forest reduction can already be seen today in Belem's increasingly extreme weather conditions [42].

An additional - but omnipresent - issue affecting the possibility of integration to the market of the Quilombola is racism. Since most Brazilians of African origin are still at the bottom of the social pyramid [12], it means being invisible, poorly paid,

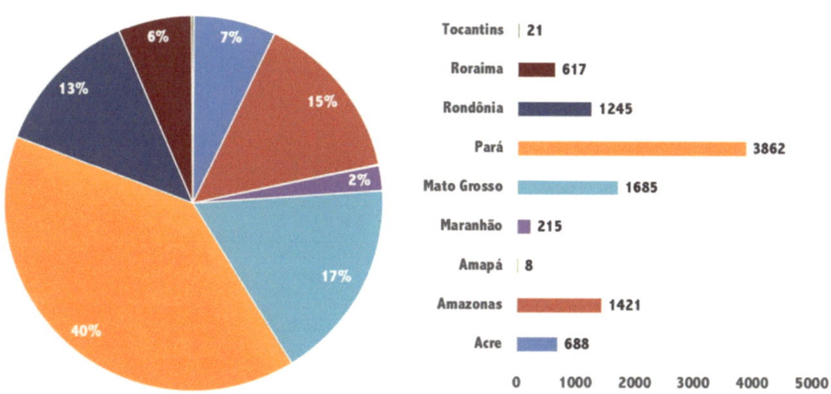

Figure 4.
Deforestation (km^2) by state in the Brazil Amazon in 2019 (from [54]).

and deprived of the most basic civil rights. To counter the racism of the Quilombola, the Brazilian government must create instruments that focus on both class and racial inequalities [55]. Brandao et al. describe that "There is a high degree of institutional distance between state and local governments and the Quilombola communities." [56], p. 10. Due to the prevailing structural racism, i.e., the social system, with its legal concepts, and its political and economic structures, causes minorities [57]. Due to the segregation often associated with this, i.e., in the context of the Quilombola, a separation of the different living spaces and the resulting social status and stigmas, e.g., black and rural poor, there are unequal life chances [34, 58]. This idea leads to the fact that it is difficult for Quilombola to sell their products on the Brazilian market. Because the population living in cities considers the people living in the Amazon (forest) as inferior [59], which can lead to a lower willingness to buy or fairly pay for their products. To bypass this barrier to access the market, one option could be to export Quilombola Honey - or any of the other traditional products such as Acai, Brazil Nuts, or fruits as Cupuassu, with similar benefits for protecting the rainforest. However, there are no reports or technical studies on assessing its viability, despite its enormous potential to provide economic inclusion and the positive consequences in protecting the rainforest and its biodiversity.

6. Analysis of the potential export market for Quilombola honey

Honey is a trendy product nowadays [60]. This trend is illustrated by the global Honey market, where Honey's international production is continuously increasing [61, 62]. Garcia (President of the International Organization of Honey Exporters)

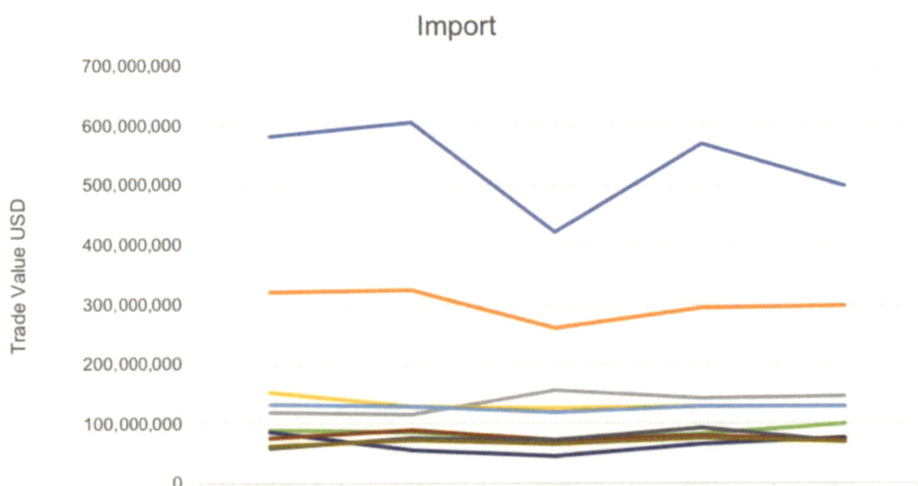

	2014	2015	2016	2017	2018
USA	582,237,980	605,042,946	422,381,731	568,734,072	497,705,051
Germany	322,024,225	324,993,295	261,881,677	292,771,537	296,993,631
Japan	120,196,038	117,643,056	157,705,833	143,010,858	145,367,382
France	153,619,194	128,473,081	126,518,253	130,491,542	129,298,226
United Kingdom	132,779,730	131,230,886	120,051,585	130,779,119	128,605,692
Italy	91,183,112	84,707,526	72,435,680	82,153,122	100,944,149
Saudi Arabia	86,191,703	56,968,565	44,448,322	67,550,125	76,976,744
Belgium	77,640,862	89,854,686	74,038,640	79,566,883	73,294,380
China	58,626,731	74,826,315	72,771,567	91,408,659	70,129,174
Spain	61,181,244	72,746,319	65,270,041	76,788,688	67,977,527

Figure 5.
Trade import value of honey 2014–2018. Top ten importing countries measured by 2018 [68].

attributes the rise in production to the increased demand. Garcia explains the increase in consumption by the increased world population and interest in natural nutrition [63]. The growing consumption of Honey can be explained to its image as a natural product that is not directly linked by society as a source of sugar and is, therefore, not associated with obesity [64]. However, Honey is rich in sugar and is not necessarily healthier [65]. Furthermore, Honey is also very popular within home medicine remedy [66, 67], contributing to increased consumption.

Many countries are no longer able to cover their domestic production needs and import Honey [63]. The trade import value from 2014 to 2018 is shown in **Figure 5** and illustrates the high demand. By far, the USA had the highest import rate in 2018, followed by Germany and Japan, France, and the UK. The chart also displays that the USA and Germany's ranking remained constant over the period shown [68].

However, to answer which export market is the most promising for the Brazilian community, no reports exist in the literature. Thus, adopting a pragmatic approach, we will concentrate on the two largest importers, the USA and Germany.

7. The US and German honey market

With imports of 197,866,579 kg, the USA offers a higher volume than Germany with 82,483,285 kg, in 2018, which makes the market opportunities in the USA appear more promising. However, comparing the revenues it in the USA (EUR 2.14/kg) was gained less than in Germany (EUR 2.85/kg) [68] and looking at the EU import prices of Brazil Honey (EUR 3.34/kg) is even higher in 2018 [69]. However, depending on the use, quality (e.g., conventional, organic), and distribution channel of the Honey, a substantial price differentiation occurs in Germany [70–72]. In contrast, the USA has little or no price differentiation between conventional and organic Honey [73]. Also, the United States Department of Agriculture shows a significant drop in organic Honey prices in 2018; it shows a price alignment within the US market [74].

In 2019, the EU concluded a free trade agreement with the Mercosur states (Argentina, Brazil, Paraguay, and Uruguay), under which around 92% of taxes on imported goods are to be abolished. Accordingly, it can be presumed that revenues from sales to the EU will increase further. However, within the agreement, the Mercosur states have committed themselves to sustainable forest management [75]. Klöckner (Federal Minister of Agriculture from Germany) emphasized under the current rainforest clearing that if Brazil does not meet its commitment to sustainable forest management, the German government will not accept it inactively [76]. Germany also supports Brazil within the framework of the "bilateral development cooperation," which (as of 20 February 2020) supports 104 projects and programs with a total of EUR 1.76 billion [77]. Nevertheless, it should be noted that the EU is in favor of stringent product safety regulations [78–85]. Thus, the barriers to selling a food product on the EU market are correspondingly high.

Guaranteeing the safety of products is also essential in the USA. The Food and Drug Administration (FDA) regulates the guidelines for food. Production facilities and dealers must be registered with the FDA [86]. The USA also has Honey specific regulations such as proper labeling "FDA-2006-P-0207" [87]. In 2014, the Agricultural Marketing Service had asked: "How a Federal standard of identity for Honey would be in the interest of consumers, the Honey industry, and US agriculture?" [88]. Moreover, it responds to various petitions that demand such clarification [88]. In principle, there is also significant economic cooperation between the USA and Brazil. The USA claims to be the second largest trading partner of Brazil, which resulted in an overall trade surplus of more than USD 20 billion on the part of the USA in 2017 [89].

On the fact sheet "Bilateral Relations between the USA and Brazil," it is pointed out that both countries have, among other things, "...respect for human rights,..." as a shared commitment and refer to the Partnership for Conservation of Amazon Biodiversity (PCAB). The PCAB intends to strengthen Brazil's vast protected area systems with USD 80 million over 2016–2024, including Indigenous territories to support the "...sustainable forest-friendly value chains..." [89]. However, because of the previously presented policy of Bolsonaro [29, 33], and Trump's trade wars, which reinforces Bolsonaro's clearing policy [90], the achievement of the project's objectives is questionable.

This paper suggests concentrating on the German market as an export country. Firstly, because Germany financially values Brazilian Honey in its conventional and organic form. Secondly, Germany's political attitude is an advantage for Quilombola, as it is the opposite of Trump's policy indirectly supporting the deforestation of the rainforest. Thirdly, Germany, together with the EU, is facilitating the export of many products through tax cuts. Lastly, concentrate on the German market on the first-hand makes it is easier to expand the export market to other EU countries such as France and Italy, which also have high imports. Therefore, the German market is for the export of Quilombola Honey more interesting.

8. Protection of the rainforest in the international trade

In the international context, Brazil is under increasing pressure concerning deforestation in the Amazon region. Since president Bolsonaro initially refused to accept the G7 countries' offer of help to fight the forest fires in the Amazon in August 2019, Brazil's governors had feared international punitive measures [76]. Within this context, afraid of the implications of the neglecting attitude from the national government and its aggressive promotion of extractivist production policies for the Amazon; Barbalho (Governor of Para) said: "I think we should now address our problems and send a signal to the world of environmental diplomacy because it is fundamental to (our) agriculture. Otherwise, we will suffer serious damage to our image, which is already a cause for concern." [76].

To understand the rate of deforestation concerning the country's total area, Brazils' forest area was around 67% of the territory in 1990 and only around 60% in 2010 [91]. According to Brazil's National Space Research Institute [92], which is publishing the official numbers of rainforest deforestation, estimates around 10,129 km^2 for the period from August 2018 to July 2019, which is 3.8% higher than the government estimated in November 2019 (**Figure 6**). This dynamic is aggravated by the fact that the rainforest's clearing will prolong the drought periods [41]. This behavior creates a vicious circle because, with increasing periods of drought, the effect of self-reinforcing forest loss would occur, estimating that the self-reinforcing forest loss could cause up to 10–13% additional rainforest death [94].

Despite these warnings, the Brazilian government does not seem to care about international pressure to reduce or even wholly refrain from clearing [76]. Evidence of this challenging - and even displacing - attitude is the fact that the Brazilian government is planning to step up its deforestation efforts. Ricardo Salles (current Brazilian Minister of the Environment) wants to take advantage of the COVIT-19 pandemic to change and simplify rainforest deforestation regulations so that mining, agroindustry, and lumberjacks would get even easier and legal access to the rainforest [30]. INPE [92] has accordingly established a record number of 1200 km^2 of deforestation from January to April 2020; this is over 50% more than in the same period in 2019.

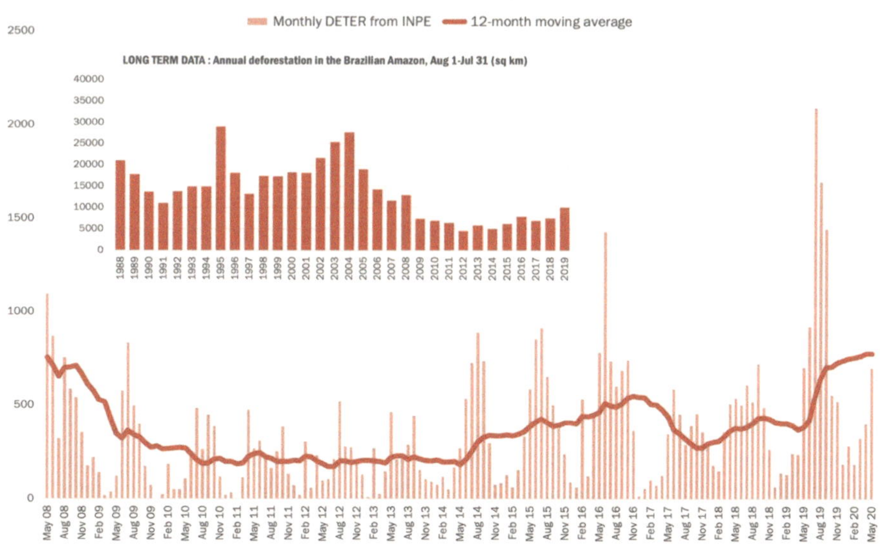

Figure 6.
Monthly deforestation alert data 2008–2020 plus yearly data 1988–2019 (from [93]). Note the peaks on the years 1995 and 2004 and their coincidence with the period of expansion of the green frontier, and the IIRSA plan for the development of big projects of infrastructure in the Amazon region.

In July 2020, 39 Brazilian companies, including some of the country's largest, have started to build up further pressure on the region - towards refraining the deforestation in the Amazon. The reason for the companies' reaction is that their international investors are threatening to withdraw their funds if the Brazilian government does not stop the rapid destruction of the country's forests [45]. Relevant to the case of the north and northeast of Brazil, Coe et al. [41] say that "...indigenous lands, sustainable-use production forests and reserves, and strictly protected forests..." are core elements of rainforest conservation. Especially Amazon's traditional peoples are rainforest protectors, as they live in and from the rainforest [25].

9. Final remarks

Some elements of the economic activity of the traditional people of the Amazon seem to be of practical use for the preservation of the rainforest. For instance, cultural practices that do not demand large extensions of territory and do not substantially affect the forest and its biodiversity. In particular, the Quilombola communities are of interest due to their focal ubication in the north of the Amazon, their family structure, and their forms of use of the land; concentrated in the production of food, the development of crops, and harvest of Amazon fruits (e.g., Acai) and the production of Honey. Out of these products, Honey seems to be the one that benefits the most from the existence of a healthy forest. Moreover, its development does not affect the forest and plays an essential role in its preservation (e.g., pollination services); Honey production does not compete with other agroforest products; it actually complements and benefits from them.

Economically, Honey has been documented in other communities as a successful product able to induce economic inclusion, even with the potential to bypass local constrains by targeting and gaining receptivity in international markets. Additionally, due to their production's biophysical characteristics, it has the potential to be certified as organic, fair trade, rainforest friendly, and many other certification schemes such as forest-friendly value chain. These certifications provide the possibility of better reception and value in some international markets. However, the development of such initiatives in Para presents some challenges: the formalization of these communities' land ownership and legal security is a significant deterrent to investment. Also, these communities' location in areas with difficult access and no infrastructure makes the logistics cost of developing a reliable provision to international markets too risky and expensive.

Politically, despite the current hostile environment under Bolsonaros' administration, it is evident that such policies are not sustainable in the middle/long-term. These practices depend on the prevalence of an ideology losing support both in the national and the international arena. Additionally, intense pressure is being made by international investors and governments from which Brazil depends to activate and support its economic development. This fact adds to the reaction of local governments in Brazil who are starting to act in opposition to Bolsonaro's policies as a consequence of the international pressure. Adding to these limitations is the fact that these communities do not have experience in international trade. However, historically they have proved to be adaptable and resilient. Also, considering the examples of other communities with similar products in Mexico and Brazil, it is reasonable to believe that this issue is not an impediment.

From the market perspective, there is an opportunity to position Honey in the international market as demand increases, particularly in the two biggest markets. In general, the case for Honey suggests that the development of commercial opportunities for traditional Amazon products, directly from the traditional people of the Amazon is not just feasible but desirable. Considering the potential for the development of socio-economic inclusion, the protection (and eventually the regeneration of) in some areas of the Amazon, and preserving the culture and biodiversity; the promotion of plans to capitalize on these productive activities should be better articulated among interested parties internationally - especially from countries interested in importing such products.

Author details

Anna Kanele and Pedro Pablo Cardoso Castro*
Leeds Beckett University, Great Britain

*Address all correspondence to: p.p.cardoso-castro@leedsbeckett.ac.uk

References

[1] Encyclopaedia Britannica (2020). Para state, Brazil. [Online]. Available from: https://www.britannica.com/place/Para-state-Brazil [Accessed 5 August 2020].

[2] IBGE (2019). Areas Territoriais. [Online]. Available from: https://www.ibge.gov.br/geociencias/organizacao-do-territorio/estrutura-territorial/15761-areas-dos-municipios.html?=&t=downloads [Accessed 5 August 2020].

[3] Pereira, L., Mendes, C., Monteiro, M. and Asp, N. (2009). Morphological and Sedimentological Changes in a Macrotidal Sand Beach in the Amazon Littoral (Vila Dos Pescadores, Pará, Brazil). Journal of Coastal Research, 56 (1), pp. 113-117, [Online]. Available from: https://www.jstor.org/stable/25737548?seq=1# metadata_info_tab_contents [Accessed 11 August 2020].

[4] Furtado Gonçalves, M., Cavalcante Blanco, C., dos Santos, V., dos Santos Oliveira, L. and Lira Pessoa, F. (2016). El Niño and La Niña and Rainfall Decrease in the state of Pará, Brazilian Amazon. Acta Scientiarum, 38 (2), pp. 209-216, [Online]. Available from: https://www.redalyc.org/pdf/3032/303245370010.pdf [Accessed 5 August 2020].

[5] IMF (2019). Population: Brazil. International Monetary Fund, October, [Online]. Available from: https://www.imf.org/external/datamapper/LP@WEO/BRA [Accessed 28 March 2020].

[6] Burns, N. (2019). Indigenous and Quilombola Land Demarcation in Brazil. Wilson Center: Brazil Institute, January, [Online]. Available from: https://www. wilsoncenter.org/article/indigenous-and-quilombola-land-demarcation-brazil [Accessed 28 March 2020].

[7] Prado, M. (2018). Quilombola Communities of Brazil. Wilson Center: Brazil Institute, August, [Online]. Available from: https://www.wilsoncenter.org/article/quilombola-communities-brazil [Accessed 28 March 2020].

[8] Google Scholar (2020a). Indigenous Communities. Available from: https://scholar.google.com/scholar?hl=de&as_sdt=0%2C5&q=indigenous+communities&btnG= [Accessed 23 July 2020].

[9] Google Scholar (2020b). Quilombola Communities. Available from: https://scholar.google.com/scholar?hl=de&as_sdt=0%2C5&q=Quilombola+communities&btnG= [Accessed 23 July 2020].

[10] Salles, V. (1988). O Negro no Pará sob o regime de escravidão. Cited in: Andrade, L. (2002) THE QUILOMBOS OF THE TROMBETAS RIVER BASIN: BRIEF HISTORY. [Online]. Available from: https://www.academia.edu/34345324/THE_ QUILOMBOS_OF_THE_TROMBETAS_RIVER_BASIN_BRIEF_HISTORY [Accessed 7 August 2020].

[11] Survival International (2020). Brazilian Indians. [Online]. Available from: https://www.survivalinternational.org/tribes/brazilian [Accessed 5 March 2020].

[12] Ruffato, L. (2013). Rassendemokratie ist ein Mythos. Festrede zur Eröffnung der Buchmesse 2013. Faust Kultur, Translation: Kegel, M. Available from: https://faustkultur.de/1456-0-Festrede-von-Luiz-Ruffato.html#.U9H_trXlpdg [Accessed 7 July 2020].

[13] Andrade, L. (2002). The Quilombolos of the Trombetas River Basin: Brief History. [Online]. Available from: https://www.academia.edu/34345324/THE_

QUILOMBOS_OF_THE_TROMBETAS_RIVER_BASIN_BRIEF_HISTORY [Accessed 7 August 2020].

[14] WCS (2020). Trombetas. Wildlife Conservation Society. [Online]. Available from: http://amazonwaters.org/basins/great-sub-basins/trombetas/ [Accessed 10 August 2020].

[15] INCRA (2019). Andamento dos Processos - Quadro Geral. Casa Civil da Presi-dência da República Instituto Nacional de Colonização e Reforma Agrária Diretoria de Ordenamento da Estrutura Fundiária Coordenação Geral de Regularização de Territórios Quilombolas – DFQ, January, [Online]. Available from: http://www.incra.gov.br/media/docs/quilombolas/andamento_processos.pdf [Accessed 11 August 2020].

[16] Gubert, M., Segall-Correa, A., Pedroso, J., Campos Coelho, S. and Perez-Escamilla, R. (2017). Household food insecurity in black-slaves descendant communities in Brazil: has the legacy of slavery truly ended?. Public Health Nutrition, 29 (8), pp. 1513-1522, [Online]. Available from: https://www.cambridge.org/core/journals/public-health-nutrition/article/household-food-insecurity-in-blackslaves-descendant-communities-in-brazil-has-the-legacy-of-slavery-truly-ended/486C22F7D836B6ECBCB16BC3AD61F98D/core-reader [Accessed 11 August 2020].

[17] Planas, R. (2017). Brazil's 'Quilombo' Movement May Be The World's Largest Slavery Reparations Program. Huffpost, updated December, Original: October, 2014, [Online]. Available from: https://www.huffpost.com/entry/brazil-quilombos_n_5572236?guccounter=1 [Accessed 7 July 2020].

[18] Campos Lima, E. (2020). Brazil is the epicenter of the Covid-19 pandemic in Latin America. Rural communities have it worse. America Magazine, May, [Online]. Available from: https://www.americamagazine.org/politics-society/2020/05/27/brazil-epicenter-covid-19-pandemic-latin-america-rural-communities-have [Accessed 4 August 2020].

[19] Dancey-Downs (2018). How beekeeping has turned around the lives of a community in Brazil. Ecologist: The Journal for the Post-Industrial Age, May, [Online]. Available from: https://theecologist.org/2018/may/04/how-beekeeping-has-turned-around-lives-community-brazil [Accessed 18 March 2020].

[20] Berr, K. and Jenal C. (2019). Landschaftskonflikte. Chapter written by Neuburger, M. and Rau, R. Die Kolonialität der brasilianischen Energielandschaft. Wiesbaden: Springer, [Online]. Available from: https://books.google.co.uk/books?id=XiaeDwAAQBAJ&pg=PA495&lpg=PA495&dq=bedrohung+quilombola&source=bl&ots=BKjpbfzhBu&sig=ACfU3U1wZkR0muq0s70P090kieDKg26CHA&hl=de&sa=X&ved=2-ahUKEwjv37a4s7_oAhVRqHEKHZfLCjYQ6AEwA3oECAoQAQ#v=onepage&q=bedrohung%20quilombola&f=false [Accessed 29 March 2020].

[21] Moloney, A. (2020). Deforestation in Brazil's Amazon two-thirds lower on titled indigenous land. Reuters, August, [Online]. Available from: https://www.reuters.com/article/us-brazil-forests-indigenous-rights/deforestation-in-brazils-amazon-two-thirds-lower-on-titled-indigenous-land-idUSKCN2562GH [Accessed 12 August 2020].

[22] CPISP (2019). Quilombolas communities in Brazil. Comissao Pro-Indio de Sao Paulo, [Online]. Available from: https://cpisp.org.br/direitosquilombolas/observatorio-terras-quilombolas/quilombolas-communities-in-brazil/ [Accessed 7 August 2020].

[23] CAISAN (2013). Direitos territoriais e patrimoniais dos povos indígenas e quilombolas – contribuições ao debate da XI Plenária do Consea – agosto de 2013. Interministerial Chamber for Food and Nutrition Security, [Online]. Available from: http://www.mds.gov.br/webarquivos/arquivo/seguranca_alimentar/caisan/caisan_nacional/documentos/terra_territorio/Subsidio_CAISAN_Plenaria_CONSEA_Direitos_Territoriais_agosto_2013.pdf [Accessed 12 August 2020].

[24] Watts, J. (2017). Brazil abolishes huge Amazon reserve in "biggest attack" in 50 years. The Guardian, August, [Online]. Available from: https://www.theguardian.com/environment/2017/aug/24/brazil-abolishes-huge-amazon-reserve-in-biggest-attack-in-50-years [Accessed 4 August 2020].

[25] Harris, B., Cowie, S., Schipani, A. and Gross, A. (2019). Powerful interests pose threat to Amazon's forest communities. Financial Times, December, [Online]. Available from: https://www.ft.com/content/355966ba-e539-11e9-b8e0-026e07cbe5b4 [Accessed 28 March 2020].

[26] Campos Lima, E. (2019). What the Catholic Church is doing to protect the Amazon in Bolsonaro's Brazil. America Magazine, October, [Online]. Available from: https://www.americamagazine.org/politics-society/2019/10/22/what-catholic-church-doing-protect-amazon-bolsonaros-brazil [Accessed 7 August 2020].

[27] Sullivan, Z. (2018). Illegale Abholzung und Bergbau bedrohen Gemeinde im Amazonasgebiet. Deutsche Welle, October, [Online]. Available from: https://www. dw.com/de/illegale-abholzung-und-bergbau-bedrohen-gemeinden-im-amazonasgebiet/a-45656372 [Accessed 7 August 2020].

[28] Cowie, S. (2018). Präsident Bolsonaro bedroht Brasiliens Regenwald - Schlechte Nachrichten für Brasiliens Umwelt: Der gerade gewählte, rechtsradikale Präsident Jair Bolsonaro möchte Umweltagenturen schließen und industriellen Großprojekten im Amazonas-Regenwald den Weg ebnen. Deutsche Welle, [Online], October. Available from: https://www.dw.com/de/präsident-bolsonaro-bedroht-brasiliens-regenwald/a-46079231 [Accessed 1 February 2020].

[29] The Intercept Brasil (2017). Bolsonaro faz discurso de ódio no Clube Hebraica. April. Available from: https://www.youtube.com/watch?v=zSTdTjsio5g [Accessed 7 July 2020].

[30] Käufer, T. and Samuel, R. (2020). Die Welt schaut weg – und Brasilien holzt ab. Welt.de, June. Available from: https://www.welt.de/politik/ausland/article2090 76003/Regenwald-Die-Welt-schaut-weg-und-Brasilien-holzt-ab.html [Accessed 5 July 2020].

[31] Boadle, A. (2019). Brazil's Bolsonaro hands indigenous land decisions back to farm sector. Reuters, June, [Online]. Available from: https://www.reuters.com/article/us-brazil-politics-indigenous/brazils-bolsonaro-hands-indigenous-land-decisions-back-to-farm-sector-idUSKCN1TK37O [Accessed 12 August 2020].

[32] Senpinar, R. (2019). Die Indigenen müssen um ihr Leben fürchten - Goldgräber und Sojafarmer verdrängen Einheimische im Amazonas. Wie die Indigenen unter den Bränden leiden, erzählt die Mitarbeiterin einer Hilfsorganisation. Zeit Online, [Online], August. Available from: https://www.zeit.de/2019-08/brasilien-indigene-voelker-waldbraende-survival-international-ngo [Accessed 1 February 2020].

[33] Russau, C. (2020). Schwerer Verbalangriff der FUNAI auf Indigenemissionsrat CIMI und

internatile Geldgeber. Kooperation Brasilien, May. Available from: https://www.kooperation-brasilien.org/de/themen/landkonflikte-umwelt/schwerer-verbalangriff-der-funai-auf-indigenenmissionsrat-cimi-und-internationale-geldgeber [Accessed 5 July 2020].

[34] Penna-Firme and Brondizio, E. (2007). The risk of commodifying poverty: rural communities, Quilombola identity and nature conservation in Brazil. Habitus, 5 (2), pp. 355-373, [Online]. Available from: http://seer.pucgoias.edu.br/index.php/habitus/article/view/541/431 [Accessed 11 August 2020].

[35] Silva, A., Alves, C., Fernandes, K. and Müller, R. (2013). Classification of Honeys from Pará State (Amazon Region, Brazil) Produced by Three Different Species of Bees using Chemometric Methods. Journal of the Brazilian chemical society, 24 (7), pp. 1135-1145, [Online]. Available from: http://static.sites.sbq.org.br/jbcs.sbq.org.br/pdf/v24n7a07.pdf [Accessed 13 August 2020].

[36] Fernandes, R., Rosa, I. and Conti-Silva, A. (2020). Honey from Tibua stingless bees (*Melipona fasciculata*) produced in different ecosystems: physical and sensory studies. Journal of the Science of Food and Agriculture, 100 (9), pp. 3748-3754, [Online]. Available from: https://onlinelibrary-wiley-com.ezproxy.leedsbeckett.ac.uk/doi/epdf/10.1002/jsfa.10415 [Accessed 13 August 2020].

[37] DW (2019a). Mexiko: Tonnenweise Bio-Honig für Deutschland. Deutsche Welle, February, [Online]. Available from: https://www.dw.com/de/mexiko-tonnenweise-bio-honig-für-deutschland/a-47406856 [Accessed 7 March 2020].

[38] Hanson, T. (2019). Why have 500m bees died in Brazil in the past three months?. The Guardian, August, [Online]. Available from: https://www.theguardian.com/commentisfree/2019/aug/29/500-million-bees-brazil-three-months [Accessed 28 March 2020].

[39] Grigori, P. (2019). Half a billion bees dead as Brazil approved hundreds more pesticides. Mongabay, August, [Online]. Available from: https://news.mongabay.com/2019/08/half-a-billion-bees-dead-as-brazil-approves-hundreds-more-pesticides/ [Accessed 28 March 2020].

[40] dos Santos, C. and de Oliveira, V. (2017). Trends in Extreme Climate Indices for Para State, Brazil. Revista Brasileira de Meteorologia, 32 (1), pp. 13-24, [Online]. Available from: https://www.scielo.br/pdf/rbmet/v32n1/0102-7786-rbmet-32-01-0013.pdf [Accessed 5 August 2020].

[41] Coe, M., Marthews, T., Costa, M., Galbraith, D., Greenglas, N., Imbuzeiro, H., Levine, N., Malhi, Y., Moorcroft, P., Muza, M., Powell, T., Saleska, S., Solorzano, L. and Wand, J. (2013). Deforestation and climate feedbacks threaten the ecological integrity of south-southeastern Amazonia. Philosophical transactions of the royal society, 368 (1619), June, [Online]. Available from: https://royalsocietypublishing.org/doi/full/10.1098/rstb.2012.0155 [Accessed 6 August 2020].

[42] Martorano, L., Vitorino, M., da Silva, B., de Moraes, J., Lisboa, L., Sotta, E. and Reichardt, R. (2017) Climate conditions in the eastern amazon: Rainfall variability in Belem and indicative of soil water deficit. African Journal of Agricultural Research, 12 (21), May, pp.1801-1810, [Online]. Available from: https://pdfs.semanticscholar.org/ 6a41/07d4b38 12f212a407fc18c7ed526a9b0f1c4.pdf [Accessed 28 March 2020].

[43] FIRMS and NASA (2019). Cited in: ZDF (2020). Zerstörung des

Amazonas-Walds nimmt stark zu. January, [Online]. Available from: https://www.zdf.de/nachrichten/panorama/amazonas-wald-zerstoerung-drastisch-zugenommen-100.html [Accessed 12 August 2020].

[44] Pendrill, F., Persson U., Gobar, J., Kastner, T., Moran, D., Schmidt, S. and Wood, R. (2019). Agricultural and forestry trade drives large share of tropical deforestation emissions. Global Environmental Change, 56, pp. 1-10, [Online]. Available from: https://reader.elsevier.com/reader/sd/pii/S0959378018314365?token=2F8B4E74FAB2F0F21CC1A14A0E619DC516B728FB569FA44ECE38523608B1C58CE45C86B96961795DDCEA7CBB19145C6F [Accessed 7 August 2020].

[45] Spiegel (2020) Brasilianische Firmen machen Druck gegen Regenbwaldzerstörung. Spiegel Wirtschaft, July, [Online]. Available from: https://www.spiegel.de/wirtschaft/unternehmen/brasilianische-firmen-machen-druck-gegen-regenwald-zerstoerung-a-10143ab9-0918-4525-aff8-52095bd85c58 [Accessed 6 August 2020].

[46] Bosma, W., Suti, S. and Deeks, P. (2017). Beekeeping as Pro-forest Income Diversification in Solomon Islands. Part of the Climate Change Management book series, Springer, February, [Online]. Available from: https://link.springer.com/chapter/10.1007/978-3-319-50094-2_23 [Accessed 18 March 2020].

[47] Lowore, J., Meaton, J. and Wood, A. (2018). African Forest Honey: an Overlooked NTFP with Potential to Support Livelihoods and Forests. Environmental Management, 62, pp. 15-28, [Online]. Available from: https://link.springer.com/article/10.1007/s00267-018-1015-8 [Accessed 18 March 2020].

[48] Watts, J. (2019). Bees in Amazon 'are greatest ally to halt rainforest destruction'. The Guardian, December, [Online]. Available from: https://www.theguardian.com/society/2019/dec/06/bees-in-amazon-are-greatest-ally-to-halt-rainforest-destruction [Accessed 18 March 2020].

[49] Mavhunga, C. (2017). Ein süßer Neuanfang in Simbabwe – Bäume retten mit Honig. Deutsche Welle, June, [Online]. Available from: https://www.dw.com/de/ein-süßer-neuanfang-in-simbabwe-bäume-retten-mit-honig/a-39430726 [Accessed 6 August 2020].

[50] Carvalho, G., Ribeiro, M., Araujo, A., Barbosa, M. Oliveira, F. and Albuquerue, P. (2016). Flora de importancia polinica utilizada por Melipona (Melikerria) fasciculata Smith, 1854 (Hymenoptera: Apidae:Meliponini) em uma area de foresta amazonica na regiao da baixada maranhense. Brasil Oecologia Australis, 20, pp. 58-68. Cited in: Fernandes, R., Rosa, I. and Conti-Silva, A. (2020) Honey from Tibu stingless bees (*Melipona fasciculata*) produced in different ecosystems: physical and sensory studies. Journal of the Science of Food and Agriculture, 100 (9), pp. 3748-3754, [Online]. Available from: https://onlinelibrary-wiley-com.ezproxy.leedsbeckett.ac.uk/doi/epdf/10.1002/jsfa.10415 [Accessed 13 August 2020].

[51] Bezerra, F., Bernardo, T., Ximenes, L. and Valente Junior A. (2015). Perfil socio-econômico do Maranhão. Banco do Nordeste do Brasil, Fortaleza. Cited in: Fernandes, R., Rosa, I. and Conti-Silva, A. (2020) Honey from Tibu stingless bees (*Melipona fasciculata*) produced in different ecosystems: physical and sensory studies. Journal of the Science of Food and Agriculture, 100 (9), pp. 3748-3754, [Online]. Available from: https://onlinelibrary-wiley-com.ezproxy.leedsbeckett.ac.uk/doi/epdf/10.1002/jsfa.10415 [Accessed 13 August 2020].

[52] Rodgers, C. (2018) Indigenous People of Brazil. Wilson Center: Brazil Institute, January, [Online]. Available from: https://www.wilsoncenter.org/article/indigenous-peoples-brazil [Accessed 28 March 2020].

[53] IBGE (2020) Projeção da população do Brasil e das Unidades da Federação. Istituto Brasileiro de Geografia e Estatistica, [Online]. Available from: https://www. ibge.gov.br/apps/populacao/projecao/index.html?utm_source=portal&utm_medium=popclock&utm_campaign=novo_popclock [Accessed 28 March 2020].

[54] INPE (2019). cited in: Butler, R. (2019) Amazon deforestation rises to 11 year high in Brazil. Mongabay, November. Available from: https://news.mongabay.com/2019/11/amazon-deforestation-rises-to-11-year-high-in-brazil/ [Accessed 7 July 2020].

[55] Silva, N. (2000) Extensão e natureza das desigualdades raciais no Brasil. In A. S. A. Guimarães & L. Huntley (Orgs.), Tirando a máscara: ensaios sobre o racismo no Brasil, pp. 33-51. São Paulo: Paz e Terra. Cited in: Costa, E. and Scarcelli, I. (2016) Psychology, public policy for quilombola populations and racism. Psicol, 27 (2), [Online]. Available from: https://www.scielo.br/scielo.php?script=sci_arttext&pid=S0103-65642016000200357&lng=pt&tlng=pt [Accessed 6 August 2020].

[56] Brandao, A., da Dalt, S. and Gouveia, V. (2010). Comunidades quilombolas no Brasil: características socioeconômicas, processos de etnogênese e políticas sociais. Cited in: Costa, E. and Scarcelli, I. (2016) Psychology, public policy for quilombola populations and racism. Psicol, 27 (2), [Online]. Available from: https://www.scielo.br/scielo.php?script=sci_arttext&pid=S0103-65642016000200357&lng=pt&tlng=pt [Accessed 6 August 2020].

[57] Hormel, U. and Scherr, A. (2010). Diskriminierung – Grundlagen und Forschungsergebnisse. Wiesbaden: Springer VS.

[58] Rotondano, R. (2019). Brazilian apartheid: racism and segregation in Salvador, Brazil. International Journal of Sociology and Social Policy, 39 (11/12), pp. 950-961, [Online]. Available from: https://www-emerald-com.ezproxy.leedsbeckett.ac.uk/insight/content/doi/10.1108/IJSSP-12-2018-0228/full/pdf?title=brazilian-apartheid-racism-and-segregation-in-salvador-brazil [Accessed 8 August 2020].

[59] Behrman, J., Gaviria, A. and Szekely, M. Social Exclusion in Latin America. In: Behrman, J., Gaviria, A. and Szekely, M. (2003). Who's in and Who's out – Social Exclusion in Latin America. Washington: IDB Publications, [Online]. Available from: https://books.google.de/books?hl=en&lr=&id=LvEzYAfbSh0C&oi=fnd&pg=PP5&dq=social+status+segregation+indigenous&ots=IJtecdMX1L&sig=8U5V9nytIn-ohb9MvVwf_tz_UFvE&redir_esc=y#v=onepage&q=social%20status%20segregation%20indigenous&f=false [Accessed 7 August 2020].

[60] Firmenich (2015) FIRMENICH NAMES HONEY THE 2015 'FLAVOR OF THE YEAR'. December, [Online]. Available from: https://www.firmenich.com/en_INT/company/news/flavor-of-the-year.html [Accessed 14 March 2020].

[61] FAO (2019). Food and Agriculture Organization of the United Nations. February. Cited in: Shahbandeh, M. (2020). Global production volume of natural honey in 2000-2017. Statista, February, [Online]. Available from: https://www-statista-com.ezproxy.leedsbeckett.ac.uk/statistics/755215/natural-honey-production-volume-worldwide/ [Accessed 6 March 2020].

[62] Orbis Research (2018). Cited in: Shahbandeh, M. (2018) Forecast value of organic honey market worldwide from 2017 to 2023. Statista, October, [Online]. Available from: https://www-statista-com.ezproxy.leedsbeckett.ac.uk/statistics/933490/global-organic-honey-market-value/ [Accessed 6 March 2020].

[63] Garcia, N. (2018). The Current Situation on the International Honey Market. International Bee Research Association, Bee World, 95 (3), July, pp.89-94, [Online]. Available from: https://www.tandfonline.com/doi/abs/10.1080/0005772X.2018.1483814?journalCode=tbee20 [Accessed 6 March 2020].

[64] Carreck, N. (2018) Special issue: Honey. Journal of Apicultural Research, 57 (1), January, [Online]. Available from: https://www.tandfonline.com/doi/full/10.1080/00218839.2017.1412565?src=recsys [Accessed 10 March 2020].

[65] Menn (n.d.) Honig: Gesund essen. Bundeszentrum für Ernährung, [Online]. Available from: https://www.bzfe.de/inhalt/honig-gesund-essen-33942.html [Accessed 8 March 2020].

[66] Pasupuleti, V., Sammugam, L., Ramaseh, N. and Gan, S. (2017) Honey, Propolis, and Royal Jelly: A Comprehensive Review of Their Biological Actions and Health Benefits. Oxidative Medicine and Celluar Longevity (Oxid Med Cell Longev), July, [Online]. Available from: https://www.ncbi.nlm.nih.gov/pmc/articles/PMC5549483/ [Accessed 8 March 2020].

[67] Hery-Moßmann, N. (2017). Honig als Heilmittel – so funktioniert's. Focus Online, August, [Online]. Available from: https://praxistipps.focus.de/honig-als-heilmittel-so-funktionierts_95273 [Accessed 8 March 2020].

[68] U.N. (2020) Comtrade. [Online]. Available from: https://comtrade.un.org/data/ [Accessed 6 March 2020].

[69] CMO (2019) Honey Market Presentation. European Commission, April, [Online]. Available from: https://ec.europa.eu/info/sites/info/files/food-farming-fisheries/animals_and_animal_products/presentations/market-presentation-honey_en.pdf/ [Accessed 6 March 2020].

[70] ProFound (2015a). Cited in: CBI (2015a) CBI Product Factsheet: Industrial honey in Germany. Center for the Promotion of Imports, November [Online]. Available from: https://www.cbi.eu/sites/default/files/market_information/researches/product-factsheet-germany-industrial-honey-2015.pdf [Accessed 14 March 2020].

[71] ProFound (2015b). Cited in: CBI (2015b) CBI Product Factsheet: Organic honey in Germany. Center for the Promotion of Imports, November [Online]. Available from: https://www.cbi.eu/sites/default/files/market_information/researches/product-factsheet-germany-organic-honey-2015.pdf [Accessed 15 March 2020].

[72] CBI (2015b) CBI Product Factsheet: Organic honey in Germany. Center for the Promotion of Imports, November [Online]. Available from: https://www.cbi.eu/sites/default/files/market_information/researches/product-factsheet-germany-organic-honey-2015.pdf [Accessed 15 March 2020].

[73] Phipps, R. (2018) International Honey Market. American Bee Journal, November. Available from: https://www.apiservices.biz/documents/articles-en/international_honey_market_report_november_2018.pdf [Accessed 13 July 2020].

[74] USDA (2018) National Honey Report. Cited in: NHB (2020) International Bulk Prices. National Honey Board, [Online]. Available from: https://www.honey.com/honey-industry/statistics/

international-bulk-prices [Accessed 13 July 2020].

[75] EC (2019). EU and Mercosur reach agreement on trade. European Commission, June, [Online]. Available from: https://ec.europa.eu/commission/presscorner/detail/en/IP_19_3396 [Accessed 14 March 2020].

[76] DW (2019b) Widerstand gegen Bolsonaros Amazonas-Politik. Deutsche Welle, August, [Online]. Available from: https://www.dw.com/de/widerstand-gegen-bolsonaros-amazonas-politik/a-50190837 [Accessed 6 August 2020].

[77] BMZ (2020) Brasilien: Bilaterale Entwicklungszusammenarbeit. Bundesministerium für wirtschaftiche Zusammenarbeit und Entwicklung, [Online]. Available from: https://www.bmz.de/de/ministerium/zahlen_fakten/transparenz-fuer-mehr-Wirksamkeit/iati/index.jsp [Accessed 7 March 2020].

[78] BMEL (2020) Sichere Lebensmittel. Bundesministerium für Ernähung und Landwirschaft [Online]. Available from: https://www.bmel.de/DE/Ernaehrung /Sichere Lebensmittel/sichereLebensmittel_node.html [Accessed 14 March 2020].

[79] EC (2002a) REGULATION (EC) No 178/2002 OF THE EUROPEAN PARLIAMENT AND OF THE COUNCIL of 28 January 2002: laying down the general principles and requirements of food law, establishing the European Food Safety. Official Journal of the European Communities, L31, pp. 1-24, February, [Online]. Available from: https://mobil.bfr.bund.de/cm/343/2002_178_en_efsa.pdf [Accessed 14 March 2020].

[80] EC (2002b) COUNCIL DIRECTIVE 2001/110/EC of 20 December 2001 relating to honey. Official Journal of the European Communities, L10, pp. 47-52, December, [Online]. Available from: https://eur-lex.europa.eu/legal-content/EN/TXT/PDF/?uri=CELEX:32001L0110&from=en [Accessed 14 March 2020].

[81] EC (2004). REGULATION (EC) No 852/2004 OF THE EUROPEAN PARLIAMENT AND OF THE COUNCIL of 29 April 2004: on the hygiene of foodstuffs. Official Journal of the European Communities, L139, April, [Online]. Available from: https://eur-lex.europa.eu/LexUriServ/LexUriServ.do?uri=OJ:L:2004:139:0001:0054:en:PDF [Accessed 14 March 2020].

[82] EC (2011). REGULATION (EU) No 1169/2011 OF THE EUROPEAN PARLIAMENT AND OF THE COUNCIL of 25 October 2011: on the provision of food information to consumers, amending Regulations (EC) No 1924/2006 and (EC) No 1925/2006 of the European Parliament and of the Council, and repealing Commission Directive 87/250/EEC, Council Directive 90/496/EEC, Commission Directive 1999/10/EC, Directive 2000/13/EC of the European Parliament and of the Council, Commission Directives 2002/67/EC and 2008/5/EC and Commission Regulation (EC) No 608/2004. Official Journal of the European Communities, L304, November, [Online]. Available from: https://eur-lex.europa.eu/LexUriServ/LexUriServ.do?uri=OJ:L:2011:304:0018:0063:EN:PDF [Accessed 14 March 2020].

[83] EU (2018) REGULATION (EU) 2018/848 OF THE EUROPEAN PARLIAMENT AND OF THE COUNCIL of 30 May 2018: on organic production and labelling of organic products and repealing Council Regulation (EC) No 834/2007. Official Journal of the European Union, L150, June, [Online]. Available from: https://eur-lex.europa.eu/legal-con-tent/EN/TXT/?uri=uriserv:OJ.L_.2018.150.01.0001.01.ENG&toc=OJ:L:20 18:150:TOC [Accessed 15 March 2020].

[84] EU (2019) COMMISSION IMPLEMENTING REGULATION

(EU) 2019/2164 of 17 December 2019: amending Regulation (EC) No 889/2008 laying down detailed rules for the implementation of Council Regulation (EC) No 834/2007 on organic production and labelling of organic products with regard to organic production, labelling and control. Official Journal of the European Union, L328, December, [Online]. Available from: https://eur-lex.europa.eu/legal-con-tent/EN/TXT/?uri=uriserv:OJ.L_.2019.328.01.0061.01.ENG&toc=OJ:L:2019:328:TOC [Accessed 15 March 2020].

[85] HonigV. (2020). Honigverordnung. Bundesministerium der Justiz und für Verbraucher, [Online]. Available from: https://www.gesetze-im-internet.de/honigv _2004/BJNR009200004.html [Accessed 7 March 2020].

[86] FDA (2018a) Importing Food Products into the United States. Food and Drug Administration, March, [Online]. Available from: https://www.fda.gov/food/food-imports-exports/importing-food-products-united-states [Accessed 18 March 2020].

[87] FDA (2018b) Proper Labeling of Honey and Honey Products: Guidance for Industry. Food and Drug Administration, February, [Online]. Available from: https://www.fda.gov/files/food/published/PDF---Guidance-for-Industry--Proper-Labeling-of-Honey-and-Honey-Products.pdf [Accessed 18 March 2020].

[88] Federal Register (2014) United States Standard of Identity for Honey. August, [Online]. Available from: https://www.federalregister.gov/documents/2014/08/20/2014-19770/united-states-standard-of-identity-for-honey [Accessed 18 March 2020].

[89] State.Gov (2019) U.S. Relations with Brazil – Bilateral Relations Fact Sheet. October. Available from: https://www.state.gov/u-s-relations-with-brazil/ [Accessed 19 July 2020].

[90] Cullen, A. (2019) How Trump's trade wars are fueling the Amazon fires. The Guardian, September. Available from: https://www.theguardian.com/commentisfree/2019/sep/01/brazil-amazon-wildfires-trump-trade-wars [Accessed 19 July 2020].

[91] Imöhl, S. (2020) Die zehn größten Länder der Welt nach Fläche. Handelsblatt, January, [Online]. Available from: https://www.handelsblatt.com/politik/international/top-ten-die-zehn-groessten-laender-der-welt-nach-flaeche/24428374.html?ticket=ST-563453-iKegeqAR5l6Z207MZtAG-ap4 [Accessed 28 March 2020].

[92] INPE (2020) cited in: Gurk (2020) Der Regenwald als Beute. Süddeutsche Zeitung, May, Available from: https://www.sueddeutsche.de/politik/brasilien-regenwald-bolsonaro-abholzung-landraub-indigene-coronavirus-1.4906376 [Accessed 8 July 2020].

[93] INPE and DETER (2020) cited in: Butler, R. (2020) Brazil revises deforestation data: Amazon rainforest loss topped 10,000 km in 2019. Mongabay, June, [Online]. Available from: https://news.mongabay.com/2020/06/brazil-revises-deforestation-data-amazon-rainforest-loss-topped-10000-sq-km-in-2019/ [Accessed 4 August 2020].

[94] Zemp, D., Schleussner, C-F., Barbosa, H., Hirota, M., Montade, V., Sampaio, G., Staal, A., Wang-Erlandsson, L. and Ramming A. (2017) Self-amplified Amazon forest loss due to vegetation-atmosphere feedbacks. Nature Communications, 8, Document nr. 14681, March, [Online]. Available from: https://www.nature.com/articles/ncom ms14681.pdf [Accessed 6 August 2020].